Wireless Sensor Systems for Extreme Environments

Wireless Sensor Systems for Extreme Environments

Space, Underwater, Underground, and Industrial

Edited by

Habib F. Rashvand
Advanced Communication Systems
University of Warwick
UK

Ali Abedi
Department of Electrical and Computer Engineering
University of Maine
Orono
USA

This edition first published 2017

Registered Offices
John Wiley & Sons, Inc., 111 River Street, Hoboken, NJ 07030, USA
John Wiley & Sons Ltd, The Atrium, Southern Gate, Chichester, West Sussex, PO19 8SQ, UK

Editorial Office
The Atrium, Southern Gate, Chichester, West Sussex, PO19 8SQ, UK

For details of our global editorial offices, customer services, and more information about Wiley products visit us at www.wiley.com.

Wiley also publishes its books in a variety of electronic formats and by print-on-demand. Some content that appears in standard print versions of this book may not be available in other formats.

Library of Congress Cataloging-in-Publication Data

Names: Rashvand, Habib F., editor. | Abedi, Ali, editor.
Title: Wireless sensor systems for extreme environments : space, underwater, underground, and industrial / [edited by] Habib F. Rashvand, Ali Abedi.
Description: Hoboken, NJ : John Wiley & Sons, 2017. | Includes bibliographical references and index.
Identifiers: LCCN 2017005391 (print) | LCCN 2017010421 (ebook) | ISBN 9781119126461 (cloth) | ISBN 9781119126478 (Adobe PDF) | ISBN 9781119126485 (ePub)
Subjects: LCSH: Wireless sensor networks. | Extreme environments.
Classification: LCC TK7872.D48 W587 2017 (print) | LCC TK7872.D48 (ebook) | DDC 004.6–dc23
LC record available at https://lccn.loc.gov/2017005391

A catalogue record for this book is available from the British Library.

Cover design by Wiley
Cover Image: © alex-mit/Gettyimages

Set in 10/12pt Warnock by SPi Global, Chennai, India
Printed and bound in Malaysia by Vivar Printing Sdn Bhd

10 9 8 7 6 5 4 3 2 1

Habib dedicates this work to his family, especially Madeleine, Laurence and the late and dearly missed Joan Elmes. Ali dedicates this work to his wife Shahrzad for her unconditional love and support.

Contents

List of Contributors

Ali Abedi
Department of Electrical and Computer Engineering
University of Maine
Orono
USA

James Agajo
Federal University of Technology Minna
Ihiagwa
Nigeria

Deepshikha Agarwal
Amity University
Lucknow
Uttar Pradesh
India

Ian F. Akyildiz
Broadband Wireless Networking Lab
Georgia Institute of Technology
Atlanta
USA

Gholamreza Alirezaei
Institute for Theoretical Information Technology
RWTH Aachen University
Aachen
Germany

Filip Barac
Business Unit Network Products
Ericsson AB
Sweden

Gerard Chalhoub
Clermont University
Aubière
France

Xuewu Dai
Northumbria University
Newcastle upon Tyne
UK

Rana Diab
Clermont University
Aubière
France

Lloyd Emokpae
U.S. Naval Research Laboratory
Code 7160 – Acoustics Division
USA

Eriza Fazli
Zodiac Inflight Innovations
Weßling
Germany

Jorge M. Finochietto
Digital Communications Research Laboratory Electronic Department
Universidad Nacional de Córdoba
Córdoba – CONICET
Argentina

Juan A. Fraire
Digital Communications Research Laboratory Electronic Department
Universidad Nacional de Córdoba
Córdoba – CONICET
Argentina

Shang Gao
Nanjing University of Aeronautical and Astronautics
China

Wolfgang H. Gerstacker
Institute for Digital Communications
Friedrich-Alexander-Universität Erlangen-Nürnberg
Germany

Mikael Gidlund
Department of Information Systems and Technology
Mid Sweden University
Sweden

Murat Gürsu
Chair of Communication Networks
Technical University of Munich
Germany

David Harrison
School of Engineering and Computer Science
Victoria University of Wellington
New Zealand

Pan Hu
College of Automation Engineering
Nanjing University of Aeronautics and Astronautics
China

Dingde Jiang
College of Information Science and Engineering
Northeastern University
Shenyang
China

Kübra Kalkan
Boğaziçi University
Istanbul
Turkey

Wolfgang Kellerer
Chair of Communication Networks
Technical University of Munich
Germany

Hendra Kesuma
Institute of Electrodynamics and Microelectronics (ITEM)
University of Bremen
Germany

Nand Kishor
Motilal Nehru National Institute of Technology
Lucknow
Uttar Pradesh
India

Steven Kisseleff
Institute for Digital Communications
Friedrich-Alexander-Universität Erlangen-Nürnberg
Germany

Lonnie Labonte
Department of Electronics and Computer Engineering
University of Maine
USA

Albert Levi
Sabancı Üniversitesi
Istanbul
Turkey

Erwan Livolant
Inria Paris
France

Pablo Madoery
Digital Communications Research
Laboratory Electronic Department
Universidad Nacional de Córdoba
Córdoba – CONICET
Argentina

Ali Mahani
Department of Electrical Engineering
Shahid Bahonar University of Kerman
Kerman Province
Iran

Rudolf Mathar
Institute for Theoretical Information
Technology
RWTH Aachen University
Aachen
Germany

Pascale Minet
Inria Paris
France

Michel Misson
Clermont University
Aubière
France

Vallipuram Muthukkumarasamy
Griffith University
Gold Coast
Australia

Kewen Pan
School of Electrical and Electronic
Engineering
University of Manchester
UK

Steffen Paul
Institute of Electrodynamics and
Microelectronics (ITEM)
University of Bremen
Germany

Jean-Francois Perelgritz
Airbus Group Innovations
Suresnes
France

Anisur Rahman
East West University
Dhaka
Bangladesh

Habib F. Rashvand
Advanced Communication Systems
University of Warwick
UK

Badr Rmili
CNES Launcher Directorate
Paris
France

Winston K.G. Seah
School of Engineering and Computer
Science
Victoria University of Wellington
New Zealand

Johannes Sebald
Airbus Safran Launchers (former EADS
Astrium) GmbH
Bremen
Germany

Praveen Shankar
Department of Mechanical and
Aerospace Engineering
California State University
California, Long Beach
USA

Saeideh Sheikhpour
Department of Electrical Engineering
Shahid Bahonar University of Kerman
Kerman Province
Iran

Emiliano Sisinni
Department of Information Engineering
University of Brescia
Italy

Ridha Soua
Inria Paris
France

Ali Imam Sunny
School of Electrical and Electronic Engineering
Newcastle University
Newcastle Upon Tyne
UK

Omid Taghizadeh
Institute for Theoretical Information Technology
RWTH Aachen University
Aachen
Germany

Chaoqing Tang
School of Electrical and Electronic Engineering
Newcastle University
Newcastle Upon Tyne
UK

Gui Yun Tian
School of Electrical and Electronic Engineering
Newcastle University
Newcastle upon Tyne
UK

Mikhail Vilgelm
Chair of Communication Networks
Technical University of Munich
Germany

Bang Wang
School of Electronic Information and Communications
Huazhong University of Science & Technology (HUST)
Wuhan
China

Haitao Wang
College of Automation Engineering
Nanjing University of Aeronautics and Astronautics
China

Wen-Qin Wang
School of Communication and Information Engineering
University of Electronic Science and Technology of China
Chengdu
China

Sherali Zeadally
University of Kentucky
Lexington
US

Tingting Zhang
Department of Information Systems and Technology
Mid Sweden University
Sweden

Jiwen Zhu
Northumbria University
Newcastle upon Tyne
UK

Preface

Just before the turn of the new millennium, the advent of short- and medium-range wireless communication technologies brought rapid developments in the field of lightweight sensors and actuators, creating a fast-moving business area in which new types of smart wireless sensors were brought to market. These smart wireless sensors benefit greatly from the use of infrastructural features of traditional telecommunications and data systems, enabling clustered interworking. New features include autonomous and adaptive coverage. Two common terminologies are 'wireless sensor networks' (WSNs) for applications of large to very large numbers of sensing nodes and 'wireless sensor systems' (WSS) for smaller numbers.

With the recent surge in commercialization of wireless technology and the availability of advanced coding, signal processing, and information technology, which enable reliable wireless connectivity at high data rates, most industries are reconsidering the need for wires. The saving on weight and cost of wires, cables, fixtures and connectors is an obvious benefit of wireless technology, but extra savings are also associated with design, testing, and modification of sensing systems and these may not always be obvious. Flexibility and scalability of wireless sensing systems are additional benefits.

Space and other extreme environments –underwater, underground – and unconventional industrial environments may significantly benefit from wireless sensing, but due to their challenging environmental conditions and requirements for high reliability, they have not yet fully taken advantage of these technologies. Our recent series of workshops entitled 'Fly-By-Wireless' and the parallel workshops called 'WiSense4Space' have come together under the IEEE's umbrella to create an IEEE International Conference on Wireless for Space and Extreme Environments. The first of these workshops was in 2013 in Baltimore, USA. This was a technical forum for scientists and engineers from space agencies, industry and academia to share knowledge in the area of wireless for space and extreme environments, and paved the way towards an eventual widespread elimination of wires in challenging environments. Underwater acoustics, optical wireless, communication and sensing inside mines, and of course the most challenging of all wireless environments – space – are among the many topics that have been discussed at these conferences. Discussions continue to date.

This book was motivated by recent advances in the area, and covers a wide variety of topics, from power allocation to battery-free communication, underwater links, and even wireless feedback control. We hope that engineers and scientists who read it will take these ideas and methods to the next level and use them in their future designs. The impact of wireless technology in extreme environments is not just about efficiency and savings due to the elimination of wires. The most important impact is enabling new applications and ways of collecting data from locations that were previously inaccessible. Networked wireless sensors can create a shared data environment that can be used to monitor complex systems, looking for anomalies and acting to prevent disasters. Data collection from structures and machines helps designers develop a better understanding of their designs, determine performance in real-world settings and avoids them having to use simulations. The transition from schedule-based maintenance to condition-based maintenance is another significant advantage of wireless sensors in harsh environments.

We would like to acknowledge the consistent support from our contributing authors, without whom this project would not have been possible. Colleagues from industry, the space agencies and academic institutions around the world – from America, to Europe and to Asia – all worked tirelessly for over two years to prepare the chapters of this book. Their contributions and dedication to this project is highly appreciated.

We would like to thank the John Wiley publishing team for their guidance and efforts from the early stages of draft proposals, to reviewing the final production stage, so this editorial volume could be presented at the required level: a high-quality book for the academic and industrial research and development community.

We sincerely hope that these efforts have created a useful book for students to learn from, to inspire engineers and scientists to use these concepts in their designs, and for academic instructors to teach these emerging concepts.

We believe that the advanced technologies of 'fly by wireless', 'drive by wireless', and 'live by wireless', apart from their convenience, effective use of smart sensors should enable our industries in space, Earth, cities, and homes to be more environmentally friendly.

The editors

Habib F. Rashvand and Ali Abedi

Part I

Wireless Sensor Systems for Extreme Environments–Generic Solutions

1

Wireless Sensor Systems for Extreme Environments

Habib F. Rashvand[1] *and Ali Abedi*[2]

[1] *Advanced Communication Systems, University of Warwick, UK*
[2] *Department of Electrical and Computer Engineering, University of Maine, Orono, USA*

Taking a new step, uttering a new word, is what people fear most
Fyodor Dostoyevsky

1.1 Introduction

The last 40 years of economic and political unrest has wrought a series of drastic changes throughout the world. Many technological trends have come to a halt as new developments have taken over, surprising the experts. Amongst the successful ones are smart sensing, flourishing as a result of promises of a higher quality of life and worries about the deterioration of the climate.

Although there have been many projects throughout the world and many successful civil and industrial applications, we are still awaiting to see a real paradigm shift. As increasing resources have expanded and increased research activity, too many research reports have somehow failed to demonstrate the eye-catching industrial applications required to justify the resources being expended. To this end, we have to judge on a global scale the performance of sensors in the last 20 years; we have looked at earlier surveys [1] and analysed the economic effectiveness of the projects described. One of the main conclusions is that too many young researchers try to make their work publishable rather than practical and useful for real applications that to help improve the quality of life. As well as the few useful research activities – such as energy conservation, optimized performance, cross layering, efficient sampling, and data management – we see many trivial patterns of common networking manipulation: routing, scheduling, node replacement, mobility, and coverage under oversimplified working conditions, where simple computer simulations can generate huge volumes of inaccurate data; they are simply creating a new black hole for consuming computer resources.

Following our series of conferences on wireless technologies for space and extreme environments (WiSEE) and the associated sensor workshops we have decided that we need to direct research towards the environments that need sensors most: space and other harsh, industrial or unconventional environments.

Wireless Sensor Systems for Extreme Environments: Space, Underwater, Underground, and Industrial.
First Edition. Edited by Habib F. Rashvand and Ali Abedi.

Following Edison's problem-solving attitude when demonstrating the use of electricity to create the light to brighten our nights, we need to encourage our youth to have strong belief and true dedication. They need to enjoy creativity and achieving their objectives so that they can engineer a better quality of life. They should be solving problems, breaking the old boundaries, opening new windows of opportunity and creating new paradigms. Applying new technologies, such as wireless and ever-improving smart sensors and actuators, gives us many possibilities for creating new and much smarter technological systems and services.

To be successfully deployed, a new technology must meet four basic measures: trust, objectivity, security, and sustainability. Here, objectivity is the demand for a product or service, which in our case means overcoming unconventional working conditions, to that the working product or a system enables new services, whether in the vacuum of space, in the oceans, underground or in places with very high, very low, and highly variable temperature, humidity, winds and pressure.

The rest of this chapter is devoted to two main summary sections. Section 1.2 describes our earlier work on wireless sensor systems (WSSs) for space and other extreme environments, while Section 1.3 provides an extended summary of the remaining twenty-one chapters of the book.

1.2 Wireless Sensor Systems for Space and other Extreme Environments

This section summarises our earlier review of our work on WSSs for space and extreme environments [1]. This was based on our WSS workshop at WiSEE 2013. Our main message is this section is to analyse how to break away from conventional wireless sensor networks (WSNs) by adopting an agile heterogeneous unconventional wireless sensing (UWS) deployment system.

1.2.1 Definitions

A comparative analysis is better than a simple definition of the terms, which often can vary upon application scenarios and its working environment.

Wireless sensor networks (WSNs) are normally complex networks of large numbers of interconnected sensor nodes and clusters. A wireless sensor system (WSS), however, is a smaller-scale system of data-oriented interconnected sensing devices for extracting well-defined sensing information. The sensor nodes in WSSs are expected to be less constrained and more flexible, and therefore more adaptive and autonomous. In WSSs, use of terms such as wireless sensor and actuator networks, wireless smart intelligent sensing, wirelessly connected distributed smart sensing, and unmanaged aerial vehicle sensor networks makes sense. However, wireless underground sensor networks, underwater wireless sensor networks, wireless body-sensor mesh networks and industrial wireless sensor networks are normally more complex, and are therefore more applicable to WSNs by definition.

As heterogeneous sensing services require UWS solutions, one way to compare WSSs and WSNs is to look objectively at the purpose for which they are designed. WSS-based solutions for self-managed heterogeneous sensing services are more dynamic and practical if kept small. This is due to our basic service principles:

- conventional WSNs, normally deployed for homogenous sensing services using generic smart sensors
- unconventional WSSs, designed for dynamic, heterogeneous, UWS services using specific sensors.

UWS solutions therefore require to be kept simple and they therefore suit smaller and less complex WSSs.

1.2.2 Networking in Space and Extreme Environments

In many WSNs, the simplicity of the data collection can allow deployment of sensors on multi-service networks, in which densely distributed sensors and actuators are used for a wide range of applications. In space and extreme environments (SEEs) smart networking is needed to make this process more efficient, and so it can benefit from the low-cost, low-power operation of networks. For example, a multi-timescale adaptation routing protocol can use multi-timescale estimation to minimize variation of packet transmission times by calculating the mean and variance. Another example is the deployment of distributed radar sensor networks (RSNs), grouped together in an intelligent cluster network in an ad hoc fashion. These can then provide spatial resilience for target detection and tracking. Such RSNs may be used for tactical combat systems deployed on airborne, surface, and subsurface unmanned vehicles in order to protect critical infrastructure.

1.2.3 Node Synchronization in SEEs

Management aspects of WSNs for time synchronization and cooperative collaboration of the nodes is important in SEEs. Techniques such as the sliding clock synchronization protocol is used for time synchronization under extreme temperatures. The key aspect of this protocol is a central node that periodically sends time synchronization signals. Then, the node measures the time between two consecutive signals as well as the locally measured time, from which it can determine and rectify any possible errors.

Another good example is creation of an ultra-reliable WSN that will never stop monitoring, even in extreme conditions, and does not require maintenance. Such a system can detect a failing sensor node through a dynamic routing protocol, enabling other nodes to take over the function being carried out by the dead node.

1.2.4 Spectrum Sharing in SEEs

In space, the demand for spectrum is huge, particularly where the safety of personnel and the reliability of control systems are heavily dependent on wireless sensors such as:

- structural health
- impact detection and location
- leak detection and localization.

Robust and reliable dynamic spectrum-sharing schemes are needed. In order to make use of spectrum-sharing in space, we need to make modify systems used in terrestrial networks, in which, for example, errors in spectrum sensing are unavoidable but which often lack incentives for primary users to allow network access to secondary users.

1.2.5 Energy Aspects in SEE

Medium access control (MAC) plays a crucial role in providing energy-efficient and low-delay communications for WSNs. Sensing systems designed for operation in space or underwater face additional challenges, notably long and potentially variable propagation delays, which severely inhibit the throughput capability and delay performance of conventional MAC schemes. Outages due to energy shortages and adverse propagation conditions also pose significant problems. We now examine similar challenges associated with reliable and efficient multiple access in SEEs, focusing on underwater sensing systems.

The use of energy-harvesting technology has important implications for medium access, since uncertainty surrounding the future availability of energy makes it difficult to arrange reliable duty-cycles, schedules or back-off times in the traditional way. The challenges associated with long propagation delays are well understood for satellite systems. Demand assignment multiple access is commonly employed as a means of achieving high channel utilization, since capacity can be allocated to nodes in response to time-varying requirements.

1.3 Chapter Abstracts

The chapters for this book come from two sources:

- expansions of previously published journal or conference papers, where the authors' work has already been peer reviewed
- original reviews to expand the scope of the book, at the choice of senior and experienced academic or industrial experts.

1.3.1 Abstract of Chapter 2

This chapter, entitled 'Feedback Control Challenges with Wireless Networks in Extreme Environments', presents a new perspective on feedback control systems that operate in a wireless fashion. Motivated by the high cost of installing a wired control system in aerospace vehicles and even automobiles and the added weight and fuel requirements that comes with it, this chapter aims at redesigning control systems by eliminating wires from sensors to the controller and eventually to actuators. Replacing wires with wireless links in a control system may be modeled in different ways. This chapter describes a delay and noise model with parameters coming from the wireless system. The performance of the control system is then studied with added delay and noise in the loop to address the feasibility of wireless control.

A case study is presented in which a launch vehicle is instrumented with several accelerometer sensors to model the vibration modes. This information will be useful for fine-tuning the trajectory of a rocket as its structure bends at high speeds. The system dynamics and controller are modeled using first- and second-order differential equations with a parameter used to determine rise time, settling time, and overshoot of the closed-loop response of the system.

A fixed delay is then added to the system and presented in rational form using the Pade approximation. The effect of the delay on the stability of a first-order system is then studied. This result is further extended to multi-sensor inputs with different delays. The effect of the delay on the transient response of a second-order system is studied too.

External disturbances affecting a wireless link are modeled using an additive white Gaussian noise model, which will slightly alter the parameters of the system's differential equation. Rise time and overshoot changes versus noise are plotted and analysed.

Although there is still a long way to go before such systems are implemented in critical applications, this chapter lays the groundwork for modeling, analysing and studying such systems and presents a framework for designing a wireless controller for sensor and actuator networks in SEEs.

1.3.2 Abstract of Chapter 3

This chapter, entitled 'Optimizing Lifetime and Power Consumption for Sensing Applications in Extreme Environments', considers power optimization in a general sensor network; that is, without specifying any particular application. An optimization problem is defined with a view to extending the network's lifetime by using the minimum power possible. The proposed problem is shown to be convex, therefore having a global solution that can be obtained by applying traditional numerical methods and convex optimization theory.

An upper and lower bound on network lifetime is derived as a guideline for designing power-constrained WSNs. A lower-complexity method to find near-optimal solutions is also presented. Several practical scenarios are also described and numerical results are presented to verify the proposed method.

A specific section in this chapter is dedicated to power optimization in applications in SEEs, which is the main focus of this book. It is noted that since sensors nodes may fail for various reasons (including battery depletion), the number of sensors in a network may change over time, hence power allocation should also be changed to follow the optimal solution for the most recent status of the network. The proposed power allocation approach performs better than uniform power allocation; in other words, using the same battery in each sensor node in sensor networks with limited and non-renewable energy sources.

The application scenarios considered in this chapter include passive multiple radar sensing, where an unknown target signal is detected or classified. Carrying out this task using a network of low-cost sensors is much more reliable and cost-effective than using a single complex radar system.

Another scenario is a solution for construction of oceanographic maps. This application uses sinking sensors to create a 3D map of the ocean bed. Using the optimal power-allocation strategies presented in this chapter eliminates the need for using large batteries and can reduce the overall cost of the network.

In summary, this chapter finds the power-allocation scheme that can best extend the network lifetime without violating signal-quality requirements.

1.3.3 Abstract of Chapter 4

This chapter, entitled 'On Improving Connectivity-based Localization in WSNs', addresses the issues of uncommon sensor-node localization methods. These are methods that are used where more common methods, such as GPS, are unavailable (most extreme environments do not have the luxury of GPS), too expensive or complicated.

A review of recent advances in localization for both single- and multi-hop networks is presented, emphasizing connectivity-based localization built upon information gathered from neighboring nodes.

For single-hop networks, centroid and improved-centroid algorithms are presented. The improved method discriminates against hearable anchor nodes among all other nodes and assigns them different weights. For multi-hop networks, the distance vector (DV) hop algorithm is presented. This method is based on the average hop distance and the calculation of a correction factor. The theoretical underpinning of the Hop-count-based localization is discussed, using probability theory and maximum-likelihood estimation. The chapter goes on to present several ways to improve the accuracy of connectivity-based localization: by adjusting the correction factor based on the expected hop progress, an approach that is then extended to exploit neighborhood information. The problem of hop distance ambiguity is noted and a neighbor-partitioning algorithm is proposed to overcome this issue. Numerical results show that this method gives more accurate localization than the original DV hop method as the number of nodes is increased.

Reading this chapter sheds light on the different definitions of the shortest path – shortest hop, shortest regulated neighborhood distance (RND) and shortest distance – and how these can affect localization strategy. The effect of packet reception rate, which captures both physical and network layer status, is also studied for different localization methods. This chapter also distinguishes between isotropic and anisotropic networks and demonstrates that many improved algorithms may emerge from shortest-path or straight-line calculations due to the nature of the network being anisotropic.

In summary, this chapter provides a comprehensive overview of connectivity-based localization (also called the range-free method) in WSNs. The simplicity and low cost of implementation of these methods suits them to many different applications. In SEEs, the lack of GPS requires use of these methods anyway. Therefore, the solutions in this chapter are potentially beneficial for many SEE applications.

1.3.4 Abstract of Chapter 5

This chapter, entitled 'Rare Events Sensing and Event-powered Wireless Sensor Networks', focuses on rare events with low probability but high impact, such as structural failure of a bridge or engine failure of an aircraft.

Detection probability and detection delay are defined as the main parameters of interest in WSNs for detecting rare events. An interesting idea to address the challenge of prolonging network lifetime and increasing the probability of detection while reducing detection delays is energy harvesting.

This chapter starts with investigation of a fully distributed sensor network with energy harvesting and duty cycling to conserve energy. The discussion continues with study of rare events that can produce enough energy to be harvested and used to power sensor networks. Examples such as earthquakes or explosions are among the many rare events that can produce enough vibration energy to trigger a sensor network to switch on and then harvest the energy.

Wireless sensor-node design is examined in two sections: on microcontroller and power-management circuitry. The concept of cluster-centric WSNs for monitoring rare events is presented next. Monitoring civil infrastructure often requires a large amount of data, which may be beyond the current limits of MAC protocols. The network processing and data-aggregation methods presented in this chapter can alleviate this issue, thus expanding the reach of WSNs to SEEs.

The system model presented in this chapter includes a personal area network (PAN) coordinator, which remains idle until an event occurs. When listening to data coming from different clusters, high priority is given to uncorrelated data from different clusters to ensure fairness. A performance analysis in terms of packet arrival times for the IEEE 802.15.4 standard is presented later in the chapter to show how cluster-centric MAC performs in comparison with traditional methods. Average and total time-to-transmit are also studied in the performance analysis section.

In summary, this chapter provides new insights into how to design a WSN for rare-event detection that is both efficient and reliable.

1.3.5 Abstract of Chapter 6

This chapter, entitled 'Batteryless Sensors for Space', describes the challenges of battery operation or replacement and maintenance in SEEs, and the difficulties that this imposes on use of WSNs. Although most examples are from space, the concepts can be generalized to other challenging environments. The benefits of using wireless technology – reduced vehicle weight and cost – are presented in detail.

A cost–benefit analysis of wired and wireless networks is presented. Two different categories of sensors, namely passive and active, are studied and relevant design considerations are presented. Other than the obvious benefits of wireless over wired systems, such as cost and weight, other challenges such as supporting structures, fixtures, cabling, cost of routing and electromagnetic interference and compatibility are noted.

A reliability analysis of wireless systems from data-collection and transmission-channel perspectives is presented. It is noted that combined-source channel-coding methods that take correlation among data streams into account may give high accuracy and reliability even when the individual sensor data may not be reliable. This result is based on well-known CEO problem, which can be extended to correlated sensor data when multiple sensors are in proximity to the same source.

The different categories of active and passive sensors, and the maintenance- and battery-free nature of passive reflective RF sensors are discussed next. The basic operation principles of these kinds of sensor are presented and the challenges in terms of sensor materials, code design and interference are noted. Several methods of interference cancellation at the source and at the receiver side are studied.

In summary, this chapter presents the benefits and design challenges of battery-free wireless sensors for SEEs. Implementation technologies covering the material and physical layer all the way to coding and higher layers such as networking and interference issues are all presented in a coherent manner. The references in this chapter range from traditional academic papers to industry and space agency reports, giving a realistic impression of the current state of the art in this field.

1.3.6 Abstract of Chapter 7

This chapter, entitled 'Contact Plan Design for Predictable Disruption-tolerant Space Sensor Networks', focuses on the design, planning and implementation of disruption-tolerant networks in space. The concept of end-to-end connectivity in the context of the highly dynamic space environment is defined. Network disruption that causes delays but not data losses is at the core of this chapter.

Methodologies to design contact plans for disruption-tolerant WSNs in a systematic way are presented next. The performance of three contract plan design methodologies (FCP, RACP and TACP) are compared.

Communications-system parameters, such as transmission power, modulation and bit error rate (BER0 are used together with orbital dynamics parameters such as position, range and antenna orientation to determine the future contacts. Other constraints include time-zone limitations and concurrent resources. Several examples, such as satellite-to-Earth and intra-satellite communications, are provided for further clarity.

The contact plan design is based on input parameters such as the number of nodes, topology states, the initialization time, interval duration, buffer capacity and traffic. A case study is presented to analyse the contact plan design and includes an experiment, 3 hours and 22 minutes in duration, with four flights over the north pole. Resulting traffic flow for all four half orbits is presented. The TACP, RACP and FCP methodologies are compared for various network loads.

The chapter continues with a discussion of safeguard margins and topology granularity of TACP, followed by the contact plan computation and the distribution and implementation considerations.

In summary, this chapter presents the benefits of using WSNs in space to enhance the performance of Earth observation missions. Since conventional Internet-based protocols fail in the space, disruption-tolerant networks are offered as a reliable alternative and the problems and challenges associated with their design and performance analysis are presented. The main goal in these systems is to ensure reliable sensor data delivery, even if some portion of the data may experience more delays than others.

1.3.7 Abstract of Chapter 8

This chapter, entitled 'Infrared Wireless Sensor Network Development for the Ariane Launcher', describes WSN development for launch vehicles. This poses a unique set of constraints and requires careful attention to reliability. One of these constraints is the strict limit on electromagnetic emission of the sensor nodes. The focus of this chapter is on infrared WSN links inside the upper stage of the rocket and finding ways to minimize packet loss.

Experimental results on the infrared transmitter–receiver link at 1–2 m without line of sight are presented. The design consideration at the device level in terms of bit-error-rate changes versus diode resistance at different illumination levels and angles is presented. Dual use of multi-layer insulation as reflector for infrared communications at low data rates is studied.

This chapter also covers ASIC development strategies for low-power infrared operations. Deciding on the optimal modulation, synthesizing and testing the receiver in a FPGA, and designing the ASIC and testing it on the sensor are discussed. Signalling methods, such as Uni-polar and Manchester codes, are compared in terms of their power spectral efficiency. The ASIC design details include an analysis for both leakage and dynamic power. A radiation-hardened ASIC with similar characteristics was ultimately used on the launch vehicle.

Time synchronization in a reliable sensor network is also examined. The objective is to achieve time synchronization with minimal hardware on the sensor nodes and with as small a load as possible. The stochastic nature of wireless communication, which yields less deterministic delay than wired communication, is noted. Techniques using visible light communication are used to synchronize all sensors. Some of the sensors used in this system are air pressure, temperature, infrared and visible light and humidity sensors.

The chapter presents a complete sensor-node system from the design to the implementation and test phases for space launch vehicles.

1.3.8 Abstract of Chapter 9

This chapter, entitled 'Multichannel Wireless Sensor Networks for Structural Health Monitoring of Aircraft and Launchers', discusses an adaptive multi-channel approach to structural health monitoring that is collision-free and addresses challenges such as latency, throughput and robustness.

Although many other books may address different aspects of structural health monitoring, this chapter considers the very complex structure of an aircraft with its many interconnected sections. Automated monitoring of such a complex system requires many different types of sensor, all connected to a central processing unit using wired or wireless systems. This chapter focuses on wireless systems, which are desirable because of their lower weight and cost. Wireless may even be the only solution, if the component under monitoring is not easily accessible or on a rotating blade, where wires cannot be used.

WSN requirements for aircraft in an industrial setting are considered. Only non-critical sensors and off-the-shelf technologies are used, to make for shorter development times. Static and dynamic sensors with various sampling rates and network sizes are considered.

After a brief overview of the existing research and development activities in this area, the challenges facing multi-channel use and data-gathering applications are presented. The solution presented in this chapter is a unified approach, which addresses both IEEE 802.15.4 requirements and the limitations imposed by use in aircraft. Issues such as signal propagation inside the aircraft cabin, mesh multi-channel wireless networks, node discovery and synchronization, channel selection and network connectivity are among the many challenges that are discussed in this chapter.

When it comes to medium access control, contention-free, contention-based and hybrid protocols are studied. Dynamic multi-hop routing and energy efficiency are discussed next. Collisions, overheating, control-packet overhead, idle listening and interference issues are discussed as part of the energy-efficiency topic. Energy-saving methods, such as data reduction, overhead reduction, energy-efficient routing, duty cycling and topology control are discussed in detail.

Finally, the robustness of adaptive WSNs to changes in environment, topology or traffic are discussed, and centralized and distributed methods are compared.

1.3.9 Abstract of Chapter 10

This chapter, entitled 'Wireless Piezoelectric Sensor Systems for Damage Detection and Localization', starts with applications of Lamb waves and their detection using piezoelectric lead zirconate titanate (PZT) sensors, and a brief overview of prior work in this area. The majority of has been based on wired systems, but this chapter presents the advantages of using wireless systems for damage detection and the challenges that need to be addressed. Transforming current WSNs to support high sampling rates is one such challenges, and is due to the high-frequency content of Lamb waves. Techniques adopted from the compressive sensing literature are presented as a possible workaround solution. This may seem to solve the problem, but embedding a compressive sensing algorithm in a sensor node is another challenge that comes with this solution, and is discussed in this chapter in detail.

An introduction to Lamb-wave-based damage detection is presented in the chapter, before WSNs aspects are addressed in more depth. Active piezoelectric-based sensing is discussed next, and issues such as frequency tuning, windowing and data processing are presented in depth.

The next part of this chapter moves to networking aspects and presents a topology with multiple regions, each with its own wireless node, relaying PZT sensor data to the base station for processing.

The detailed architecture of the sensor nodes, including analogue-to-digital converters, digital signal-processing cores, ARM processors and wireless radio communications, are presented. Distributed data processing for the proposed method is described next. This discussion covers data synchronization issues and consideration of the required buffers and direct memory access.

The chapter concludes with some remarks on synchronization and the scalability and reliability of this method. It is noted that damage detection using structural-borne ultrasonics is a recently revisited and hot topic of research. This chapter presents a concise but comprehensive overview of this area of research that can be used by academics to develop further theories and by professional engineers to implement new systems with more capabilities and higher efficiency.

1.3.10 Abstract of Chapter 11

This chapter, entitled 'Navigation and Remote Sensing using Near-space Satellite Platforms', deals with navigation in near-space (20–100 km altitude). This is a special area of the atmosphere, which is unsuitable for aircraft and satellites. With recent developments in microelectronics and propulsion systems, however, it has gained some attention mainly due to the lower cost of implementing near-space platforms compared to aeroplanes or satellites.

Some advantages of near-space are the ability to control the platform and keep it stationary over the region of interest or to move it on demand to other areas. Two examples of NASA-designed platforms for near-space use are HELIOS and Pathfinder, which are solar powered. Since near-space platforms are much closer to Earth and potential observation-area wireless sensors can detect much weaker signals, which are not otherwise detectable by low-Earth orbit satellites. It is also important to note the line-of-sight advantage of communications using these platforms.

In addition to detailed discussions of the NASA-developed platforms, this chapter also provides a information on ESA-developed projects such as HeliNet, CAPANINA, UAVNET, CAPECON, and USICO.

Other applications of near-space platforms, such as radar and navigation sensors, are also discussed. Specifically, storage aspect ratio imaging is highlighted as one of the potential areas that can benefit from these closer-to-Earth imaging platforms.

An integrated framework that combines near-space platforms with satellites and aircraft into one coherent network with sensor systems at different levels is depicted, illustrating the potential benefits and the coordination challenges. The design considerations for satellite-to-Earth augmentation using near-space platforms and optimal placement and coverage analysis are also presented in this chapter. The limitations and vulnerabilities, as well as legal and implementation issues, are also discussed.

In summary, this chapter presents an alternative platform for wireless sensor networking, which enables new applications but can also enhance the performance of existing applications. The design challenges in such environments will be interesting for professional engineers who are interested in designing such systems.

1.3.11 Abstract of Chapter 12

This chapter is entitled 'Underwater Acoustic Sensing: An Introduction', and introduces readers to underwater networked sensors, their limitations, and potential applications. The authors argue that the underwater environment is far more hostile than it normally sounds, making it a true extreme environment. This is mainly due to its very poor communication channels under existing technologies. Whilst exploring a wide range of transmission possibilities they look at options such as free-space optics, magnetic induction, radio with upper and lower frequencies, and acoustics in the form of audio, sonar and ultrasonic waves. The analysis and modelling of acoustic waves for their characteristics and multi-path solutions highlight important features of underwater acoustic transmission and its associated networking.

They expand this analysis with an in-depth consideration of acoustics, as the most popular technology, used for centuries in marine applications, along with its well-known problems including:

- long distance requirements
- heavy signal loss
- node losses
- poor localization due to a lack of GPS (taken for granted in terrestrial sensor networks)
- poor accessibility
- problems of biological fouling
- high cost of experimentation
- high cost of maintenance.

The authors survey some recent developments to highlight the advances made. They describe small mobile platform technologies, such as smart and autonomous vehicles for monitoring and exploration: remotely operated underwater vehicles, autonomous underwater explorers, unmanned underwater vehicles, underwater drones, and underwater gliders. They also look at the possibility of larger complex platforms that might be the basis of a smart underwater environment, with multifunctional complexes as pilot deployment centers for the demonstration of new multi-technological developments.

1.3.12 Abstract of Chapter 13

This chapter, entitled 'Underwater Anchor Localization using Surface-reflected Beams', addresses the localizing of underwater objects near the surface of the water using a new mathematical model. This is applicable to most immersing vehicles and well over 90% of the sensing and underwater traffic today. Then, the author extends the model to make use of bottom-reflected acoustic beams, an approach applicable to shallow waters and some special cases of object localization in deep water. A lab-based prototype has been set up to validate some parts of the model in a scaled 3D underwater environment.

The localization model uses line-of-sight and surface-reflected non-line-of-sight links for locating a node that has been lost or drifted away from the network-controlled system, using reference points shown to be very effective in accuracy and speed of the process. One of the main features of the method is estimation of the angle of arrival (AOA) at the lost node; this has made a significant contribution and involves combining the surface-reflected non-line-of-sight signal arrays using directional acoustic transducers. The simulation results demonstrate the method's localization performance and the advantages of the proposed scheme when more reference nodes are available; the localization error is halved. Further observations bring the benefits of combining both the line-of-sight and non-line-of-sight signals. This is significant when the water's surface is rough, and most of the effective signal power of the acoustic channel comes from indirect rather than direct paths.

1.3.13 Abstract of Chapter 14

This chapter, entitled 'Coordinate Determination of Submerged Sensors with a Single Beacon using the Cayley–Menger Determinant', introduces a new dynamic method for locating sensor-enabled nodes, objects and vehicles with higher precision.

In this measurement system, a closed-form solution is used to determine the coordinates and associated specifications using beacon nodes at the surface; these can give continuous updates, with the coordinates claimed to be achieved instantly, without a need for any complex infrastructure or use of reference points. The method uses the Cayley–Menger determinant and linearized trilateration. An interesting feature is associated with using the Cayley–Menger determinant: the six edges of the planar quadrilateral are not independent and need to satisfy certain equality constraints. This constraint then can be exploited to reduce the impact of distance measurement errors, and this insight can be extended to give TDOA and AOA localization measurements.

The Cayley–Menger determinant gives the volume of a tetrahedron created by one beacon at the surface and three sensors at the bottom of the water column to determine the coordinates of each of the sensors with respect to one of them, where the determinant is nonlinear. Then, applying the degree-of-freedom property to expand the determinant, a linearized system is solved once the coordinates of the sensor are found with respect to known distances, trilateration and linear transformation of the reference point. This is used to determine the coordinates of the sensors with respect to the beacon node.

At first, the coordinates of the submerged sensors are determined assuming the sensors are stationary; voluntary or involuntary mobility of the sensors is incorporated into the mathematical model at a later stage.

1.3.14 Abstract of Chapter 15

This chapter, entitled 'Underwater and Submerged Wireless Sensor Systems Security Issues and Solutions', addresses the security aspects of underwater wireless sensor systems (UWSSs), which for most applications – scientific exploration, commercial, surveillance and defence – is essential. It looks at the specific characteristics of the underwater sensing environment and sets out the need for a new and more suitable network architecture that can overcome the most critical drawbacks of UWSSs, including the unavailability of commonly used features of open-air and terrestrial facilities, such as GPS localization and the maintenance of security requirements.

Security considerations of underwater projects are related to issues such as adversary nodes eavesdropping on traffic and also can interrupting and modifying messages, creating more critical vulnerabilities for data confidentiality, integrity, and authentication processes.

Because of the impracticality of asymmetric encryption mechanisms, the authors have looked for *symmetric* encryption mechanisms suitable for UWSSs, and have therefore tried to find a suitable key-management technique too. Aspects considered include:

- Connectivity security requirements and issues for the new key-distribution approach, considering nomadic and meandering current mobility models.
- Denial of service attacks: jamming by reducing duty cycles, power exhaustion by limiting retransmission, multiple identities using location verification techniques and sinkhole attacks by adopting more secure routing protocols.
- Secure localization approaches, which are mostly under development.
- Secure cross layer techniques for more efficient communications; these are to be designed and adopted as they mature.
- Secure time synchronization in scheduling and TDMA protocols normally caused by long propagation-delay transmission and highly mobile nodes.

1.3.15 Abstract of Chapter 16

This chapter, entitled 'Achievable Throughput of Magnetic induction-based Sensor Networks for the Underground Communications', addresses one of the most difficult communication environments. Underground communication for sensing and actuating is truly challenging due to the unique channel variability and complexity involved. Wireless underground sensor networks (WUSNs) are required for industrial applications ranging from soil monitoring, earthquake prediction, inspection of mines, oil reservoirs, tunnels, structural and construction health monitoring, and agriculture. They can also be used for many other common purposes, such as underground object identification and localization.

Communication in such environments represents a bottleneck for WUSNs. Due to the heavy path loss, traditional electromagnetic radio wave propagation techniques can be used only over very short distances and degrade extensively with increase in soil moisture. The authors therefore consider alternative communication technologies and, due to its popularity in short-depth applications, they suggest magnetic induction (MI) as a possible solution. MI has already been used for near-field communication and wireless power transfer. The MI-WUSN has been tried in various ways, including MI waveguides with passive relaying, in which multiple magnetic relays implemented as resonant coils are combined in a waveguide structure for making connections between the active nodes of the networks. The practical round figure of a 3-m relay system with no passive relays is an interesting solution for many underground industrial applications.

Following their underground transmission model for MI-WUSNs the chapter explores some details of modelling and maximization, including network specifications, throughput maximization, direct transmission throughput and waveguide style throughput, and compares the methods used for various applications.

1.3.16 Abstract of Chapter 17

This chapter, entitled 'Agricultural Applications of Underground Wireless Sensor Systems', is a technical review addressing underground WSSs for agriculture. There is a technological challenge between the established terrestrial wireless sensor network (TWSN) technology and the new and immature technology of WUSN. Comparing underground technologies with TWSNs is unrealistic. Combination of WUSN and TWSN technology can provide breakthroughs. Bringing these new solutions to dry lands and mountains where the soil is extremely rich but where water efficiency is vital may become more important than just a contribution to the global economy.

The authors have made significant efforts to unearth the recent history of sensor technology development from decades of scattered research and less-known developments and real projects in remote areas. At a time where pressure from global businesses and expanding towns threatens the future of traditional farming, and while Internet trading and self-sufficiency could save many small villages from disappearance, there could be a big market for combined TWSNs and WUSNs: mass production of the parts and systems by larger industries, and SMEs providing a whole range of services and systems including parts, devices, sensors, expertise and advice, training, and sharing or hiring out new and advanced agricultural tools, machineries, and systems.

1.3.17 Abstract of Chapter 18

At the dawn of the Internet of Things (IoT) and WSNs this chapter, 'Structural Health Monitoring with WSNs', shows us how smart sensors bring new safety measures to our monuments, buildings, tunnels and bridges, reducing casualties and saving the structures.

A statistical literature review shows that whilst research into the IoT and WSNs has begun to fall, interest in structural health monitoring (SHM) is still on the rise, demonstrating its resistance to the hype. The authors examine SHM technologies and associated breakthroughs including compressed sensing and energy consumption. Then, under the title of 'WSN-enabled SHM applications', they look into:

- integrating the IoT and SHM
- commercially available acoustic emission sensors
- RFID-assisted SHM.

The network topology and associated network overlay part of the chapter introduces the new idea of multifunctional overlaying network management.

Then the authors consider the power requirements and energy consumption of WSNs, trying to ensure the maximum lifetime achieved. Choices include use of the correct processors. For a global architecture, rapid development of IoT-SHM systems will enable mass production of the parts at extremely low prices, following global Internet and sensor interworking standards.

SHM is unusual; although it sounds as though it should be an industrial application of WSNs, because of its use in, for example, internal measurement of walls, buildings, bridges, tunnels, and in embedded, and movement-enabled monitoring, we therefore group it with underground and confined applications of the WSSs.

1.3.18 Abstract of Chapter 19

This chapter, entitled 'Error Manifestation in Industrial WSN Communications and Guidelines for Countermeasures', addresses the problem of insufficient reliability in classic industrial wireless sensor networks (IWSNs). The authors suggest that this is due to the existing de facto standards' inability to cope with the physical and electromagnetic properties of industrial environments.

As the compromising factors in industrial WSN communications, they list physical factors as follows: sensor placement, reflective surfaces, open-space layout and moving objects. They also consider electromagnetic interference and signal distortion.

To meet the high reliability requirements of wireless sensor applications in the harsh propagation conditions of industrial environments, they propose a resilient lightweight solution based upon the nature of errors, follow their footprints on the bit-, symbol- and DSSS chip-level of IWSN signals, trying to distinguish the differences made by WLAN interference to those caused by multipath fading and attenuation. Because signal waveforms and physical layer properties are defined by WSN communication standards, the solutions for higher reliability are to be found in the design space of the data link layer and its medium access control sublayer.

1.3.19 Abstract of Chapter 20

This chapter is entitled 'Medium Access Approach to Wireless Technologies for Reliable Communication in Aircraft', addresses wireless communication systems and their impact on sensor and actuator networks and the safety aspects of controlling the air craft. After a preliminary investigation and discussion of practical wiring problems and outlining passenger concerns about aeroplane safety, the authors set out a procedure to meet the required objectives. To do this they introduce their fault-tree analysis, as the framework of a reliability assessment. Then they introduce their performance metrics and analyse feasible wireless technologies. The chosen candidate wireless system is evaluated against the performance metrics in the reliability framework.

The reliability model is illustrated through an example of a passenger heat sensor application, building upon the errors from four main contributing components of power, sensor, control, and communication systems. Then they expand these elements to the hardware, application message delivery, and security components, which are all converted into an accumulative set of packet-level failures.

The reliability assessment model takes the requirements from aircraft safety standards and merges them down to the level of packet transmission probability failures. Then there is a lengthy analysis and comparison of six associated standards:

- Wireless Interface for Sensors and Actuators
- ECMA-368
- IEEE 802.11e
- IEEE 802.15.4 (IoT)

- IEEE 802.15.4 (WirelessHART)
- 3GPP LTE

for a typical aircraft application. There is benchmarking of the unmodified capabilities, and plots and tables to be used by designers for engineering the best possible solution for a wireless enabled sensor based aircraft communication system.

The authors claim their method is practical and note that is makes use of commercial off-the-shelf hardware and that their work gives communication engineers better insights into designing more reliable MAC systems.

1.3.20 Abstract of Chapter 21

This chapter, entitled 'Application of Wireless Sensor Systems for Monitoring of Offshore Windfarms', addresses the way WSSs can be used in offshore windfarms. Wind is everywhere and abundant. In coastal areas, the intensity of wind energy is higher and present everywhere, and wind turbines can go onshore or offshore. Onshore wind turbines are easier to maintain, monitor and control than offshore wind turbines. Offshore turbines do not occupy valuable land, but suffer from corrosion because of the salt water and humid air.

The authors discuss the various applications of WSNs in windfarm monitoring systems. A unique energy-efficient application-specific routing protocol called the network lifetime enhancing tri-level clustering and routing protocol (NETCRP) is discussed. A fault detection method is also discussed. Suitable enhancements can be incorporated to further decrease the energy consumption and hence increase network lifetime. Examples include the introduction of sleep periods. Moreover, the effects of

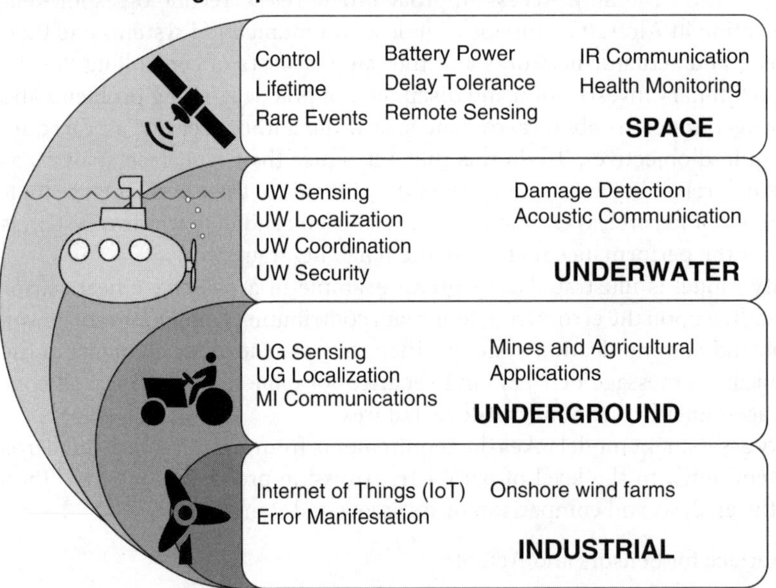

Figure 1.1 An illustration of the four harsh, and most difficult areas for using the WSS and its new applications.

increasing and decreasing the number of quantization levels are studied, and equations for the optimal number are established.

The above Chapter summaries demonstrate a unique blend of R&D and Implementation activities in the field of Wireless Sensor Systems from today's academic research and industrial deployments. Penetrating and breaking the existing limiting barriers in the very challenging environments, under four topical applications of Space, Underwater, Underground, and Industrial, with considerable potentials to enhance our lives and prosper our future industries, as shown in Figure 1.1.

Reference

1 Rashvand, H., Abedi, A., Alcaraz-Calero, J., Mitchell, P. D., and Mukhopadhyay, S. (2014). Wireless sensor systems for extreme environments: a review. *IEEE Sensors Journal*, **14**(11), 3955–3970.

2

Feedback Control Challenges with Wireless Networks in Extreme Environments

Lonnie Labonte[1], *Ali Abedi*[2] *and Praveen Shankar*[3]

[1] *Department of Electronics and Computer Engineering, University of Maine, USA*
[2] *Department of Electrical and Computer Engineering, University of Maine, Orono, USA*
[3] *Department of Mechanical and Aerospace Engineering, California State University, California, Long beach, USA*

2.1 Introduction

The efficient control of complex systems with improved reliability and operability depends in part on the use of new technologies. Wireless networks provide an opportunity to improve control systems in terms of their cost, performance, implementation and maintenance. The high cost of installing a wired control system in an application is a key motivation for introducing wireless control systems. The modularity that is possible with a wireless control system simplifies maintenance and lowers maintenance costs. These wireless networked control systems (WNCSs) can be used in a wide range of industries. Industries that have benefitted from WCNSs include but are not limited to: aerospace [1], transportation, biomedical engineering and civil engineering [2].

There many challenges in the design of WNCSs. The wireless link between the sensors and the controller–actuator system inherently creates a delay in the relay of the information from sensor to controller. The overall delay between sampling and eventual decoding at the receiver can be highly variable because both the network-access delays and the transmission delays depend on highly variable network conditions, such as congestion and channel quality. Delay can be constant, time-varying or even random, and is frequently a source of instability and performance deterioration of control systems. The conditions that are required for keeping a system stable with respect to signal-to-noise ratio have been discussed in the literature [3, 4]. The effects of time delay and data packet loss have been analysed by considering a data packet loss as a special kind of time delay [5]. Time delay in wireless networks is caused primarily by buffering and propagation delays as well as the time spent processing information: encoding or decoding for example [6]. These delays are significant compared to the pre- and post-processing delays.

This chapter addresses some of the issues of wireless sensors networks when they are used in control system design. The primary focus is the effect of delays, but a brief description of the effect of noise is also presented. Section 2.2 describes the motivations for using wireless control networks and current wired control theory as it pertains to launch-vehicle control. Section 2.3 provides brief overview of system dynamics

Wireless Sensor Systems for Extreme Environments: Space, Underwater, Underground, and Industrial.
First Edition. Edited by Habib F. Rashvand and Ali Abedi.

and control system design relevant to the rest of the chapter. Section 2.4 describes the effect of constant delay on control systems when only one or more wireless sensors are implemented in a first-order system. Additionally, this section also addresses the effect of the combination of delay and noise on the transient response characteristics of a second-order system. Section 2.6 provides a summary of the chapter and Section 2.7 briefly discusses the direction that research is going in order to make wireless control networks a reality in the aerospace and other fields.

2.2 Controllers in Extreme Environments

Using wireless sensors and actuators in a feedback control system can eliminate the extra weight and cost required by a wired system. Fly-by-wire was proposed in the 1970s to replace heavy hydraulic control systems with wires. Ever since, the aerospace and sensor communities have been pushing the limits of energy and fuel efficiency by developing advanced control systems [1]. To create more advanced and reliable control systems, additional sensors and actuators are necessary. The fabrication and installation of this equipment is estimated to cost a several thousand dollars per kilogram [7]. Fly-by-wireless is a vision of wireless feedback control networks for active flight [8]. Wireless sensor networks have several benefits, such as reduced costs of installation and maintenance, additional redundancy and the ability to use advanced control algorithms. For example, as will be described in Section 2.2.1, it has been proposed that wireless sensor networks be used for estimating the bending-mode characteristics of a nanosat launch vehicle in real time, allowing for adaptive and reconfigurable control strategies to be implemented. However, with the implementation of wireless sensors and actuators, the existence of network dynamics and packet drops will cause additional complexities in designing the control system. The designer will not only require novel control strategies to mitigate these effects but also tools to analyse the effectiveness of the approaches.

In this chapter, the effect of a wireless network on the control system is reduced to a simple constant delay. Through the rest of the chapter, the effect of this delay on the stability and performance of the system will be discussed.

2.2.1 Case Study: Wireless Sensor Networks in Extreme Environments

Next-generation launch vehicles that are designed to be reconfigurable and reusable will require control strategies that actively minimize oscillations about the three rotational axes: roll, pitch and yaw. The pitching and yawing dynamics of the vehicle are significantly degraded by the elasticity of the structure [9], which manifests itself as lateral vibrations. While optimal design of the launch vehicle may reduce some of these effects, active control systems are required. The performance of the control system, however, relies on accurate knowledge of the vibration characteristics. Modeling the vibration characteristics from first principles is the initial design step, followed by extensive structural testing [10]. Ultimately, even with structural testing, it is very difficult to determine which of the vibration modes will get excited during a particular phase of the flight. While modern launch-vehicle controllers rely on a priori knowledge of these modes, the best solution is to determine the vibration characteristics in real time, and cancel its effects using a feedback control algorithm. This new approach requires many sensors, which have to be networked wirelessly because it becomes rapidly impractical to have all of them 'wired' together.

Rezaei et al. [11] conducted a simulation study for the real-time determination of the elastic-mode characteristics (frequency and displacement mode shape) of a launch vehicle using wirelessly networked sensors. They assumed that an array of sensors to measure the local strain at different locations along the length of the vehicle was deployed (see Figure 2.1). The determination of the frequency of the excited modes is accomplished using a real-time fast Fourier transform (FFT) of the buffered strain data. The algorithm to determine the mode shape is based on the strategy employed by Jiang et al. [12]. The locally measured strain for a pre-determined length of time is used to determine the cross-correlation function matrix of the excited modes (see Figure 2.2). The eigenvectors of this matrix provide the strain mode shape, and can also be double-integrated to obtain the displacement mode shape. While Jiang et al. [12] provides comprehensive simulation and experimental evidence of the validity of the methodology, the data from multiple sensors is assumed to be available without any delay between the sensor arrays. In Rezaei et al. [11], the data is obtained from a wirelessly networked array of sensors that brings in additional complexity, which is simply modeled as a constant delay between each of the sensors.

The results of the study—the accuracy of the reconstructed modes—consists of the comparison between the cases with and without delay in the network. For this study, the launch vehicle is modeled as a free–free beam with five wireless strain sensors located equidistant from each other on the beam. The launch vehicle is assumed to exhibit the first three modes of vibration. Each of the three modes are excited independently and

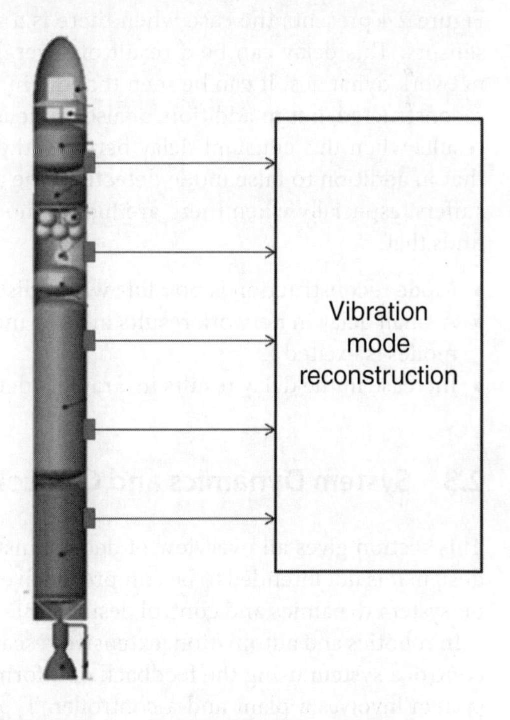

Figure 2.1 Launch vehicle with wirelessly networked strain sensors [11].

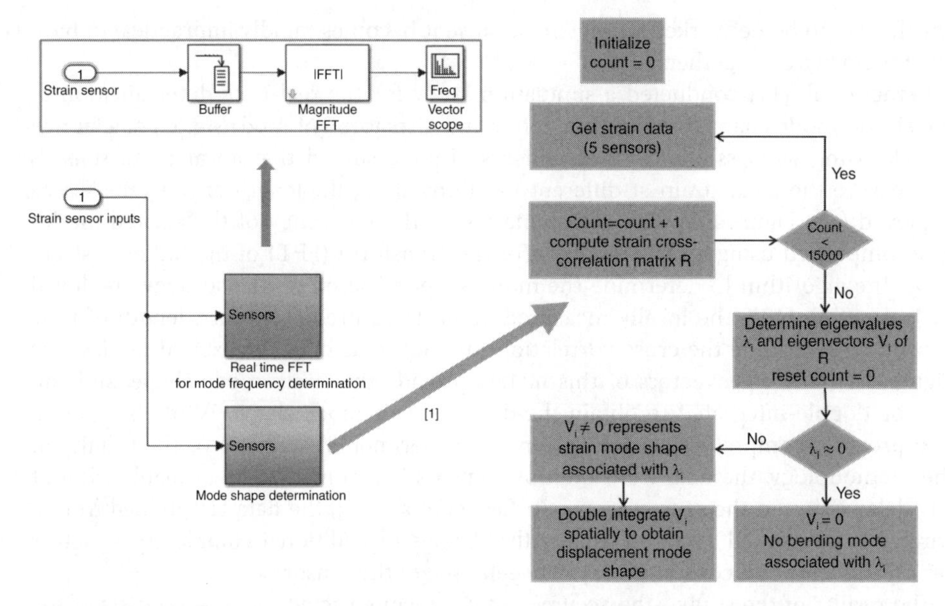

Figure 2.2 Elastic mode reconstruction with wirelessly networked strain sensors [12].

the mode-reconstruction algorithm is executed. Figure 2.3 presents the reconstructed modes when each of the three modes are independently excited and there is no delay between the sensors. It can be seen that each of the modes is accurately reconstructed. Figure 2.4 presents the case when there is a small constant delay between each of the sensors. This delay can be a result of several complexities, including packet loss and network dynamics. It can be seen that the independently excited modes are accurately reconstructed, but in addition, a false mode is detected as well. Figure 2.5 presents the results when this constant delay between the sensors is increased. It can be seen that that in addition to false mode detection, the accuracy of the mode reconstruction also suffers, especially when there are higher modes being excited. In summary, the study finds that:

- Mode reconstruction is possible with a distributed sensor network
- A small delay in network results in 'false mode reconstruction' when any of the three modes is excited
- Increase in the delay results in a rapidly deteriorating reconstruction of the mode.

2.3 System Dynamics and Control Design Fundamentals

This section gives an overview of deterministic system dynamics and classical control design. It is not intended to be comprehensive and the reader is directed to the literature on system dynamics and control design [13].

In robotics and automation, extensive research has provided designs that can actively control a system using the feedback of information from sensors. A very basic control system involves a plant and a controller. Typically the plant outputs information that is read by the sensors and this information is given to the controller to actively control

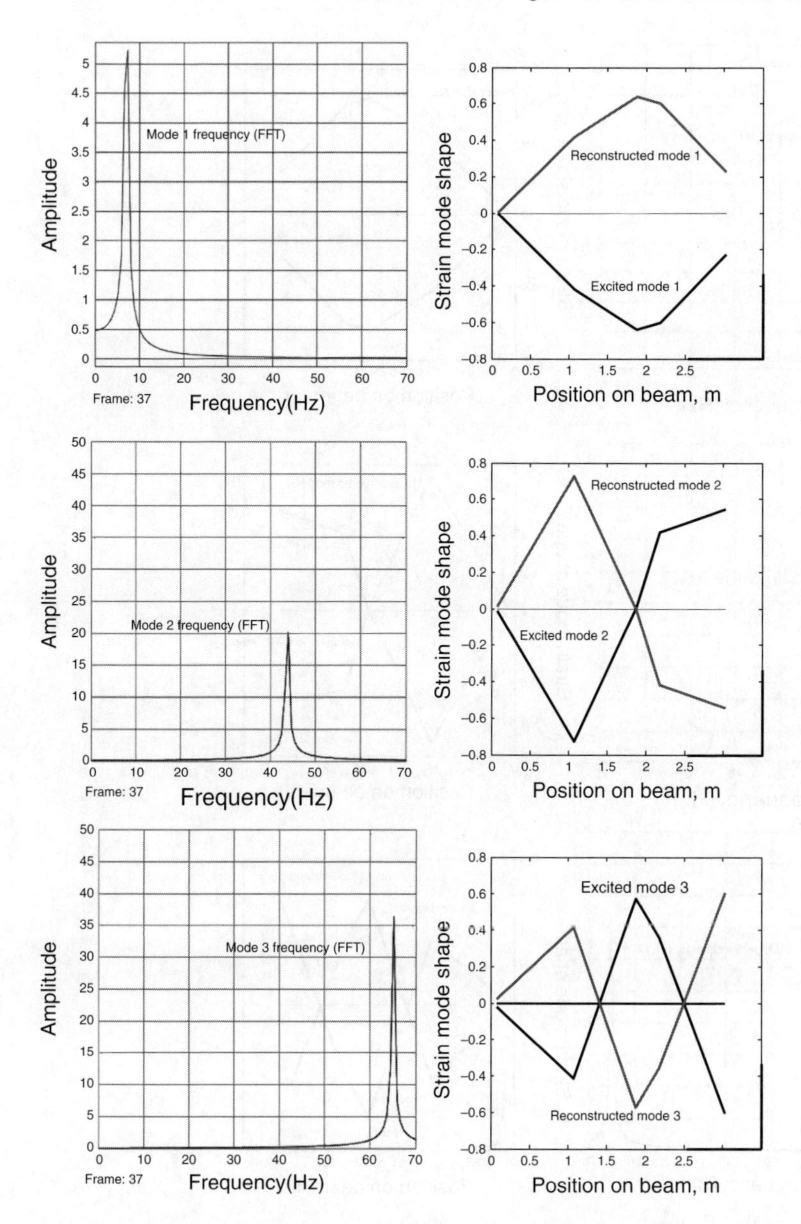

Figure 2.3 Mode reconstruction results with no delay between sensors.

the plant as it operates. It is important to note that any control system model is only a representation of how the system behaves; in reality, the system may act differently due to changing parameters and inaccurate values in the system model. With that being said, most systems can be represented well enough for the purposes of practical plant design and analysis of the system parameters. Figure 2.6 shows the component block diagram of an elementary feedback control system.

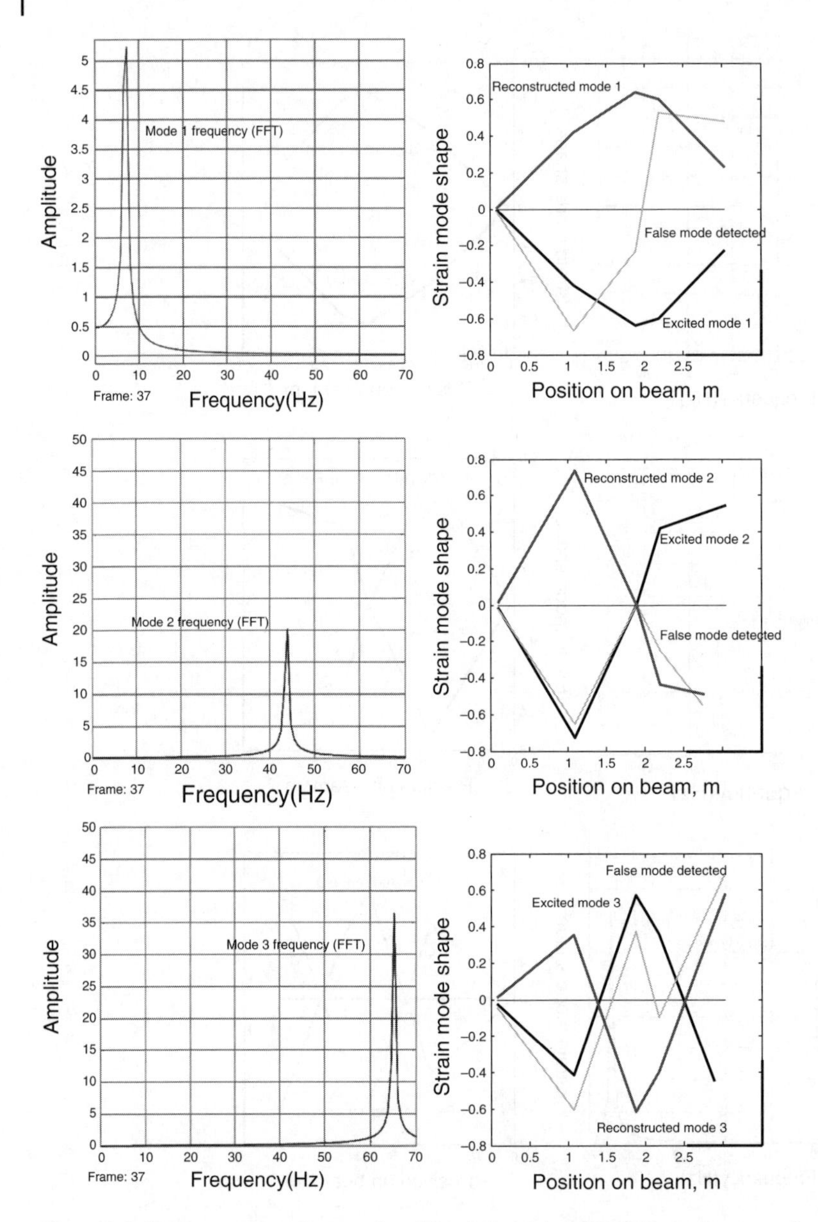

Figure 2.4 Mode reconstruction results with relative delay of 0.0015 s between each sensor.

2.3.1 System Dynamics

Every element in the block diagram of Figure 2.6 is a dynamic system; that is, it has characteristics that vary with time. A typical deterministic dynamic system is represented by an ordinary differential equation. The order of the differential equation represents the number of variables (also called the 'state' variables) that vary with time. While a dynamic system may have several state variables, a first-order system (one state) and

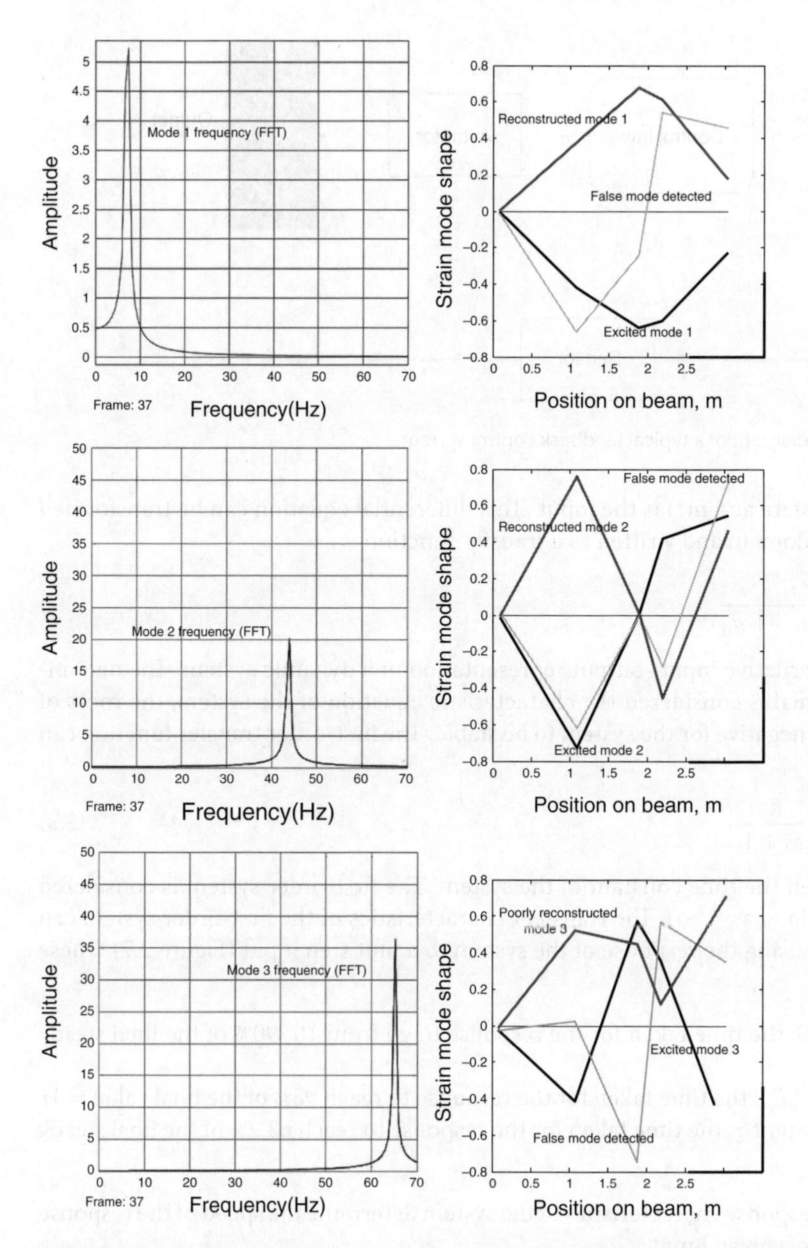

Figure 2.5 Mode reconstruction results with relative delay of 0.015 s between each sensor.

second-order system (two states) provide a simple system for analysing novel control strategies.

A first-order system can be represented by a differential equation such as

$$a_1 \frac{dx}{dt} + a_0 x = u(t) \tag{2.1}$$

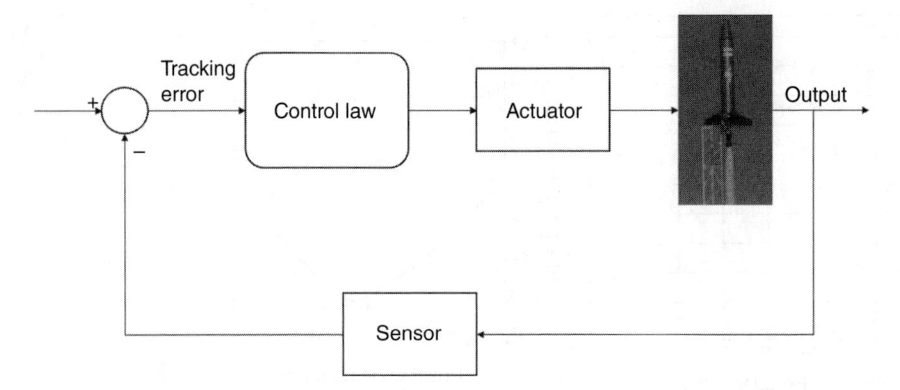

Figure 2.6 Block diagram of a typical feedback control system.

where x is the state and $u(t)$ is the input. This differential equation can be transformed to the Laplace domain and written as a transfer function

$$\frac{X(s)}{U(s)} = \frac{1}{a_1 s + a_0} \tag{2.2}$$

which is an alternative input–output representation of a dynamic system. The denominator polynomial is considered the characteristic equation of the system, the roots of which must be negative for the system to be stable. The first-order transfer function can be rewritten as

$$\frac{X(s)}{U(s)} = \frac{K}{\tau s + 1} \tag{2.3}$$

where τ is called the time constant of the system. The first-order system is considered to be stable as long as $\tau > 0$. The transient characteristics of the first-order system can be determined using the response of the system to a unit step input (Figure 2.7). These include

- rise time (T_r): the time taken for the response to go from 10–90% of the final steady value $= 2.2\tau$
- settling time (T_s): the time taken for the response to reach 98% of the final value $= 4\tau$
- the time constant τ, the time taken for the response to reach 63.2% of the final steady value.

The transient response characteristics of the system determine the speed of the response of the system to a given input.

A second-order system can be written as a differential equation of the form

$$a_2 \frac{d^2x}{dt^2} + a_1 \frac{dx}{dt} + a_0 x = u(t) \tag{2.4}$$

where $x(t)$ and $\frac{dx}{dt}$ are the two states and $u(t)$ is the input. This differential equation can be transformed to the Laplace domain and written as a transfer function

$$\frac{X(s)}{U(s)} = \frac{1}{a_2 s^2 + a_1 s + a_0} \tag{2.5}$$

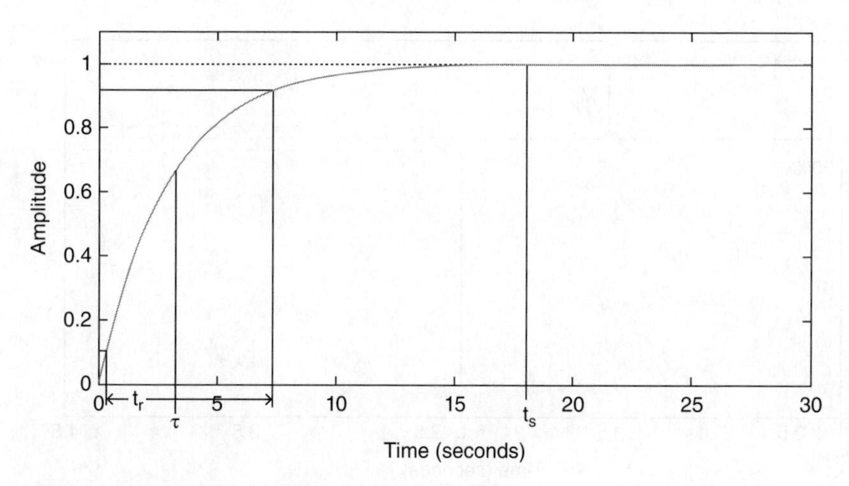

Figure 2.7 First-order transient response.

This can be alternatively written as

$$\frac{X(s)}{U(s)} = \frac{K}{s^2 + 2\zeta\omega_n s + \omega_n^2} \tag{2.6}$$

where ζ is the damping ratio and ω_n is the natural frequency. The second-order system is stable as long as $a_i > 0, i = 0, 1, 2$. This will result in the characteristic equation having roots that are negative or having a negative real part. Stable second-order systems can be of four types depending on the value of the damping ratio:

$$\begin{array}{lll} \text{Case 1} & \zeta = 0 & \text{Undamped} \\ \text{Case 2} & 0 < \zeta < 1 & \text{Underdamped} \\ \text{Case 3} & \zeta = 1 & \text{Critically damped} \\ \text{Case 4} & \zeta > 1 & \text{Overdamped} \end{array} \tag{2.7}$$

For the undamped case, because ζ is zero, all the roots of the characteristic equation are purely imaginary. For the underdamped case the roots are complex conjugate. For a system to be critically damped the roots must be real and equal to each other. The overdamped case is caused by the roots of the characteristic equation being real and distinct.

The transient response characteristics of an underdamped second-order system, as obtained through the unit step response of the transfer function, are given by the following quantities:

- rise time (T_r): the time taken for the response to go from 10–90% of the final steady value

$$T_r = \frac{\pi}{\omega_n\sqrt{1-\zeta^2}} \tag{2.8}$$

- settling time (T_s): the time taken for the response to reach and stay between 98% and 102% of the final steady value

$$T_s = \frac{4}{\zeta\omega_n} \tag{2.9}$$

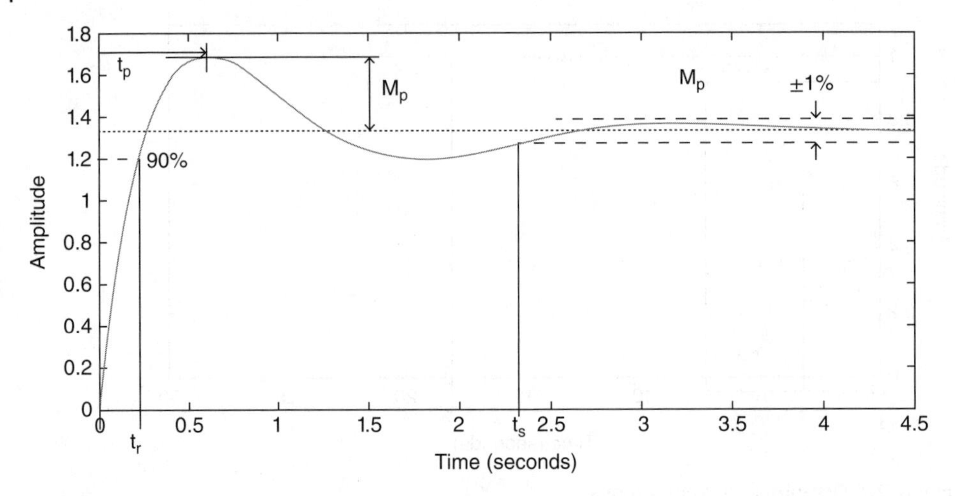

Figure 2.8 Underdamped transient response.

- percentage overshoot (%OS): the time taken for the response to reach the maximum (peak) value

$$\%OS = e^{\frac{-\pi\zeta}{\sqrt{1-\zeta^2}}} \times 100 \tag{2.10}$$

The transient response of an underdamped second-order system can be seen in Figure 2.8.

2.3.2 Classical Control System Design

A simplified block diagram of a feedback control system with *series compensation* is shown in Figure 2.9. The transfer function $G_p(s)$ represents the dynamics of the system being controlled and is known as the *plant transfer function*. As compared to Figure 2.6, the plant transfer function also includes the dynamics of the actuators that influence the plant and the sensors that measure the outputs of the plant. The transfer function $G_c(s)$ is referred to as the *compensator* and incorporates the logic of the control design. In its elementary form, the compensator consists of poles (roots of the denominator polynomial) and zeros (roots of the numerator polynomial) that are chosen by the control designer to meet the performance requirements. In the figure, R(s) is the reference input for the system and E(s) is the error in the current state. The error is calculated by subtracting the current output from the reference signal. U(s) is the control input signal, which is used to drive the plant to react to the reference signal and make the error go to zero, resulting in an output, Y(s), that matches the reference input. There are three different reference inputs that are typically used to characterize a control system.

The performance of the controller is evaluated by looking at the stability and response characteristics of the closed-loop transfer function. The closed-loop transfer function for the system in Figure 2.9 is given by

$$G_{cl}(s) = \frac{G_c(s)G_p(s)}{1 + G_c(s)G_p(s)} \tag{2.11}$$

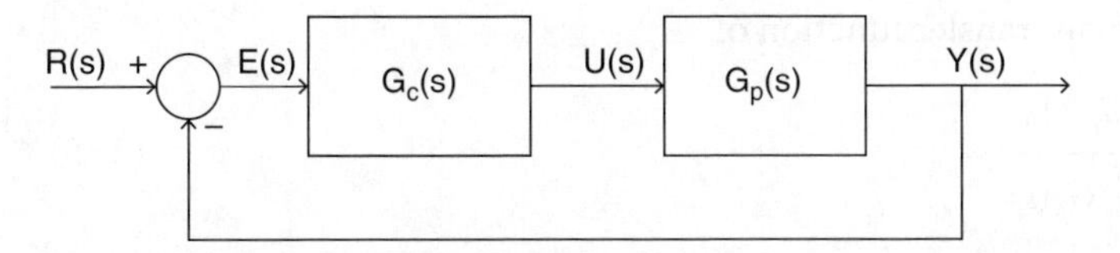

Figure 2.9 Elementary feedback control system.

When designing a controller for a system, the designer needs to determine what the system should do and how it should do. These are referred to as the design specifications. Next, the engineer should determine the controller configuration and how it is connected to the controlled process. Then the parameter values of the controller can be determined to achieve the design goals.

Design specifications for a typical controller design should begin with ensuring the output of the system follows the reference input. Stability is a crucial factor in control system design; the controller should be designed to stabilize an unstable system. The steady-state accuracy must always be close to zero, meaning that the output of the system eventually matches the reference input. It is also important to meet desired transient response specifications, such as maximum overshoot, settling time and rise time. Robustness is important to ensure that small inaccuracies in the dynamic model do not cause the system to become unstable. Disturbance rejection is important in many control systems, to prevent the system from becoming unstable due to small disturbances in the output.

Series compensators can be generalized into four types of configuration:

- proportional controller (P)
- proportional integral controller (PI)
- proportional derivative controller
- proportional integral derivative controller (PID).

A proportional controller is the most basic of the four types of controllers. Proportional control is simply a gain value that is applied to the error of the system. In Figure 2.9 the controller block is replaced with a constant K_p.

$$G_c(s) = K_p \tag{2.12}$$

A proportional controller results in a closed-loop transfer function given by

$$G_{cl}(s) = \frac{K_p G_p(s)}{1 + K_p G_p(s)} \tag{2.13}$$

A PI controller involves not only a gain but also an integral block. It takes the integral of the error as well and removes what is known as steady-state error from the system by placing a pole at the origin. Steady-state error is the difference between the final output value and the reference that was put into the system. A PI control transfer function is given by:

$$G_c(s) = K_p + \frac{K_I}{s} \tag{2.14}$$

and results in a closed-loop transfer function of

$$G_{cl}(s) = \frac{\frac{K_p s + K_I}{s} G_p(s)}{1 + \frac{K_p s + K_I}{s} G_p(s)} \quad (2.15)$$

where K_p is the proportional gain and K_I is the integral gain value.

A PD controller has a transfer function given by

$$G_c(s) = K_p + K_D s \quad (2.16)$$

and the resulting closed-loop transfer function

$$G_{cl}(s) = \frac{(K_p + K_D s)G_p(s)}{1 + (K_p + K_D s)G_p(s)} \quad (2.17)$$

where K_p is the proportional gain and K_D is the derivative gain value. A PD controller improves the transient response characteristics of the closed-loop system.

A PID controller includes the terms that a PI controller has plus a derivative term that takes the rate of change of the error into account. A PID controller is given by the transfer function

$$G_c(s) = K_p + \frac{K_I}{s} + K_D s \quad (2.18)$$

The closed-loop transfer function is given by

$$G_{cl}(s) = \frac{\frac{K_p s + K_I + K_D s}{s} G_p(s)}{1 + \frac{K_p s + K_I + K_D s}{s} G_p(s)} \quad (2.19)$$

2.4 Feedback Control Challenges when using Wireless Networks

Wireless sensor and actuator networks have several advantages in the design of control systems, including distributed computation, large-scale redundancy and localized error detection and mitigation. However, with the implementation of such networks, the control designer is faced with additional challenges, some of which are exaggerated in extreme environments. Typical issues with wireless networks include

- packet drops resulting in incomplete information
- complex network dynamics that cannot be modeled accurately
- external noise that can be significantly amplified by the network characteristics
- the presence of delays that might be time-varying.

In this section, we focus on the effect of constant delays on the closed-loop control system. While research in the area of delays and their effects on control system design is very rich and encompasses many different applications, this section only provides a very simple analysis of the effect of delays (both single and multiple) on the stability and transient response of first and second-order systems; the analysis is intended to provide the reader with a starting point for more in-depth reading on wireless networks for control design.

2.4.1 Approximated Model of Delay

To perform an accurate analysis of the effect of delay on a control system, it is important to properly model the delay, which allows it to be seen as a transfer function. It can then be included in the overall transfer function of the control system. One of the methods used to model delay as a transfer function is known as the Padé approximation. The original delay transfer function Eq. (2.43) is irrational, so when performing frequency-response-based analysis, it is important to replace it with a rational approximation [14]:

$$e^{-Ds} \approx \frac{1 - k_1 s + k_2 s^2 + \ldots \pm k_n s^n}{1 + k_1 s + k_2 s^2 + \ldots + k_n s^n} \tag{2.20}$$

where n is the order of the approximation. A first-order approximation of delay has been defined as

$$k_1 = \frac{D}{2} \tag{2.21}$$

Plugging k_1 and $n = 1$ into (2.20) gives

$$H_{\text{Pade}}(s) \approx \frac{1 - \frac{D}{2}s}{1 + \frac{D}{2}s} \tag{2.22}$$

which can be transformed into its final, most simplified version of

$$H_{\text{Pade}}(s) \approx \frac{2 - Ds}{2 + Ds} \tag{2.23}$$

Figure 2.10 shows a unit step input passing through a MATLAB transport delay block and the Padé approximation transfer function. While it appears that the Padé approximation is a good match for the delay block, it must be noted that when the approximation is used in a feedback loop, it becomes sensitive to the frequency at which the root locus of the open-loop transfer function crosses the imaginary axis. In other words, the validity of the approximation depends on the transfer function of the system parameters.

The rationalization of the transcendental form of the delay given by $\exp^{-Ds}$ can also be performed more accurately using the Rekasius substitution [15]. The Rekasius substitution has been discussed in several articles that specifically address the stability of systems with delays [16–18]. A brief discussion is provided below.

Consider the delay representation e^{-Ds}. The variable s is the complex number $i\omega$. Now

$$e^{-iD\omega} = \cos D\omega - i \sin D\omega \tag{2.24}$$

The magnitude of this complex number is 1 and the phase is given by

$$\angle e^{-iD\omega} = -D\omega \tag{2.25}$$

Consider an approximation of this delay by the transfer function

$$G(s) = \frac{1 - \overline{D}s}{1 + \overline{D}s} \tag{2.26}$$

Considering $s = i\omega$, the magnitude of the above transfer function is 1, while the phase angle is given by

$$\angle G(s) = -2\tan^{-1}(\overline{D}\omega) \tag{2.27}$$

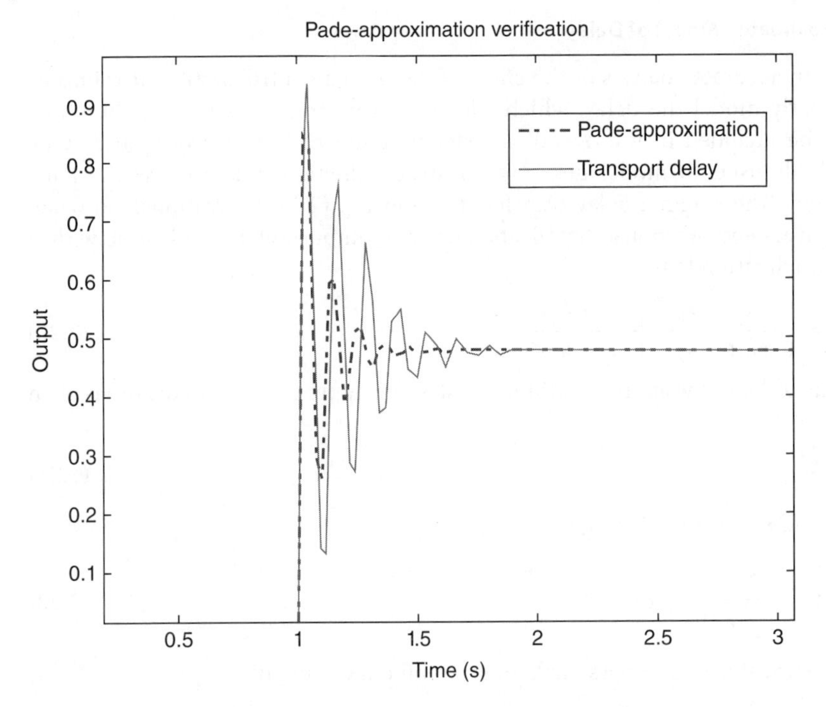

Figure 2.10 Delay verification.

It is evident that both the exact and approximate transfer functions of the delay have the same magnitude. If a $\overline{D}$ can be chosen such that it makes the phase obtained from (2.25) and (2.27) equal, then the transfer function in (2.26) is an equivalent substitute for e^{-Ds}. This value of $\overline{D}$ is computed from

$$D = \frac{2}{\omega}\tan^{-1}(\overline{D}\omega) \tag{2.28}$$

Note that, if the magnitude of ω is sufficiently small:

$$\frac{2}{\omega}\tan^{-1}(\overline{D}\omega) = \frac{2}{\omega}(\overline{D}\omega) \tag{2.29}$$

This implies that $D = 2\overline{D}$ and hence the approximate rational transfer function in (2.26) can be rewritten as

$$G(s) = \frac{1 - \frac{D}{2}s}{1 + \frac{D}{2}s} = \frac{2 - Ds}{2 + Ds} \tag{2.30}$$

This is similar to the Padé approximation of the delay block shown in (2.23).

2.4.2 Effect of Delay on the Stability of a First-order System

As outlined earlier, wireless sensor networks have the ability to simplify implementation and increase redundancy, but have the drawback of introducing delays into the system. To understand the effect of a delay on the stability of a closed-loop system, consider the

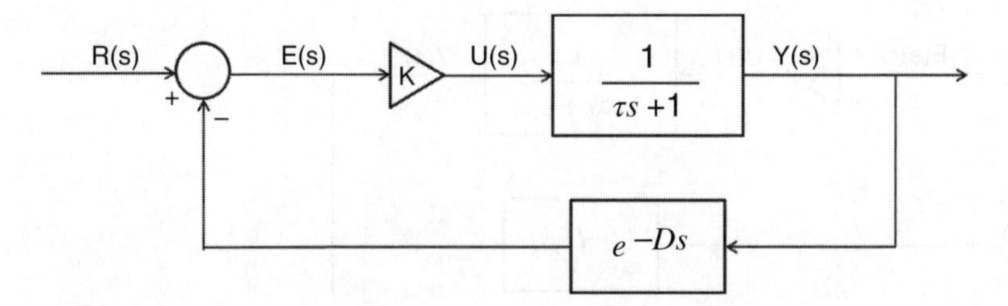

Figure 2.11 First-order system with single delay in feedback path.

example of a first-order system with a single wireless sensor in the feedback path, where τ is the time constant (see Figure 2.11). Consider the delay in the feedback path given by D and a simple proportional controller gain given by K. If the delay in the feedback path was zero, then the closed-loop system will be given by

$$G_{cl}(s) = \frac{K}{\tau s + (1 + K)} \tag{2.31}$$

It is evident that the system is stable for all values of $K > -1$. This implies that there is no upper bound on the gain K to stabilize the closed-loop system.

Now consider a non-zero delay D. The rationalization of the delay block using the Rekasius substitution and without an assumption about the value of ω is given by

$$G(s) = \frac{1 - \overline{D}s}{1 + \overline{D}s} \tag{2.32}$$

where

$$\overline{D} = \frac{1}{\omega}\tan(\frac{D\omega}{2}) \tag{2.33}$$

The closed-loop system with a proportional controller K is given by

$$G_{cl}(s) = \frac{K(1 + \overline{D}s)}{\tau\overline{D}s^2 + (\tau + \overline{D} - K\overline{D})s + K + 1} \tag{2.34}$$

For stability of the closed-loop system, the following conditions need to be met

- $\tau\overline{D} > 0$, which is implied since we are considering a stable first-order plant ($\tau > 0$) and delay is positive
- $K + 1 > 0 \Longrightarrow K > -1$, similar to the case with no delay
- $\tau + \overline{D} - K\overline{D} > 0 \Longrightarrow K < 1 + \frac{\tau}{\overline{D}}$.

It should be noted that the introduction of a non-zero delay in the feedback path not only increases the order of the system but also produces an upper bound on the gain K that can be used to stabilize the closed loop. This upper bound on K can be termed the *gain margin* of the control system and is a typical measure of closed-loop performance. It is evident from the above analysis that delay decreases the performance of a system to a point that can even result in instability [19].

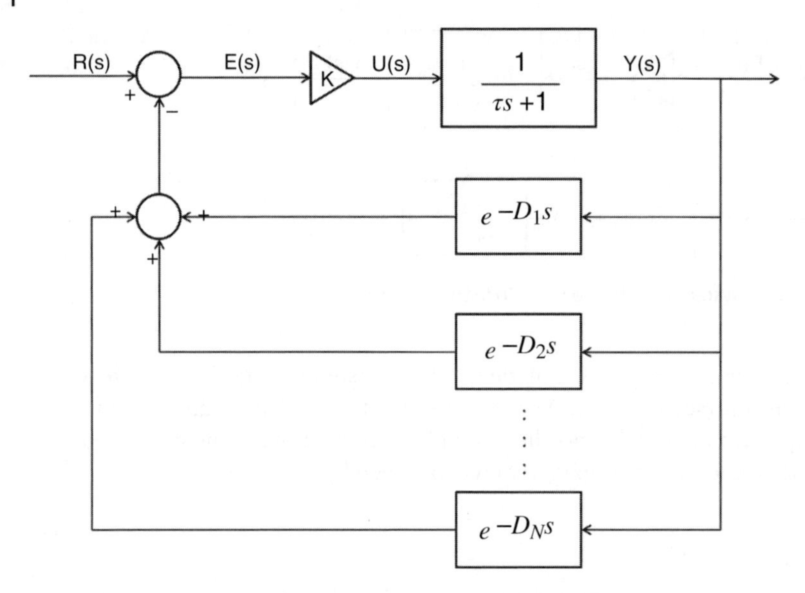

Figure 2.12 First-order system with multiple delays in the feedback path.

2.4.2.1 Multi-sensor Systems

When a network of sensors is used for a control system application, delay may be present in each of the nodes. Consider the first-order system with multiple delays in the feedback path considered earlier and shown in Figure 2.12. This is a more practical control system than the basic one in Figure 2.9, because in many cases in industrial and aerospace applications it is necessary to have multiple wireless sensors in order to accurately measure the output of a system.

The approximation of each delay in the feedback path (the feedback transfer function) using Rekasius substitution is given by

$$H_i(s) = \frac{1 - \overline{D}_i s}{1 + \overline{D}_i s} \tag{2.35}$$

Using the closed-loop analysis process, a closed-loop transfer function is created. This represents a system with N sensors. It is important to first note that the order of this closed-loop system is dependent on the number of sensors used. The characteristic equation of closed-loop transfer function of the system is given by

$$G_{Cl}(s) = (\tau s + 1)\prod_{i=1}^{N}(1 + \overline{D}_i s) + K\sum_{i=1}^{N}\left[(1 - \overline{D}_i s)\prod_{j=1}^{N}(1 + \overline{D}_j s)\right] \quad j \neq i \tag{2.36}$$

The characteristic equation of the closed-loop system is a polynomial of degree N+ 1. The stability of the system can be determined by constructing the Routh array and applying the Routh–Hurwitz criterion [20]. Let us assume that the characteristic equation is written as

$$a(s) = a_{N+1}s^{N+1} + a_N s^n + a_{N-1}s^{n-1} + \ldots\ldots a_1 s + a_0 \tag{2.37}$$

The primary requirement for stability when using the Routh–Hurwitz criterion is that all the coefficients (a_i) of the characteristic polynomial be of the same sign. Assuming this is true, a Routh array is arranged as follows

$$\begin{array}{c|c|c|c|}
s^{N+1} & a_{N+1} & a_{N-1} & \cdots \\
s^{N} & a_{N} & a_{N-2} & \cdots \\
s^{N-1} & b_1 & \cdots & \cdots \\
s^{N-2} & c_1 & \cdots & \cdots \\
\vdots & \vdots & \cdots & \cdots \\
s^{N-k} & \vdots & \cdots & \cdots \\
\vdots & \vdots & \cdots & \cdots \\
s & \vdots & \cdots & \cdots \\
s^{0} & \vdots & \cdots & \cdots
\end{array}$$

where the coefficients a_{N+1}, a_N and a_0 can be determined as

$$a_{N+1} = \tau \sum_{i=1}^{N} \overline{D}_i \tag{2.38}$$

$$a_N = \prod_{i=1}^{N}(\overline{D}_i) + \tau \sum_{i=1}^{N} \frac{\left(\prod_{j=1}^{N} \overline{D}_j\right)}{\overline{D}_i} - KN\prod_{i=1}^{N}(D_i) \tag{2.39}$$

$$a_0 = N + NK \tag{2.40}$$

A system can be stable if and only if all the elements in the first column of the Routh array are of the same sign. Equations to find the b_i, c_i coefficients can be found by using the a coefficients; a description of this process can be found in the literature [13]. The coefficients of the s^{N+1} and s^N terms are the first elements of the first two rows. It is also apparent that the coefficient of s^{N+1} is always positive for an inherently stable first-order system with positive delay. This implies that all the elements in the first column must be positive for stability. However, the coefficient of s^N need not necessarily be positive and its value depends on the control gain K. It is possible to determine the limit on K based on the requirement that $a_N > 0$. This limit is termed K_u and it is given by

$$K < \frac{1}{N} + \frac{\tau \sum_{i=1}^{N} \frac{\left(\prod_{j=1}^{N} \overline{D}_j\right)}{\overline{D}_i}}{N\prod_{i=1}^{N}(D_i)} \tag{2.41}$$

Note that when there is only one delay (D_1) in the feedback path, the above equation reduces to

$$K_u = 1 + \frac{\tau}{D_1} \tag{2.42}$$

It should be noted that if the proportional control gain K exceeds K_u, the system will become unstable. However, it is not possible to confirm that K_u represents the gain margin of the system unless each of the quantities in the first column of the Routh array is analysed for the limits on K that will make the system unstable ($a_{i1} \leq 0$). One way to determine the gain margin is to construct the Routh array for different values of K. This can be accomplished by starting with $K = K_u$ and iteratively reducing its value until the system is no longer stable. It should be noted that an analytical limit to the gain margin cannot be obtained for an N-sensor problem since the Routh array construct depends closely on the order of the system and cannot be generalized.

The above analysis provides an example of the important role that delays play in the design of a control system that uses wireless sensor networks.

2.5 Effect of Delay on the Transient Response of a Second-order System

A system with wireless feedback always features delay and noise, which affect several parameters of the transient response of a second-order system, such as rise time and overshoot [21]. Consider a second-order system with a proportional controller of unity gain, as shown in Figure 2.13.

The block diagram includes both a delay in the feedback path and an external disturbance, which is modeled as additive white Gaussian noise (AWGN). The delay in the feedback path is modeled as

$$A(s) = e^{-Ds} \tag{2.43}$$

The noise variance in this system is represented by σ^2. The noise term can be expressed as

$$n(t) = \frac{1}{\sqrt{2\pi\sigma^2}} e^{\frac{-t^2}{2\sigma^2}} \tag{2.44}$$

with a frequency-domain representation of

$$N(s) = \frac{1}{\sqrt{2\pi\sigma^2}} \frac{\sqrt{\pi}}{2} e^{\frac{s^2}{4}} erfc\left(\frac{s}{2}\right) \tag{2.45}$$

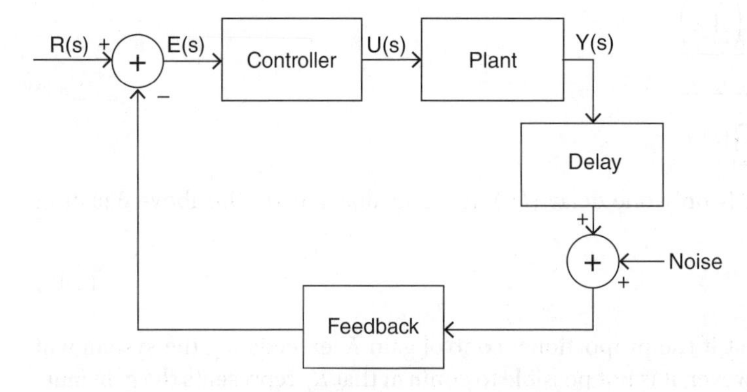

Figure 2.13 Single-sensor feedback loop with AWGN-CD model.

where the complementary error function *erfc* is

$$erfc(z) = \frac{2}{\sqrt{\pi}} \int_{z}^{\inf} e^{-t^2} dt \tag{2.46}$$

Therefore, the overall transfer function of the second-order system becomes

$$\tilde{H}(s) = e^{-Ds} \frac{kw^2}{s^2 + 2s\zeta\omega_n + \omega_n^2} + \frac{1}{\sqrt{2\pi\sigma^2}} \frac{\sqrt{\pi}}{2} e^{\frac{s^2}{4}} erfc\left(\frac{s}{2}\right) \tag{2.47}$$

A Taylor series expansion has been used in performance evaluation of coded communication systems [22]. A similar approach is now followed in order to view the effects of noise and delay on the system. The Taylor series expansion of the transfer function evaluated at 0 is examined. The first two terms provide the following second-order approximation of the transfer function including both delay and noise:

$$\tilde{H}(s) = k + \frac{1}{2\sigma\sqrt{2}} - s\left(kD + \frac{2\zeta}{\omega_n} - \frac{1}{2\sigma\sqrt{2\pi}}\right) \tag{2.48}$$

To properly analyse this equation with respect to the rise time it is necessary to be able to compare (2.48) to the original second-order system in (2.6). In order to do this, the Taylor series expansion of (2.6) is compared with (2.48).

$$H(s) \approx k - s(2k\zeta\omega_n) \tag{2.49}$$

The first term of (2.48) and (2.49) represent the DC gain of the system, so they have negligible effect on rise time and overshoot. The second terms, however, can be compared to each other to create an equation to calculate the rise time and overshoot of the wireless feedback network. If $2k$ is factored out of the second term of (2.48), one can find the wireless system's $\zeta\omega_n$ equivalent. The second term coefficient becomes

$$2k\left(\frac{D}{2} + \frac{\zeta}{k\omega_n} - \frac{1}{4k\sigma\sqrt{2\pi}}\right) \tag{2.50}$$

Replacing $\zeta\omega_n$ in (2.51) with the coefficient term from (2.50), an equation to calculate rise time ($\tilde{T}_r$) of the wireless feedback network is

$$\tilde{T}_r \approx \frac{2.2}{\frac{D}{2} + \frac{\zeta}{k\omega_n} - \frac{1}{4k\sigma\sqrt{2\pi}}} \tag{2.51}$$

This equation can now be used to predict the rise time of the system while varying the noise and delay. This equation is used to create the theoretical rise time curves in Figures 2.14 and 2.15.

The feedback loop is analysed by observing the rise time and overshoot of the system when noise variance increased and delay is held constant. The case when the delay is varied and the noise variance held constant is also examined.

Observing the rise time is the proper way to observe how fast the system will respond to the changing input signal. In Figure 2.14, the delay is set as 0.01 s and the noise variance is varied from 1–10% of the input signal. When comparing these results to typical results from a wired system, some differences and similarities can be noted. Figure 2.14 shows that the rise time decreases slightly as the noise variance increases. The rise time

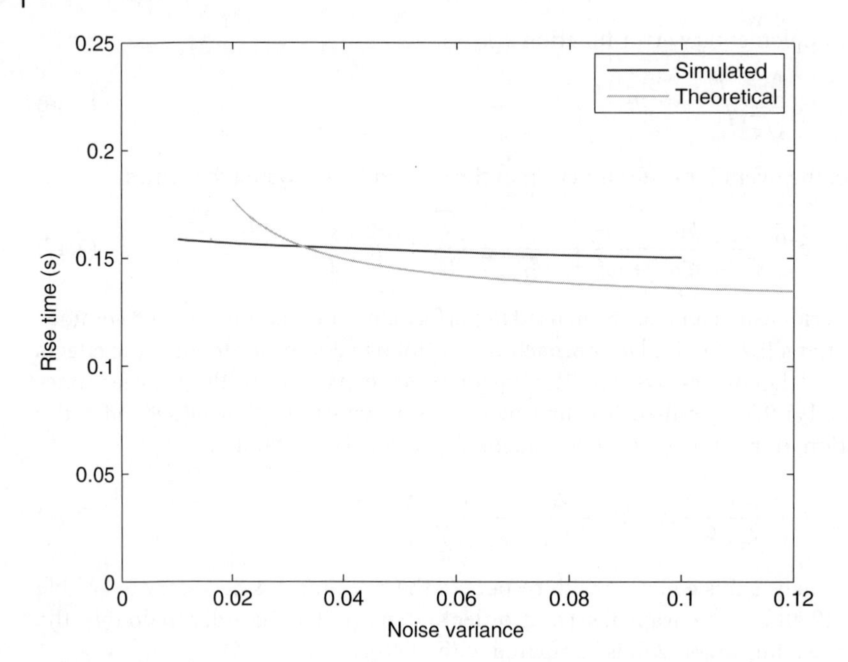

Figure 2.14 Rise time vs noise variance.

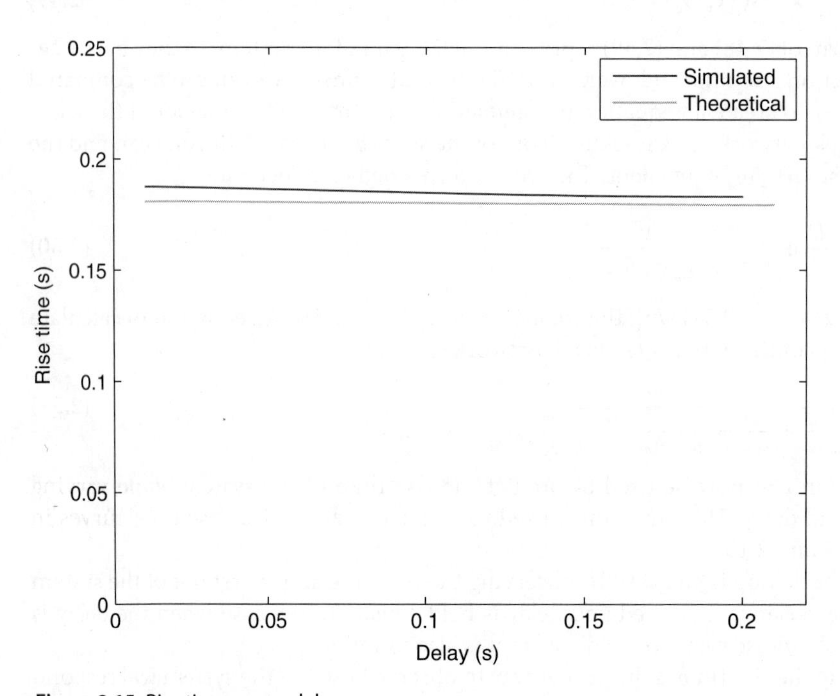

Figure 2.15 Rise time versus delay.

only changes by 9 ms as noise variance increases from 1% to 10%. Figure 2.14 shows that the rise-time response to noise variance may seem different to what is expected; one might expect the rise time to increase as noise variance increases. This inverse proportionality is caused by adding the noise to the signal, causing the signal to rise faster. It can be seen that the theoretical rise-time values calculated from (2.51) match the simulated values relatively well. The rate of change of these values matches particularly well.

It is also important to study how delay can affect the step response. It is important to observe the effects of a delay in the feedback path because any system using wireless feedback will have some latency that will affect the behaviour of the system. To produce meaningful plots with varying delay, the noise variance is held to zero and the delay is varied from 0.01 s to 0.2 s. Figure 2.15 shows the effect of delay on the rise time of the step response. It shows that the theoretical equation is a linear approximation of the effect that delay has on the rise time of the system. The figure also shows that rise time depends on delay until the delay reaches a certain value, after which rise time is constant for all delay values until it becomes unstable. It can be seen that the rise time decreases with delay as long as the delay is less than or equal to the rise time. This is an interesting observation; it means if the delay is lower than the rise time then it has an effect on the rise time, but if the delay is greater than the rise time, then the rise time is unaffected by the amount of delay in the system. This can be used as a design criterion.

Figure 2.16 shows that an increase of noise variance in the system causes the overshoot to increase. The values increase from 10% to 12.5% as noise variance changes from 1% to 10%. This is a small change in overshoot and would not noticeably affect the system response time, but given a higher signal-to-noise ratio could do so significantly.

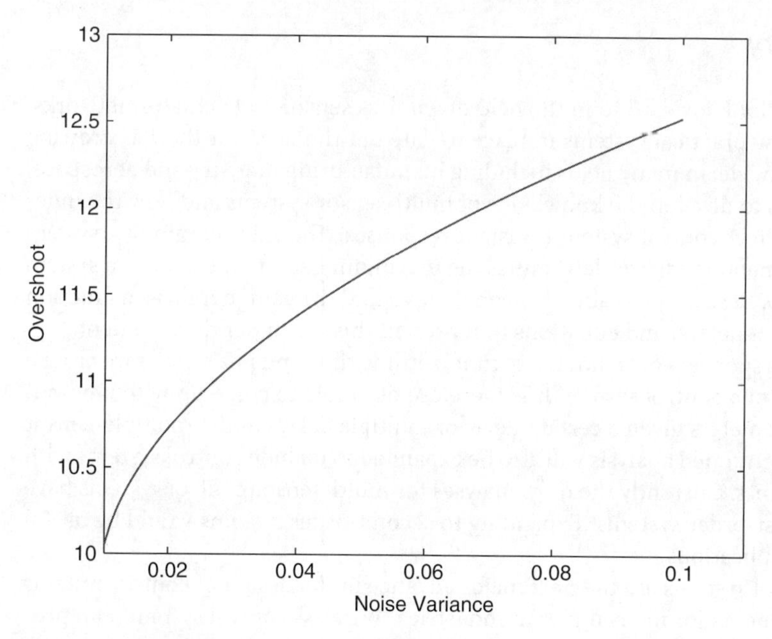

Figure 2.16 Overshoot versus noise variance.

Figure 2.16 is an example of why noise in a wireless feedback system should be minimized. In combination with Figures 2.14 and 2.15, it verifies that a wireless system with noise can be implemented.

2.6 Discussion

Transient characteristics of a control system are the basis of how the control system behaves. Wireless control systems in particular have a significantly larger delays and noise than traditional wired systems due to the additional signal processing and interference between transmission signals that occurs in wireless networks. It is important to know how significant delay and noise in a system will affect these characteristics, in order to properly analyse wireless control systems.

It has been shown that (2.51) is a viable approximation for calculating the rise time of a second-order single-sensor system given the delay values and the noise variance in the system. While it is true that these values will not always be known in a practical system, the upper bounds of of the delay and noise variance can be used during the design of the system to ensure a rise time that is acceptable for a given application.

Two methods of delay equation approximation were considered when carrying out the analysis for the wireless control system. This is crucial for a simulation of a wireless control system with realistic behavior when delay is included. A method for finding the gain margin of a wireless control system with a given delay has been described.

2.7 Summary

There is much to look forward to in the field of wireless sensor and actuator networks. There are very few practical systems in place to date but the need for them is growing immensely worldwide, in many fields including manufacturing industry and aerospace.

There are plans to develop the knowledge of multi-sensor systems and how the independent delays affect control system transient responses. The gain margin of a system with known independent delays determines the maximum delay that can push a system toward instability. Research has already been done to find the gain margin as a function of the independent delays, and equations to represent this are under development.

The transient responses–overshoot, rise time, settling time and peak time–are crucial to the behaviour of a control system. It is therefore desirable to come up with new values for these parameters given a certain delay or multiple delays in the multiple-sensor case. The aforementioned analysis will also be expanded to include control systems with second-order plants. Currently the only analyses for multi-sensor control systems have been done on first-order systems. Expanding to second-order systems would be useful for real-world applications.

Wireless control systems are the next major advance in the design of control systems and will bring benefits for many different industries. Wireless control systems can provide more redundant performance through the ability to have more sensors, and will simplify installation and maintenance of systems by increasing modularity and removing wires.

References

1 Creech, G. (2003) Fly by wire, *Tech. Rep.*, NASA.

2 Swartz, R., Lynch, J., and Loh, C.H. (2009) Near real-time system identification in a wireless sensor network for adaptive feedback control, in *American Control Conference, 2009. ACC '09.*, pp. 3914–3919, doi:10.1109/ACC.2009.5160493.

3 Braslavsky, J., Middleton, R., and Freudenberg, J. (2007) Feedback stabilization over signal-to-noise ratio constrained channels. *Automatic Control, IEEE Transactions on*, **52** (8), 1391–1403, doi:10.1109/TAC.2007.902739.

4 Rojas, A., Braslavsky, J., and Middleton, R. (2006) Output feedback stabilisation over bandwidth limited, signal to noise ratio constrained communication channels, in *American Control Conference, 2006*, doi:10.1109/ACC.2006.1656646.

5 Zhang, Y., Zhong, Q., and Wei, L. (2008) Stability of networked control systems with communication constraints, in *Control and Decision Conference, 2008. CCDC 2008. Chinese*, pp. 335–339, doi:10.1109/CCDC.2008.4597325.

6 Topakkaya, H. and Wang, Z. (2008) Effects of delay and links between relays on the min-cut capacity in a wireless relay network, in *Communication, Control, and Computing, 2008 46th Annual Allerton Conference on*, pp. 319–323, doi:10.1109/ALLERTON.2008.4797574.

7 Dang, D.K., Mifdaoui, A., and Gayraud, T. (2012) Fly-by-wireless for next generation aircraft: Challenges and potential solutions, in *Wireless Days (WD), 2012 IFIP*, pp. 1–8, doi:10.1109/WD.2012.6402820.

8 Studor, G. (2008) Fly-by-wireless: A less-wire and wireless revolution for aerospace vehicle architectures, *Tech. Rep.*, NASA.

9 Swaim, R. (1969) Control system synthesis for a launch vehicle with severe mode interaction. *Automatic Control, IEEE Transactions on*, **14** (5), 517–523.

10 Templeton, J.D., e. (2010) Ares I-X launch vehicle modal test measurements and data quality assessments, *Tech. Rep. TR-LF99-9048*, NASA.

11 Rezaei, R., Ghabrial, F., Besnard, E., Shankar, P., Castro, J., Labonte, L., Razfar, M., and Abedi, A. (2013) Determination of elastic mode characteristics using wirelessly networked sensors for nanosat launch vehicle control, in *Wireless for Space and Extreme Environments (WiSEE), 2013 IEEE International Conference on*, pp. 1–2, doi: 10.1109/WiSEE.2013.6737572.

12 Jiang, H., Van Der Week, B., Kirk, D., and Guiterrez, H. (2013) Real-time estimation of time-varying bending modes using fiber Bragg grating sensor arrays. *AIAA Journal*, **51** (1), 178–185.

13 Franklin, G., Powell, J.D., and Emami-Naeini, A. (2009) *Feedback Control of Dynamic Systems*, Pearson Higher Education.

14 Gibson, J. and Hamilton-Jenkins, M. (1983) Transfer-function models of sampled systems. *IEE Proceedings G—Electronic Circuits and Systems*, **130** (2), 37–44.

15 Rekasius, Z. (1980) A stability test for systems with delays, in *Joint Automatic Control Conference*, ASME, San Francisco, CA, TP-9A.

16 Sipahi, R. and Olgac, N. (2005) Complete stability robustness of third-order LTI multiple delay systems. *Automatica*, **41**, 1413–1422.

17 Sipahi, R. and Olgac, N. (2006) A unique methodology for stability robustness of multiple delay systems. *Automatica*, **55**, 819–825.

18 Fazelinia, H., Sipahi, R., and Olgac, N. (2007) Stability robustness analysis of multiple time-delay systems using building-block concept. *Automatic Control, IEEE Transactions on*, **52** (5), 799–810.
19 Cloosterman, M., van de Wouw, N., Heemels, M., and Nijmeijer, H. (2006) Robust stability of networked control systems with time-varying network-induced delays, in *Decision and Control, 2006 45th IEEE Conference on*, pp. 4980–4985, doi:10.1109/CDC.2006.376765.
20 Bothwell, F.E. (1950) Nyquist diagrams and the Routh–Hurwitz stability criterion. *Proceedings of the IRE*, **38** (11), 1345–1348, doi:10.1109/JRPROC.1950.234428.
21 Labonte, L., Castro, J., Razfar, M., Abedi, A., Rezaei, R., Ghabrial, F., Shankar, P., and Besnard, E. (2013) Wireless sensor and actuator networks with delayed noisy feedback (WiSAN), in *Wireless for Space and Extreme Environments (WiSEE), 2013 IEEE International Conference on*, pp. 1–5, doi:10.1109/WiSEE.2013.6737575.
22 Abedi, A. and Khandani, A. (2004) An analytical method for approximate performance evaluation of binary linear block codes. *Communications, IEEE Transactions on*, **52** (2), 228–235, doi:10.1109/TCOMM.2003.822704.

3

Optimizing Lifetime and Power Consumption for Sensing Applications in Extreme Environments

Gholamreza Alirezaei, Omid Taghizadeh and Rudolf Mathar

Institute for Theoretical Information Technology, RWTH Aachen University, Aachen, Germany

3.1 Introduction

Due to the rise of Internet of things (IoT) and certain applications of 5G wireless systems, sensor networks are quickly gaining importance. Since large-scale sensor networks usually utilise cheap and weak sensor nodes (SNs), complex signal processing within each SN is not applicable. Thus amplify-and-forward techniques are mostly used to relay an observed signal from each SN to a centralised unit, such as a fusion center. The task of the fusion center is to process all of the transmitted signals in order to create a reliable global observations at the fusion center based on the collection of individual, independent and unreliable observations.

Several publications show that the reliability increases with the transmission power of each SN. Naturally, the question arises of how to guarantee a minimum signal quality at the fusion center while having an energy-efficient sensor network. Especially for sensor networks with huge numbers of SNs, energy efficiency is essential, because the overall power consumption can be drastically reduced, prolonging the network lifetime.

In this chapter, we consider a common sensor network that is used for sensing applications: target signal detection, localization, classification and tracking. Figure 3.1 shows the target emitter, the sensing channel, the independent and distributed SNs, the communication channel and a fusion center. This scenario has been examined in many publications and will also serve as our framework in the present chapter.

Raghunathan et al. [1] considered a sensor network composed of microsensors and described general architectural and algorithmic approaches for enhancing the energy awareness of wireless sensor networks. Muruganathan et al. [2] used a cluster-based approach and a centralised routing protocol to improve network lifetime. A theoretical upper bound for the network lifetime was investigated by Bhardwaj et al. [3], but this bound is in practice not achievable. A further notable publication, by Cardei et al. [4] used different heuristics for lifetime maximization. The corresponding optimization problems were subsequently solved by numerical methods. In contrast, an analytical solution in closed form to the power allocation problem was presented for several power constraints by Alirezaei et al. [5], an extension of their earlier work [6]. This investigation was later further extended [7–15]. In the present chapter, we aim to extend some of our other earlier works [16, 17], and we will answer the question: 'How much energy

Wireless Sensor Systems for Extreme Environments: Space, Underwater, Underground, and Industrial.
First Edition. Edited by Habib F. Rashvand and Ali Abedi.

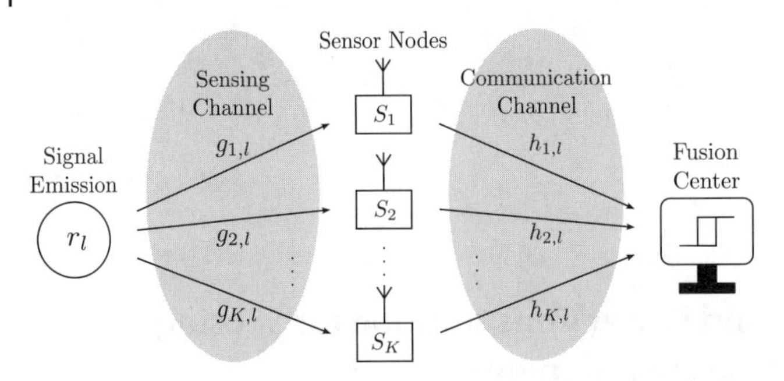

Figure 3.1 A distributed sensor network.

is needed to achieve a given lifetime of a sensor network?' This question has rarely been investigated in the past, because of the mathematical challenges involved.

In the present chapter, we will first briefly introduce the system model and formulate a power minimization problem for a given lifetime under several power constraints. Subsequently, the optimization problem will be rewritten to show its convex nature and to enable its computation using standard numerical methods. Although the optimization problem is of a theoretical nature and is, in addition, computationally intensive to solve, it provides sharp lower and upper bounds for the network power consumption and its lifetime, respectively. Second, we develop some practical methods that require less computational effort but will give nearly optimal solutions to the power minimization problem in more realistic scenarios. Finally, selected numerical results are visualised to show the performance of the new methods and to discuss the power consumption of the entire sensor network.

3.1.1 Mathematical Notation

Throughout this chapter, we denote the sets of natural, real and complex numbers by $\mathbb{N}$, $\mathbb{R}$ and $\mathbb{C}$, respectively. Note that the set of natural numbers does not include the element zero. Moreover, $\mathbb{R}_+$ denotes the set of non-negative real numbers. Furthermore, we use the subset $\mathbb{F}_N \subseteq \mathbb{N}$, which is defined as $\mathbb{F}_N := \{1, \dots, N\}$ for any given natural number N. We denote the absolute value of a real or complex-valued number z by $|z|$. The expected value of a random variable v is denoted by $\mathcal{E}[v]$. Moreover, the notation $V^{\star}$ stands for the value of an optimization variable V where the optimum is attained.

3.2 Overview and Technical System Description

In this chapter, we use an extension of the system model that was introduced by Alirezaei et al. [5]. The extended system model is depicted in Figure 3.2 and is briefly presented in the following. An overview of all system parameters is given in Table 3.1.

We assume a discrete time system and denote the index of each observation process by $l \in \mathbb{F}_L$, where $L \in \mathbb{N}$ describes the lifetime of the sensor network. Therefore, the network under consideration can only perform L consecutive observations with optimal performance and will be out of use afterwards. We consider a sensor network consisting of $K \in \mathbb{N}$ independent and spatially distributed SNs, which receive random observations

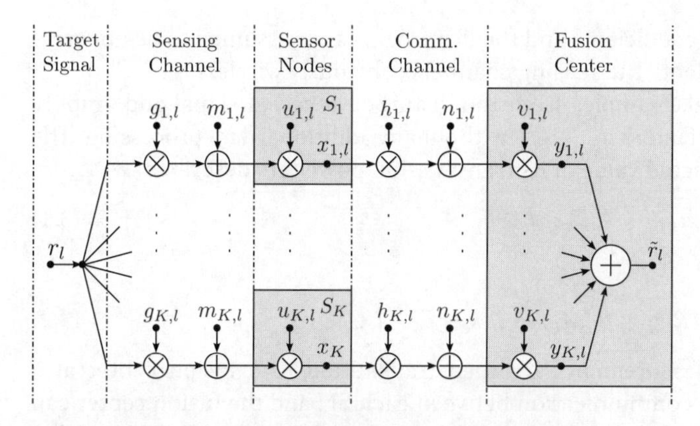

Figure 3.2 System model of the distributed sensor network.

Table 3.1 Symbols for the description of each observation process.

Notation	Description
K	Number of sensor nodes.
k	The kth sensor node.
L	Number of consecutive observation processes, equivalent to the service lifetime of the network.
l	The lth observation process.
r_l, R	Present target signal in the lth observation process and its quadratic absolute mean.
$\tilde{r}_l$	Estimate of r_l.
$g_{k,l}$, $h_{k,l}$	Complex-valued channel coefficients.
$m_{k,l}$, $n_{k,l}$	Complex-valued zero-mean AWGN.
M_k, N_k	Variances of $m_{k,l}$ and $n_{k,l}$.
$u_{k,l}$, $v_{k,l}$	Non-negative amplification factors and complex-valued weights.
$\vartheta_{k,l}$	Phase of $v_{k,l}$.
$\phi_{k,l}$	Phase of the product $g_{k,l}h_{k,l}$.
$x_{k,l}$, $X_{k,l}$	Output signal and output power of kth sensor node.
$y_{k,l}$	Input signals of the combiner.
P_{min}, P_{max}	Lower and upper output power limitations of each sensor node.
$P_{k,l,\text{bud}}$	Available power budget for the lth observation process at the kth sensor node.
P_{over}	Overall power consumption of the network.

in each observation process. If a target signal $r_l \in \mathbb{C}$ with $R := \mathcal{E}[|r_l|^2]$ and $0 < R < \infty$ is present, then the received power at the SN S_k is a part of the emitted power from the actual target. Each received signal is weighted by the corresponding channel coefficient $g_{k,l} \in \mathbb{C}$ and is disturbed by additive white Gaussian noise (AWGN) $m_{k,l} \in \mathbb{C}$ with $M_k := \mathcal{E}[|m_{k,l}|^2] < \infty$. We assume that the coherence time of all sensing channels is much longer than the whole length of the observation process. Therefore, the expected value and the quadratic mean of each coefficient during each observation step can be assumed to be equal to their instantaneous values: $\mathcal{E}[g_{k,l}] = g_{k,l}$ and $\mathcal{E}[|g_{k,l}|^2] = |g_{k,l}|^2$.

Furthermore, the channel coefficients and the disturbances are assumed to be uncorrelated and jointly independent. The sensing channel is obviously wireless.

All SNs continuously take samples from the disturbed received signal and amplify them by the amplification factors $u_{k,l} \in \mathbb{R}_+$ without any additional data processing. The output signal and the expected value of its transmission power are described by

$$x_{k,l} := (r_l g_{k,l} + m_{k,l}) u_{k,l}, \quad k \in \mathbb{F}_K, \quad l \in \mathbb{F}_L, \tag{3.1}$$

and

$$X_{k,l} := \mathcal{E}[|x_{k,l}|^2] = (R|g_{k,l}|^2 + M_k) u_{k,l}^2, \quad k \in \mathbb{F}_K, \quad l \in \mathbb{F}_L, \tag{3.2}$$

respectively. The local measurements are then transmitted to a fusion center at a remote location. The data communication between each SN and the fusion center can be either wired or wireless. In the latter case, a distinct waveform for each SN is used to distinguish the communication of different SNs and to suppress inter-user (inter-node) interference at the fusion center. Hence, all K received signals at the fusion center are pairwise uncorrelated and are assumed to be conditionally independent. Each received signal at the fusion center is also weighted by the corresponding channel coefficient $h_{k,l} \in \mathbb{C}$ and disturbed by additive white Gaussian noise $n_{k,l} \in \mathbb{C}$ with $N_k := \mathcal{E}[|n_{k,l}|^2] < \infty$. We also assume that the coherence time of all communication channels is much longer than the whole length of the observation process. Thus the expected value and the quadratic mean of each coefficient during each observation step can be assumed to be equal to their instantaneous values: $\mathcal{E}[h_{k,l}] = h_{k,l}$ and $\mathcal{E}[|h_{k,l}|^2] = |h_{k,l}|^2$. Furthermore, the channel coefficients and the disturbances are assumed to be uncorrelated and jointly independent.

The noisy received signals at the fusion center are weighted by the fusion weights $v_{k,l} \in \mathbb{C}$ and combined in order to obtain a single reliable observation $\tilde{r}_l$ of the actual target signal r_l. In this way, we obtain

$$y_{k,l} := ((r_l g_{k,l} + m_{k,l}) u_{k,l} h_{k,l} + n_{k,l}) v_{k,l}, \quad k \in \mathbb{F}_K, \quad l \in \mathbb{F}_L, \tag{3.3}$$

and hence,

$$\tilde{r}_l := \sum_{k=1}^{K} y_{k,l} = r_l \sum_{k=1}^{K} g_{k,l} u_{k,l} h_{k,l} v_{k,l} + \sum_{k=1}^{K} (m_{k,l} u_{k,l} h_{k,l} + n_{k,l}) v_{k,l}. \tag{3.4}$$

Note that the fusion center can separate the input streams because the data communication is either wired or performed by distinct waveforms for each SN.

In order to obtain a single reliable observation at the fusion center, the value $\tilde{r}_l$ should be a good estimate of the present target signal r_l. Thus the amplification factors $u_{k,l}$ and the weights $v_{k,l}$ should be chosen so as to minimize the average absolute deviation between $\tilde{r}_l$ and the true target signal r_l. The corresponding optimization program is elaborated in the next section.

3.3 Power and Lifetime Optimization

In this section, we first introduce the power minimization problem and subsequently develop its solution. Since the corresponding optimization problem is non-convex in its general form, we solve it by subsequent applications of the Lagrangian multipliers method with equality constraints, Karush–Kuhn–Tucker conditions, and straightforward usage of mathematical analysis, see [18], pp. 323–335 and [19], pp. 243–244.

3.3.1 The Optimization Problem

As mentioned in the last section, the quantity $\tilde{r}_l$ should be a good estimate for the present target signal r_l. In particular, we aim at finding estimators $\tilde{r}_l$ of minimum mean-squared error in the class of unbiased estimators for each r_l.

The estimate $\tilde{r}_l$ is unbiased simultaneously for each r_l if $\mathcal{E}[\tilde{r}_l - r_l] = 0$; in other words, from Equation (3.4) we obtain the identity

$$\sum_{k=1}^{K} g_{k,l} u_{k,l} h_{k,l} v_{k,l} = 1, \quad l \in \mathbb{F}_L. \tag{3.5}$$

This identity is our first constraint in what follows. Note that the mean of the second sum in (3.4) vanishes since the noise is zero-mean. Furthermore, we do not consider the impact of either random variable $g_{k,l}$ or $h_{k,l}$ or their estimates in our calculations because the coherence time of both channels is assumed to be much longer than the target observation time. Note that Equation (3.5) is complex-valued and may be separated as

$$\sum_{k=1}^{K} u_{k,l} |v_{k,l} g_{k,l} h_{k,l}| \cos(\vartheta_{k,l} + \phi_{k,l}) = 1, \quad l \in \mathbb{F}_L, \tag{3.6}$$

and

$$\sum_{k=1}^{K} u_{k,l} |v_{k,l} g_{k,l} h_{k,l}| \sin(\vartheta_{k,l} + \phi_{k,l}) = 0, \quad l \in \mathbb{F}_L, \tag{3.7}$$

where $\vartheta_{k,l}$ and $\phi_{k,l}$ are the phases of $v_{k,l}$ and the product $g_{k,l} h_{k,l}$, respectively.

The average power consumption of each node is approximately equal to its average output power $X_{k,l}$, if the input signal is negligible in comparison to the output signal and if the nodes have smart power components with low-power (dissipation) loss. We assume that equality between $X_{k,l}$ and the average power consumption of each node is ensured. In this chapter, we assume that the average output power range of each SN is limited by $P_{\min} \in \mathbb{R}_+$ and $P_{\max} \in \mathbb{R}_+$ with $0 \leq P_{\min} < P_{\max}$. The lower limit $P_{\min}$ denotes the minimum power that is needed to guarantee the awareness and presence of the SN; the upper limit $P_{\max}$ denotes the maximum allowed transmission power per SN due to power regulation standards or due to the functional range of the integrated circuit elements. In addition, each SN is usually powered by weak energy supplies, such as batteries, such that the operation time of the kth SN is limited by an available power budget $P_{k,l,\text{bud}} \in \mathbb{R}_+$. Note that $l = 0$ describes the time point at which each SN has a full-power budget; after each observation process the new power budget $P_{k,l+1,\text{bud}}$ is equal to $P_{k,l,\text{bud}} - X_{k,l}$, with $X_{k,0} = 0$ for all $k \in \mathbb{F}_K$. In this way, the sensor network operates under the constraints

$$P_{\min} \leq X_{k,l} \leq P_{\max} \Leftrightarrow P_{\min} \leq (R|g_{k,l}|^2 + M_k) u_{k,l}^2 \leq P_{\max}, \quad k \in \mathbb{F}_K, \quad l \in \mathbb{F}_L, \tag{3.8}$$

and

$$\sum_{l=1}^{L} X_{k,l} \leq P_{k,0,\,\mathrm{bud}} \Leftrightarrow \sum_{l=1}^{L} (R|g_{k,l}|^2 + M_k)u_{k,l}^2 \leq P_{k,0,\,\mathrm{bud}}, \quad k \in \mathbb{F}_K\,. \tag{3.9}$$

In order to guarantee a certain signal quality at the fusion center, the mean squared error $\mathcal{E}[|\tilde{r}_l - r_l|^2]$ should not exceed a given maximum value $V_{\mathrm{max}} \in \mathbb{R}_+$. By using (3.4) and the identity (3.5) we may write the next constraint as

$$\mathcal{E}[|\tilde{r}_l - r_l|^2] = \sum_{k=1}^{K} (M_k u_{k,l}^2 |h_{k,l}|^2 + N_k)|v_{k,l}|^2 \leq V_{\mathrm{max}}, \tag{3.10}$$

which must hold for all $l \in \mathbb{F}_L$.

The objective is now to minimize the overall power consumption of the sensor network for a given lifetime L, that is:

$$P_{\mathrm{over}}^{\star} := \underset{u_{k,l},v_{k,l}}{\mathrm{minimize}} \sum_{k=1}^{K}\sum_{l=1}^{L} X_{k,l} = \underset{u_{k,l},v_{k,l}}{\mathrm{minimize}} \sum_{k=1}^{K}\sum_{l=1}^{L} (R|g_{k,l}|^2 + M_k)u_{k,l}^2\,. \tag{3.11}$$

In summary, the optimization problem is to minimize the overall power consumption in (3.11) with respect to $u_{k,l}$ and $v_{k,l}$, subject to constraints (3.6)–(3.10). Note that the optimization problem is a *signomial program*, which is a generalisation of *geometric programming*, and is thus non-convex in general [20].

In order to avoid misunderstandings, we note that the minimization of the overall power consumption for any given lifetime L is in general not equivalent to the maximization of the lifetime for a corresponding given overall power P_{over}, since the lifetime is of a discrete nature while the power consumption is usually continuous. However, the solution difference between both optimization methods is at most only a single count in the lifetime and hence can be neglected in practice. Since the mathematical description of the lifetime maximization needs considerably more effort than the overall power minimization, we thus investigate the minimization of the power consumption in order to obtain insights about the maximization of the network lifetime.

3.3.2 Theoretical and Practical Solutions

For the sake of brevity, we define two new quantities $\alpha_{k,l}$ and $\beta_{k,l}$ by

$$\alpha_{k,l} := \sqrt{\frac{|g_{k,l}|^2}{M_k}} \quad \text{and} \quad \beta_{k,l} := \sqrt{\frac{N_k(R|g_{k,l}|^2 + M_k)}{M_k|h_{k,l}|^2}}\,. \tag{3.12}$$

The above optimization problem is closely related to an optimization problem considered by Taghizadeh et al. [15]. An optimization over $v_{k,l}$ leads to the problem

$$\underset{X_{k,l}}{\text{minimize}} \sum_{k=1}^{K} \sum_{l=1}^{L} X_{k,l} \tag{3.13a}$$

$$\text{subject to } P_{\min} \leq X_{k,l} \leq P_{\max}, \quad k \in \mathbb{F}_K, \quad l \in \mathbb{F}_L, \tag{3.13b}$$

$$\sum_{l=1}^{L} X_{k,l} \leq P_{k,0,\,\text{bud}}, \quad k \in \mathbb{F}_K, \tag{3.13c}$$

$$\sum_{k=1}^{K} \frac{X_{k,l}\alpha_{k,l}^2}{X_{k,l} + \beta_{k,l}^2} \geq V_{\max}^{-1}, \quad l \in \mathbb{F}_L, \tag{3.13d}$$

where (3.2) for the relation between $u_{k,l}$ and $X_{k,l}$ is used. It is easy to show that (3.13) is a convex optimization problem [5], and it can be solved using standard convex optimization tools.

The solution of (3.13) yields the overall power consumption $P_{\text{over}}^{\star}$ and all allocated powers $X_{k,l}^{\star}$ for a given lifetime L, signal quality $V_{\max}$, minimum and maximum allowed transmission powers $P_{\min}$ and $P_{\max}$, and power budgets $P_{k,0,\,\text{bud}}$. However, this solution needs advance knowledge about all channel realizations $g_{k,l}$ and $h_{k,l}$ and all noises $m_{k,l}$ and $n_{k,l}$, which are mostly unknown at the starting time of the sensor network. Nevertheless, the solution of (3.13) provides theoretical limits for the overall power consumption $P_{\text{over}}^{\star}$ and the network lifetime L, and furthermore enables comparisons of more practical methods. To provide a more realistic method, we highlight the following heuristic.

One possible approach is to optimize the power consumption per observation time. This means that at the beginning of the lth observation process, a relaxed version of (3.13) is solved which neglects the impact of all upcoming observation steps. The relaxed version of (3.13) for the lth observation process is described by

$$\underset{X_{k,l}}{\text{minimize}} \sum_{k=1}^{K} X_{k,l} \tag{3.14a}$$

$$\text{subject to } P_{\min} \leq X_{k,l} \leq P_{\max}, \quad k \in \mathbb{F}_K, \tag{3.14b}$$

$$X_{k,l} \leq P_{k,l,\,\text{bud}}, \quad k \in \mathbb{F}_K, \tag{3.14c}$$

$$\sum_{k=1}^{K} \frac{X_{k,l}\alpha_{k,l}^2}{X_{k,l} + \beta_{k,l}^2} \geq V_{\max}^{-1}. \tag{3.14d}$$

The solution of the relaxed optimization problem (3.14) is well-investigated and has even been solved analytically in [15] (compare also [5]). Since this solution is available in closed form, the computation of the power allocation needs less effort and can simply be performed online for each observation process. However, because of the relaxation of (3.13), the solution of (3.14) is suboptimal. If we denote the suboptimal solution of (3.14) by $\tilde{P}_{\text{over}}^{\star}$ with the suboptimal powers $\tilde{X}_{k,l}^{\star}$, then the inequality

$$\begin{aligned} \tilde{P}_{\text{over}}^{\star} &= \sum_{l=1}^{L} \left(\sum_{k=1}^{K} \tilde{X}_{k,l}^{\star} \right) = \sum_{l=1}^{L} \underset{X_{k,l}}{\text{minimize}} \sum_{k=1}^{K} X_{k,l} \\ &\geq \underset{X_{k,l}}{\text{minimize}} \sum_{k=1}^{K} \sum_{l=1}^{L} X_{k,l} = \sum_{l=1}^{L} \sum_{k=1}^{K} X_{k,l}^{\star} = P_{\text{over}}^{\star} \end{aligned} \tag{3.15}$$

obviously holds. This relation simply shows that due to the shortage of information, a sensor network that is optimized stepwise will consume more power and thus will have a shorter lifetime than when there is an overall optimization by (3.13). Another difference is the development of the available power budget $P_{k,l,\,\text{bud}}$ over time. Recall that for each observation process the new available power budgets must be updated as $P_{k,l+1,\,\text{bud}} = P_{k,l,\,\text{bud}} - \tilde{X}^{\star}_{k,l}$; in contrast, for an optimization by (3.13) the development would be $P_{k,l+1,\,\text{bud}} = P_{k,l,\,\text{bud}} - X^{\star}_{k,l}$. Fortunately, we will see later that both developments converge in many scenarios. The convergence speed is certainly a function of all parameters, especially dominated by P_{min}, P_{max}, $P_{k,0,\,\text{bud}}$ and V_{max}.

It is self-evident that both optimizations (3.13) and (3.14) represent two extreme cases for a variety of optimization methods. Based on the optimization method in (3.14), other heuristics can be proposed to obtain better performance. For example, the optimization in (3.14) can be extended by a robust method, in which information about the channel states and noise values of upcoming observation steps is not needed. Another approach is to extend (3.14) by channel and noise estimation methods, for example based on Kalman filtering, in order to obtain sufficient information about the unknown parameters of upcoming observation steps. Compared to the the method proposed in (3.14), these and other smart optimization methods will give improved performance at the cost of higher complexity. The discussion of two such methods is the focus of the next subsection.

3.3.3 More Practical Solutions

In general, an accurate suboptimal solution of (3.14), for which the upcoming channel states are not needed, is preferable in order to reduce the calculation complexity and to overcome unknown parameters. A smart approach would be the calculation of

$$\underset{X_{k,l}}{\text{minimize}} \sum_{k=1}^{K} w_{k,l}\, X_{k,l} \tag{3.16a}$$

$$\text{subject to } P_{\text{min}} \le X_{k,l} \le P_{\text{max}}, \quad k \in \mathbb{F}_K, \tag{3.16b}$$

$$X_{k,l} \le P_{k,l,\,\text{bud}}, \quad k \in \mathbb{F}_K, \tag{3.16c}$$

$$\sum_{k=1}^{K} \frac{X_{k,l}\alpha^2_{k,l}}{X_{k,l} + \beta^2_{k,l}} \ge V^{-1}_{\text{max}}. \tag{3.16d}$$

for the lth observation process with some predefined quantities $w_{k,l} \ge 0$. Each outcome $\tilde{X}^{\star}_{k,l}$ of (3.16) is then a function of these quantities $w_{k,l}$. In order to achieve a high performance by the approach in (3.16), the quantities $w_{k,l}$ may be optimized in various ways. For example, we can formulate a stepwise optimization by

$$\underset{w_{k,l}}{\text{minimize}} \sum_{k=1}^{K} \sum_{l=1}^{L} \mathcal{E}[|X^{\star}_{k,l} - \tilde{X}^{\star}_{k,l}|^2] \tag{3.17a}$$

$$\text{subject to } w_{k,l} \ge 0, \quad k \in \mathbb{F}_K, \quad l \in \mathbb{F}_L, \tag{3.17b}$$

or an overall optimization by

$$\underset{w_{k,l}}{\text{minimize}}\ \mathcal{E}\left[\left|\sum_{k=1}^{K}\sum_{l=1}^{L} X^{\star}_{k,l} - \tilde{X}^{\star}_{k,l}\right|^2\right] \tag{3.18a}$$

$$\text{subject to } w_{k,l} \geq 0,\ \ k \in \mathbb{F}_K,\ \ l \in \mathbb{F}_L, \tag{3.18b}$$

where the expectation should be properly chosen to exclude unwanted and unknown parameters. Since the optimization of $w_{k,l}$ by both above approaches is challenging and would not fit in the framework of this chapter, we provide the heuristic

$$w_{k,l} = f\left(\frac{P_{k,l,\text{ bud}}}{\sum_{k=1}^{K} P_{k,l,\text{ bud}}}\right),\ \ k \in \mathbb{F}_K,\ \ l \in \mathbb{F}_L, \tag{3.19}$$

where f should be a decreasing function of its argument (in particular, convex functions are better than others). For example, we can choose

$$w_{k,l} = \exp\left(\kappa - \frac{\kappa P_{k,l,\text{ bud}}}{\sum_{k=1}^{K} P_{k,l,\text{ bud}}}\right),\ \ k \in \mathbb{F}_K,\ \ l \in \mathbb{F}_L\,. \tag{3.20}$$

As readily seen, $w_{k,l}$ is a function of all $P_{k,l,\text{ bud}}$, which in turn include all channel states and the selection strategies of the past observation processes. The parameter κ is scenario dependent and should be chosen carefully. The solution of (3.16) with the aid of (3.20) gives good performance, particularly in scenarios with time-dependent channel states. A big advantage of (3.16) in comparison to (3.14) is its faster convergence speed; the computation effort nearly remains the same.

Another approach that benefits greatly from time-dependent channels is obtained if a long-term channel estimator is used. Assuming that channel estimation can be perfectly performed for I consecutive observations, we deduce the optimization problem

$$\underset{X_{k,l},\ldots,X_{k,l+I}}{\text{minimize}} \sum_{k=1}^{K}\sum_{i=l}^{l+I} X_{k,i} \tag{3.21a}$$

$$\text{subject to } P_{\min} \leq X_{k,i} \leq P_{\max},\ \ k \in \mathbb{F}_K,\ \ i = l, \ldots, l+I, \tag{3.21b}$$

$$\sum_{i=l}^{l+I} X_{k,i} \leq P_{k,l,\text{ bud}},\ \ k \in \mathbb{F}_K, \tag{3.21c}$$

$$\sum_{k=1}^{K} \frac{X_{k,i}\alpha_{k,i}^2}{X_{k,i} + \beta_{k,i}^2} \geq V_{\max}^{-1},\ \ i = l, \ldots, l+I\,. \tag{3.21d}$$

Note that in practice the channel estimation is imperfect. However, the imperfectness is not really decisive, since the calculation of (3.21) only leads to a suboptimal solution. In order to achieve accurate solutions and keep the computation effort as low as possible, the estimation depth should rather be small, say $I = 2$ or $I = 3$. The optimization by (3.21) is stated here only for the sake of completeness and will not be discussed further in the remaining part of this chapter.

3.4 Visualization and Numerical Results

In this section, we set out to compare the solution of (3.13) with (3.14) and (3.16) via numerical simulations. We consider a scenario with randomly distributed SNs so as to demonstrate the performance of (3.13) and (3.14). Afterwards, we use the same scenario to compare the solution of (3.16) with (3.13) and (3.14).

3.4.1 Comparison of (3.14) with (3.13)

In order to evaluate the performance of the optimization problems (3.13) and (3.14), we perform four simulations of the same scenario and for the same network. For all four simulations the values in Table 3.2 are kept constant. In particular, all channel and noise realizations remain the same in all simulations to simplify subsequent comparisons. These realizations are drawn randomly from independent Gaussian distributions. The available power budget $P_{k,0,\,\mathrm{bud}}$ is assumed to be equal for all SNs and thus denoted in short-form by P_{bud}. Figure 3.3 shows simulation results with the parameter values

Table 3.2 Values of fixed parameters for all plots.

Parameter	R	$\mathcal{E}[\lvert g_{k,l}\rvert^2]$	$\mathcal{E}[\lvert h_{k,l}\rvert^2]$	M_k	N_k	P_{min}
Value	1	1	1	1	1	0

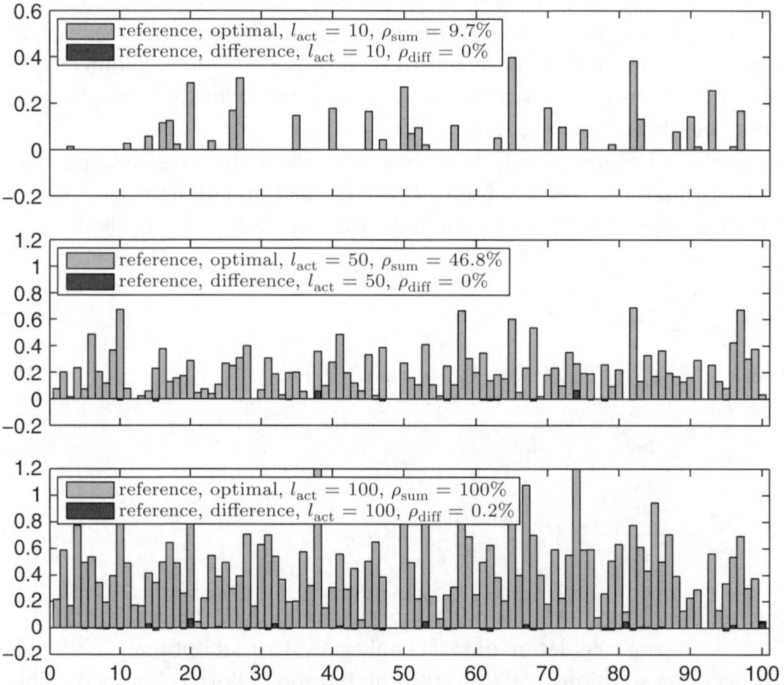

Figure 3.3 Reference bars for comparisons with other simulations. The values for the minimum overall power consumption are $P^{\star}_{\mathrm{over}} = 44.12$ and $\tilde{P}^{\star}_{\mathrm{over}} = 44.20$, and and the expected number of active SNs in each observation step is 4.

Table 3.3 Default parameter values for all plots.

Parameter	K	L	P_{max}	P_{bud}	V_{max}
Default value	100	100	0.36	1.2	1.2

given in Table 3.3. This figure will serve as a reference for all other figures. In all other figures, only one of the parameters P_{max}, P_{bud} or V_{max} is changed and the corresponding results are shown for comparison against the reference. The specific new value of the changed parameter is noted in the legend of the corresponding figure. In each legend three other values are given. The first value is the actual observation process l_{act}, which shows in which observation step the power distribution within the sensor network is illustrated. The second and third values are defined by

$$\rho_{sum} := \sum_{l=1}^{l_{act}} \sum_{k=1}^{K} \frac{X_{k,l}^{\star}}{P_{over}^{\star}} \tag{3.22}$$

and

$$\rho_{diff} := \sum_{l=1}^{l_{act}} \sum_{k=1}^{K} \frac{\tilde{X}_{k,l}^{\star} - X_{k,l}^{\star}}{P_{over}^{\star}}, \tag{3.23}$$

respectively. The value of ρ_{sum} describes the percentage amount of $P_{over}^{\star}$ that is already consumed by the network at the observation step l_{act} by utilisation of the optimization (3.13). The value of ρ_{diff} reflects the percentage amount of $P_{over}^{\star}$ that is additionally consumed at the observation step l_{act} due to the utilisation of the suboptimal optimization (3.14). Hence, the absolute value of ρ_{diff} should be as small as possible for describing an accurate fit of the suboptimal solution to the global one. The index of all SNs is shown on the abscissa of each plot. In each figure, the sum $\sum_{l=1}^{l_{act}} X_{k,l}^{\star}$ of consumed powers for each SN is visualised in a grey bar. These bars together show how the actual power distribution over the SNs is. Note that an equilibrated height of the grey bars over all SNs is preferential to achieve a long lifetime. In addition, the differences $\sum_{l=1}^{l_{act}} \tilde{X}_{k,l}^{\star} - X_{k,l}^{\star}$ are depicted as black bars. In contrast to the grey bars, the black bars show how well the theoretical and practical methods match. Thus smaller deviations of each black bar from zero is more favourable. Moreover, the captions include the minimum achieved values $P_{over}^{\star}$ and $\tilde{P}_{over}^{\star}$ for each simulation. Furthermore, the expected number of active SNs (the number of SNs that are allocated with positive transmission power) in each observation step is stated in the caption. Fortunately, in all our simulations the same number of active SNs always results for both optimization methods, which substantiates the accurate accordance of both methods.

In each figure, three states of the network lifetime are presented. The upper, the middle, and the lower illustrations represent the power distribution after $l_{act} = 10$, $l_{act} = 50$ and $l_{act} = 100$ observation processes, respectively.

In Figure 3.3, we observe that with the suboptimal optimization (3.14) the overall power consumption is slightly increased from $P_{over}^{\star} = 44.12$ to $\tilde{P}_{over}^{\star} = 44.20$. This small increment of the power consumption depends on the specific values of the parameters P_{max}, P_{bud} and V_{max} and is in practice negligible for most scenarios. Additionally,

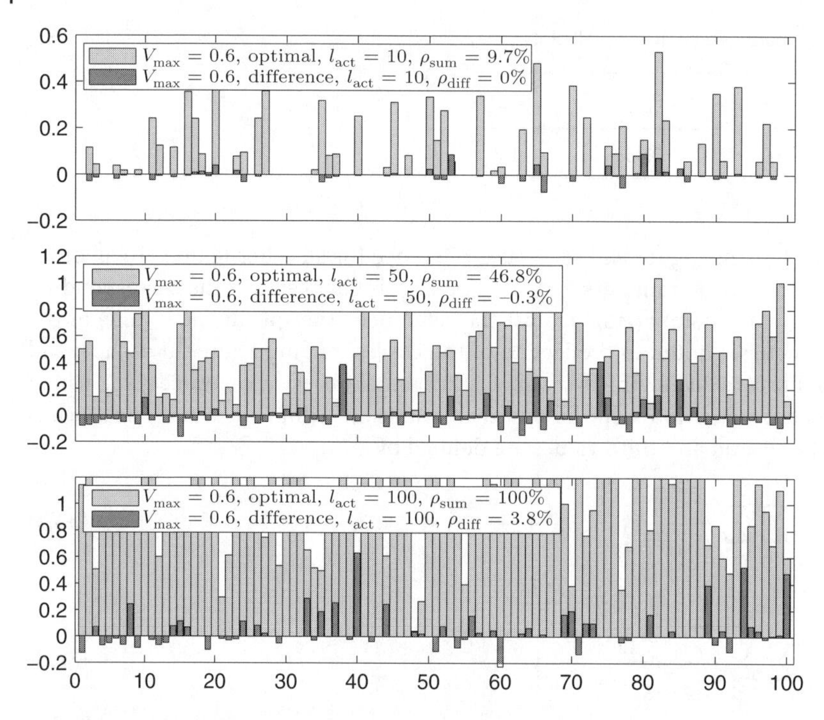

Figure 3.4 Decreasing V_{max} worsens the lifetime, the power consumption, and the convergence speed between both optimization results. The values of the minimum overall power consumptions are $P^{\star}_{over} = 95.15$ and $\tilde{P}^{\star}_{over} = 98.72$, and the expected number of active SNs in each observation step is 7.

it can be seen from the few black bars, that both optimization methods match. These observations are generally valid and they can be verified in the other figures.

Figure 3.4 shows the variation of V_{max} and its impact on the performance for both optimization methods. This parameter has the most impact on the performance. Both the number of active SNs and the overall power consumption highly fluctuate with the variation of V_{max}.

In Figure 3.5 the performance change of both optimization methods due to the parameter P_{bud} is depicted. The available power budget has the least influence on both the number of active SNs and the overall power consumption. The reason is that either the available power budget is large enough to achieve the given lifetime or it is so small that the corresponding optimization problem becomes infeasible.

Figure 3.6 illustrates the effect of P_{max} on the performance. As expected, the lifetime increases with P_{max} while the power consumption decreases. Thus any limitation of the transmission power by P_{max} and/or by P_{min} always has a negative effect on the performance of sensor networks.

In conclusion, the overall power consumptions $P^{\star}_{over}$ and $\tilde{P}^{\star}_{over}$ are monotonically decreasing with the parameters P_{max}, P_{bud} and V_{max} while the lifetime L is an increasing function of these parameters. Another important observation is that by decreasing P_{bud} or V_{max} while the other parameters are kept constant, the value of ρ_{diff} increases. This shows that the stepwise optimization by (3.14) converges more slowly to the global solution (3.13) by decreasing P_{bud} or V_{max}. The reason is that the gap between

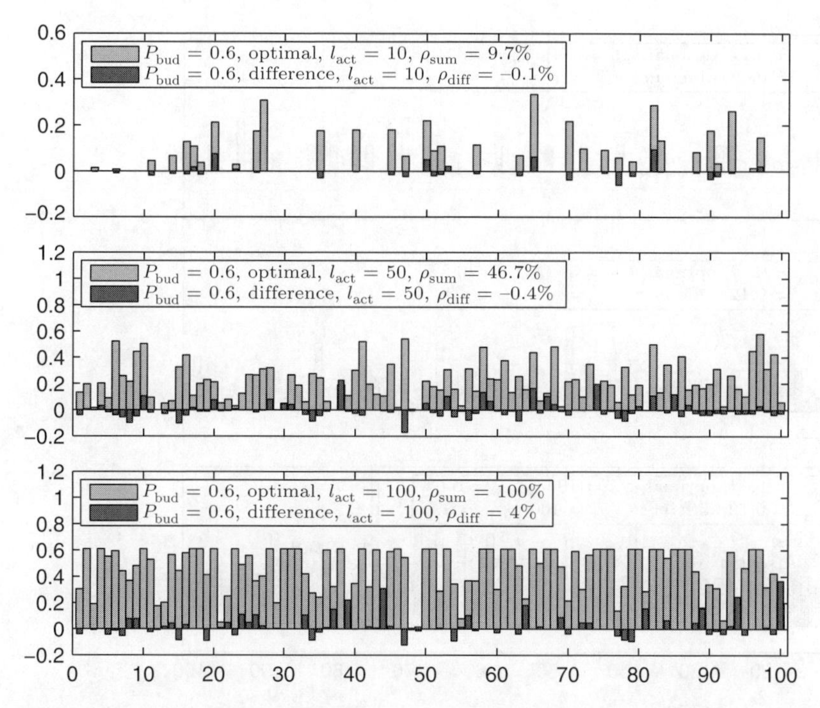

Figure 3.5 Decreasing P_{bud} can worsen lifetime, power consumption, and convergence speed. The values for the minimum overall power consumptions are $P^{\star}_{over} = 44.64$ and $\tilde{P}^{\star}_{over} = 46.41$, and the expected number of active SNs in each observation step is 4.

the solutions (3.13) and (3.14) increases since sharper constraints are to be handled, especially by the relaxed optimization method. Conversely, decreasing P_{max} while the other parameters are kept constant will result in a decreased ρ_{diff}, which in turn shows the increased convergence speed of the solution using (3.14) compared to the global one using (3.13). By decreasing P_{max} the power allocation engages more SNs and the activation of SNs becomes more distributed.

Since the evaluation of (3.13) is highly computation intensive, we unfortunately were not able to simulate networks with larger values of L that are more relevant in practice. Furthermore, a sensitivity analysis of both lifetime and power consumption under variation of other network parameters is important and will be investigated in future studies.

3.4.2 Comparison of (3.16) with (3.13) and (3.14)

In order to ensure fair comparisons between all methods, we perform the optimization method (3.16) in the same scenario with the same parameters as used in the previous subsection. In particular, the same channel and noise realizations are considered. We again illustrate four simulation results, analogous to the previous ones. The free parameter κ, which is needed for (3.16), is experimentally chosen and is set equal to ten. Note that an optimization over κ would lead to a better performance, but such a numerical optimization is computationally highly intensive.

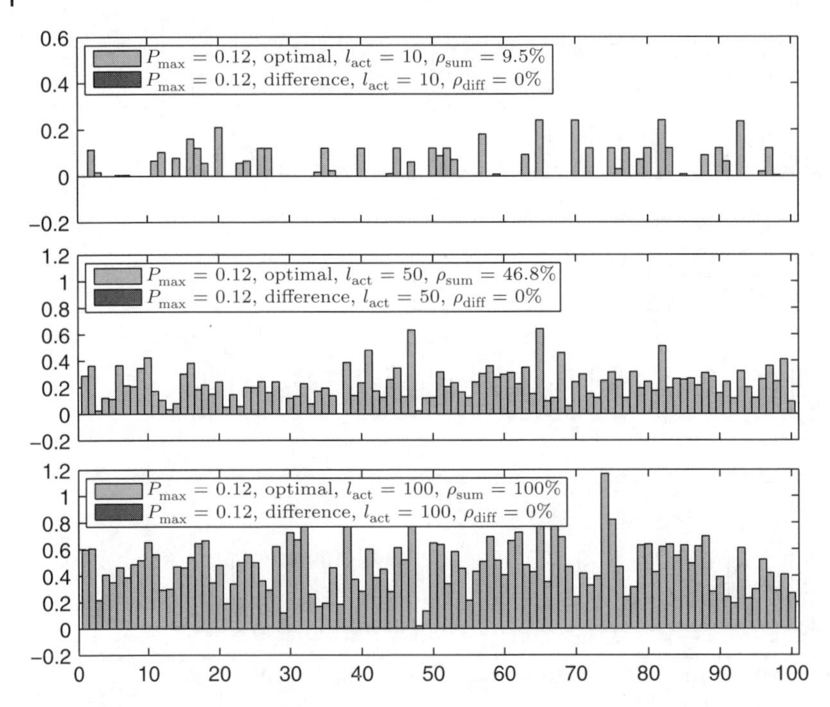

Figure 3.6 Decreasing P_{max} always worsens both the lifetime and the power consumption, but in contrast it increases the convergence speed. The values for the minimum overall power consumption are $P^{\star}_{over} = 47.39$ and $\tilde{P}^{\star}_{over} = 47.39$, and the expected number of active SNs in each observation step is 6.

In Figure 3.7, we show the reference plots, which are nearly the same as Figure 3.3. Only a small improvement is to seen by comparing the values of ρ_{diff} in both figures. The reason is that the results in Figure 3.3 already exhibit good convergence.

A better improvement is seen by comparing Figure 3.8 with Figure 3.4. Both ρ_{diff} and the overall power consumption $\tilde{P}^{\star}_{over}$ are smaller in Figure 3.8. The improvement is in general better for larger lifetimes L and the number K of SNs. Similar behavior is seen when comparing Figure 3.9 to 3.5. The comparison of Figure 3.10 with 3.6 shows that even in good scenarios, where the convergence speed is high, the solution of (3.16) behaves better or at least similarly to the solution of (3.14).

In summary, the power allocation method, which is relevant in practice and described by the optimization problem (3.16), provides performance close to the theoretical limits. This performance is robust over a wide range of parameters and is better for larger lifetimes L and number K of SNs.

3.5 Application of Power Control in Extreme Environments

The implementation of smart power allocation in sensor networks has various advantages. First, designing, planning and deploying the sensor network will not be customised for specific scenarios and they thus can be performed time- and

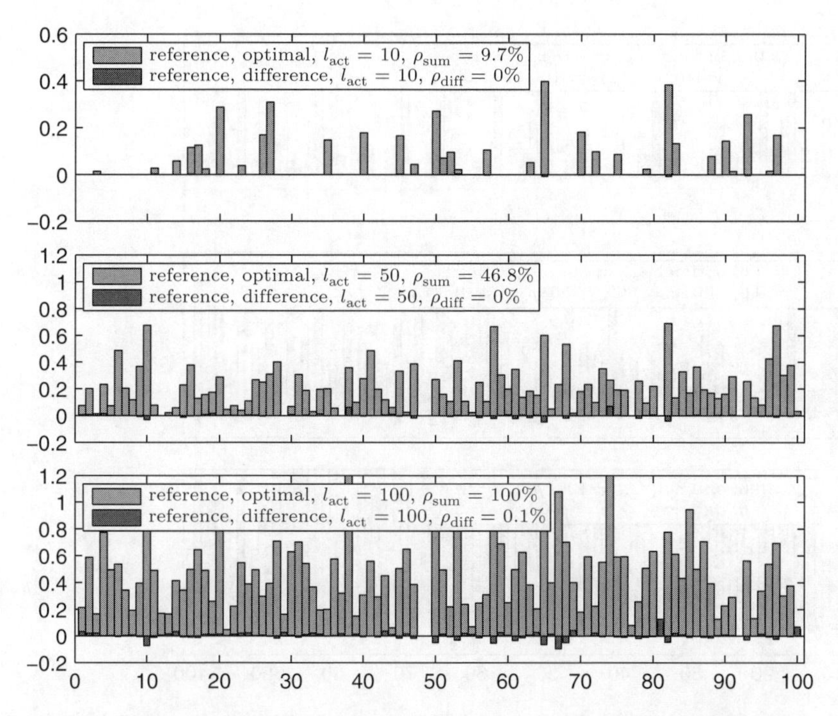

Figure 3.7 Reference bars for comparisons with other simulations. The values for the minimum overall power consumptions are $P^{\star}_{over} = 44.12$ and $\tilde{P}^{\star}_{over} = 44.17$, and the expected number of active SNs in each observation step is 4.

cost-efficiently. Second, the sensor network will be able to adapt its sensing strategy whenever the environment changes, by altering the number of active SNs and the power allocation to them to achieve maximal system performance. Finally, the lifetime of the sensor network will be prolonged in scenarios with limited energy resources.

The main concept behind the energy- and resource-awareness by applying a smart power allocation is as follows. Since the SNs are passive and cannot control or improve the sensing quality, the performance of each SN is dominated by the sensing channel. In contrast, the communication quality to the fusion center is determined by the transmission power of the SNs and can thus be regulated as needed. Since the effective quality of each observation process is compounded by the series qualities of the sensing and communication parts, the communication quality only needs to be a little better than the sensing quality to ensure the best effective observation quality. By optimal regulation of the transmission powers of the SNs we hence can enable an energy-aware sensor network and save more power compared to a network in which the power allocation is uniform or cannot be changed. Naturally, the greatest benefits of power optimization and lifetime maximization in sensor networks are found in scenarios in which energy resources are scarce or in which the exhausted power supply cannot be renewed. In particular in wireless sensor networks for extreme environments, power allocation and energy optimization techniques are indispensable. Two such environments are elucidated in the following in order to

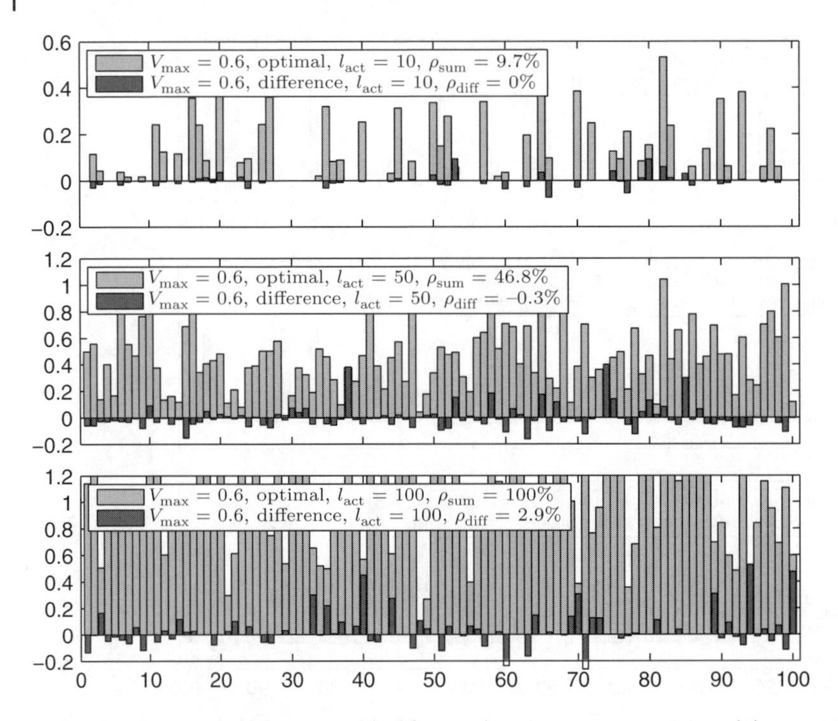

Figure 3.8 Decreasing V_{max} worsens the lifetime, the power consumption, and the convergence speed between both optimization results. The values for the minimum overall power consumptions are $P^{\star}_{over} = 95.15$ and $\tilde{P}^{\star}_{over} = 97.91$, and the expected number of active SNs in each observation step is 7.

demonstrate wireless sensor systems for space, underwater, underground and industrial applications.

A potential application for power minimization methods is passive multiple-radar sensing, where an unknown target signal is estimated, detected or classified. Instead of using a complex single-radar system, this task is carried out by a network of cheap and energy-efficient SNs: so-called *distributed passive multiple-radar systems* (DPMRSs). DPMRSs have several practical applications nowadays. Physicists use them to detect or to determine specific characteristics of particles, for example in the IceCube Neutrino Observatory, a neutrino telescope at the Amundsen–Scott South Pole Station in Antarctica, where a network with over 5000 nodes is deployed 2500 m under the icecap [21, 22]. DPMRSs are also used in radio astronomy, for instance in the Karl G. Jansky Very Large Array of the National Radio Astronomy Observatory in Socorro County, New Mexico [23]. In both these scenarios the fusion center is connected to the SNs via cables, to communicate the sensing information and to supply the SNs with energy. This means that the SNs have unlimited energy for operation and thus the entire sensor network has an unbounded lifetime. In this way, there is no need to maximize the network lifetime based on available resources. But note that the transportation of fossil fuels to Antarctica and the generation of electrical energy at South Pole are extremely difficult and very expensive. Hence any reduction of the energy consumption is of high interest. As we have seen, the power consumption of the entire network can drastically be

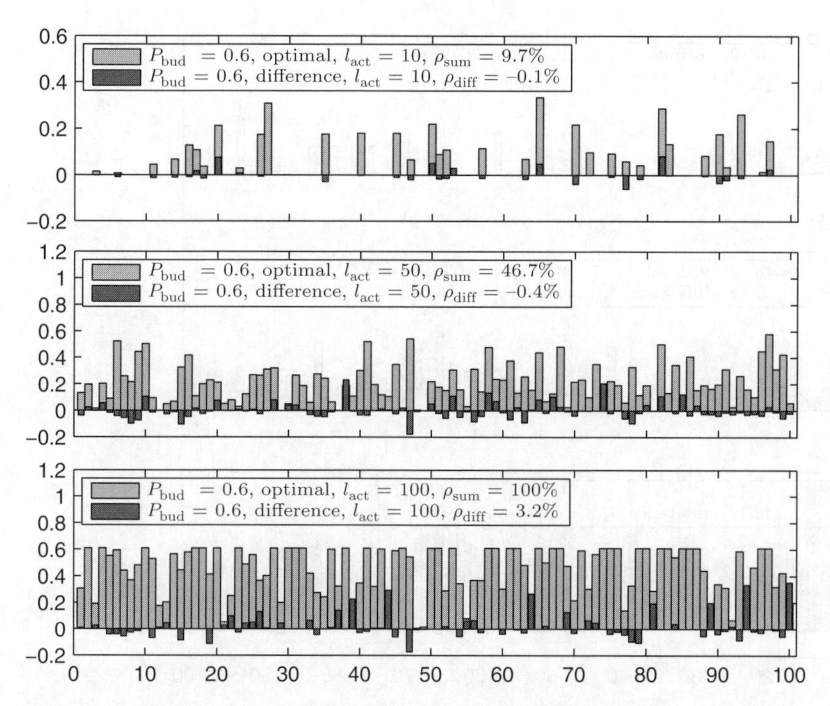

Figure 3.9 Decreasing P_{bud} can worsen the lifetime, the power consumption, and the convergence speed. The values for the minimum overall power consumptions are $P^{\star}_{over} = 44.64$ and $\tilde{P}^{\star}_{over} = 46.09$, and the expected number of active SNs in each observation step is 4.

reduced by a smart power allocation method without any performance loss [11]. In this way, sensor networks can be designed to be economical and eco-friendly.

An application for lifetime maximization methods is the construction of oceanographic maps, for which battery-based SNs are used. The utilised SNs are heavy and robust so that they can resist the pressures at great depths. The SNs are turned on and thrown into the water, where they will sink and communicate their information to a fusion center. By specific signaling techniques between the SNs, it is possible to construct a 3D map of the ocean bed (around 10 000 m deep) at the fusion center by applying sophisticated methods, for example multi-dimensional scaling [24, 25], and refs therein. For this technique a powerful communication link between the SNs as well as between each SN and the fusion center is required. Since each SN needs a certain time to reach the floor of the ocean and be in a stable position (the required time is usually unknown beforehand), the batteries are often oversized to guarantee a successful mission. In contrast, oversized batteries lead to pollution of waters and can cause fish die-off after the service time has ended. To avoid these problems, an investigation of energy resources is essential. The lifetime of the sensor network plays an important role in achieving accurate observation results and determining the power supplies by optimizing the number of batteries and the size of each. The methods set out in Section 3.3 enable the discussion of power consumption versus lifetime for scenarios in which access to or maintenance of SNs is infeasible.

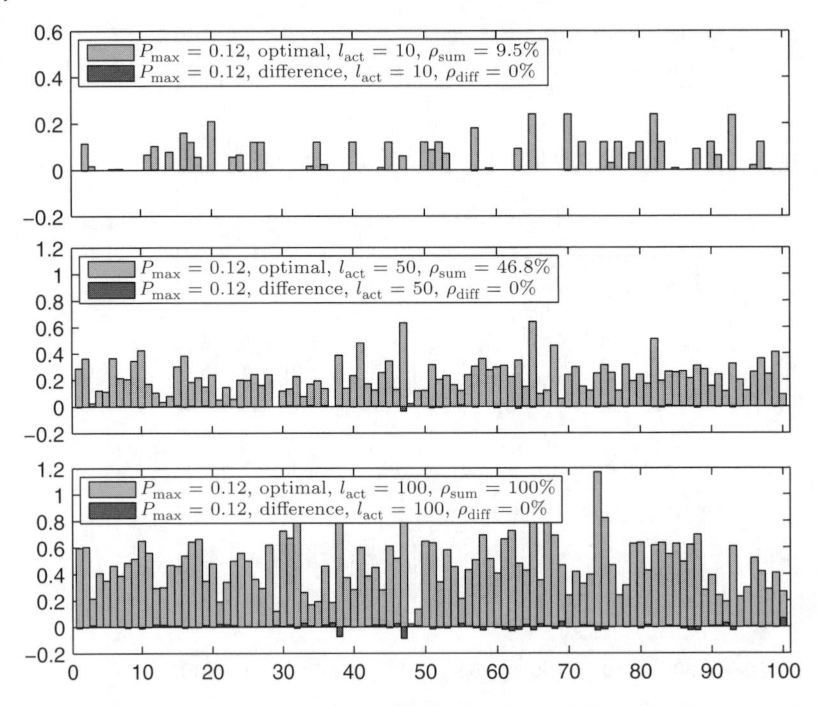

Figure 3.10 Decreasing P_{max} always worsens both the lifetime and the power consumption, but in contrast it increases the convergence speed. The values for the minimum overall power consumptions are $P^{\star}_{over} = 47.39$ and $\tilde{P}^{\star}_{over} = 47.40$, and the expected number of active SNs in each observation step is 6.

3.6 Summary

Power consumption and lifetime are essential features of sensor networks. On the one hand, the power consumption should be as low as possible to ensure an energy-efficient system. On the other hand, the lifetime should be as long as possible to ensure comprehensive coverage. For the application of sensor networks in extreme environments, it is also necessary to achieve high reliability over the whole lifetime. However, these features are contrary and they must be optimized simultaneously to achieve optimal performance. In this chapter, we have studied the minimization of the overall power consumption for any given lifetime and any required signal quality. Our goal has been the provision of insights into the power distribution when certain constraints on both the transmission power and the sum-power of the sensor nodes are prescribed. We have provided a theoretical approach and several practical methods to optimize the power consumption in common sensor networks.

First, we have seen that the theoretical approach is challenging and very computationally intensive to solve, although the corresponding optimization problem belongs to the class of convex problems. Furthermore, the theoretical approach is not suitable in practice due to the lack of channel-state information. However, the theoretical approach describes the feasible boundaries for power reduction and energy-awareness in sensor networks. Second, we have shown different, more practical approaches, which can nearly achieve the optimal solution and match the results of the theoretical approach well. Their main advantage is the low computational effort, albeit coming at the cost of a small increment in the overall power consumption. Another advantage of such practical methods is their modest complexity, since sophisticated methods for channel estimation are not needed. We have also seen that the convergence speed of the practical approaches is parameter dependent, but usually high enough for most scenarios. Especially in scenarios where the channel realizations are time-dependent, the practical methods will achieve their best performance. Finally, selected results via extensive simulations have shown both the performance of the methods and the power distribution among the sensor nodes over the lifetime of the sensor network. It becomes apparent that the required signal quality V_{max} has the major impact on both power consumption and lifetime, while a limitation of the transmission power by P_{min} and/or P_{max} has minor effects. The power budget P_{bud} of each sensor node definitely influences the overall power consumption and can even prevent the sensor network from lasting for the required lifetime.

References

1 Raghunathan, V., Schurgers, C., Park, S., and Srivastava, M. (2002) Energy-aware wireless microsensor networks. *Signal Processing Magazine, IEEE*, **19** (2), 40–50, doi:10.1109/79.985679.

2 Muruganathan, S., Ma, D., Bhasin, R., and Fapojuwo, A. (2005) A centralized energy-efficient routing protocol for wireless sensor networks. *Communications Magazine, IEEE*, **43** (3), S8–13, doi:10.1109/MCOM.2005.1404592.

3 Bhardwaj, M., Garnett, T., and Chandrakasan, A. (2001) Upper bounds on the lifetime of sensor networks, in *The IEEE International Conference on Communications (ICC'01)*, vol. 3, pp. 785–790, doi:10.1109/ICC.2001.937346.

4 Cardei, M., Thai, M., Li, Y., and Wu, W. (2005) Energy-efficient target coverage in wireless sensor networks, in *The 24th Annual Joint Conference of the IEEE Computer and Communications Societies (INFOCOM'05)*, vol. 3, pp. 1976–1984, doi:10.1109/INFCOM.2005.1498475.

5 Alirezaei, G., Reyer, M., and Mathar, R. (2014) Optimum power allocation in sensor networks for passive radar applications. *Wireless Communications, IEEE Transactions on*, **13** (6), 3222–3231, doi:10.1109/TWC.2014.042114.131870.

6 Alirezaei, G., Mathar, R., and Ghofrani, P. (2013) Power optimization in sensor networks for passive radar applications, in *The Wireless Sensor Systems Workshop (WSSW'13), co-located with the IEEE International Conference on Wireless for Space and Extreme Environments (WiSEE'13)*, Baltimore, Maryland, USA, doi:10.1109/WiSEE.2013.6737565.

7 Alirezaei, G. and Mathar, R. (2013) Optimum power allocation for sensor networks that perform object classification, in *Australasian Telecommunication Networks and Applications Conference (ATNAC'13)*, Christchurch, New Zealand, pp. 1–6, doi:10.1109/ATNAC.2013.6705347.

8 Alirezaei, G., Taghizadeh, O., and Mathar, R. (2014) Optimum power allocation with sensitivity analysis for passive radar applications. *Sensors Journal, IEEE*, **14** (11), 3800–3809, doi:10.1109/JSEN.2014.2331271.

9 Alirezaei, G. and Mathar, R. (2014) Optimum power allocation for sensor networks that perform object classification. *Sensors Journal, IEEE*, **14** (11), 3862–3873, doi:10.1109/JSEN.2014.2348946.
10 Taghizadeh, O., Alirezaei, G., and Mathar, R. (2014) Complexity-reduced optimal power allocation in passive distributed radar systems, in *International Symposium on Wireless Communication Systems (ISWCS'14)*, Barcelona, Spain.
11 Alirezaei, G. (2014) *Optimizing Power Allocation in Sensor Networks with Application in Target Classification*, Shaker Verlag. ISBN: 978-3-8440-3115-7.
12 Alirezaei, G. and Schmitz, J. (2014) Geometrical sensor selection in large-scale high-density sensor networks, in *The IEEE International Conference on Wireless for Space and Extreme Environments (WiSEE'14)*, Noordwijk, Netherlands, doi:10.1109/WiSEE.2014.6973064.
13 Alirezaei, G., Taghizadeh, O., and Mathar, R. (2015) Optimum power allocation in sensor networks for active radar applications. *Wireless Communications, IEEE Transactions on*, **14** (5), 2854–2867, doi:10.1109/TWC.2015.2396052.
14 Alirezaei, G. and Mathar, R. (2015) Sensitivity analysis of optimum power allocation in sensor networks that perform object classification. *Australian Journal of Electrical and Electronics Engineering*, **12** (4), 267–274, doi:10.1080/1448837X.2015.1093679.
15 Taghizadeh, O., Alirezaei, G., and Mathar, R. (2015) Optimal energy efficient design for passive distributed radar systems, in *IEEE International Conference on Communications (ICC'15)*, London, UK, pp. 6773–6778, doi:10.1109/ICC.2015.7249405.
16 Alirezaei, G., Taghizadeh, O., and Mathar, R. (2015) Lifetime and power consumption analysis of sensor networks, in *The IEEE International Conference on Wireless for Space and Extreme Environments (WiSEE'15)*, Orlando, Florida, USA, doi:10.1109/WiSEE.2015.7392980.
17 Alirezaei, G., Taghizadeh, O., and Mathar, R. (2016) Comparing several power allocation strategies for sensor networks, in *The 20th International ITG Workshop on Smart Antennas (WSA'16)*, Munich, Germany, pp. 301–307.
18 Luenberger, D.G. and Ye, Y. (2008) *Linear and Nonlinear Programming*, 3rd edn, Springer Science & Business Media.
19 Boyd, S. and Vandenberghe, L. (2004) *Convex Optimization*, Cambridge University Press.
20 Chiang, M. (2005) *Geometric Programming for Communication Systems*, Now Publishers.
21 University of Wisconsin-Madison and National Science Foundation (2010), IceCube Neutrino Observatory webpage. URL http://icecube.wisc.edu/.
22 Abbasi, R. (2010) IceCube neutrino observatory. *Int. J. Mod. Phys.*, **D19**, 1041–1048, doi:10.1142/S021827181001697X.
23 The National Radio Astronomy Observatory (2012), The Very Large Array webpage. URL http://www.vla.nrao.edu/.
24 Mathar, R. (1997) *Multidimensionale Skalierung: Mathematische Grundlagen und algorithmische Aspekte*, Teubner Verlag.
25 Xu, G., Shen, W., and Wang, X. (2014) Applications of wireless sensor networks in marine environment monitoring: A survey. *Sensors*, **14** (9), 16 932–16 954, doi:10.3390/s140916932.

4

On Improving Connectivity-based Localization in Wireless Sensor Networks

Bang Wang

School of Electronic Information and Communications, Huazhong University of Science and Technology (HUST), Wuhan, China

4.1 Introduction

Wireless sensor networks, which consist of a large number of small sensor nodes, each capable of sensing, processing and transmitting environmental information, have many applications in various fields, especially in space and extreme environments [1]. In these applications, the sensors' measurements may become useless without information about the nodes' positions. Furthermore, location information is also important in networking protocols for deployment, coverage, routing and so on.

Although equipping each sensor node with a GPS unit give its location easily, it incurs prohibitive hardware costs for large-scale sensor networks. As an alternative, localization can be used, with the coordinates for most of unknown nodes being determined without GPS from only a few anchor nodes. An *anchor node* is a node that knows its own coordinates; an *unknown node* is a node that does not. Localization, as one of the key issues in wireless sensor networks, has been intensively researched in recent years.

In general, localization algorithms for large-scale wireless sensor networks can be divided into two categories: range-based and range-free approaches. Mao et al. [2] have reviewed localization algorithms, mainly focusing on range-based ones, in which each node is assumed to be able to measure the distances between itself and its neighboring nodes using some ranging technique, such as

- the angle of arrival (AOA) [3]
- time of arrival (TOA) [4]
- time difference of arrival (TDOA) [5]
- received signal strength (RSS) [6].

The AOA method uses special antennas to detect the direction of the received signal or an antenna array to estimate its phase. The TOA and TDOA techniques estimate the distance between two nodes through the common relation of the signal transmission time, speed and distance. For the light transmission velocity, if the flight time of the received signal can be obtained, then the distance between a receiver and a transmitter can be calculated. The RSS approach estimates the distance by measuring the signal strength, based on the assumption of an a-priori known radio transmission model. These

Wireless Sensor Systems for Extreme Environments: Space, Underwater, Underground, and Industrial.
First Edition. Edited by Habib F. Rashvand and Ali Abedi.

techniques, however, often require sophisticated hardware for measurements, or suffer from inaccurate transmission models.

After obtaining the distances between neighboring nodes, an unknown node can estimate the distance to an anchor from the shortest cumulative travelling distance [3]. If the unknown node can estimate the distances to three or more anchor nodes, it can locate itself by trilateration or multilateration [7]. The range-based approaches can achieve high localization accuracy. On the other hand, their performance depends on the accuracy of the pairwise distance measurement.

In range-free localization, nodes generally have no ability to measure the distances to their neighbors, having only connectivity information. In extreme environments, such as monitoring in virgin forests, exact distance information might not be available due to obstructions. Also, using a sophisticated ranging unit is cost prohibitive. Range-free localization is therefore of particular importance as an alternative for extreme environments, because of its less demanding hardware and computational requirements. In many range-free localization algorithms, the *hop* is often used as the proximity measure between two neighboring nodes that have a direct communication link. For two non-neighboring nodes, the accumulated hop count for the shortest hop path between them can be used for distance estimation. Range-free localization methods are suitable for large-scale wireless ad hoc and sensor networks due to their less demanding hardware requirements.

In this chapter, we first review some recent advances in range-free localization; for one-hop networks in Section 4.2 and for multi-hop networks in Section 4.3. We then focus on improvements in classic connectivity-based localization approaches via exploitation of neighborhood information in Section 4.4. Finally, we conclude the chapter in Section 4.5.

4.2 Connectivity-based Localization in One-hop Networks

4.2.1 The Centroid Algorithm

The centroid localization algorithm [8] is a basic one-hop range-free localization algorithm. Its core concept is to use the geometric center of several anchor nodes in the communication range of an unknown node as its estimated position. This algorithm can be divided into two steps. In the first step, each anchor node broadcasts information maintained in a table $A_i(id; x_i, y_i)$ to all the neighbor nodes, where (x_i, y_i) are the coordinates of the anchor node A_i. After an unknown node S has heard this information from n anchors, it then calculates its location (s_x, s_y) as the arithmetic average of the coordinates of the anchors. That is,

$$\hat{s}_x = \frac{1}{n}\sum_{i=1}^{n} x_i; \quad \hat{s}_y = \frac{1}{n}\sum_{i=1}^{n} y_i. \tag{4.1}$$

The centroid algorithm is easy to implement. However, it generally requires a high anchor density for successful localization. When the anchor density is insufficient, it is possible that some unknown nodes will not hear even from one anchor, and therefore cannot directly apply the centroid algorithm for localization. Furthermore, the distribution of anchor nodes also greatly impacts on the localization performance. It is desirable that all anchor nodes be uniformly distributed.

4.2.2 Improved Centroid Algorithms

The centroid algorithm averages the coordinates of hearable anchor nodes to estimate the location of an unknown node, without differentiating these anchor nodes in any way. Some authors have proposed discriminating hearable anchor nodes by assigning different weights, so as to give a weighted centroid localization [9–11]. A general weighted centroid localization is as follows:

$$\hat{s}_x = \sum_{i=1}^{n} \omega_i x_i \bigg/ \sum_{i=1}^{n} \omega_i; \quad \hat{s}_y = \sum_{i=1}^{n} \omega_i y_i \bigg/ \sum_{i=1}^{n} \omega_i, \tag{4.2}$$

where ω_i is the weight for anchor A_i.

Blumenthal et al. [9] suggested making the weight for an anchor proportional to its received signal strength at the unknown node. The weight is estimated as $\omega_{ij} = 1(d_{ij})^g$, where where the attenuation coefficient g stands for the environmental impacts. The distance between two nodes d_{ij} can be obtained through the RSS value.

The authors assumed that a radio transmission model with a fixed path-loss attenuation exponent. However, this may not be the case if the field is not homogenous, with varying environment conditions. To address this problem, an adaptively weighted centroid localization approach has been proposed [10]. This uses least square error identification and maximum likelihood identification to identify the rational attenuation exponent for different environments.

Instead of using RSS, the improved weighted centroid localization algorithm (IWCA) exploits the time of arrival to compute the weights [11]. The distance between two nodes can be calculated as $d = c \times t$, where c is the speed of light and t is the signal on-flight time. The weight parameter can be obtained from $\sum_{i=1}^{n} t_i / n \times t_i$. For the IWCA, exploiting time of arrival to estimate the weight, the result is easily affected by the environment and it needs powerful hardware.

The centroid algorithm is a rough localization method, which may cause large localization errors in some cases. Suppose that there are three anchor nodes A_1, A_2 and A_3 with coordinates (0, 0), (7, 0) and (7, 2). For the disk communication model (see Section 4.3.1), an unknown node that can hear from the three anchor nodes only should be located in the intersection area of three circles centred on the three anchor nodes. However, the position $\hat{S}$ estimated using the centroid algorithm is $(14/3, 2/3)$, which is outside the intersection area. To address this problem, the localization algorithm proposed by Chen et al. [12] uses the intersection of the radical lines as the estimated location. The radical line is the line joining two intersection points of two intersecting circles.

4.3 Connectivity-based Localization in Multi-hop Networks

In a multi-hop network, it is possible that some unknown nodes cannot directly hear from any anchor node. In this case, we need to resort to other approaches to estimate the distance between an unknown node and an anchor node. In this section, we introduce the DV-hop algorithm and its variants, which use hop-count and hop-distance information to this end.

4.3.1 The DV-hop Algorithm

The core idea of the DV-hop algorithm [13] is to compute a correction factor, namely, the average hop distance (with the unit of meters per hop), for multi-hop distance estimation. The original DV-hop algorithm is based on the disk communication model, in which a node can communicate directly only with those nodes located within a disk centered at itself and with a radius of communication range r. For nodes beyond the disk, the node needs to use relay nodes for multi-hop communication.

The original DV-hop algorithm is divided into three steps:

- correction factor computation
- multi-hop distance estimation
- multi-lateration localization.

To compute the correction factor, each anchor node needs to know its hop count to another anchor node, which can be found by simple controlled flooding [14]. The hop count h_{ij} between an anchor i and another anchor j indeed measures the least number of multi-hop relays required to send one packet from i to j. Since anchors know their own coordinates, the Euclidean distance between the two anchors d_{ij} can be easily computed. Then the correction factor of anchor i can be computed by:

$$\phi_i = \frac{\sum_{j=1,j\neq i}^{n} d_{ij}}{\sum_{j=1,j\neq i}^{n} h_{ij}}, \tag{4.3}$$

where n is the total number of anchor nodes in the network. An unknown node k can estimate its Euclidean distance $\hat{d}_{ki}$ to an anchor i by

$$\hat{d}_{ki} = \phi_i \times h_{ki}, \tag{4.4}$$

where h_{ki} is the hop count between k and i.

Let $(x_i, y_i), i = 1, 2, \ldots, n$, denote the coordinates of the anchors, and $\hat{d}_i, i = 1, 2, \ldots, n$, the estimated distances from an unknown node to these anchors. After obtaining the estimated distances to these n anchors, an unknown node can then estimate its coordinates $(\hat{x}, \hat{y})$ using multi-lateration localization as follows:

$$\begin{bmatrix} \hat{x} \\ \hat{y} \end{bmatrix} = (A^T A)^{-1} A^T B \tag{4.5}$$

where

$$A = 2 \times \begin{bmatrix} x_2 - x_1 & y_2 - y_1 \\ x_3 - x_1 & y_3 - y_1 \\ \vdots & \vdots \\ x_n - x_1 & y_n - y_1 \end{bmatrix} \tag{4.6}$$

and

$$B = \begin{bmatrix} x_2^2 - x_1^2 + y_2^2 - y_1^2 + \hat{d}_1^2 - \hat{d}_2^2 \\ x_3^2 - x_1^2 + y_3^2 - y_1^2 + \hat{d}_1^2 - \hat{d}_3^2 \\ \vdots \\ x_n^2 - x_1^2 + y_n^2 - y_1^2 + \hat{d}_1^2 - \hat{d}_n^2 \end{bmatrix}. \tag{4.7}$$

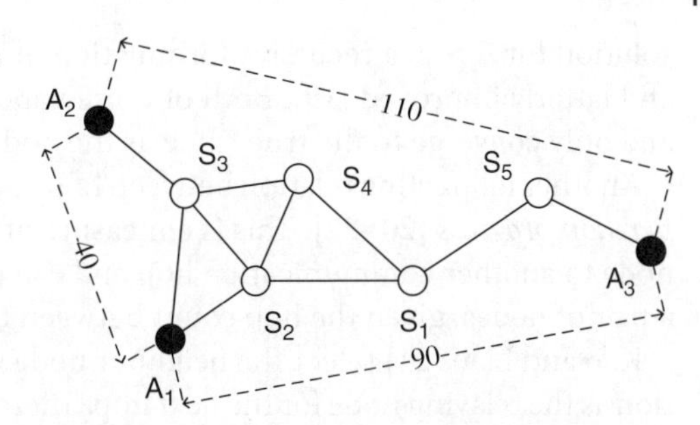

Figure 4.1 Illustration of the DV-hop algorithm.

Figure 4.1 illustrates the basic idea of the DV-hop algorithm. The network contains three anchors (A_1, A_2, A_3) and five unknown nodes ($S_1, S_2, \ldots, S_5$). The Euclidean distances between any pair of anchor nodes are also shown in the figure. The correction factor ϕ_{A_1} of anchor A_1 is calculated as $\phi_{A_1} = (40 + 90)/(2 + 4) = 21.67$. Similarly, the correction factors of A_2 and A_3 are $\phi_{A_2} = (40 + 110)/(2 + 5) = 21.43$ and $\phi_{A_3} = (90 + 110)/(4 + 5) = 22.22$, respectively. And the estimated distance between the anchor A_1 and the unknown node S_1 is computed by $\hat{d}_{A_1S_1} = 21.67 \times 3 = 65.01$. Similarly, the estimated distances $\hat{d}_{A_2S_1}$ and $\hat{d}_{A_3S_1}$ are $\hat{d}_{A_2S_1} = 21.43 \times 2 = 42.86$ and $\hat{d}_{A_3S_1} = 22.22 \times 2 = 44.44$, respectively.

4.3.2 Mathematics of Hop-count-based Localization

The philosophy of the DV-hop is to transform from an observed hop count h between two nodes to their Euclidean distance d through a heuristically obtained correction factor ϕ. Indeed, in a large network with uniformly distributed nodes, the relationship between h and d can be expressed by the probability density function of the distance d between two nodes conditioned on their hop-count; that is, $f(d|h)$. The minimum mean-square error estimate of d, given only the observation h, is simply the expectation of d conditioned on h; that is, $\mathbb{E}[d|h]$. Therefore, if we can compute $f(d|h)$, then $\mathbb{E}[d|h]$ can be used as an estimation of the true distance d.

The DV-hop algorithm applies a heuristic method to obtain ϕ and computes the mean distance as $\phi \times h$ for two nodes separated by h-hops from each other. Obviously, it is not an exact expression for $\mathbb{E}[d|h]$. Much effort has been devoted to deriving the statistical relation between d and h. For one-dimensional networks, the solution of $f(d|h)$ has been provided [15], as have the related conditional expectation $\mathbb{E}[d|h]$ and conditional variance $\mathbb{V}[d|h]$ [16]. In two-dimensional networks, for $h = 1$,

$$f(d|h = 1) = \frac{2d}{r^2} \text{ and } \mathbb{E}[d|h = 1] = \frac{2}{3}r, \tag{4.8}$$

where r is the radius of the communication disk model. However, for $h > 2$, the exact closed-form solution of $f(d|h)$ in two-dimensional networks is very hard to derive, and still remains an open problem.

A dual problem is to derive $P(h|d)$, the probability of a node being h-hops away from another node, given their Euclidean distance d. Given the conditional probability $P(h|d)$, the maximum-likelihood estimator can be formulated to estimate d based on the observation of h. Bettstetter and Eberspacher [17] derived the exact analytical solution of $P(h|d)$ for $h = 1$ and $h = 2$. However, conditional probabilities $P(h|d)$ for $h > 2$ are estimated by analytical bounds and extensive simulations. In order to obtain an analytical solution for $h > 2$, a recursive formulation of $P(h|d)$ for $h > 2$ has been proposed [18], and latterly improved [19]. Both of these papers are approximations of the true $P(h|d)$ and only converge to the true value as the node density in a network tends to infinity.

Another interesting relation between hop count and distance is known as the *expected per-hop progress* [20–22]. This is a measure of progress in Euclidean distance from one node to another communication hop and can be used to estimate the distance between a pair of nodes, given the hop count between them.

Kuo and Liao [21] select the neighbor node with the shortest distance to the destination as the relaying node for the next hop. There are three arcs centered at the destination node D with radii of l_0, x and $l_0 - a$, respectively, where l_0 denotes the initial distance between nodes S and D. Let $\text{Cov}(S, r)$ denote the transmission coverage of the node S with radius r, and let $L_{b,m}$ denote the remaining distance to the destination node after m hops, given that the effective node density is b. Thus the conditional cumulative distribution function for $L_{1,1}$ is given by:

$$F_{X_1}(x) = \Pr\{d(N_1) \leq x\} = \frac{\text{Cov}(S, r) \cap \text{Cov}(D, x)}{\text{Cov}(S, r) \cap \text{Cov}(D, l_0)}, \tag{4.9}$$

and $F_{X_b}(x)$ of $L_{b,m}$ is obtained by

$$F_{X_b}(x) = \Pr\{d(N_1) \leq x \cup d(N_2) \leq x \cdots d(N_2) \leq x\}. \tag{4.10}$$

This gives the conditional probability density function $f_{X_b}(x)$ of $L_{b,m}$ and the hop progress $\mathbb{E}[L_{b,m}]$.

4.4 On Improving Connectivity-based Localization

Many algorithms have been proposed to improve the DV-hop algorithm by adjusting the correction factor. The original DV-hop algorithm only uses average hop-distance for multi-hop distance estimation. However, information such as neighborhood information can be used too.

4.4.1 Improvements by Adjusting Correction Factor

In the original DV-hop algorithm, the correction factor ϕ_i is computed by averaging the hop distance, an approach which only functions well with enough uniformly deployed sensors. An alternative approach, based on hop progress, can be used to calculate the correction factor with randomly deployed sensors and arbitrary node density [22–25].

Wang et al. [22] used the expected hop progress to estimate the optimal path distance between sensors. Sensors are distributed randomly according to a two-dimensional homogeneous Poisson point process of density λ. The probability of m sensors located within a section between $(-\theta, \theta)$ of a sensor's transmission range πr_0^2 can be obtained as

$$P(m, \theta, r_0) = (\lambda\theta r_0^2)^m e^{-\lambda\theta r_0^2}/m!. \tag{4.11}$$

The hop-distance between S and its next forwarding sensor is denoted as L. The probability of L being less than l can be given as

$$P[L \leq l] = \sum_{m=0}^{N} P(m, \theta, r_0)P(m, \theta, l) = e^{-\lambda\theta(r_0^2 - l^2)}. \tag{4.12}$$

This gives the probability density function $f_L(l)$ of $P[L \leq l]$. Finally, the correction factor of sensor s and the expected hop progress can be derived as follows:

$$\phi_s = \mathbb{E}[R] = \frac{1}{\theta}\int_0^{\theta}\int_0^{r_0} f_L(l) l \cos\omega \, dl \, d\omega. \tag{4.13}$$

And the distance between two nodes is calculated as $\hat{d} = h \times \mathbb{E}[R]$.

Similarly, Vural and Ekici [23] select a node with the maximum distance to the current one as the next relaying node. They also analyse the maximum hop progress of a single hop, and calculate the multi-hop progress with these results.

We et al. [24] analyse the hop distance of different hop counts to reduce the linear approximation error of the correction factor. The authors assume that nodes are uniformly distributed with the probability distribution function

$$f(l) = 2l/r^2. \tag{4.14}$$

Then the hop progress of a 1-hop is given by

$$\mathbb{E}[R] = \int_0^{r} l f(l) dz = 2r/3, \tag{4.15}$$

and this is confirmed by experimental results. Thus when the communication range is known, the distance can be obtained by

$$\hat{d}_{ki} = \phi_i \times (h_{ki} - 1) + 2r/3, \tag{4.16}$$

where

$$\phi_i = \sum_{j=i,j\neq i}^{m} (d_{ij} - 2r/3)(h_{ij} - 1) \Big/ \sum_{j=i,j\neq i}^{m} (h_{ij} - 1)^2. \tag{4.17}$$

Some other approaches have been proposed for adjusting correction factors in non-uniformly distributed networks. The *density-aware hop-count localization* algorithm adjusts the correction factor according to local node densities [25]. For each pair of anchors, they also can use the computed correction factor to estimate their distance, However, there may be some estimation error, especially in highly asymmetric networks. To address such problems, the DDV-hop algorithm [7] adjusts the correction factor for an anchor according to its regulated distance estimation error.

4.4.2 Improvements by Exploiting Neighborhood Information

The DV-hop algorithm is easy to implement, but it suffers from the *hop-distance ambiguity* problem: those nodes with the same hop-count to an anchor will have the same estimated Euclidean distance, although they have different distances to this anchor. For example, as shown in Figure 4.2, nodes i and k are both h hops away from the source node. According to the DV-hop algorithm, they will have the same estimated distance to the source node. However, as seen from the figure, the distance d_i to the source is

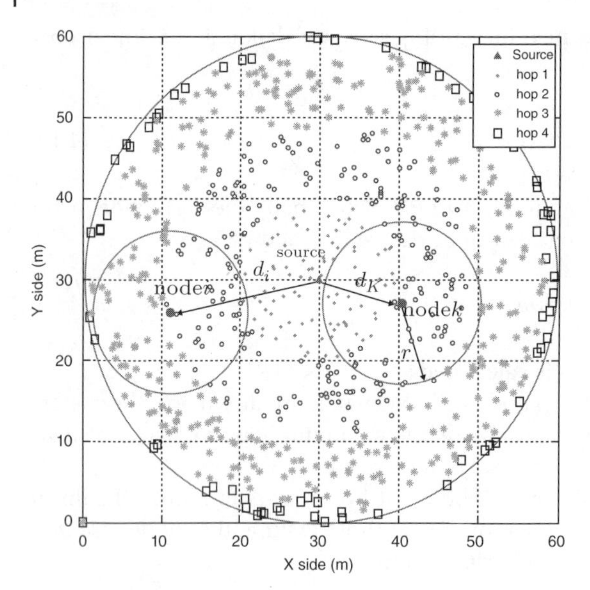

Figure 4.2 The hop-distance ambiguity problem [26].

larger than the distance d_k. To address the hop-distance ambiguity problem, some have proposed using neighborhood information to discriminate two nodes of the same hop count [26–35].

Ma et al. [26] proposed a *hop-count-based neighbor partition algorithm* (HCNP) to exploit neighborhood information to address the hop-distance ambiguity problem. For example, as shown in Figure 4.2, different shapes represent nodes with different hop counts to the source node. The neighbors of a node with hop count h can be generally divided into three disjoint sets: S^{h-1}, S^h and S^{h+1}, namely, neighbors with hop counts $h-1$, h and $h+1$, respectively. Also from Figure 4.2, node i has more neighbors with hop count $h-1$ than node k, and node k has more neighbors with hop-count $h+1$ than node i. It should be noted that such neighborhood information can easily be obtained via local exchange of information.

The HCNP algorithm first assumes that all nodes of hop-count h are located within a ring centered at the source and with inner radius γ_{h-1} and outer radius γ_h. This is called the *perfect-hopping assumption*, and set $0 < \gamma_1 < \cdots < r_h < \cdots$. Obviously, this assumption only holds when the number of total sensors N tends to infinity, and in such a case, $\gamma_h \times r$ and $\gamma_h - \gamma_{h-1} = r$ for all hop rings $h \geq 1$, as illustrated in Figure 4.3.

For a node with hop-count h to the source, the coverage disk intersects with hop rings $h \pm 1$ and creates two intersection regions, denoted by $a^{h\pm1}$. Let $A^{h\pm1}$ denote the areas of $a^{h\pm1}$. The geometric relations between the distance between the node to the source d and the intersection areas $A^{h\pm1}$ can be computed by

$$A_{h-1} = r^2\cos^{-1}\left(\frac{d^2 + r^2 - \gamma_{h-1}^2}{2dr}\right) + \gamma_{h-1}^2\cos^{-1}\left(\frac{d^2 + \gamma_{h-1}^2 - r^2}{2d\gamma_{h-1}}\right)$$
$$-\frac{1}{2}\sqrt{4d^2\gamma_{h-1}^2 - (d^2 - r^2 + \gamma_{h-1}^2)^2}. \quad (4.18)$$

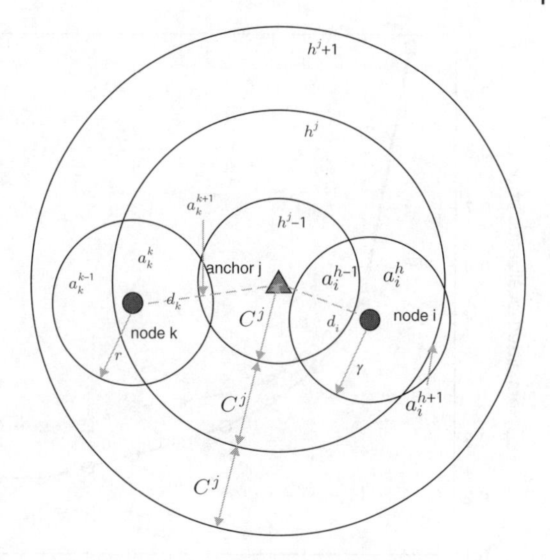

Figure 4.3 Perfect-hopping scenario and its geometric relation [26].

$$A_{h+1} = \pi r^2 - r^2 \cos^{-1}\left(\frac{d^2 + r^2 - \gamma_h^2}{2dr}\right) - \gamma_h^2 \cos^{-1}\left(\frac{d^2 + \gamma_h^2 - r^2}{2d\gamma_h}\right)$$
$$+\frac{1}{2}\sqrt{4d^2\gamma_h^2 - (d^2 - r^2 + \gamma_h^2)^2}. \tag{4.19}$$

Assuming that the sensors are uniformly distributed in the whole sensor field with density λ, the intersection areas can also be estimated through the node density by

$$\curlywedge^{h\pm1} = \frac{|S^{h\pm1}|}{\lambda}, \tag{4.20}$$

where $|S^{h\pm1}|$ is the number of nodes with hop-count $h \pm 1$. Let $A^{h\pm1} = f_{h\pm1}^{-1}(A^{h\pm1})$. We can substitute the estimated areas $\curlywedge^{h\pm1}$ into function $f_{h\pm1}^{-1}(A^{h\pm1})$. Then the distance can be obtained by

$$\hat{d}^{h\pm1} = f_{h\pm1}^{-1}(A^{h\pm1}). \tag{4.21}$$

Because we can get two distance estimations—in other words, $\hat{d}^{h\pm1}$—based on the two intersection areas, the final multi-hop estimation is computed as their average; that is,

$$\hat{d}_{\text{HCNP}} = (\hat{d}^{h-1} + \hat{d}^{h+1})/2. \tag{4.22}$$

Figure 4.4 compares the normalised localization errors of the HCNP and the DV-hop approach. We randomly deploy N nodes, each with communication range $r = 10$, into a disk field of radius of 60. It can be seen that the performances of the two algorithms are similar when the number of nodes is small, but the HCNP algorithm performs better than DV-hop when the number of nodes increases. The amount of improvement also increases with the number of nodes. Recall that the HCNP algorithm is based on the perfect-hopping assumption. The larger the sensor density, the more accurate the hop ring estimates and so the localization performance.

The HCNP algorithm is based on the computation of hop-ring boundaries γ_h. In order to compute γ_h, it implicitly assumes a global uniform distribution for the whole network,

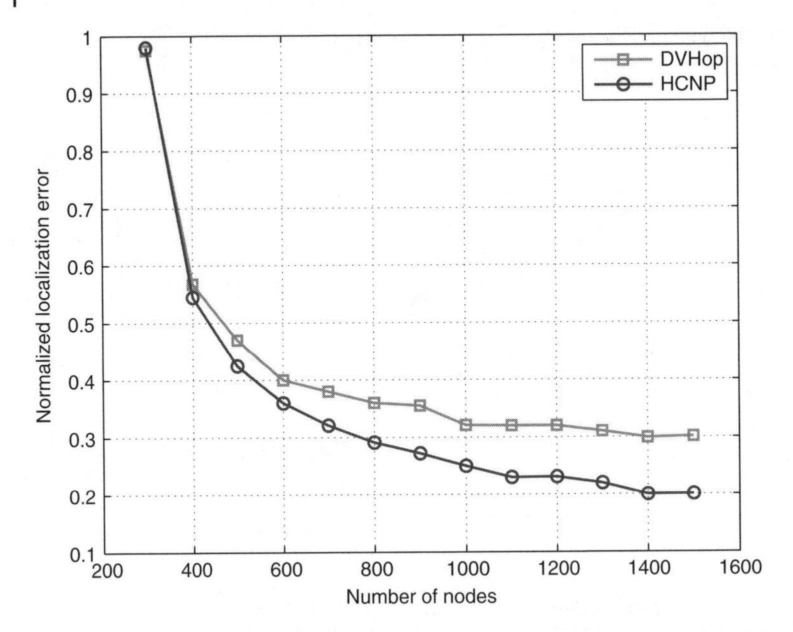

Figure 4.4 Comparison of the normalised localization error and the number of deployed nodes [26].

and the algorithm needs the network-wide node density λ also. Wu et al. [30] consider only local neighborhood information for multi-hop distance estimation. They define the *regulated neighborhood distance* (RND) as a proximity measure for two neighbors.

Let $\mathcal{M}_i$ denote the set of neighbors of a node i, which is defined as

$$\mathcal{M}_i = \{j | j \neq i \text{ and } d_{ij} \leq r\}, \tag{4.23}$$

where d_{ij} is the Euclidean distance between node i and node j, and r the communication radius. We can observe from Figure 4.5 that the distance d_{ij} between a node i and its neighbor j determines the size of the intersection area, $A(d_{ij})$, as the communication radio is constant. The larger the d_{ij}, the smaller the $A(d_{ij})$. Furthermore, $A(d_{ij})$ can be computed by,

$$A(d_{ij}) = 2r^2 \arccos\left(\frac{d_{ij}}{2r}\right) - \frac{d_{ij}}{2}\sqrt{4r^2 - d_{ij}^2} \tag{4.24}$$

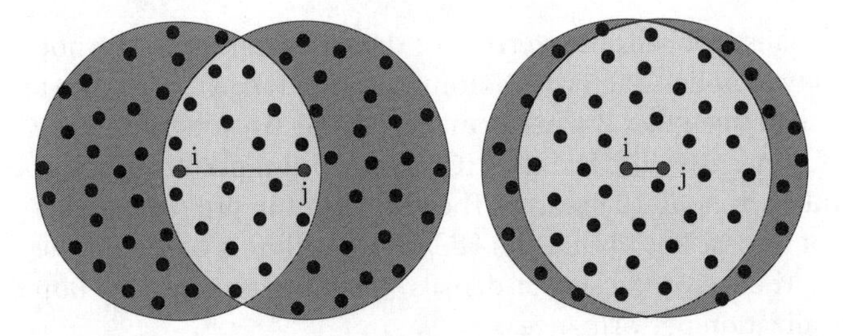

Figure 4.5 The neighborhood distance model: the distance between *i* and *j* is dependent on the number of common neighbors between *i* and *j* [30].

As the distance between two nodes, d_{ij}, is not known, $A(d_{ij})$ can be estimated by,

$$\lambda(d_{ij}) = \frac{m_{ij}}{M_i} \cdot \pi r^2 \tag{4.25}$$

where m_{ij} is the number of nodes within the intersection area and $M_i = |\mathcal{M}_i| + 1$ the number of nodes within the communication disk. Note that m_{ij} actually counts the number of nodes in the intersection area formed by i and j; that is, $m_{ij} = |\mathcal{M}_i \cap \mathcal{M}_j| + 2$. In practice, m_{ij} can be easily obtained by exchanging the neighbor information between the two nodes. The node i then can use $\lambda(d_{ij})$ to estimate d_{ij}. The *neighborhood distance* from node i to its neighbor j is defined by,

$$\text{ND}(i,j) = 1 - \frac{m_{ij}}{M_i}, j \in \mathcal{M}_i. \tag{4.26}$$

Note that the neighborhood distance from node j to i, ND(j, i), may not equal to ND(i, j). The RND is then defined by

$$\text{RND}(i,j) = \frac{1}{2}(\text{ND}(i,j) + \text{ND}(j,i)). \tag{4.27}$$

Instead of using the hop-count, the shortest-RND path is then found from the accumulated RND of a multi-hop path between two nodes. The shortest RND distance between two nodes is defined by

$$\text{RND}_{\min}(s_1, s_n) \doteq \min\left\{\sum_{i=1}^{n-1} \text{RND}(s_i, s_{i+1}\right\}. \tag{4.28}$$

Similar to the hop-count-based correction factor, then the RND-based correction factor is computed as:

$$\phi_{\text{rnd}} = \frac{\sum_{j \neq i} d_{ij}}{\sum_{j \neq i} \text{RND}_{\min}(i,j)}. \tag{4.29}$$

And an unknown node s estimates its Euclidean distance to anchor i by

$$\hat{d}_{si} = \phi_{rnd} \times \text{RND}_{\min}(s,i). \tag{4.30}$$

We next use an example to illustrate the DV-RND and DV-hop algorithms for distance estimation. As shown in Figure 4.6, in a randomly deployed network, we can find the shortest-RND path marked by the blue line and the shortest-hop path marked by the red line from node s to node t by applying the Floyd–Warshall [36] algorithm. We can observe that compared with the shortest-hop path, the shortest-RND path is closer to the direct link marked by the black dotted line between s and t, although the shortest-RND path has more hops (11 hops) than the shortest-hops (10 hops) path. The idea behind DV-hop and DV-RND is to use the shortest-hop or shortest-RND path to approximate the distance between two nodes. As seen from this example, the DV-RND algorithm gives a better approximation than the DV-hop algorithm.

We next compare the localization performance with the DV-hop [3], the DV-distance [3] and the DV-CNED [37] algorithms. The DV-distance algorithm assumes that each node can measure the true Euclidean distance to its neighbors. As shown in Figure 4.7, in a uniform network deployment, DV-RND outperforms the DV-CNED, DV-distance and DV-hop algorithms in terms of lower localization error, if the node degree is smaller

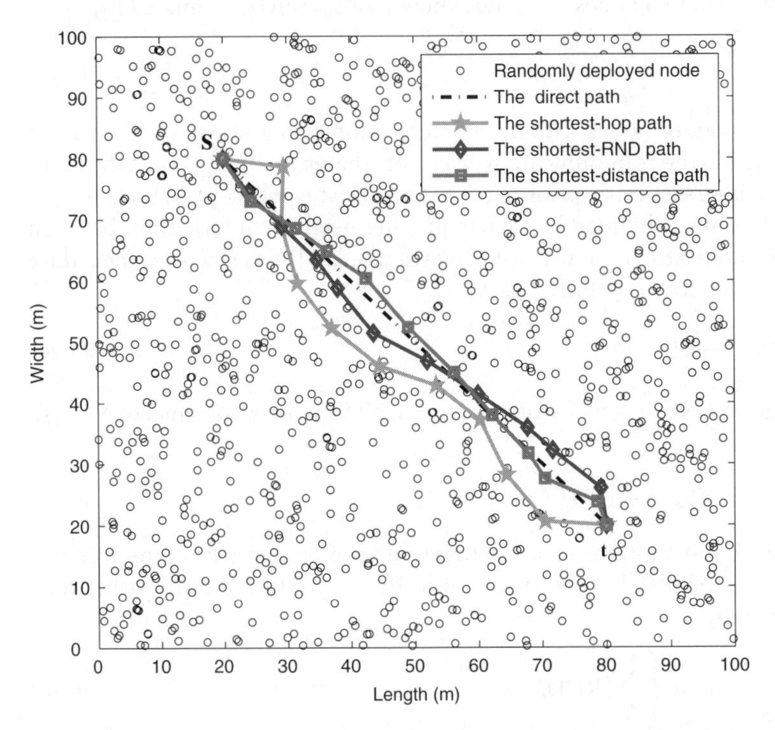

Figure 4.6 Illustration of the shortest-RND path, the shortest-hop path and the shortest-distance path [30].

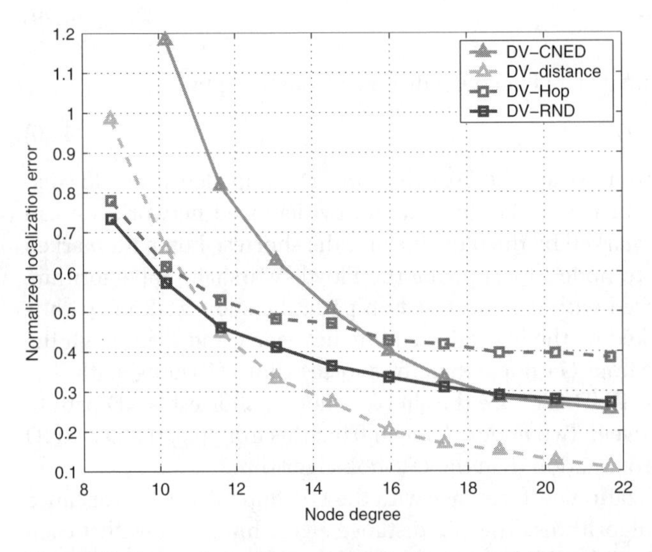

Figure 4.7 Comparison of average localization error of different connectivity-based localization algorithms [30].

than 11.5; it outperforms DV-CNED and DV-hop if the node degree is smaller than 17.5. We can also see that our proposed DV-RND algorithm will give a larger improvement over DV-hop as the node degree increases. This is because DV-RND successfully combines the advantages of DV-hop and DV-CNED.

Many connectivity-based range-free localization techniques have assumed the disk communication model. However, radio propagation in the real world is very complicated and far from the ideal disk model. For example, the well-known log-normal shadowing model [38] is given by (in dB)

$$P_r(d) = P_r(d_0) - 10\beta \log\left(\frac{d}{d_0}\right) + X_\sigma, \tag{4.31}$$

where $P_r(d)$ and $P_r(d_0)$, respectively, are the received power at the distance d and the reference distance d_0 to the transmitter, and β is the distance attenuation factor. X_σ is a factor to model the shadowing effect, which follows a Gaussian distribution (in dB) with zero mean and σ standard deviation.

Observing the shadowing model, the received power is distance-dependent and its mean is attenuated with the distance. Furthermore, due to the randomness of the shadowing term, the received power may be different at different receptions for a transmitter–receiver pair. On the other hand, the received power determines whether a receiver can correctly receive a packet sent from a transmitter. Under such a shadowing model, a transmitter communicates with a receiver in a probabilistic way. Therefore, the definition of 'hop' and 'neighboring node' in the disk model should be revisited in realistic environments.

Wang et al. [39] consider a rather general radio propagation model, where the RSS is modeled as a random variable and its expectation is a non-increasing function of the distance between a transmitter–receiver pair. For wireless networks with such a general model, the hop and neighboring-node definitions are revisited based on the packet reception rate (PRR). The performance of the RND-based localization scheme is dependent on the choice of an appropriate PRR threshold to define the hop and the neighboring nodes. For the general propagation model, Wang et al. [39] use the PRR to define neighborhood as follows:

> Suppose that T localization packets are transmitted for neighborhood determination. Let γ denote the PRR threshold and $\frac{1}{T} \leq \gamma \leq 1$. A node j is a neighbor of node i if the PRR at node i from the node j is not smaller than the PRR threshold. That is, if $P_{ij}(T) \geq \gamma$, then node j is a neighbor of node i.

It is possible that a node j is a neighbor of node i, while node i is not a neighbor of node j. According to the definition, the neighbors of a node i are dependent on the choices of T and γ. The hop is defined based on the mutual neighborhood; that is, two nodes i and j are considered to be neighboring nodes (one-hop neighbors of each other), if $i \in \mathcal{M}_j(T,\gamma)$ and $j \in \mathcal{M}_i(T,\gamma)$. Similarly, a node i is two hops away from a node j if i is not a one-hop neighbor of j, and both node i and j have a common one-hop neighbor of another node. In this way, we can construct a shortest-hop path from one node to another, and define the hop count accordingly.

Rappaport [39] proposed an adaptive RND selection algorithm to selects the best PRR threshold γ, if T localization packets are to be used for neighborhood determination.

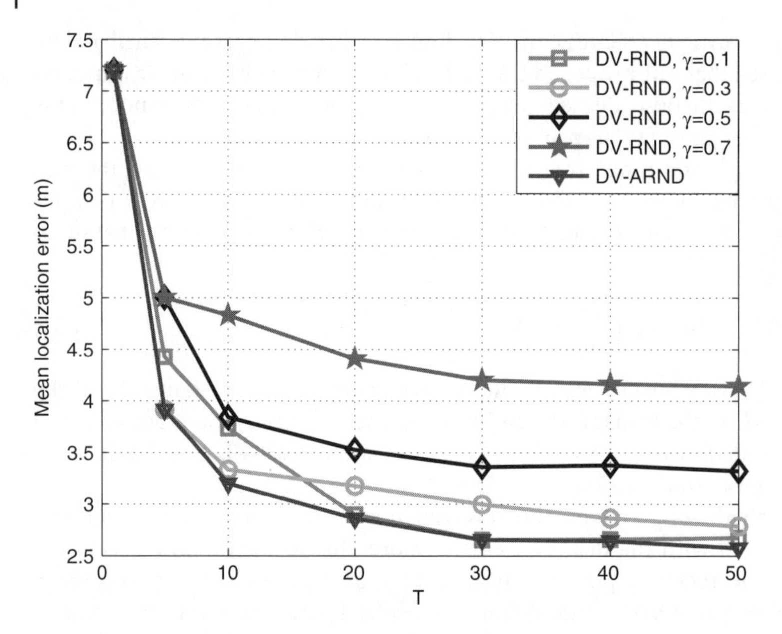

Figure 4.8 Localization error against the value of the PRR threshold [39].

The actual PRR threshold is only from the threshold set $\{\frac{1}{T}, \frac{2}{T}, \dots, \frac{T}{T}\}$. The basic idea is to find the best threshold γ^* from the threshold set; the one that leads to the smallest distance estimation error in between all anchors. As the optimal RND is adaptive to the number of localization packets T, this is called the DV-ARND localization algorithm.

Figure 4.8 compares the localization error for the DV-RND and DV-ARND algorithms under the log-normal shadowing model. It is observed that the DV-ARND algorithm consistently outperforms the DV-RND algorithm with the fixed settings of $\gamma = 0.1, 0.3, 0.5, 0.7$ in terms of lower localization error. This is not unexpected, as the DV-ARND algorithm adaptively selects an optimal PRR threshold γ, according to the number of localization packets T.

4.5 Summary

In this chapter, we have reviewed connectivity-based range-free localization in wireless sensor networks. The main advantage of such localization techniques is their simplicity of implementation, which is important to applications for space and extreme environments. The centroid and DV-hop algorithms are two cornerstone algorithms that can be applied in one-hop and multi-hop networks, respectively. We have also reviewed many improvements of the two algorithms, especially those that exploit neighborhood information. These improvements have been shown to greatly improve the localization accuracy over the primitive connectivity-based algorithms.

We need to note that there are some other issues in the design of connectivity-based localization algorithms. For example, the sensor field to deploy a network may be in diverse shapes and even with obstacles and holes. The localization algorithms discussed actually make an implicity assumption that the networks are isotropic; that is, that the shortest path between two nodes is only impacted by the node density. However, in *anisotropic networks*, obstacles and holes may cause there to be a distorted shortest path and therefore an impact on distance estimation performance. Many improved methods deal with detoured shortest path problems from two approaches: straight-line distance calculation [40–42] and anchor nodes selection [43–47]. The former mainly focuses on reducing the distance estimation error, while the latter excludes the detoured shortest path via judicious anchor selection.

References

1 Rashvand, H.F., Abedi, A., Alcaraz-Calero, J.M., Mitchell, P.D., and Mukhopadhyay, S.C. (2014) Wireless sensor systems for space and extreme environments: A review. *IEEE Sensors Journal*, **14** (11), 3955–3970.

2 Mao, G., Fidan, B., and Anderson, B.D. (2007) Wireless sensor network localization techniques. *Elsevier Computer Networks*, **50** (10), 2529–2553.

3 Niculescu, D. and Nath, B. (2003) Ad hoc positioning system (APS) using AOA, in *IEEE INFOCOM*, pp. 1734–1743.

4 Venkatraman, S., Caffery, J., and You, H.R. (2004) A novel ToA location algorithm using LoS range estimation for NLoS environments. *IEEE Transactions on Vehicular Technology*, **53** (5), 1515–1524.

5 Kovavisaruch, L. and Ho, K.C. (2005) Alternate source and receiver location estimation using TDOA with receiver position uncertainties, in *IEEE International Conference on Acoustic, Speech, and Signal Processing*, pp. 1065–1068.

6 Patwari, N., Hero, A.O., Perkins, M., Correal, N.S., and O'Dea, R.J. (2003) Relative location estimation in wireless sensor networks. *IEEE Transactions on Signal Processing*, **51** (8), 2137–2148.

7 Hou, S., Zhou, X., and Liu, X. (2010) A novel DV-Hop localization algorithm for asymmetry distributed wireless sensor networks, in *IEEE International Conference on Computer Science and Information Technology (ICCSIT)*, pp. 243–248.

8 Bulusu, N., Heidemann, J., and Estrin, D. (2000) Gps-less low-cost outdoor localization for very small devices. *IEEE Personal Communications*, **7** (5), 28–34.

9 Blumenthal, J., Grossmann, R., Golatowski, F., and Timmermann, D. (2007) Weighted centroid localization in zigbee-based sensor networks, in *IEEE International Symposium on Intelligent Signal Processing*, pp. 1–6.

10 Chen, Y., Pan, Q., Liang, Y., and Hu, Z. (2010) AWCL: Adaptive weighted centroid target localization algorithm based on RSSI in WSN, in *IEEE International Conference on Computer Science and Information Technology (ICCSIT)*, pp. 331–336.

11 Zhou, Y., Qiu, T., Xia, F., and Hou, G. (2011) An improved centroid localization algorithm based on weighted average in WSN, in *IEEE International Conference on Electronics Computer Technology (ICECT)*, pp. 351–357.

12 Chen, H., Chan, Y.T., Poor, H.V., and Sezaki, K. (2010) Range-free localization with the radical line, in *IEEE International Conferenc on Communications (ICC)*, pp. 1–5.

13 Niculescu, D. and Nath, B. (2001) Ad hoc positioning system (APS), in *IEEE Global Telecommunications Conference (Globecom)*, pp. 2926–2931.

14 Wang, B., Fu, C., and Lim, H.B. (2007) Layered diffusion based coverage control in wireless sensor networks, in *IEEE the 32rd Conference on Local Computer Networks (LCN)*, pp. 504–511.

15 Cheng, Y. and Robertazzi, T.G. (1989) Critical connectivity phenomena in multihop radio models. *IEEE Transactions on Communications*, **37** (7).

16 Vural, S. and Ekici, E. (2007) Probability distribution of multi-hop-distance in one-dimensional sensor networks. *Elsevier Computer Networks*, **51** (3), 3727–3749.

17 Bettstetter, C. and Eberspacher, J. (2003) Hop distances in homogeneous ad hoc networks. *IEEE Vehicular Technology Conference (VTC)*, **4**, 2286–2290.

18 Chandler, S. (1989) Calculation of number of relay hops required in randomly located radio network. *IEEE Electronic Letters*, **25**, 1669–1671.

19 Ta, X., Mao, G., and Anderson, B.D. (2007) On the probability of k-hop connection in wireless sensor networks. *IEEE Communication Letters*, **11** (9).

20 Kleinrock, L. and Silvester, J. (1978) Optimum transmission radii for packet radio networks or why six is a magic number. *IEEE National Telecommunications Conference*.

21 Kuo, J. and Liao, W. (2007) Hop count distribution of multihop paths in wireless networks with arbitrary node density modeling and its applications. *IEEE Transactions on Vehicular Technology*, **56** (4).

22 Wang, Y., Wang, X., Wang, D., and Agrawal, D.P. (2009) Range-free localization using expected hop progress in wireless sensor networks. *IEEE Transactions on Parallel and Distributed Systems*, **20** (10), 1540–1552.

23 Vural, S. and Ekici, E. (2010) On multihop distances in wireless sensor networks with random node locations. *IEEE Transactions on Mobile Computing*, **9** (4), 540–552.

24 We, Q., Han, J., Zhong, D., and Liu, R. (2012) An improved multihop distance estimation for DV-Hop localization algorithm in wireless sensor networks, in *IEEE Vehicular Technology Conference Fall (VTCFall)*, pp. 1–5.

25 Wong, S.Y., Lim, J.G., Rao, S., and Seah, W.K. (2005) Density-aware hop-count localization (DHL) in wireless sensor networks with variable density, in *IEEE Wireless Communications and Networking Conference (WCNC)*, pp. 1848–1853.

26 Ma, D., Er, M.J., and Wang, B. (2010) Analysis of hop-count-based source-to-destination distance estimation in wireless sensor networks with applications in localization. *IEEE Transactions on Vehicular Technology*, **59** (6), 2998–3011.

27 Ma, D., Er, M.J., Wang, B., and Lim, H.B. (2010) A novel approach toward source-to-sink distance estimation in wireless sensor networks. *IEEE Communications Letters*, **14** (5), 384–386.

28 Ma, D., Er, M.J., Wang, B., and Lim, H.B. (2012) Range-free wireless sensor networks localization based on hop-count quantization. *Springer Telecommunication Systems*, **50** (3), 199–213.

29 Wu, G., Wang, S., Wang, B., and Dong, Y. (2012) Multi-hop distance estimation method based on regulated neighbourhood measure. *IET Communications*, **6** (13), 2084–2090.

30 Wu, G., Wang, S., Wang, B., Dong, Y., and Yan, S. (2012) A novel range-free localization based on regulated neighborhood distance for wireless ad hoc and sensor networks. *Elsevier Computer Networks*, **56** (16), 3581–3593.

31 Cao, Y., Chen, X., Yu, Y., and Kang, G. (2009) Range-free distance estimate methods using neighbor information in wireless sensor networks, in *IEEE Vehicular Technology Conference Fall (VTCFall)*, pp. 1–5.
32 Huang, B., Yu, C., Anderson, B., and Guoqiang (2010) Connectivity-based distance estimation in wireless sensor networks, in *IEEE Global Telecommunications Conference (Globecom)*, pp. 1–5.
33 Buschmann, C., Pfisterer, D., and Fischer, S. (2006) Estimating distances using neighborhood intersection, in *IEEE Symposium on Emerging Technologies and Factory Automation (ETFA)*, pp. 314–321.
34 Villafuerte, F.L., Terfloth, K., and Schiller, J. (2008) Using network density as a new parameter to estimate distance, in *IEEE International Conference on Networks (ICN)*, pp. 30–35.
35 Zhong, Z. and He, T. (2011) RSD: A metric for achieving range-free localization beyond connectivity. *IEEE Transactions on Parallel and Distributed Systems*, **22** (11), 1943–1951.
36 Floyd, R.W. (1962) Algorithm 97: Shortest path. *Communications of the ACM*, **5** (6), 345.
37 Aslam, F., Schindelhauer, C., and Vater, A. (2009) Improving geometric distance estimation for sensor networks and unit disk graphs, in *International Conference on Ultra Modern Telecommunications and Workshops*, pp. 1–5.
38 Rappaport, T.S. (2001) *Wireless Communications, Principles and Practice*, 2nd edn., Prentice Hall.
39 Wang, B., Wu, G., Wang, S., and Yang, L.T. (2014) Localization based on adaptive regulated neighborhood distance for wireless sensor networks with a general radio propagation model. *IEEE Sensors Journal*, **14** (11), 3754–3762.
40 Li, M. and Liu, Y. (2010) Rendered path: Range-free localization in anisotropic sensor networks with holes. *IEEE/ACM Transactions on Networking*, **18** (1), 320–332.
41 Wang, Y., Li, K., and Wu, J. (2010) Distance estimation by constructing the virtual ruler in anisotropic sensor networks, in *IEEE INFOCOM*, pp. 1–9.
42 Xiao, Q., Xiao, B., Cao, J., and Wang, J. (2010) Multihop range-free localization in anisotropic wireless sensor networks: A pattern-driven scheme. *IEEE Transactions on Mobile Computing*, **9** (11), 1592–1607.
43 Liu, X., Zhang, S., Wang, J., Cao, J., and Xiao, B. (2011) Anchor supervised distance estimation in anisotropic wireless sensor networks, in *IEEE Wireless Communications and Networking Conference (WCNC)*, pp. 938–943.
44 Xiao, B., Chen, L., Xiao, Q., and Li, M. (2010) Reliable anchor-based sensor localization in irregular areas. *IEEE Transactions on Mobile Computing*, **9** (1), 60–72.
45 Zhang, S., Wang, J., Liu, X., and Jiannong (2012) Range-free selective multilateration for anisotropic wireless sensor networks, in *IEEE Annual Communications Society Conference on Sensor, Mesh and Ad Hoc Communications and Networks (SECON)*, pp. 299–307.
46 Fan, Z., Chen, Y., Wang, L., Shu, L., and Hara, T. (2011) Removing heavily curved path: Improved DV-hop localization in anisotropic sensor networks, in *IEEE International Conference on Mobile Ad-hoc and Sensor Networks (MSN)*, pp. 75–82.
47 Zhong, C., Wang, B., and Yang, L.T. (2015) On improvement of the DV-RND localization in wireless sensor networks, in *The 12th IEEE International Conference on Ubiquitous Intelligence and Computing (UIC)*, pp. 1–5.

5

Rare-events Sensing and Event-powered Wireless Sensor Networks

Winston K.G. Seah and David Harrison

School of Engineering and Computer Science, Victoria University of Wellington, New Zealand

The emergence of wireless sensor networks (WSNs) can be traced back to an initiative by the National Research Council [1]. Development has been motivated by military applications, such as battlefield surveillance, where sensor nodes are small battery-powered devices typically consisting of a microcontroller, a modest quantity of random access memory, some non-volatile storage capacity, one or more sensors, and a low-power radio transceiver. The finite charge-storage capacity of batteries shapes WSN research to the extent that minimizing energy consumption becomes a preoccupation; the less energy consumed, the longer the network will continue to operate. Once the stored energy is exhausted, the sensor is deemed to be useless, and often the first sensor node to exhaust its stored energy has been used as the determining factor of the entire wireless sensor network's lifetime or operational effectiveness. Acknowledging this rather unrealistic measure of a sensor network's effectiveness, researchers have proposed various other measures of network lifetime that take into consideration other factors as well as operational scenarios [2].

Since then, new WSN applications, such as global-scale environmental monitoring and structural health monitoring (SHM) of critical infrastructure, such as bridges, have emerged [3] where deployment is expected to last much longer periods. The simplest way for a sensor node to conserve energy, and in doing so maximize its operational life, is to power down for extended periods. For star-topology networks, where sensor nodes are connected to a permanently powered base station, each node can adopt an independent duty cycle and a medium access control protocol based on unslotted carrier sense multiple access with collision avoidance. Nodes deployed to periodically sample data that are relayed via multi-hop transmission to a base station synchronize their activity [4] to ensure network connectivity, and employ algorithms that minimize overuse of individual routing nodes [5]. Similar techniques can be adopted for networks where data collection is initiated by a request from the base station [6]. Sensing rare events, however, introduces additional complexity.

A rare event can be defined as 'an event occurring with a very small probability, the definition of "small" depending on the application domain' [7]; some typical examples of rare events include a civil aircraft failing during a typical eight-hour flight and a high-speed network node experiencing a buffer overflow where the probability of occurrence is expected to be less than 10^{-9}. In other words, we can regard rare events

Wireless Sensor Systems for Extreme Environments: Space, Underwater, Underground, and Industrial.
First Edition. Edited by Habib F. Rashvand and Ali Abedi.

axiomatically as infrequent occurrences, which may also be short-lived, unpredictable (that is, random) and leave no trace of their presence when complete. Successfully sensing a rare event requires consideration of the extent to which it is ephemeral (lasting for a short time) and transitory (transient and leaving no evidence afterwards.) A scheme that proves effective for sensing forest fires in progress may not be as successful for detecting the starting instant of the same fires. Similarly, a scheme for detecting perimeter intrusion on a battlefield where events last fractions of a second and leave no discernible trace may prove inappropriate for sensing landslides that last for comparatively extended periods and leave significant physical evidence in their wake, yet occur less frequently.

Rare-event WSNs are found in a variety of situations, including the battlefield [8] where low unit costs allow high densities, short time-frames, and disposable deployments. In industrial settings [9], WSNs are cost efficient when compared to fixed wiring, and their robust self-organizing characteristics make them suitable for monitoring hazardous machinery and protecting high value assets. It is entirely practical to perform periodic data collection and rare-event detection in one WSN, but quality-of-service-aware routing protocols [10] should be implemented to prevent unacceptable delays when urgent messages relating to rare events have to queue up behind more mundane traffic. Whatever the deployment scenario, two metrics emerge as fundamental to rare-event sensing [11–13]:

- *Detection probability*: the likelihood of an event being detected.
- *Detection delay*: the time taken for notification to reach a network sink, where the necessary action can be taken.

WSNs can be considered part of the Internet of Things (IoT) [14], and continue to benefit from active research and regular publication of surveys targeting specific topics. Recently, statistical, probabilistic, artificial-intelligence, and machine- learning methods for event detection have been surveyed [15], but the majority of WSN surveys in the last five years have focused on areas unrelated to rare events. One exception does consider event sensing as it relates to energy-efficient routing [16], with an emphasis on how the criticality of an event informs protocol design. Most recently, a survey specifically targeted rare-event detection and propagation in WSNs [17].

The challenge of maximizing network longevity and detection probability, and minimizing detection delay can be moderated by harvesting environmental energy to augment or replace batteries in sensing nodes. Networks powered by energy harvesting have the potential to stay active for longer than those relying on batteries. However, energy availability may be unpredictable, leading to unexpected node outages and temporary reductions in detection probability. Energy harvesting has been extensively investigated for data-capture applications [18], but comparatively little research exists on the successful application of this technology to rare-event WSNs.

In this chapter, we present research that addresses the two metrics listed above. Firstly, we discuss an asynchronous, fully distributed duty-cycling algorithm for preserving full coverage in energy-harvesting-powered WSNs for rare-event sensing [19]. This duty-cycling algorithm aims to maximize the likelihood of a rare event being detected by ensuring the coverage of the area being sensed does not drop below the level when the network began operation. We then consider rare events that, when they occur, produce enough energy to be harvested and used to power the sensor nodes.

For example, a major seismic event like an earthquake causes buildings and structures to shake, and sensor nodes that are mounted on them can be powered from the vibrations. We describe a sensor that has been designed to do just that, using the energy harvested to measure the level of stress and vibration that the building has been subject to [20]. We follow with a networking protocol that has been designed to work specifically with such event-powered sensors [21], and conclude with a discussion of ongoing and future research directions.

5.1 Coverage Preservation [19]

Establishing and maintaining sensing coverage in WSNs for rare events can be regarded as that ensuring every point in the sensing area is covered by at least one sensor node at all times whilst maintaining energy efficiency [22]. WSNs with a sufficiently dense overpopulation of sensing nodes can self-organize such that a given node is in a position to power down if a subset of its neighbors are willing to take temporary responsibility for its sensing area. In battery-powered WSNs, this duty-cycling extends the operational lifetime of a subset of deployed nodes and in doing so can both extend the period during which the network maintains its initial coverage and the period during which there is at least some coverage [23].

When energy harvesting replaces battery power, duty-cycled nodes not only preserve their stored energy but also have the opportunity to replenish it more rapidly than when active. The design of solar energy harvesting systems for WSNs is non-trivial [24], yet if the energy harvested and stored is sufficient, networks have the opportunity to maintain initial sensing coverage for extended periods, and with the potential for indefinite operation. For solar energy harvesting, the physical dimensions, output voltage, maximum current and efficiency of the installed solar panels determine the harvestable energy for a given incident radiation. Once harvested, energy is lost though inefficiencies in storage componentry, yet recent low-power management systems demonstrate charging efficiencies in excess of 90% [25].

The equitable sleep coverage algorithm for rare geospatial occurrences (ESCARGO) [19] is designed to address both coverage and connectivity requirements of rare-events sensing. While similar to ESCARGO in seeking to address coverage, the role alternating coverage preserving coordinated sleep algorithm (RACP) [23] for battery-powered WSNs specifically excludes connectivity from consideration. Furthermore, detectable events are assumed to occur in a finite set of known locations, with coverage of only these locations being required; ESCARGO aims to cover the entire network sensing area, as events can occur anywhere within it. In addition, assessment of RACP's efficacy assumes error-free transmission and focuses on dense random node distributions with few similarities to the sparse planned placements featured here. Although RACP has been shown to extend the lifetime of battery-powered WSNs when compared to random sleep cycling and the coverage algorithm proposed by Tian and Georganas [26], the sensing nodes, whilst location-aware, do not know the bounds of the sensing area; this leads to abrupt drops in coverage as nodes situated at or close to the perimeter of the sensing area simultaneously exhaust their energy stores. For discussion purposes, we assume a two-dimensional rectangular distribution of location and network sensing area aware nodes. Node sensing areas are assumed identical and exactly circular, and the node communication range is at least twice that of the sensing range to preserve network connectivity [27].

5.1.1 Overview

Nodes are in one of four states:

- *sponsored*, where sensing responsibility is delegated to at least one other node
- *sponsoring*, where they are taking responsibility for the sensing area of one or more neighboring nodes
- *seeking*, where they are actively attempting to find neighboring nodes willing to sponsor them
- *passive*, where they are unsponsored, not seeking sponsorship, and are not sponsoring any of their neighbors.

A state transition diagram for the algorithm is shown in Figure 5.1.

All nodes start simultaneously and enter the passive state, rapidly broadcasting periodic status ('STAT') messages containing their position and cumulative run-time statistics. During this start-up period, nodes listen for STAT messages from other nodes and determine who their neighbors are. Two nodes are deemed to be neighbors if they are separated by a distance less than or equal to their common sensing range. Pseudocode for the passive state is given in Algorithm 1.

A node is eligible to enter sponsored state if and only if it determines its sensing area is fully contained by the union set of some combination of its passive, seeking or sponsoring neighbors' sensing areas. Combinations of neighboring nodes that could sponsor a given node are known as *sponsor groups*. Sponsor group membership is determined by geometric calculations and taking into consideration the boundary of the sensing area,

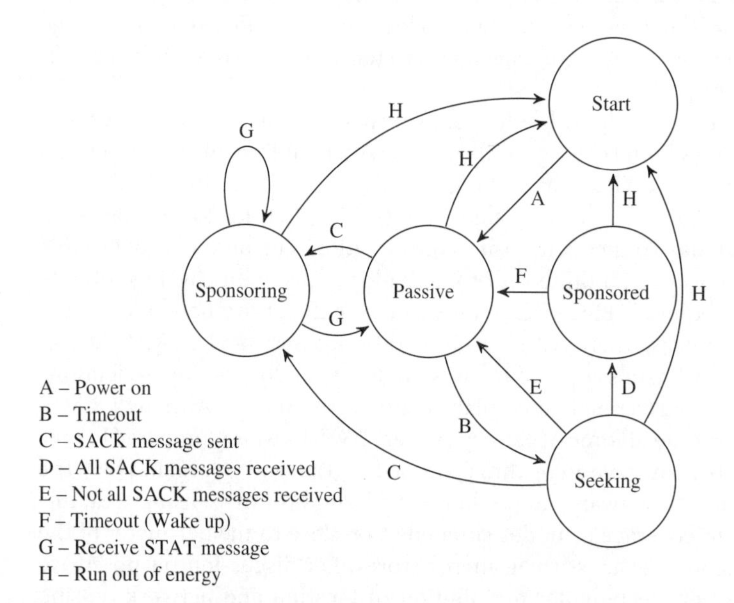

Figure 5.1 ESCARGO state transitions.

Algorithm 1 Passive state

```
/* Listening thread */
while not interrupted do
  listen for message
  if STAT from unknown neighbor then
    update neighbor list
    recalculate sponsor groups
  else if SREQ then
    transition to sponsoring state
  end if
end while
/* Main thread */
for a predefined period do
  periodically broadcast STAT messages
end for
interrupt listening thread
transition to seeking state
```

as detailed in Section 5.1.2. Nodes adjust their sponsor-group lists each time a STAT from a previously unknown neighbor is received.

Nodes that deem themselves eligible for sponsorship cycle though their sponsor groups in a round-robin fashion, seeking for sponsors, starting with the group that comes after the previous sponsor group. For each sponsor group selected, a sponser request ('SREQ') message is sent to each node in the group. Nodes in receipt of an SREQ add the requesting node to their sponsored list and return a Sponsorship Acknowledgment ('SACK') message. Once a node has agreed to be a sponsor it will no longer attempt to gain its own sponsors until it is notified by the nodes it is sponsoring that they no longer require assistance. Pseudocode for the seeking state is given in Algorithm 2.

When a seeking node wishing to be sponsored has received a SACK from each of the nodes it sent an SREQ to, it enters the sponsored state for a predetermined period. In the sponsored state, nodes enter a low-energy mode in which all sensors and the radio transceiver are powered down. Pseudocode for the sponsored state is given in Algorithm 3.

If fewer SACKs are received than SREQs sent, the requesting node reverts to the passive state and broadcasts a burst of STATs. Neighboring nodes that had already agreed to be sponsors remove the requesting node from their sponsored lists on receipt of any of these STATs. On wake-up from a period of sponsorship, nodes similarly broadcast a small number of STATs and their sponsoring nodes adjust their sponsored lists accordingly. Pseudocode for sponsoring state is given in Algorithm 4.

SREQs contain the stored charge of the requesting node. Potential sponsor nodes will respond to the SREQ with a SACK if and only if the sponsoring node has more stored charge than the requester. Whilst waiting for SACKs, a seeking node that receives an SREQ from a node with less stored energy than itself will abandon waiting for any remaining inbound SACKs, send a SACK of its own to the requesting node and enter the sponsoring state.

Algorithm 2 Seeking state

```
/* Listening thread */
while not interrupted do
  listen for message
  if SREQ then
    send SACK
    interrupt requesting thread
    transition to sponsoring state
  else if SACK then
    add sender to list of acknowledged sponsors
  end if
end while
/* Requesting thread */
while not interrupted do
  for all sponsor groups do
    for all members of sponsor group do
      send SREQ
    end for
    pause
    if SACKs received from all members of sponsor group then
      interrupt listening thread
      transition to sponsored state
    else
      for all members of sponsor group do
        send STAT
      end for
    end if
  end for
  transition to passive state
end while
```

Algorithm 3 Sponsored state

```
shutdown sensors and transceiver
sleep for a predefined period
energise sensors and transceiver
transition to passive state
```

5.1.2 Sleep Eligibility

When a node goes to sleep, it takes no part in the sensing or communication links of the WSN until it wakes up again. To maintain sensing coverage in the WSN, the entire sensing area a sleeping node was responsible for must be covered by one or more neighboring nodes. Only when a node is confident its sensing area is covered by a set of awake neighbors does it become sleep eligible.

Wang et al. [27] performed a geometric analysis of the relationship between connectivity and coverage in networks made up of nodes with uniform sensing and

Algorithm 4 Sponsoring state

while sponsoring one or more neighbors **do**
  listen for message
  if STAT **then**
    remove sender from list of sponsored neighbors
  else if SREQ **then**
    send SACK
    add sender to list of sponsored neighbors
  end if
end while
transition to passive state

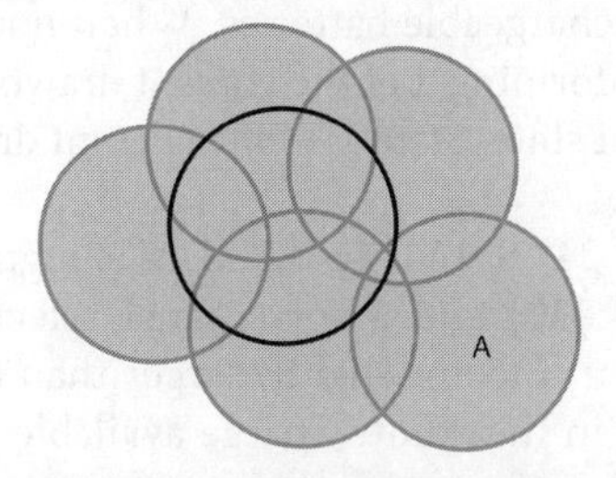

(a) 1-coverage eligibility; sensing node A is not required to provide coverage for sensing node with area of darkest circle.

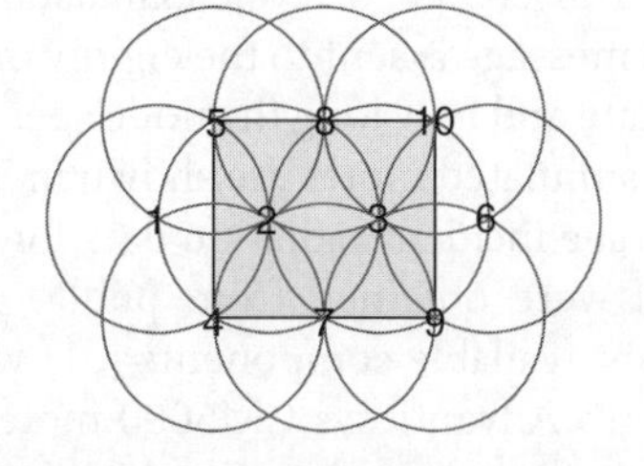

(b) Ideal placement of 10 nodes relative to sensing area; node identifiers shown, sensing area shaded.

Figure 5.2 Node placements. (a) 1-coverage eligibility; sensing node A is not required to provide coverage for sensing node with area of darkest circle. (b) Ideal placement of 10 nodes relative to sensing area; node identifiers shown, sensing area shaded.

transmission ranges. For sensing range R_s and communication range R_c it is shown that a convex region A is k-covered by a set of sensors if all of the following conditions are met:

- there exist, in A, intersection points between sensors or between sensors and A's boundary
- all intersection points between any sensors are at least k-covered
- all intersections points between any sensor and A's boundary are at least k-covered.

This is the k-coverage eligibility rule. ESCARGO only requires nodes to exhibit 1-coverage eligibility, an example being given in Figure 5.2a, where the node with the highlighted sensing area is eligible to enter sleep mode because its sensing area is completely covered by neighboring nodes. Note that the node with sensing area marked A, whilst overlapping slightly with the sensing area of the eligible node (shown by the darkest circle), is not required to act as a sponsor for the eligible node.

5.1.3 Performance Evaluation

ESCARGO's efficacy has been evaluated using simulations of a representative ten-node network, as shown in Figure 5.2b, operating over a full calendar year. Four operational configurations were studied:

- battery only—no synchronized sleep scheduling, no energy harvesting
- battery power with ESCARGO
- energy harvesting replaces battery but ESCARGO is disabled
- ESCARGO with energy harvesting.

The nodes are modeled as solar-powered devices. Energy available to sensing nodes is modeled as a linear charge store. As the nodes consume energy, charge is removed from the store; when harvesting energy during daylight hours, charge is added, up to a maximum capacity. In practice, dual capacitor storage systems are required to facilitate simultaneous charge and discharge of the storage system [28]. When the charge held by a node is entirely depleted, it shuts down and the energy model waits until the store achieves 10% of initial charge before waking the node up to resume its sensing responsibilities. The model does not consider practical storage issues such as supercapacitor leakage and self-discharge, or cycle exhaustion in rechargeable batteries. When nodes change state, a message is sent to the energy model informing it of the current drawn by the previous state and how long the node spent in that state. State-specific current draw figures for the simulated motes are shown in Table 5.1.

Historic average incident radiation data for 41° 19′ 24″ S, 174° 46′ 12″ E (Wellington, New Zealand) were obtained from public records [30] and a solar panel selected from commonly available components [31] with a form factor slightly larger than the 81.90 × 32.50 mm Advanticsys CM5000 motes used in the study. Charge available for storage in a given time period s seconds starting at time T is:

$$Q = \sum_{t=T}^{T+s} MIN \left\{ I_p, \left(\frac{\lambda_t \times l_p \times w_p \times \eta_p}{V_p} \right) \right\} \times \eta_c \tag{5.1}$$

where λ_t is the incident radiation at time t, η_p and η_c are the efficiencies of the solar panel and charging circuit respectively, I_p and V_p are the peak current and peak voltage of the solar panel, of length l_p and width w_p.

Mean stored charge for all nodes during the first month of operation is shown in Figure 5.3a. On battery power alone, stored charge depletes entirely in around eleven days at which point all coverage is lost. Adding ESCARGO to battery power extends network lifetime significantly. However, original coverage is not maintained for the entire period. As nodes' stored charge become exhausted, they die and coverage is compromised. Original coverage is maintained until around day 17 when the first nodes die. Coverage is restored briefly as sponsored nodes return to passive state then tails off quickly as further nodes die.

During the first month of operation (the height of summer) adding ESCARGO to energy harvesting maintains a higher mean stored charge, but the practical benefits are negligible as more energy is available for harvesting than can be stored and used. During

Table 5.1 Advanticsys CM5000 current draw by state.

Sleep	Idle listen	Receive	Transmit
mA	mA	mA	mA
0.0001	18.4	19.2	19.9

Source: Advanticsys website [29].

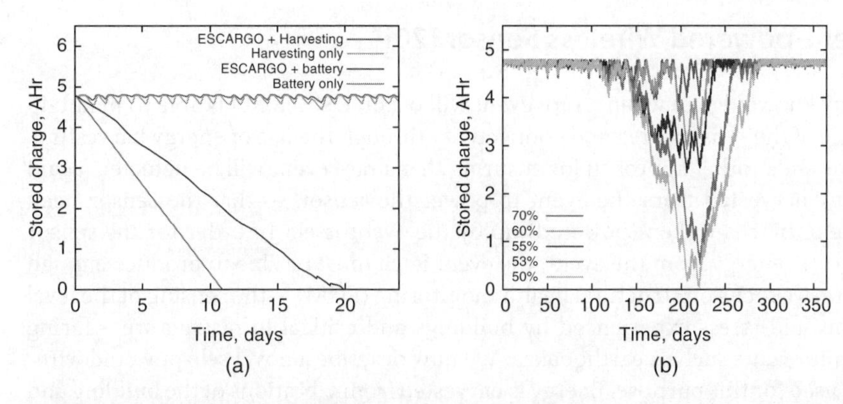

Figure 5.3 Stored charge. (a) Mean stored charge during first month of operation (b) Mean stored charge over time by charging efficiency. (*See color plate section for the color representation of this figure.*)

Table 5.2 Duration of coverage maintenance.

Configuration	Battery (days)	Energy harvesting (days)
Power source only	11	152
With ESCARGO	17	Indefinite

the winter months, when less energy is available, a harvesting only solution exhibits a sharp degradation in average stored charge as more energy is used each day than can be replenished. Combining ESCARGO with energy harvesting results in significantly improved mean stored charge. When ESCARGO and energy harvesting are combined, no nodes die, a high proportion of nodes are asleep at any time and initial sensing coverage is maintained throughout the entire year. The number of days that initial coverage was maintained for each configuration is shown in Table 5.2.

Khosropour et al. [25] propose a low-power charging circuit with 90% efficiency and reference other similar systems with charging efficiencies of 67%, 70%, and 86%. Running simulations based on the planned ten-node network (Figure 5.2b) and varying the charging efficiency η_c in Eqn (5.1) shows that a charging efficiency of 55% is sufficient to maintain mean stored charge close to 50% of the original. Figure 5.3b shows mean stored charge over a full year for charging efficiencies from 50–70%. For planned placements, ESCARGO assures equitable discharge and recharge for all nodes in the WSN, hence any charging efficiency that realizes a non-zero average stored charge across all nodes will ensure 100% of original coverage is maintained. Figure 5.3b shows a charging efficiency of 50% is insufficient to maintain sensing coverage throughout the year, as the charge drops to zero at some point.

As full connectivity is also ensured by ESCARGO, any WSN routing protocol can be used to transmit the rare-event detection data to the sink. However, to ensure that detection delay is minimized, a priority-based routing protocol, such as the scalable priority-based multi-path routing protocol [32], should be used.

5.2 Event-powered Wireless Sensor [20]

Without any knowledge of when a rare event will occur, ESCARGO is able to maintain full coverage of the sensing area and connectivity through the use of energy harvesting. We now present another approach for ensuring that a rare event will be detected, again using energy harvested from the event to power the sensor, so that the sensor need not operate until it is woken (powered) up by the event itself. In order for the sensor node to harvest energy from the event, the event itself must be able to produce enough energy. Here, we focus on structural health monitoring (SHM) —the sensing of the level of vibrations and stress experienced by buildings and critical infrastructure—during major seismic events such as earthquakes. We now describe a novel self-powered wireless sensor used for this purpose. Energy is harvested from vibrations of the building and sensor data are transmitted wirelessly to collection/access points for further analysis by structural engineers.

5.2.1 Earthquakes and Structures

The magnitude of an earthquake is typically measured and reported using the Richter scale. This is an accurate measure of the energy released from the tremor, but it does not give an accurate measure of the acceleration felt on the ground [33]. The ground acceleration is critical because there is a distinct correlation between that and the damage caused to buildings [34]. This makes the peak ground acceleration (PGA) is a more practical measurement for an earthquake when determining the structural damage sustained on the surface, as it is the maximum acceleration felt on the ground in the place of interest. The peak ground velocity and distance can be determined from the acceleration, and are sometimes considered instead of PGA when evaluating structural damage in extreme earthquakes; when large tremors above $1.2g$ strike earthquake-flexible buildings, damage is proportional to velocity not acceleration [33]. The overall structural damage index (OSDI) is a numbering system used to give an indication of the extent of the damage to a structure caused by an earthquake. This index gives a single value between 0 and 1 that summarises all existing damage on columns and beams in a structure, to give a representation of the significance of the earthquake. The destruction extent of the earthquake is considered low if OSDI < 0.3, medium if $0.3 <$ OSDI < 0.6, great if $0.6 <$ OSDI < 0.8, and total if OSDI > 0.8. A list of notable earthquakes with their corresponding OSDI and PGA values is shown in Table 5.3, and plotted in Figure 5.4 to show the positive correlation between structural damage and the PGA [34]. It is seen that earthquakes classified as 'low destruction' on the OSDI index have a PGA of 0.6–$1.0g$, with higher-OSDI earthquakes having PGA of up to $1.4g$. Therefore, a good acceleration rate to target for structural health monitoring caused by earthquake tremors is at least $0.6g$.

5.2.2 Vibration Energy Harvesting

A transducer is used to transform the energy from movements into usable energy in the form of an electric current that will power the WSN nodes. Vibration energy is normally generated by a mechanical component attached to an inertial frame acting as a fixed reference. The inertial frame transmits the vibrations to a suspended inertial mass that produces a relative displacement between them [35]. These types of system have a resonant

Table 5.3 OSDI/PGA of notable earthquakes.

Earthquake	OSDI	PGA Max
Alkion (L)	0.081	0.603
Alkion (T)	0.082	0.575
Big Bear (270°)	0.071	0.702
Big Bear (360°)	0.103	0.799
Erzincan (N–S)	0.397	0.991
Erzincan (E–W)	0.169	0.834
Izmir (N–S)	0	0.309
Izmir (E–W)	0	0.14
Hyogo-Ken Nanbu	0.55	1.149
Kalamata	0.094	0.582
Montenegro	0.198	1.049
Landers (0°)	0.129	0.622
Landers (90°)	0.151	0.714
Cape Mendocino (0°)	0.222	1.476
Cape Mendocino (90°)	0.098	0.757
Naghloo	0.15	0.963
San Salvador (0°)	0.106	0.794
San Salvador (90°)	0.096	0.889
Strazhitsa	0	0.298
Whittier	0.116	0.945

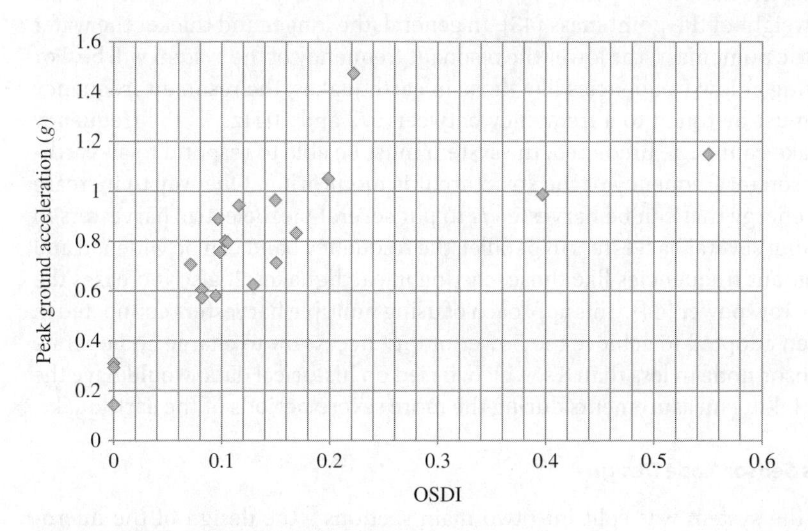

Figure 5.4 OSDI and corresponding PGA values.

frequency that needs be matched with the characteristic frequencies of the application environment, which in this case are the frequencies found in earthquakes. Several mechanisms have been investigated in the past for creating vibration-based energy harvesters. The main mechanisms used are piezoelectric, electromagnetic and electrostatic [36]. Piezoelectric energy harvesters rely on the piezoelectric effect, in which charge is generated on an active material when mechanically stressed. Electromagnetic harvesting utilises Faraday's law of induction to induce an electric field from a changing magnetic field caused by the vibrations. An electrostatic generator utilises the relative movement between electrically isolated charged capacitor plates to generate energy, where the work done against the electrostatic force between the plates provides the harvested energy [35]. A measurement study to select a vibration-based transducer for low frequencies of less than 10 Hz has found that piezoelectric-based harvesters give the best power output at these frequencies [37].

5.2.3 Piezoelectric Harvesting for Structural Monitoring during Earthquakes

The challenge with using a piezoelectric transducer to provide energy is that the duration of the quake needs to be long enough to generate adequate power to operate the wireless sensor node. For example, the earthquake that struck Christchurch, New Zealand on 22 February 2011 had only 12 s worth of intense shaking [38]. In the case of the vibrations produced by earthquakes, the waves occur at very low frequencies of around 0.5–10 Hz [39]. Most commercially available vibration-based energy harvesters have been designed to work on industrial machinery, which typically vibrates at much higher frequencies around 50–300 Hz [40–42]. Since the amount of energy harvested depends on the frequency of oscillations, the amplitude (PGA) and the duration, gathering energy from building motion during earthquakes proves to be a difficult task.

For successful energy harvesting, the harvester must be tuned to the natural frequency of the environment. For a cantilevered beam configuration, the most significant parameters influencing the natural frequency of the system are the length and thickness of the wafer and the weight of the point mass [43]. In general, the longer and thicker the wafer and the larger the point mass, the lower the resonant frequency of the system will be. For energy harvesting at low frequencies like those in earthquakes, the resonant frequency of the system must be tuned to a frequency between 0.5 and 10 Hz. As the frequency of the earthquake cannot be predicted, the system must be able to respond to an earthquake at the resonant frequency of the structure it is monitoring. One way to increase the amount of energy that can be harvested is to put several piezoelectric harvesters in parallel. By having several harvesters in parallel, the frequency band can be widened and shifted to dominant frequencies like those existing in earthquakes; it also increases the total obtainable raw power [44]. This approach of using multiple harvesters connected in parallel has been adopted to achieve the power output necessary to charge and operate the wireless sensor node in less than 8 s, which, based on historical data, would have the node ready for taking measurements during the more severe periods of the earthquake.

5.2.4 Wireless Sensor Node Design

The design of the system was split into two main sections - the design of the microcontroller circuitry (referred to as microcontroller board or MCB) and the design of the power management circuitry (referred to power management board or PMB). Each

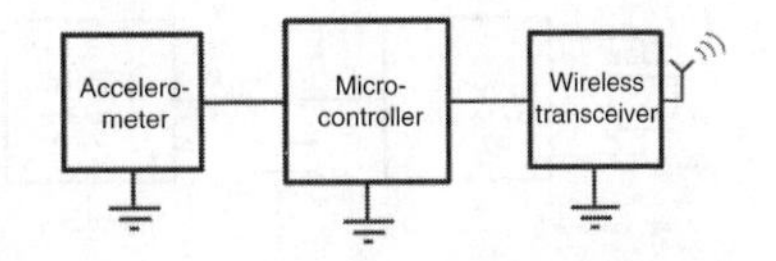

Figure 5.5 Microcontroller board.

has its own circuit board; the aim was to keep the design modular and easy to manage, since creating two separate boards created clear goals to aim for and made for easier debugging.

5.2.4.1 Microcontroller Board

The main features of the MCB, as shown in Figure 5.5, include the processor unit or microcontroller needed to initialize the other components and create the packet of data, the wireless transceiver which sends the packets (via an antenna) and an accelerometer to read the acceleration of the building during an earthquake. With low power being the most important consideration in the design of our WSN node, the Texas Instruments MSP430F2619 microcontroller (TI 2013b) was chosen for the processor unit to ensure ultra-low power consumption. It also has an input voltage that the PMB can produce, provides at least two communication ports to connect to the transceiver and the accelerometer, and has at least 10 kB of flash memory. The choice of accelerometer is a compromise between accuracy and power consumption; the range and sensitivity (related to accuracy) and the number of axes were also considered. The LIS331HH accelerometer (ST 2013) was chosen for its low power consumption (as low as 10 μW at 1.8 V), and a suitable range and sensitivity of ±6 g and 3 mg respectively. This was deemed adequate since the earthquakes being measured would be unlikely to exceed 6g. The other major component of the MCB is the wireless transceiver. The Texas Instruments CC2520 was selected because it is a low-power transceiver using IEEE 802.15.4 technology that transmits on the 2.4 GHz unlicensed band.

As the main design goal of the current prototype is to achieve the lowest possible energy usage, the accelerometer has been programmed to update its values in the registers once every 200 ms, which is the lowest usable energy-consuming state. Using this configuration, the IEEE 802.15.4 packets sent are 19 bytes in length, comprising the preamble header, timestamp and the payload data (6 bytes in length). Each payload contains three accelerometer values (x, y and z) in raw format from the LIS331HH. Considering the IEEE 802.15.4 packet has a maximum payload size of 127 bytes, there remains space for up to 20 more sets of raw accelerometer data; more data can be sent if data compression is utilised. The accelerometer also uses significantly less energy than other components, such as the wireless transceiver, and this enables the sampling rate to be increased easily to acquire more acceleration data—up to the level that is required for structural health monitoring. A cluster-based medium access control protocol has also been developed for a network of such sensing devices to transmit their data and this will be discussed in the next section.

5.2.4.2 Power Management Board

The basic components of the PMB are shown in Figure 5.6. It comprises an energy transducer (in this case the piezoelectric harvester, which in its raw form is a high- voltage, low-current AC signal) being converted into DC form for energy storage on a capacitor. This is then converted into a usable low-voltage, high-current signal by a voltage regulator to be used by the MCB. As our aim was to build a low-cost wireless sensor node

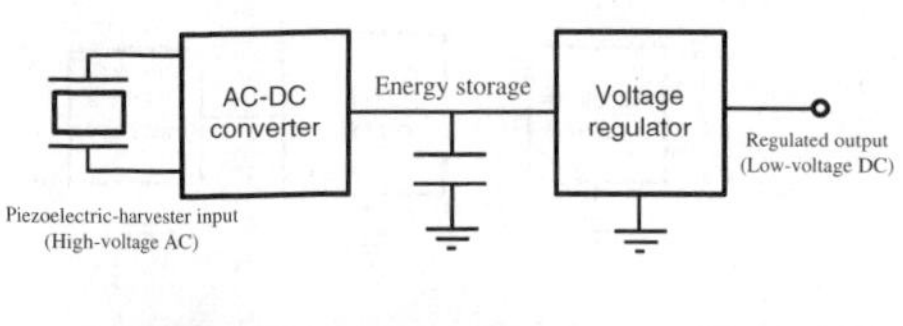

Figure 5.6 Power management board.

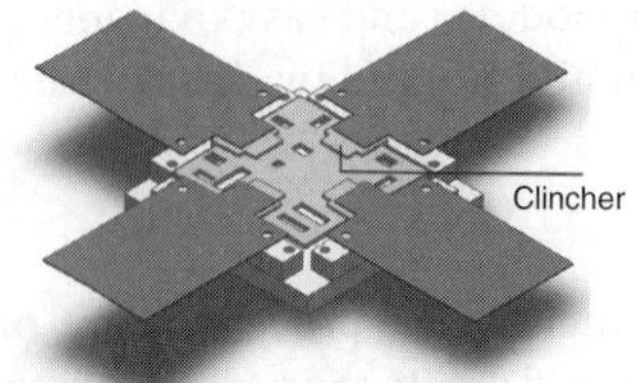

Figure 5.7 Base platform.

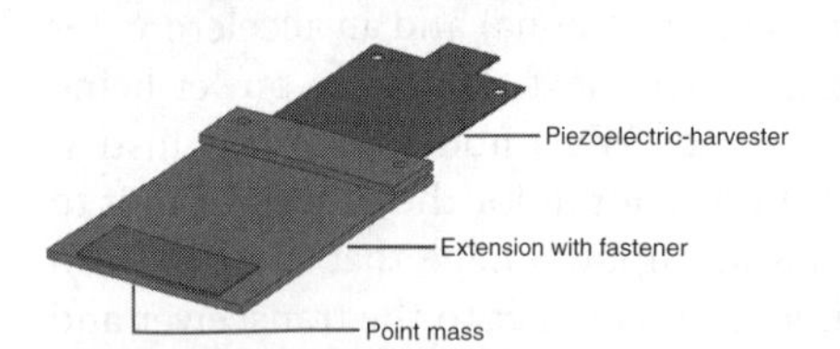

Figure 5.8 Extension and point mass.

that could be deployed in large numbers, we picked the Midé V25W for our prototype. It provided the flexibility to tune the resonant frequency to below the 10 Hz that we needed, at a cost of less than US$100, considerably less than customised solutions, that might cost over US$5000.

Four harvesters connected in parallel are needed to produce enough energy to power up the WSN within 5 s [44]. To accommodate the PMB printed circuit board layout, a base platform is used to secure the four harvesters with the PMB in the middle, as shown in Figure 5.7 The resonant frequency of the harvester was lowered using an extension with a point mass, as shown in Figure 5.8. As this was only a proof-of-concept prototype, a crude approach was used to lower the resonant frequency to match those measured during earthquakes. Custom-built vibration-energy harvesters with low resonant frequencies in the range of 0.5–10 Hz can be used when the system is refined for actual deployment purposes.

5.2.5 System Test and Evaluation

The time needed to charge the capacitors on the PMB is the most critical element in the design of the system; due to the momentary nature of earthquakes, the charge time must be as short as possible. This addresses the (energy) supply side of the problem. However, we also need to know the demand side, which is how much energy the MCB requires to start operating and continue operating until the earthquake has passed. While datasheets provide information on the power consumption of the various components used to implement the system, many other aspects of the system cannot be easily determined, except through careful measurements. Most importantly, how the system performs in an actual earthquake is a very challenging evaluation task.

Firstly, the voltage at the input to the MCB is sampled at 10 kHz as it transmits packets to a packet sniffer (a SmartRF board from Texas Instruments.) The results of the test, as shown in Figure 5.9, indicate that the MCU consumes a small amount of power in the first 12 ms for initialization and powering up, while the majority of the power

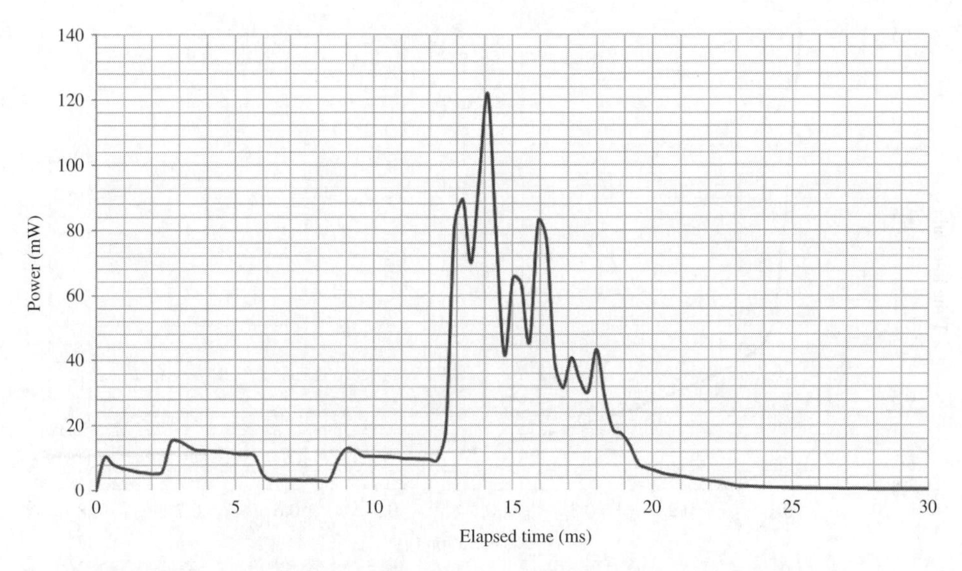

Figure 5.9 Power consumed by the MCB.

is consumed by the transceiver in order to send packets. In this test, the number of packets observed by the packet sniffer was three. Three distinct peaks can be seen, corresponding to the three separate packets sent. From this, we computed the energy required to send a single packet as 0.30 mJ, which is the minimum needed before the system can operate. Further experimentation and optimization of the system showed that the optimal capacitance of the PMB is 94 μF, which could yield 0.49 mJ of energy. With the system mounted on a mechanical oscillator that can be tuned to shake at a specified frequency and constant acceleration, we measured the charging times over varying acceleration rates at a set frequency of 7 Hz to obtain the results shown in Figure 5.10. The charging time indicates the time taken for the buck converter (voltage regulator) to turn on and provide energy to the MCB. Various other tests that were conducted, which showed that with an acceleration rate of 0.6g, we are able to send the first packet 0.8 s after the onset of shaking.

5.2.6 Earthquake Simulator Test

'Awesome Forces: Earthquake House' is a permanent exhibition at the Museum of New Zealand 'Te Papa' [45] that simulates the horizontal movement of the 1987 Edgecumbe Earthquake. This was chosen for more realistic testing of the system, as the frequencies and accelerations are modeled closely on a real earthquake; a mechanical oscillator has a single frequency and keeps the same acceleration over time. The earthquake that struck Edgecumbe just after noon on 2 March was reported to be of magnitude 6.5 with an average acceleration rate of 0.261g [46]. The completed wireless sensor node, comprising the PMB with four vibration energy harvesters plus extensions, MCB and antenna, mounted in an enclosure for protection, is shown in Figure 5.11. When deployed, the enclosure is fully covered to protect the wireless sensor node from damage from environmental causes or due to tampering. The system was configured to send as many packets (containing acceleration rates measured by the accelerometer) as possible for as long as it

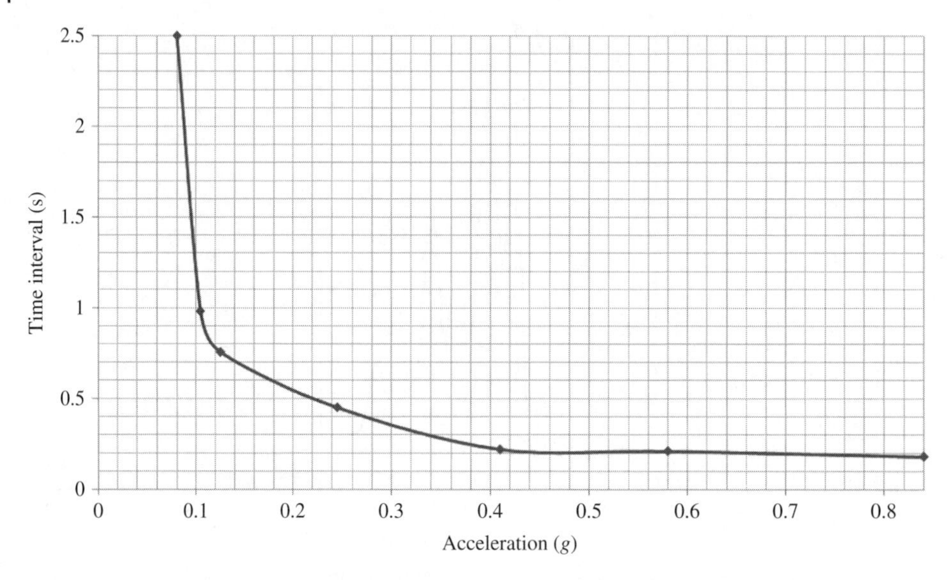

Figure 5.10 Charge times versus acceleration.

Figure 5.11 Assembled wireless sensor node inside enclosure (patent pending).

Figure 5.12 Assembled system being installed by Te Papa technicians on Earthquake House.

had power. After the system was mounted in an unobtrusive location above the exit (Figure 5.12), a receiving station comprising a SmartRF packet sniffer connected to a laptop computer was located nearby to receive and log the transmitted packets.

With the harvesters tuned and receiving station set up, the system was left to run every day during the opening hours of the museum, from 10:00 to 18:00, for a period of one week. A portable accelerometer (a Midé Slam Stick) was attached to the Earthquake House to sample the acceleration for the duration of simulated shakes, and to record the

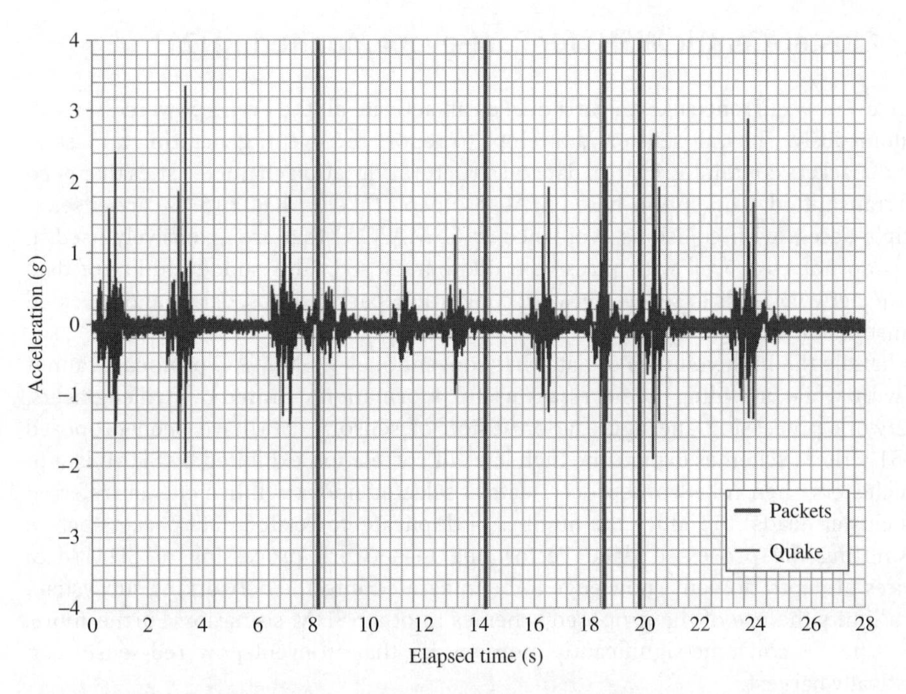

Figure 5.13 Acceleration over time showing when each packet was sent.

jolts together with the times they occurred. As shown in Figure 5.13, from a cold start (that is, with empty capacitors) it took 8 s for enough energy to be harvested before the first packet was sent. The second packet was sent about 6 s later, which implied that there was some remaining charge in the capacitors. The next two jolts, which appeared to be stronger, were able to produce enough energy to send the next packet, and another soon after; this is possible because the capacitors were able to hold enough charge to send one extra packet. Averaging over all the data collected, the time needed to send the first packet from the instance shaking started was 7.2 s. While the acceleration rates have been measured by the accelerometer, the system can still provide an approximate estimation of the shaking severity by simply noting the time instances when consecutive packets are generated, and then work out the amounts of energy needed to produce those packets. That is, measuring how fast the packets are generated and sent can give an indication of the shaking severity.

5.2.7 Implications for Networking Protocol Design

Using the energy from the event itself to power the sensor ensures that the event will not be missed. While solving the detection problem, it creates a problem for the wireless communication protocol because all the sensors will be activated almost simultaneously. They will attempt to transmit concurrently, resulting in serious channel contention. In the next section, we describe a medium access protocol that is specifically designed for such event-powered wireless sensors in rare-event detection.

5.3 Cluster-Centric WSNs for Rare-event Monitoring [21]

A rare event occurrence like an earthquake can activate all the event-powered sensors simultaneously. They all sense and attempt to report the event, generating a massive burst of packets containing critical data on structural vibration characteristics that need to be transmitted. Contention-based MAC protocols (like IEEE 802.15.4's carrier sense multiple access with collisions avoidance or CSMA/CA) that are extensively used in WSNs do not cope well with massive traffic bursts [47]. This sudden influx of data into the network leads to severe network congestion, packet drops, delays and repeated retransmission attempts.

To handle the high volumes of data that conventional WSN MAC protocols cannot deal with, while providing the necessary useful information required by civil engineers, in-network processing and aggregation/fusion of sensory data have been proposed [48–51]. These schemes exploit the high level of correlation exhibited by SHM data to form clusters, then collect and process data at selected nodes within a cluster (referred to as cluster heads) to reduce the number of duplicate/correlated data packets before transmitting the processed data. Performing in-network processing of SHM data requires in-depth domain knowledge to be integrated into the networking subsystem, which limits the use of the proposed schemes in other SHM scenarios. Furthermore, these schemes consume significantly more energy that an event-powered sensor can realistically harvest.

The MAC protocol for an event-powered WSN also adopts clustering, whereby sensors that generate highly correlated data are grouped into a cluster. The protocol arbitrates access to the wireless channel in such a way that every cluster has a fair opportunity to transmit their data; this is achieved by viewing each cluster as a supernode. In this way, contention is reduced and data is sent quickly and fairly. Ultimately, the aim is to provide civil and structural engineers the (raw) data that they need and at the same time to address WSN constraints.

5.3.1 System Model

All nodes are within a one-hop transmission range of a personal area network (PAN) coordinator that receives all the data from the sensors under its charge. For example, a PAN coordinator might be installed on a lamp post that is next to a building where SHM sensors are deployed, as shown in Figure 5.14. Exploiting the high degree of correlation in SHM data, more emphasis is put on transmitting uncorrelated data from different clusters evenly rather than allowing all nodes to compete for channel access individually. In practice, the clusters of sensors are defined by domain experts, such as structural engineers, who have the knowledge to best identify and organize nodes into clusters, such that each cluster is a set of correlated data points.

Each cluster is viewed as a *supernode*, and as soon as one node of a cluster has successfully transmitted its data, the cluster is deemed to have succeeded in the current cycle; the rest of the nodes in that cluster refrain from transmitting further until all the other clusters in the network have also succeeded in the current cycle. When the PAN coordinator successfully receives a packet from a node, it simply broadcasts an acknowledgment (ACK) packet containing the identifier of the successful node, and the other nodes in the cluster upon receiving this ACK get to know that a node in their cluster

Figure 5.14 Deployment scenario (adapted from a figure on the Wireless Building Automation website at https://wlba.wordpress.com).

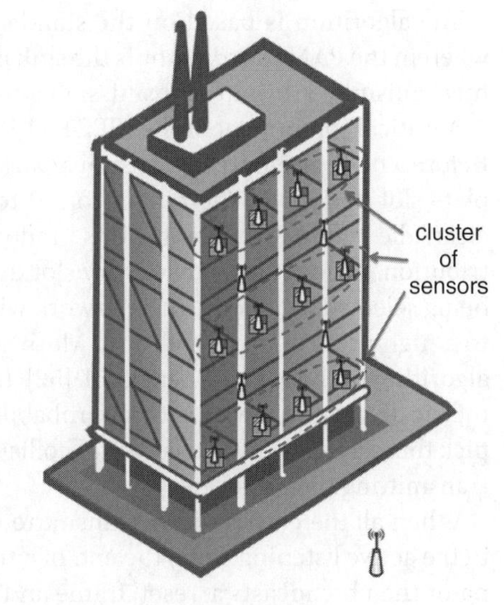

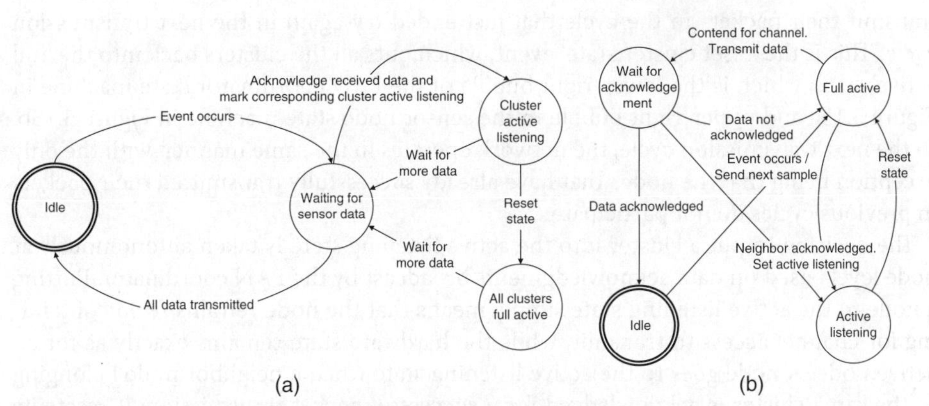

Figure 5.15 State transition diagrams. (a) PAN coordinator (b) Sensor node.

has succeeded. The nodes in this successful cluster will transition into 'active listening' mode, and the cluster as a whole is said to be in an active listening state when all the member nodes refrain from transmitting but actively listen to broadcast signals from the PAN coordinator. Figure 5.15a shows the state transition diagram for the PAN coordinator in the IEEE 802.15.4 context.

For rare events (like earthquakes), the PAN coordinator typically remains idle until an event occurs, after which it waits for data from the sensor nodes. However, the PAN coordinator can remain in listening mode all the time (waiting for any transmission from the sensors), if it is assumed to have a sustained power source. Minor tremors can also cause a low degree of structural vibrations to trigger an event and activate the sensors. This cluster-centric approach has two advantages:

- Higher priority is given to transmission of uncorrelated data from different clusters; this is an implicit 'round-robin' that ensures fairness among clusters.
- Nodes within a cluster have a fair chance to transmit during each transmission cycle.

The algorithm is based on the standard IEEE 802.15.4 non-beacon mode protocol, wherein the PAN coordinator is the sink node and solely relies on the slotted CSMA/CA mechanism to arbitrate transmission attempts by the nodes.

A critical component of the IEEE 802.15.4 MAC protocol is the *backoff exponent* (BE). Before a node attempts to transmit a packet, it first delays for a random number of complete slot periods in the range from 0 to $2^{BE} - 1$ and then checks that the channel is clear/idle before it transmits. This random number is selected based on a uniform distribution, which means that every slot in the range 0 to $2^{BE} - 1$ has an equal chance of being selected by a node. In a network with many nodes wanting to transmit, this leads to a high probability of collision. Motivated by the optimal backoff time slot selection algorithm proposed by Cheng et al. [52], the nodes randomly select a slot based on a geometric distribution, with a lower probability of selecting an early slot so that fewer nodes pick these, reducing the chances of collisions, and giving a higher chance of successfully transmitting their packets.

When all the clusters have transmitted one data packet each and all the clusters are in the active listening state, the end of a transmission cycle is reached. The PAN coordinator then broadcasts a 'reset' frame and the nodes that did not manage to successfully transmit their packets in the cycle that just ended try again in the next transmission cycle. This is the 'teset cluster state' event, which puts all the clusters back into the 'full active' state, which is the lower-right bubble of the PAN coordinator state machine in Figure 5.15a and upper-right bubble in the sensor node state machine in Figure 5.15b. In the next transmission cycle, the network operates in the same manner, with the only exception being that the nodes that have already successfully transmitted their packets in previous cycles do not participate.

The decision to put a cluster into the active listening state is taken autonomously at node level based on data acknowledgments broadcast by the PAN coordinator. Putting a node to the active listening state simply means that the node refrains from contending for channel access to transmit, while the hardware state remains exactly as for an active node. A node goes to the active listening state when a neighbor node belonging to the same cluster is acknowledged for a successful packet transmission; it resets its state back to full active only when the PAN coordinator sends a reset signal, as shown in Figure 5.15b. Once all the clusters are in the active listening state, the PAN coordinator broadcasts a 'reset' frame to reset all the nodes' states for next round of transmissions. To reduce the nodes' energy consumption, the PAN coordinator can also include in the ACK packet the number of remaining clusters to transmit; 'active listening' nodes can estimate the quickest time required for these remaining clusters to successfully transmit, and then go to sleep for this period of time.

5.3.2 Performance Evaluation

The design is simple, with minor changes to the IEEE 802.15.4 MAC algorithm, yet able to achieve significant performance improvements and eliminate network bias. The scheme was evaluated using simulations and compared against other IEEE 802.15.4 variants as well as the WSN approach proposed by Liu et al. [49], which employs in-network processing of SHM data. The evaluation uses the standard IEEE 802.15.4 protocol, and the cluster-centric MAC is built on top of that. Varying cluster sizes (5, 10, 15, 20 and 25) and network sizes (100, 150, 200 and 250) are used for the evaluation. To obtain accurate averages, the result for each combination is an average of ten different runs, with each run using a different seed value.

The sensor nodes are placed as they would be in a real-life SHM system. The PAN coordinator is usually outside the building and not very high above the ground; the sensor nodes are in the building, with the ground-floor sensor nodes closest to the PAN coordinator and the highest-floor sensor nodes furthest away. Nodes in cluster #1 are closest to the PAN coordinator and the cluster numbers increase with increasing distance from the PAN coordinator. This deployment scenario is similar to that shown in Figure 5.14. The assumed data in the evaluation scenario are generated by the occurrence of an event (time t_0) that warrants attention; say a strong tremor or earthquake. After all the data generated by that event has been transmitted, the WSN goes back to sleep until another event activates it. In our targeted scenario, we assume a fixed amount of data are generated at each node as a result of an event.

5.3.2.1 Time to Completion in a Cluster

Figure 5.16 shows the *packet delivery characteristics* of the IEEE 802.15.4 MAC and the cluster-centric MAC protocols for the 250-node network, which is the largest network evaluated. The vertical plots for each cluster show the time duration (since t_0) at which consecutive packets within a cluster are successfully transmitted and received by the

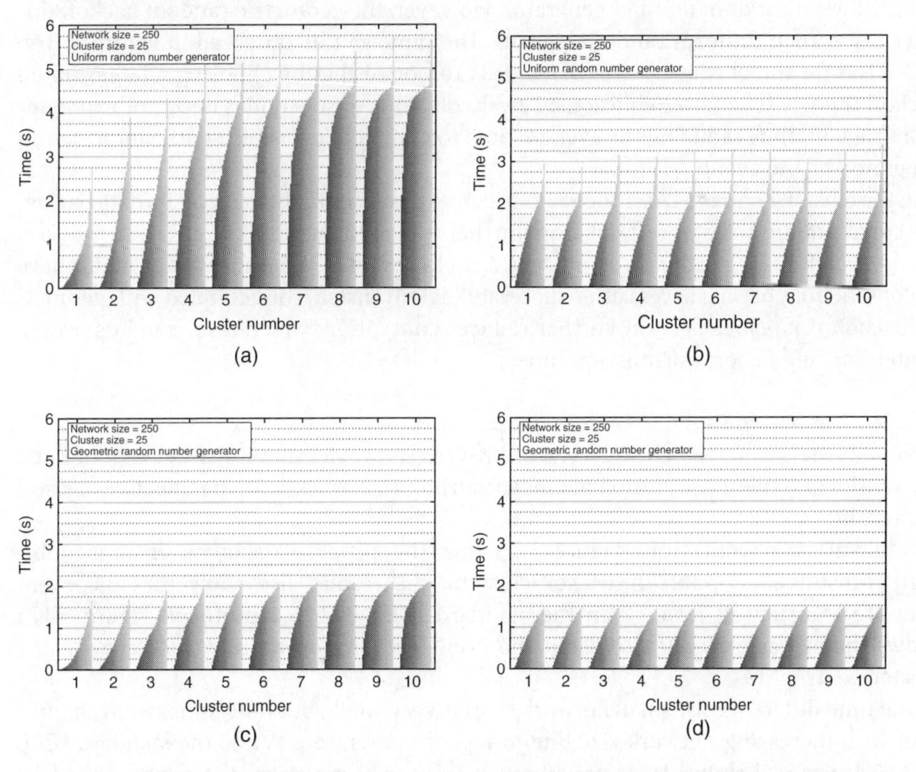

Figure 5.16 Packet arrival time ($N = 250, C = 25$). (a) IEEE 802.15.4 with uniform random backoff slot selection (b) Cluster-centric MAC with uniform random backoff slot selection (c) IEEE 802.15.4 with geometric random backoff slot selection (d) Cluster-centric MAC with geometric random backoff slot selection. (*See color plate section for the color representation of this figure.*)

PAN coordinator; for example the blue bar on the left shows the time needed for the first successfully transmitted packet of a cluster (irrespective of which sensor within the cluster it came from) and the next bar shows the second successful packet, and so forth.

In Figures 5.16a and 5.16c, the standard IEEE 802.15.4 MAC protocol produces a bias towards nodes and clusters closer to the PAN coordinator node. This skewed performance for nodes/clusters closer to the PAN coordinator can be attributed to a capture effect [53], which has been observed and studied in IEEE 802.15.4 networks [54, 55]. This leads to clusters that are closer to the PAN coordinator node being able to transmit all their data much sooner than the clusters that are further away.

The cluster-centric MAC eliminates this bias by ensuring that each cluster gets a fair chance rather than individual nodes. This is achieved by taking a cluster (and all the nodes therein) out of the contention for the channel once it is successful in the current cycle. No bias results are observed in Figure 5.16b and 5.16d, which show every cluster evenly sending packets to the PAN coordinator. Since there is no bias among clusters, all the clusters finish transmitting their data around the same time. This is a favourable consequence of the cluster-centric approach, which reduces overall network contention and improves the entire network's performance.

Both Figure 5.16a and 5.16c use the same IEEE 802.15.4 CSMA/CA MAC algorithm but a different random number generator. However, the geometric random backoff finishes faster than uniform random backoff. The same can be observed in Figures 5.16b and 5.16d for the cluster-centric MAC. It is reasoned that by choosing a larger initial backoff there will be fewer collisions. Fewer collisions means nodes need not exponentially backoff, thus reducing the average time to complete transmitting the information generated by the event.

In the cluster-centric MAC, once a node is able to successfully transmit its data packet, the corresponding cluster refrains from further transmission until the next transmission round, which helps produce a significant drop in contention because fewer nodes have to contend for channel access after successful data transfer. This, coupled with geometric backoff time-slot selection, further reduces contention, which results in less overall contention and faster transmission times.

5.3.2.2 Average and Total Time to Transmit

Although the aim of the cluster-centric MAC is to provide unbiased delivery of data, conventional network performance metrics from a network-wide perspective are still important.

While it is evident from Figure 5.16 that the proposed scheme ensures unbiased transmission opportunities for clusters, the results also show that the event data are transmitted faster than the standard IEEE 802.15.4 protocol. Figure 5.17a shows the average times to transmit the event data are on average shorter with the cluster-centric MAC.

The time difference for small network sizes is very small, but it becomes more significant with increasing network size due to higher contention. While the standard IEEE 802.15.4 protocol shows large variations in time to completion, the proposed model gives consistent times to complete transmissions, irrespective of network size.

To further understand the time to transmit all packets, the total time to transmit all packets was measured at fixed packet error probabilities. Increased

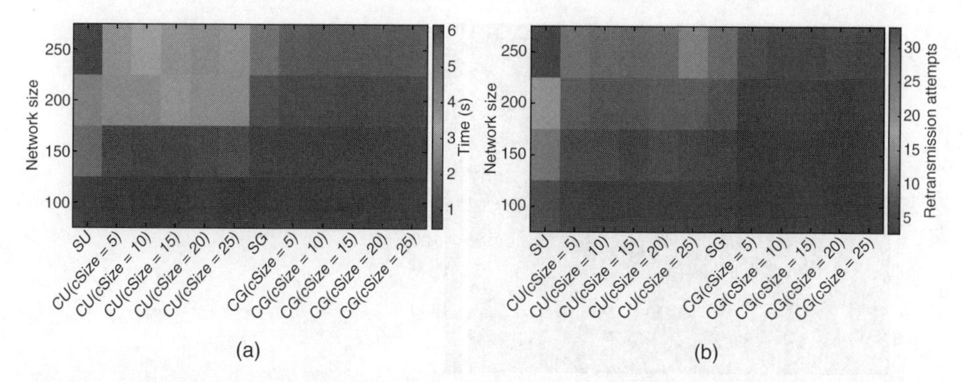

Figure 5.17 Performance for different network and cluster sizes to transmit all packets: S, standard IEEE 802.15.4 MAC; C, cluster-centric MAC; U, G, uniform/geometric random number generator; cSize, cluster size. (a) Average packet transmission time (b) Average number of retransmissions. (*See color plate section for the color representation of this figure.*)

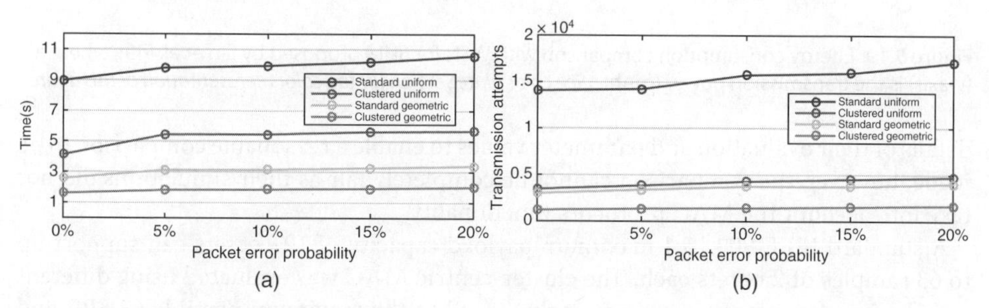

Figure 5.18 Performance in lossy environment ($N = 250$, $C = 10$). (a) Total time to transmit all packets (b) Total retransmission attempts. (*See color plate section for the color representation of this figure.*)

packet loss probability triggers higher retransmission rates, and a vicious cycle of successive retransmissions may develop due to the large number of nodes transmitting/retransmitting
simultaneously.

Figure 5.18a shows that standard IEEE 802.15.4 does suffer from high delays, while the cluster-centric approach experiences a very gradual increase in time needed as the loss probability increases. Retransmissions are costly in terms of time and energy, hence one of the aims of WSN protocols is to minimize the retransmission count. The proposed design is able to significantly reduce network congestion by taking nodes out of the contention for the channel promptly. It is therefore able to transmit all its data packets much sooner than the standard IEEE 802.15.4 protocol. Less contention in return means fewer retransmissions, and the same effect of fewer transmissions can be observed in Figures 5.17b and 5.18b.

5.3.2.3 Energy Consumption

To put the energy consumption of the proposed design into perspective, we compare our results with those of Liu et al. [49], an example of a WSN approach designed to reduce energy consumption by performing in-network processing of SHM data. This approach has been selected from among many others, because the authorshave provided sufficient

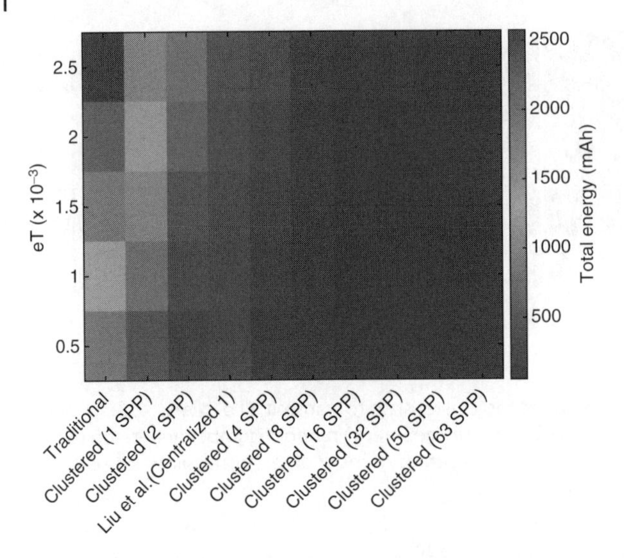

Figure 5.19 Energy consumption comparison with WSN for SHM proposed by Liu et al. [49], where e^T (y-axis) is the transmission power (mAh). (*See color plate section for the color representation of this figure.*)

details of their evaluation and parameter values to enable a reasonable comparison to be made; however, the comparison cannot be completely fair as their simulations did not take into account the MAC protocol's functionality.

A standard IEEE 802.15.4 maximum payload capacity of 127 octets can support up to 63 samples of 2 octets each. The cluster-centric MAC was evaluated using different sample sizes from one sample per packet (SPP) to the maximum possible 63 SPP, and assuming a total of 10752 samples and energy values as specified in Liu's paper. The simulations produced similar results for cluster sizes 5, 6, 7 and 8, so only the results for cluster size 5 are shown. Since a smaller SPP means a higher number of packets, intuitively it should mean that using 63 SPP not only results in fewer packets but also less contention, fewer retransmissions and lower energy consumption.

On the other hand, sending larger payloads requires more energy than shorter payloads, but there are energy and throughput gains due to the amortisation of the transmission overheads, and it has been shown that payload sizes of around 100 octets are near optimal [56]. Hence, a scenario with 50 SPP, to represent a 100-octet payload, is also included. Simulation results as presented in Figure 5.19 also show that better performance is achievable with as few as 4 SPP, without having to perform in-network processing.

5.4 Summary

Rare events present unique challenges to energy-constrained systems designed for long-term sensing of their occurrence or effects. Unlike periodic sampling or query-based sensing systems, longevity can not be achieved simply by adjusting the sensing nodes' duty cycle until an equitable balance between data density and network lifetime is established. The low probability of occurrence and random nature of rare events makes it difficult to guarantee that duty-cycled battery-powered sensing nodes will be energised when events occur. Equally, it is usually considered impractical to leave the sensing nodes energised at all times if the network is to have an acceptably long operational life. In the past decade and a half, wireless sensor network research has addressed this aspect of rare-event sensing by investigating techniques such as Synchronized duty-cycling of redundant nodes, passive sensing, duplicate message suppression and energy-efficient network protocols. Researchers have also demonstrated the efficacy of harvesting energy from the environment to extend operational life.

References

1 Committee on Networked Systems of Embedded Computers, National Research Council (2001) *Embedded Everywhere: A Research Agenda for Networked Systems of Embedded Computers*, National Academies Press.
2 Mak, N.H. and Seah, W.K.G. (2009) How long is the lifetime of a wireless sensor network?, in *Proceedings of the International Conference on Advanced Information Networking and Applications (AINA)*, Bradford, UK, pp. 763–770.
3 Stankovic, J.A., Wood, A.D., and He, T. (2011) Realistic applications for wireless sensor networks, in *Theoretical Aspects of Distributed Computing in Sensor Networks*, Springer, pp. 835–863.
4 Ye, W., Heidemann, J., and Estrin, D. (2004) Medium access control with coordinated adaptive sleeping for wireless sensor networks. *IEEE/ACM Transactions on Networking*, **12** (3), 493–506.
5 Schurgers, C. and Srivastava, M. (2001) Energy efficient routing in wireless sensor networks, in *Proceedings of the IEEE Military Communications Conference (MILCOM)*, McLean, VA, USA, pp. 357–361.
6 Yao, Y. and Gehrke, J. (2003) Query processing in sensor networks, in *Proceedings of the First Biennial Conference on Innovative Data Systems Research (CIDR)*, Asilomar, CA, USA.
7 Rubino, G. and Tuffin, B. (2009) *Rare event simulation using Monte Carlo methods*, John Wiley & Sons.
8 Arora, A. *et al.* (2004) A line in the sand: a wireless sensor network for target detection, classification, and tracking. *Computer Networks*, **46** (5), 605–634.
9 Low, K.S., Win, W.N.N., and Er, M.J. (2005) Wireless sensor networks for industrial environments, in *Proceedings of International Conference on Computational Intelligence for Modeling, Control and Automation, and International Conference on Intelligent Agents, Web Technologies and Internet Commerce*, Vienna, Austria, pp. 271–276.
10 Gelenbe, E. and Ngai, E.H. (2008) Adaptive QoS routing for significant events in wireless sensor networks, in *Proceedings of 5th IEEE International Conference on Mobile Ad Hoc and Sensor Systems*, Atlanta, GA, USA, pp. 410–415.
11 Liu, C., Wu, K., Xiao, Y., and Sun, B. (2006) Random coverage with guaranteed connectivity: joint scheduling for wireless sensor networks. *Transactions on Parallel and Distributed Systems*, **17** (6), 562–575.
12 Cao, Q., Abdelzaher, T., He, T., and Stankovic, J. (2005) Towards optimal sleep scheduling in sensor networks for rare-event detection, in *Proceedings of the 4th*

International Symposium on Information Processing in Sensor Networks (IPSN), Los Angeles, CA, USA.

13 Cao, Q., Yan, T., Stankovic, J., and Abdelzaher, T. (2005) Analysis of target detection performance for wireless sensor networks, in *Distributed Computing in Sensor Systems*, Springer, pp. 276–292.

14 Atzori, L., Iera, A., and Morabito, G. (2010) The Internet of Things: A survey. *Computer Networks*, **54** (15), 2787–2805.

15 Nasridinov, A., Ihm, S.Y., Jeong, Y.S., and Park, Y.H. (2014) Event detection in wireless sensor networks: Survey and challenges, in *Mobile, Ubiquitous, and Intelligent Computing*, Springer, pp. 585–590.

16 Pantazis, N., Nikolidakis, S.A., and Vergados, D.D. (2013) Energy-efficient routing protocols in wireless sensor networks: A survey. *Communications Surveys & Tutorials, IEEE*, **15** (2), 551–591.

17 Harrison, D.C., Seah, W.K.G., and Rayudu, R. (2016) Rare event detection and propagation in wireless sensor networks. *ACM Computing Surveys (CSUR)*. (Accepted for publication.).

18 Seah, W.K.G., Eu, Z.A., and Tan, H.P. (2009) Wireless sensor networks powered by ambient energy harvesting (WSN-HEAP)—survey and challenges, in *Proceedings of the 1st International Conference on Wireless Communication, Vehicular Technology, Information Theory and Aerospace & Electronic Systems Technology*, Aalborg, Denmark.

19 Harrison, D.C., Seah, W.K.G., and Rayudu, R. (2015) Coverage preservation in energy harvesting wireless sensor networks for rare events, in *Proceedings of the 40th Annual IEEE Conference on Local Computer Networks (LCN)*, Clearwater Beach, FL, USA.

20 Tomicek, D., Tham, Y.H., Seah, W.K.G., and Rayudu, R. (2013) Vibration-powered wireless sensor for structural monitoring during earthquakes, in *Proceedings of the 6th International Conference on Structural Health Monitoring of Intelligent Infrastructure*, SHMII, Hong Kong, China.

21 Singh, S., Seah, W.K.G., and Ng, B. (2015) Cluster-centric medium access control for WSNs in structural health monitoring, in *Proceedings of the 13th International Symposium on Modelling and Optimization in Mobile, Ad Hoc, and Wireless Networks (WiOpt)*, Mumbai, India, pp. 275–282.

22 Cardei, M. and Wu, J. (2006) Energy-efficient coverage problems in wireless ad-hoc sensor networks. *Computer Communications*, **29** (4), 413–420.

23 Hsin, C.F. and Liu, M. (2004) Network coverage using low duty-cycled sensors: random and coordinated sleep algorithms, in *Proceedings of the 3rd International Symposium on Information Processing in Sensor Networks (IPSN)*, Berkeley, CA, USA, pp. 433–442.

24 Raghunathan, V., Kansal, A., Hsu, J., Friedman, J., and Srivastava, M. (2005) Design considerations for solar energy harvesting wireless embedded systems, in *Proceedings of the 4th International Symposium on Information Processing in Sensor Networks (IPSN)*, Los Angeles, California, USA.

25 Khosropour, N., Krummenacher, F., and Kayal, M. (2012) Fully integrated ultra-low power management system for micro-power solar energy harvesting applications. *Electronics Letters*, **48** (6), 338–339.

26 Tian, D. and Georganas, N.D. (2002) A coverage-preserving node scheduling scheme for large wireless sensor networks, in *Proceedings of the 1st ACM international workshop on wireless sensor networks and applications*, ACM, Atlanta, GA, USA, pp. 32–41.
27 Wang, X., Xing, G., Zhang, Y., Lu, C., Pless, R., and Gill, C. (2003) Integrated coverage and connectivity configuration in wireless sensor networks, in *Proceedings of the 1st International Conference on Embedded Networked Sensor Systems (SenSys)*, Los Angeles, California, USA, pp. 28–39.
28 Alippi, C., Camplani, R., Galperti, C., and Roveri, M. (2008) Effective design of WSNs: from the lab to the real world, in *Proceedings of the 3rd International Conference on Sensing Technology (ICST)*, Tainan, Taiwan, pp. 1–9.
29 Advanticsys (2015), Solar cells. URL http://www.advanticsys.com/shop/mtmcm5000msp-p-14.html.
30 NIWA (2015), Solarview. URL http://solarview.niwa.co.nz.
31 Futurlec (2015), Solar cells. URL http://www.futurlec.com/Solar_Cell.shtml.
32 Liu, Y. and Seah, W.K.G. (2005) A scalable priority-based multi-path routing protocol for wireless sensor networks. *International Journal of Wireless Information Networks*, **12** (1), 23–33.
33 Wald, D.J., Quitoriano, V., Heaton, T.H., and Kanamori, H. Relationships between peak ground acceleration, peak ground velocity, and modified Mercalli intensity in California, volume = 15, year = 1999, bdsk-url-1 = http://dx.doi.org/10.1193/1.1586058. *Earthquake Spectra*, (3), 557–564.
34 Elenas, A. and Meskouris, K. (2001) Correlation study between seismic acceleration parameters and damage indices of structures. *Engineering Structures*, **23** (6), 698–704.
35 Beeby, S.P., Tudor, M.J., and White, N.M. (2006) Energy harvesting vibration sources for microsystems applications. *Measurement Science and Technology*, **17** (12), R175–R195.
36 Roundy, S., Wright, P.K., and Rabaey, J. (2003) A study of low level vibrations as a power source for wireless sensor nodes. *Computer Communications*, **26** (11), 1131–1144.
37 Raj, P.S. (2012) Vibration Energy Harvesting using PEH25W, *ECS Technical Report ECSTR2012-05*, Victoria University of Wellington.
38 Clifton, C. (2011) Christchurch Feb 22nd Earthquake: A Personal Report, *Tech. Rep.*, NZ Heavy Engineering Research Association (HERA).
39 Elvin, N.G., Lajnef, N., and Elvin, A.A. (2006) Feasibility of structural monitoring with vibration powered sensors. *Smart Materials and Structures*, **15** (4), 977.
40 Arms, S., Townsend, C., Churchill, D., Galbreath, J., Corneau, B., Ketcham, R., and Phan, N. (2008) Energy harvesting, wireless, structural health monitoring and reporting system, in *Proceedings of the 2nd Asia Pacific Workshop on SHM*, Melbourne, Australia.
41 Torah, R., Glynne-Jones, P., Tudor, J., O'Donnell, T., Roy, S., and Beeby, S. (2008) Self-powered autonomous wireless sensor node using vibration energy harvesting. *Measurement Science and Technology*, **19** (12), 125 202.
42 Park, J.H. (2010) *Development of MEMS Piezoelectric Energy Harvesters*, Ph.D. thesis, Auburn University.

43 Ahmad, M.A. and Alshareef, H.N. (2011) Modeling the power output of piezoelectric energy harvesters. *Journal of Electronic Materials*, **40** (7), 1477–1484.

44 Xue, H., Hu, Y., and Wang, Q.M. (2008) Broadband piezoelectric energy harvesting devices using multiple bimorphs with different operating frequencies. *IEEE Transactions on Ultrasonics, Ferroelectrics, and Frequency Control*, **55** (9), 2104–2108.

45 Te Papa, Museum of New Zealand, Awesome Forces (Te Papa exhibition)—Earthquake House, http://collections.tepapa.govt.nz/theme.aspx?irn=1364.

46 Dowrick, D. (1988) Edgecumbe earthquake: some notes on its source, ground motions, and damage in relation to safety. *Bulletin of the New Zealand National Society for Earthquake Engineering*, **21** (3), 198–203.

47 Kleinrock, L. and Tobagi, F. (1975) Packet switching in radio channels: Part I—Carrier sense multiple-access modes and their throughput-delay characteristics. *IEEE Transactions on Communications*, **23** (12), 1400–1416.

48 Zimmerman, A., Shiraishi, M., Swartz, R., and Lynch, J. (2008) Automated modal parameter estimation by parallel processing within wireless monitoring systems. *Journal of Infrastructure Systems*, **14** (1), 102–113.

49 Liu, X., Cao, J., Lai, S., Yang, C., Wu, H., and Xu, Y.L. (2011) Energy efficient clustering for WSN-based structural health monitoring, in *Proceedings of the 30th IEEE International Conference on Computer Communications (INFOCOM)*, Shanghai, China, pp. 2768–2776.

50 Jindal, A. and Liu, M. (2012) Networked computing in wireless sensor networks for structural health monitoring. *IEEE Transactions on Networking*, **20** (4), 1203–1216.

51 Hackmann, G., Guo, W., Yan, G., Sun, Z., Lu, C., and Dyke, S. (2014) Cyber-physical codesign of distributed structural health monitoring with wireless sensor networks. *IEEE Transactions on Parallel and Distributed Systems*, **25** (1), 63–72.

52 Cheng, M.Y., Chen, Y.B., Wei, H.Y., and Seah, W.K.G. (2013) Event-driven energy-harvesting wireless sensor network for structural health monitoring, in *Proceedings of the IEEE 38th Conference on Local Computer Networks (LCN)*, Sydney, Australia, pp. 364–372.

53 Leentvaar, K. and Flint, J. (1976) The capture effect in FM receivers. *IEEE Transactions on Communications*, **24** (5), 531–539.

54 Gezer, C., Buratti, C., and Verdone, R. (2010) Capture effect in IEEE 802.15.4 networks: Modelling and experimentation, in *Proceedings of the 5th IEEE International Symposium on Wireless Pervasive Computing (ISWPC)*, Modena, Italy, pp. 204–209.

55 Lu, J. and Whitehouse, K. (2009) Flash flooding: exploiting the capture effect for rapid flooding in wireless sensor networks, in *Proceedings of the 28th IEEE Conference on Computer Communications (INFOCOM)*, Rio de Janeiro, Brazil, pp. 2491–2499.

56 Noda, C., Prabh, S., Alves, M., and Voigt, T. (2013) On packet size and error correction optimizations in low-power wireless networks, in *Proceedings of the 10th Annual IEEE Communications Society Conference on Sensor, Mesh and Ad Hoc Communications and Networks (SECON)*, New Orleans, LA, USA, pp. 212–220.

Part II

Space WSS Solutions and Applications

6

Battery-less Sensors for Space

Ali Abedi

Department of Electrical and Computer Engineering, University of Maine, Orono, USA

6.1 Introduction

Space is one of the most extreme environments known to mankind, posing design challenges for engineering reliable systems. Any structure or machine operating in space requires some kind of sensor to monitor its integrity, reliable operation and, for manned vehicles or habitats, to ensure the safety of those living inside. There are several problems with designing sensors and electronics that can withstand temperature fluctuations from cryogenic to very high, as well as radiation and launch vibration, and still operate reliably. Going with traditional engineering standards, the solution might be too expensive or too heavy to be lifted into space. Thinking outside the box, however, innovative solutions eliminating unnecessary parts of the system or integrating other parts to save on mass and volume might be a more reasonable approach.

For instance, instead of designing a wired system, a wireless system would seem to save weight by eliminating wires and associated supporting hardware. Figure 6.1 illustrates the interior section of the wing leading edge of the orbiter Columbia [1]. It is noteworthy that wires are not the only added mass; bundling and attachment hardware also count when assessing the added mass of a wired system.

The price to pay is that the redundancy and reliability of wired systems may not be matched by wireless systems. However, if wireless systems are designed properly, this challenge can be solved. One of the major issues with wireless systems is power delivery. Using wires to deliver power may defeat the whole purpose, but using batteries that need to be changed often is also challenging, particularly in hard-to-reach areas.

This chapter presents an overview of wireless sensing in space and extreme environments, with an emphasis on battery-less sensors. Section 6.2 presents cost–benefit analyses for wired and wireless sensors. Active and passive sensors are compared in Section 6.3. Design considerations for battery-less sensors are presented in Section 6.4. The chapter is concluded in Section 6.5.

Wireless Sensor Systems for Extreme Environments: Space, Underwater, Underground, and Industrial.
First Edition. Edited by Habib F. Rashvand and Ali Abedi.

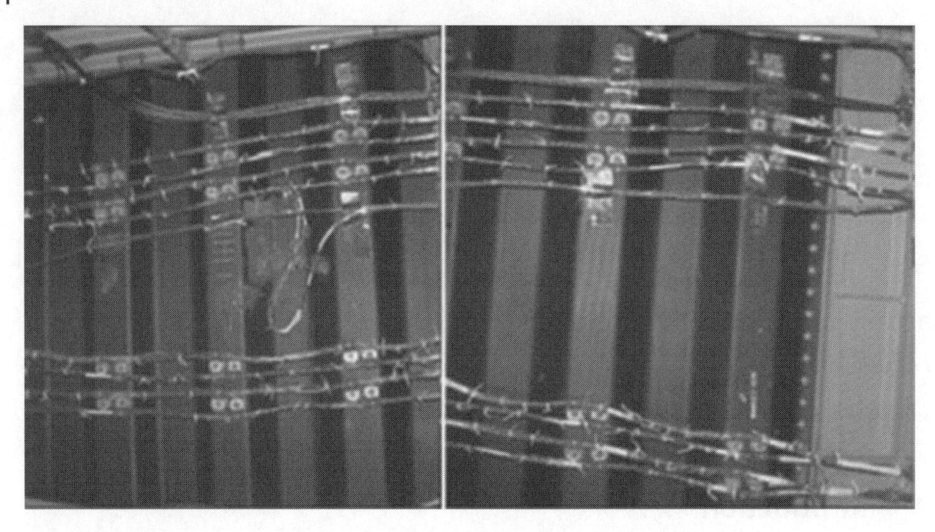

Figure 6.1 Wired sensors inside the wing leading edge of the orbiter Columbia [1].

6.2 Wired or Wireless Sensing: Cost–Benefit Analysis

6.2.1 Wired Sensing Systems

Wired sensing systems have been used for a long time and technologies surrounding them have matured to the point of acceptable reliability for space applications – the golden standard for harsh environments. Redundant wiring of multiple sensors ensures data delivery at all times, even when a wire or two have broken. The cost of wired systems increases with the number of sensors, the distance between the sensors and readers, and associated support hardware such as cable cladding, fixtures, the cost of routing design and electromagnetic interference (EMI) analysis and tests. Another issue with wired systems in addition to the weight and cost is the flexibility and scalability, which adds additional cost for adding a single wire to a current design or moving one sensor to a new location. This includes changes of routes, drawings and EMI tests.

6.2.2 Wireless Sensing Systems

Wireless sensing, on the other hand, is flexible and scalable; adding a new sensor in any new desired location is possible without adding any of the costs associated with wired sensors. Moving a sensor in a current system is also not as costly as for a wired system. This is based on the assumption that the initial wireless sensor network design can accommodate these sensors and any interference associated with them. Once the initial design is done and the number of supported sensor nodes is determined, the flexibility and scalability is achieved at no extra cost and without adding more weight.

6.2.3 Reliability Analysis

Reliability remains an issue to be discussed and compared. As seen in Figure 6.2, multiple sensors measuring various parameters and perhaps using different communication protocols send their data through a mesh network to a sink or fusion center. In some

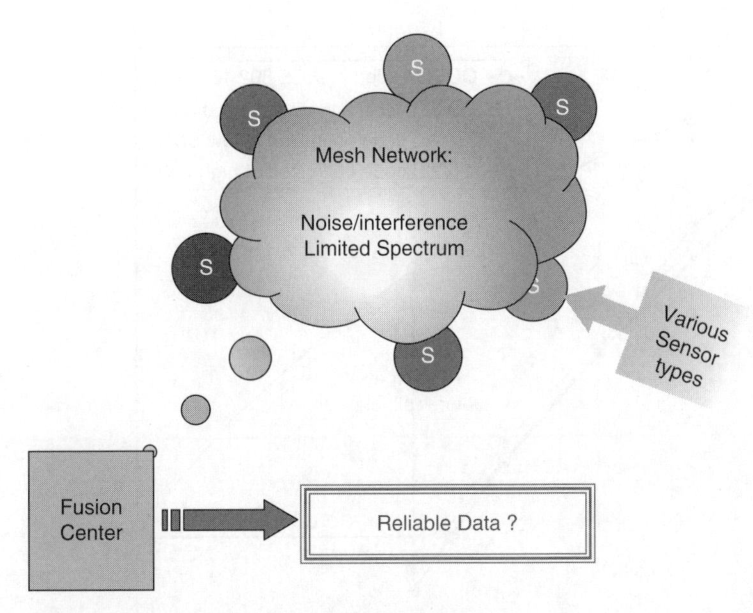

Figure 6.2 Are wireless sensor networks reliable?

texts, these two terms are used interchangeably, while an accurate distinguishing factor is that data may only be collected and not processed at the sink, while processing is an integral role of the fusion center. The stochastic nature of wireless channels may mean that corrupted packets arrive at the fusion center, but does this make the whole system less reliable than wired systems?

Two points to consider here include:

- error correction per sensor data stream
- error correction for a group of sensors with correlated data.

For the first item, even a simple error-correction code can provide thousands of times more reliability than an uncoded sensor network. One example is provided in Figure 6.3, where the IEEE 802.15.4 standard is modified to include a simple convolutional code in order to achieve bit error rates (BER) that are a thousand times lower [2].

The second item is even more important if we look at a cluster of sensors in close vicinity to each other, resulting in several correlated streams of data noting their proximity to a source. For instance, if multiple sensors are measuring the temperature of a blade inside a jet engine, while placed a various locations around the blade, they all sense a similar temperature but with different noise and estimation errors (modeled as E_1, E_2, E_n in Figure 6.4). It can be shown that combining all these unreliable estimates will yield a more reliable result if distributed coding is applied properly [3]. The homogeneous configuration shown in Figure 6.4 is scalable and only includes simple encoding/modulation at the sensor (transmitter) side. A more complicated joint decoder would reside at the sink (receiver side).

The decoder structure is depicted in Figure 6.5 and includes multiple turbo decoders with a correlation extraction block. The log likelihood a-priori values (reliability of each bit) are fed back to the decoders for future iterations, in a similar way to regular turbo-codes.

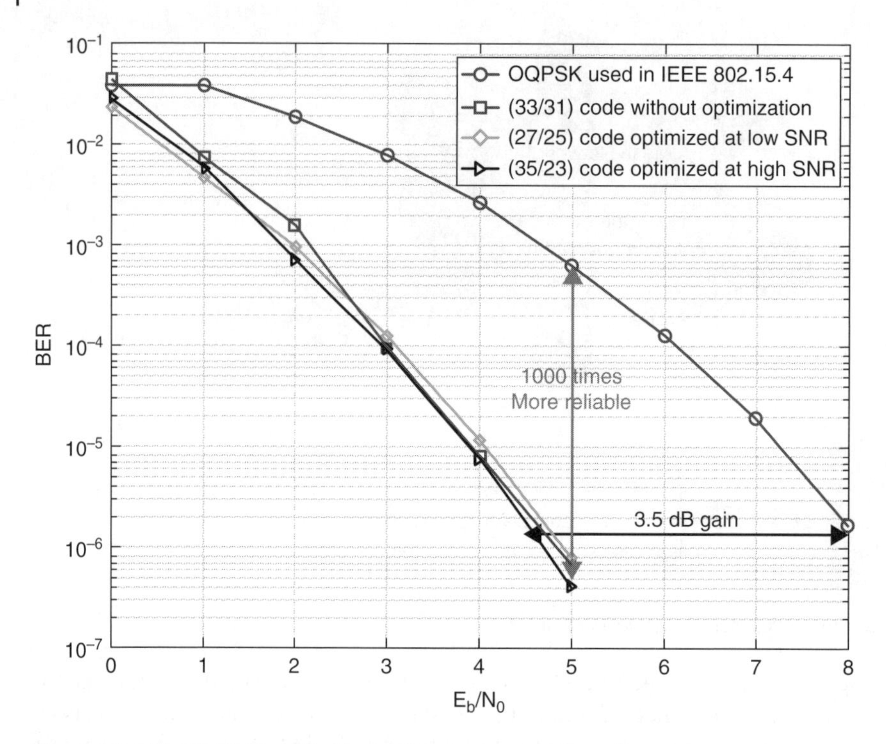

Figure 6.3 IEEE 802.15.4 modified by adding error-correction codes.

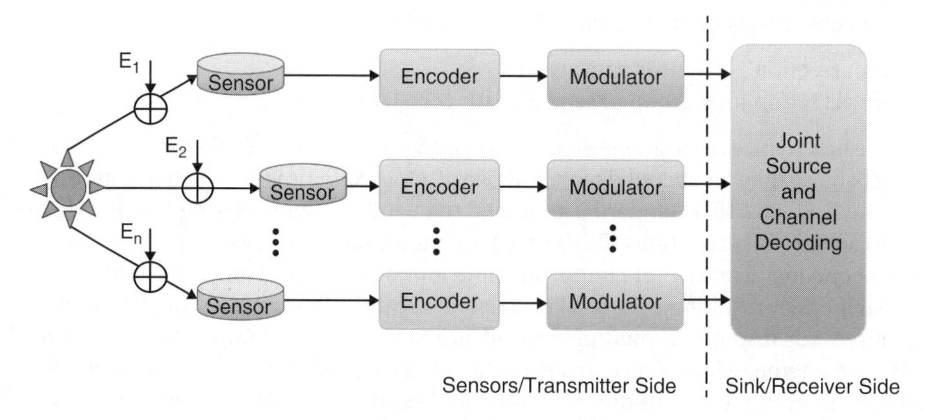

Figure 6.4 Multiple correlated observations from simple sensor nodes decoded jointly [3].

Another obvious benefit of a mesh wireless network is its immunity to wire failure (multiple links from sink to source are always available) and its robustness with respect to one or more faulty sensors as long as sufficient sensors in the network are still operating properly. This concept is illustrated in Figure 6.6, with white representing noisy sensors with erroneous estimations, and black representing faulty sensors with no communication link (due to low power or hardware failure).

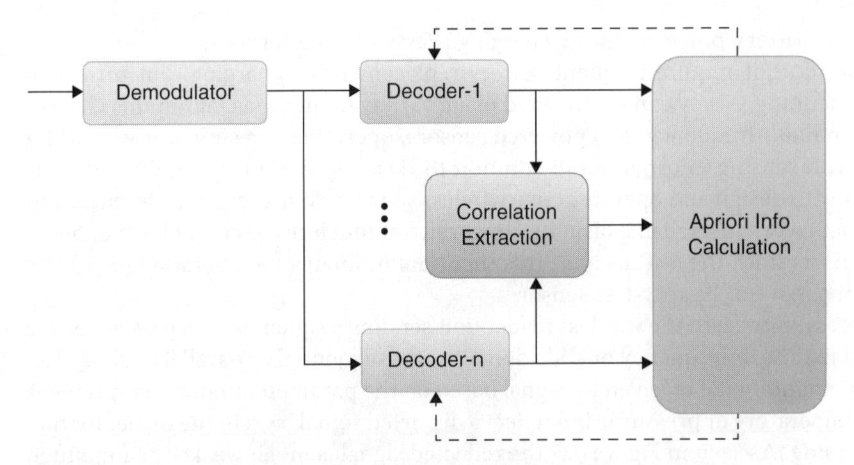

Figure 6.5 Joint decoding of correlated sensor data streams.

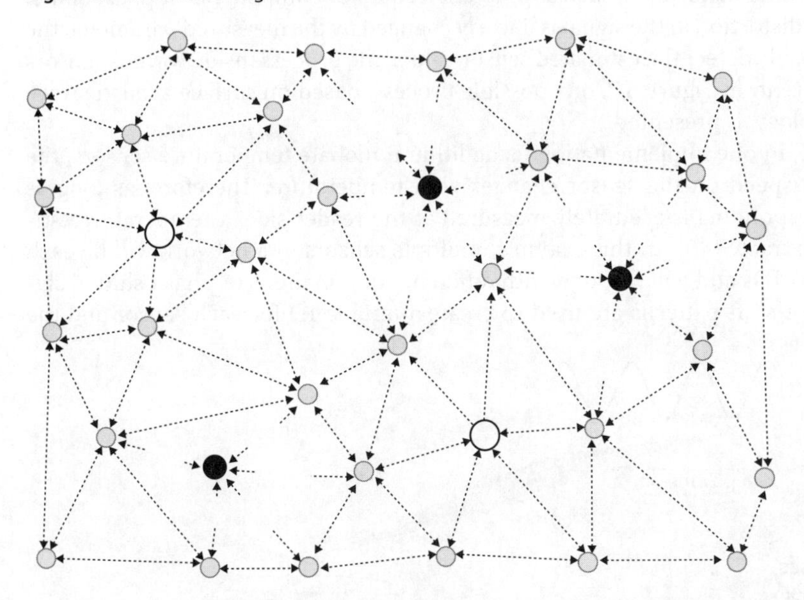

Figure 6.6 Noisy sensors (white) and faulty sensors (black) in a mesh sensor network.

Although it is hard to quantify the reliability of wired and wireless systems to accurately compare them, there is enough evidence to promise potentially highly reliable wireless networks as feasible alternatives for space applications. The savings in weight and volume, and hence cost, are attractive enough to entice researchers to move in that direction.

6.3 Active and Passive Wireless Sensors

Wireless sensors can be placed into two main categories: active (battery-operated) and passive (battery-less) [4]. Active sensors provide higher range and support more wireless standards, for example WiFi, Bluetooth, or ZigBee, but come up short in terms of the need for battery replacement or charging. Passive or battery-less sensors, on the other hand, do not require frequent battery replacement or charging, but provide a much smaller range. They can be powered using variety of methods, but in this chapter we focus on radio-frequency (RF) powered sensors, operating in a similar way to RFID tags, but with sensing information in addition to IDs. There is a large body of knowledge on how to design and operate active wireless sensors, to manage interference, and to make them work in a reliable fashion. However, not much research has been done on passive sensors until the last decade. This section illuminates the operating principles of a typical passive or battery-less sensor.

The process starts with the wireless reader unit sending a signal to each passive sensor (separated in time or frequency or code domains) for sequential or parallel reading. The sensors then modify the received RF signal based on the parameter under measurement (such as temperature or pressure) and reflect a distorted signal back to the reader for further processing. As seen in Figure 6.7, the reflected signal is much weaker in amplitude and may have other distortions, such as different frequency components or phase shifts. As long as the distortion in the signal is linearly changed by the measured parameter, the reader can reliably detect the estimated sensor value. The process inside various sensors might be different. In Figure 6.7, one possible process, based on surface acoustic wave (SAW) technology, is presented.

For instance, in one implementation using lithium niobate temperature sensors, the acoustic wave speed on the sensor changes with temperature. Therefore, as long as the delay in responding is accurately measured at the reader side, a temperature estimate can be extracted [5]. In this example, multiple sensors in a network will be each have a different IDs and will respond with different combinations of phase-shifted signals. The phase-shift patterns are used to create a unique ID for each sensor and the

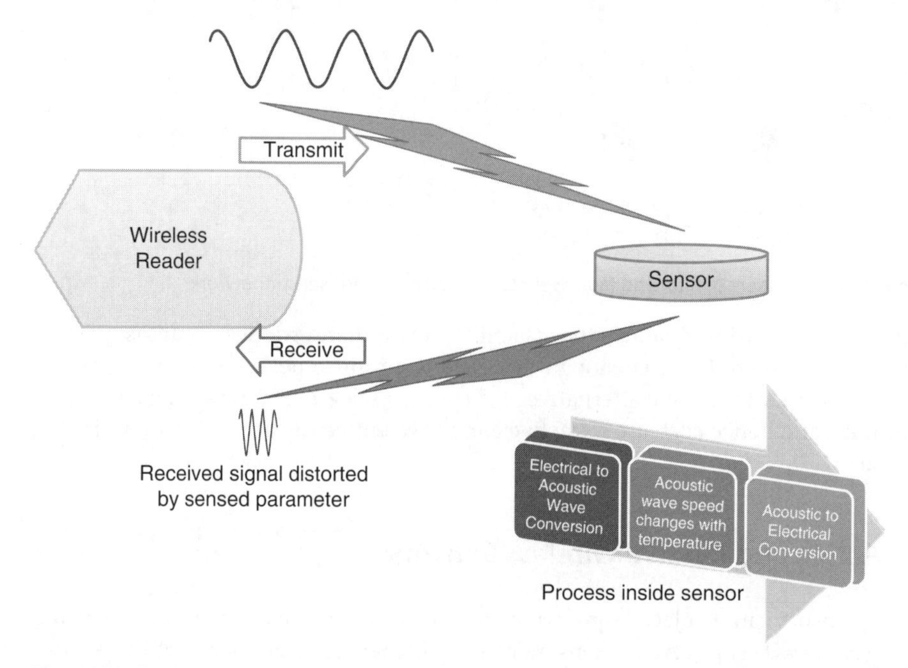

Figure 6.7 Passive sensor operation principle.

temperature-dependent delays represent sensed temperature. Several other implementations of passive sensors and how to design distinguishing codes for them can be found in the literature [6].

6.4 Design Considerations for Battery-less Sensors

This section is organized as follows. We first describe the material science behind sensor design. This is important in how the sensor substrate is designed and how the physical operation is carried out. The next part details code design to distinguish various sensor responses in a network. Principles of orthogonal code design from the digital domain are adopted for this analogue problem. Finally, we discuss the interference caused by multiple neighbor sensors on a single sensor and discuss some design techniques to prevent interference.

6.4.1 Sensor Material

Various materials may be used for sensor design. The foremost consideration is the sensitivity of the chosen material to the parameter under measurement. For instance, the velocity of acoustic waves in a SAW device built on a lithium niobate substrate changes with temperature if operated in the room-temperature range. If the same sensor needs to be used at high temperatures (above 1000 °C), the substrate needs to be changed to Langasite and coated to ensure its integrity at that temperature range [7]. If the same sensor is going to be used for chemical or biological sensing, a specific type of thin-film mask may be deposited on its surface to change the reflecting wave, based on absorption of the chemical or biological molecules under consideration. For instance, a humidity sensor based on polyvinyl alcohol and polyvinyl pyrrolidone thin films has been used [8]. It is noted that the sensor response should be reversible. In other words, if the chemical or biological parameter under measurement or the physical temperature or humidity affects the sensors and impairs its future measurements, the sensor will not be reliable for practical applications.

6.4.2 Code Design

A single sensor is usually not sufficient and cannot fulfil the engineering requirements. Multisensor networks require clear distinguishable differences between various sensor responses for obvious reasons. Traditional time- and frequency-domain separation of sensor responses, although powerful and effective, has a limiting effect on the number of sensors supported. For example, in a network with 100 sensors and a required sampling rate of 1 KHz (1000 samples per second), we only have 1 ms to sample all the sensors, meaning that delay differences between each sensor pair has to be around 10 μs, which is very difficult to implement in practice. Frequency-domain separation is also limited by available spectrum and the bandwidth of each sensor response. A block of spectrum with 6 MHz of free bandwidth can accommodate only six sensors with 1 MHz bandwidth requirements.

Adding a code or ID to each sensor adds yet another degree of freedom to a sensor network and multiplies the number of possible sensors by yet another large number. These digital codes will indirectly modulate the sensor response in such a way that each

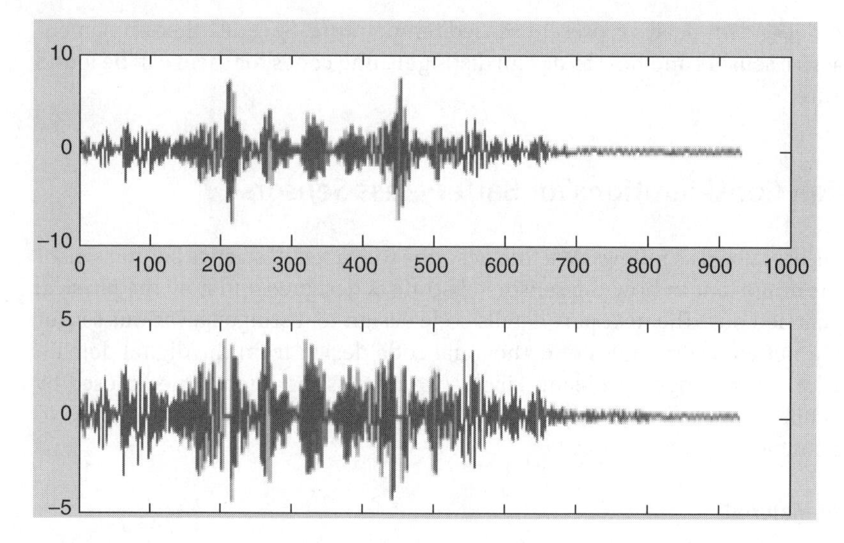

Figure 6.8 Response with 1 (top) and 5 (bottom) interfering sensors.

analogue response is uncorrelated to another one on the same frequency band and time slot. One example of such a design method is called quasi-orthogonal code design [9]. Two criteria are considered in this design methodology:

- the autocorrelation of each sensor response is maximized
- the cross correlation with other sensor responses in minimized.

These criteria are then used to solve a peak-to-side-lobe ratio maximization problem to assure simple separation of the sensor responses at the receiver upon reading all sensors. In this method and other variants of it in the literature, the main problem is interference from other sensors. The next section looks at this problem and presents some methods to avoid interference at source.

6.4.3 Interference Management

Interference from small side lobes of other sensors accumulates and eventually masks the peak of the main sensor signal. As seen in Figure 6.8, with one sensor interfering, the peaks of the response from the desired sensor are still detectable, but that is not the case when as few as five sensors start to interfere [10]. There are two different methods that can be used to minimize this interference effect. The first one [10] is called the *iterative interference method* at the receiver. The process involves a multi-filter bank that detects multi-sensor responses, sorts the output of all filters to determine the strongest signal, then removes that signal, repeating the process until all interference is eliminated.

An alternative method to alleviate the interference at source is presented in this chapter for the first time. This patent-pending method [11] applies a simple design feature to each sensor to reduce its effect on other sensors before the aggregate interference gets too large and unmanageable. The idea is to block the small cross-correlation response before leaving each sensor and only allow large peaks of autocorrelation to leave the sensor under interrogation. A simple diode or transistor operating as a hard limiter can modify each sensor at low cost and prevent all those small responses from reaching the main receiver.

As demonstrated in Figure 6.9, two simple PN-diodes with opposite polarities used in parallel will only allow signals above a specific threshold voltage to pass through the sensor output. This easily cleans up the response signal for each sensor. This method is low cost and is effective in controlling interference in battery-less sensors.

6.5 Summary

This chapter presents the benefits and design challenges of battery-less wireless sensors in space applications. Cost–benefit analyses of wireless and wired systems reveal the high potential for wireless systems to be a reliable alternative to current wired systems, while providing significant cost savings due to their reduced weight and their scalability. The main focus of this chapter is on passive or battery-less sensors for extreme environments such as space, where using batteries may not be feasible due to environmental conditions or cost.

Principles of operation in wireless systems are discussed in detail and their reliability from both physical and network perspectives are analysed. Error-correction codes and distributed decoding methods are presented as possible ways to enhance reliability in sensor networks in harsh and extreme environments where noisy sensor data is combined to extract reliable estimates.

The benefits of passive sensors with no batteries are described. Design challenges relating to materials science and code design are discussed. The main problem in most networks is interference. Two methods for interference elimination at source (sensor) and reader are presented.

This chapter presents some sample technologies, such as SAW, for implementing the proposed ideas, while a wide array of similar technologies using the proposed ideas and system-level perspectives can be used for developing applications for space and extreme environments.

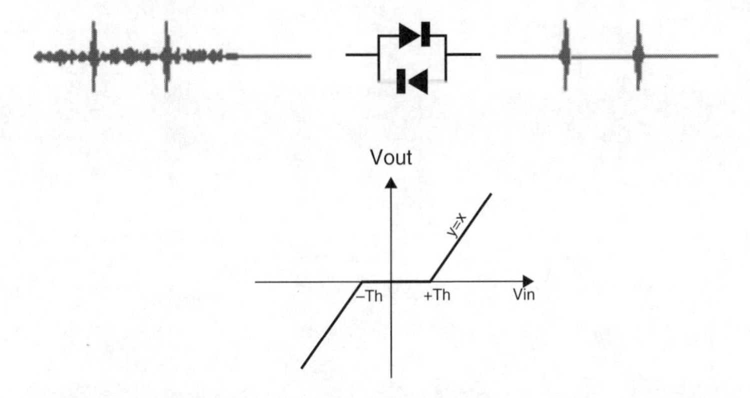

Figure 6.9 Eliminating interference at source [11].

References

1 G. Studor (2007) 'Fly-by-wireless: a revolution in aerospace vehicle architecture for instrumentation and control', *NASA Technical Report.*

2 B. Shen, A. Abedi (2007) 'A simple error correction scheme for performance improvement of IEEE 802.15.4,' *ICWN'07*, June 2007, Las Vegas, NV, pp. 387–393.

3 A. Razi, A. Abedi (2014) 'Convergence analysis of iterative decoding for binary CEO problem,' *IEEE Transactions on Wireless Communications*, 13 (5), 2944–2954.

4 A. Abedi (2012) 'Wireless sensors without batteries,' *High Frequency Electronics Magazine*, **Nov**, 22–26.

5 E. Dudzik, A. Abedi, D. Hummels, M. P. da Cunha (2008) 'Wireless multiple access surface acoustic wave coded sensor system,' *IET Electronics Letters*, 44 (12), 775–776.

6 A.T. Hines, D.Y.G. Tucker, J.H. Hines, J. Castro, A. Abedi (2012) 'Techniques for optimal DSSS code selection for SAW multi-sensor systems,' *IEEE International Frequency Control Symposium*, Baltimore, MD, May 2012.

7 M.P. da Cunha, R.J. Lad, P. Davulis, A. Canabal, T. Moonlight, S. Moulzolf, D.J. Frankel, T. Pollard, D. McCann, E. Dudzik, A. Abedi, D. Hummels, G. Bernhardt (2011) 'Wireless acoustic wave sensors and systems for harsh environment applications,' *IEEE WiSNet'11*, January 2011, Phoenix, AZ, pp. 41–44.

8 A Buvailo, Y Xing, J Hines, E Borguet (2011) 'Thin polymer film based rapid surface acoustic wave humidity sensors,' *Sensors and Actuators B: Chemical*, 156 (1), 444–449.

9 E. Dudzik, A. Abedi, M.P. da Cunha, D. Hummels (2008) 'Orthogonal code design for passive wireless sensors,' *QBSC'08*, June 2008, Kingston, Canada, pp. 316–319.

10 A. Abedi, K. Zych (2013) 'Iterative interference management in coded passive wireless sensors,' *Proceedings of IEEE Sensors 2013*, November 2013, Baltimore, MD.

11 A. Abedi (2013) *Systems and methods for interference mitigation in passive wireless sensors, US Patent 61/871,511, Filed: Aug 2013,* Patent Pending.

7

Contact Plan Design for Predictable Disruption-tolerant Space Sensor Networks

Juan A. Fraire, Pablo Madoery and Jorge M. Finochietto

Digital Communications Research Laboratory Electronic Department, Universidad Nacional de Córdoba, Córdoba - CONICET, Argentina

7.1 Introduction

Today, optical and radar images are acquired continuously from orbit. They have become a powerful scientific tool to enable better understanding and improved management of the Earth and its environment. Traditionally, a single space-based satellite sensor would gather data from sites across the world, including places too remote or otherwise inaccessible for ground-based data acquisition. This makes Earth observation from space an effective means of providing coverage across both space and time. Therefore, an orbiting network of spatially distributed autonomous wireless sensors (a space sensor network; SSN [1]) could open up new possibilities of unprecedented applications by significantly extending coverage in both dimensions. To achieve this, the nodes would need to rely on efficient protocols and algorithms to cooperatively pass sensed data through the network to the final destination on the ground [2].

However, over the past 20 years, space communications technologies have shown limited progress in comparison to Internet-based networks on Earth. Only recently, have NASA and other space agencies begun moving towards a packet-switched space communications architecture using appropriate protocols [3]. Unlike the Internet, the harsh conditions of space in which satellites have to operate, mean that designers face several situations non-existent in Earth communications: physical inaccessibility, a highly changing orbital dynamic, power limitations and hardware instability [1, 4]. Since existing protocols were not thought to perform under these conditions, the Consultative Committee for Space Data Systems [5] and a specific working group of the Internet Engineering Task Force [6] promises to throw light on these particular networks, which are known as delay-tolerant networks (DTNs). However, disruption is a particular case of delay, in which the delay is infinite. It is the prevalent effect in the low-Earth orbits (LEOs) where Earth-observation sensors are generally located. As a result, henceforth, we will refer to DTNs as disruption-tolerant networks.

Despite the recent and extensive research on DTNs, several challenges are to be overcome before operative DTN-based SSNs can be deployed in orbit. Here, we analyse the state of the art of design, planning, and implementation of the network communications opportunities (that is, the contact plan; CP). In general, the topological information

Wireless Sensor Systems for Extreme Environments: Space, Underwater, Underground, and Industrial.
First Edition. Edited by Habib F. Rashvand and Ali Abedi.

imprinted in a CP can be exploited by DTN nodes to optimize routing. To achieve this, we discuss typical spacecraft resource constraints, different contact-plan modeling techniques, possible contact selection criteria and existing contact plan design (CPD) schemes [7–10]. However, none of these exploits traffic information, which is fairly predictable for SSN applications. Indeed, in space missions data download and data acquisitions are generally scheduled in advance. Therefore, we integrate this information in a traffic-aware contact plan design (TACP): an overcoming mixed integer linear programing (MILP) model to solve the CPD for predictable disruption-tolerant wireless sensor networks operating in space environments [11]. In the chapter, we characterise, analyse and evaluate TACP performance and compare it with existing schemes, discussing the challenges of implementing it in practical SSN applications.

This chapter is organized as follows. In Section 7.1 we provide an overview of the evolution of traditional Internet communications towards disruption-tolerant schemes, and discuss the specific characteristics of these when implemented in space environments. Next, in Section 7.2, we introduce the problem of resource limitations and the derived need for CPD. An overview of existing methodologies is provided and TACPs are presented and described as an appealing CPD scheme for SSNs. A case study is set out in Section 7.3. This is used as a CPD benchmark to inform the discussion of the challenges of implementing efficient planning strategies in operative SSNs systems in Section 7.4. Finally, we draw the chapter conclusions in Section 7.5.

7.1.1 On the End-to-End Connectivity Paradigm

The Internet has enabled seamless, transparent and heterogeneous communication, thus allowing migration of centralised functions towards scalable and efficient distributed systems in fields such as banking or education. Back in the 1960s, the concept of the Internet grew out of military and academic studies on how to build robust networks. As a result, addressing and routing were decentralised but, most important, the primary purpose of the network was to remain connected after any catastrophe. Unknowingly, the Internet inherited an *end-to-end connectivity* paradigm that has shaped modern networked communications, including the popular TCP/IP [12] protocol stack currently supporting wireless sensor networks (WSNs). Despite WSNs being applied in an unprecedented range of sensing and actuating fields during the last decade [13], they might fail to operate efficiently in space and extreme environments [1]. This is because they assume:

- a persistent connection between the origin of the data and its destination
- low data-packet loss rate
- low end-to-end delay
- short maximum round-trip times between nodes
- low error rates.

Generating these Internet-like conditions in SSNs has led the space industry to have to face several problems in different orbital scenarios.

On the one hand, traditional geostationary (GEO) satellite relay systems implement bent-pipe repeaters to transmit from one location on Earth to the satellite, and back to another location on Earth. Therefore, GEO relays are appealing for broadcasting information to a large geographical area, but when considered for bidirectional and interactive data communication, challenges such as long round-trip times and frequent channel disruption must be addressed [14]. This effect is even more dramatic in deep-space (DS) systems, where longer distances lead to more delay and severe disruption, and planetary rotation making permanent and two-way Internet-like communications unfeasible [15]. In this context, and due to its conversational nature, the TCP protocol [12] simply cannot cope with the long delays, low bandwidth and errors typical of the space environment [16].

On the other hand, disruption is the prevalent problem in LEO satellite systems. For example, in order to provide voice services, the Iridium [17] satellite constellation system had to be designed with enough link margin to avoid disruption and sustain stable end-to-end multi-hop paths in a highly dynamic and extensive topology. To this end, permanent connectivity was achieved at the expense of a highly complex, expensive, and controversial system based on inter-satellite hand-off mechanisms. Most recently, the ambitious DARPA F6 [18] distributed spacecraft architecture project, aimed at deploying an Internet-like LEO mesh network, was cancelled due to significant increases in overall complexity and cost.

Whether by system complexity and cost in LEO systems, or physical unfeasibility in GEO and DS satellite systems, the end-to-end connectivity paradigm has proven hard to adapt to the highly dynamic and sparse space environment. This is a consequence of the fact that intermittent connectivity causes loss of packets, which in Internet-based protocols should be prevented by all means. In Internet communications, a TCP scheme might retransmit lost packets with lower data rates, but if the drop rate is severe, the session is lost and an error is reported to the user. As a result, TCP/IP [12] communications will always represent a technological frontier for disruptive WSNs such as those operating in space [19, 20]. Nevertheless, these types of challenged networks has become a recognised research area in computer networks and space communications, where they are known as DTNs [21]. DTN architecture [22] and the Bundle protocol derived from it [23] are designed to overcome the limitations of a persistent end-to-end connectivity. Despite DTNs promising to become the enabling technology for future sensor networks in space and others extreme environments, several challenges are stil to be addressed before they can be successfully deployed.

7.1.2 Disruption-tolerant Wireless Sensor Networks Overview

DTNs have received much attention during the last years as they have been proposed for several environments where communications can be challenged by either latency, bandwidth, errors or stability issues [21]. Originally developed as a network architecture for the Interplanetary Internet [15], DTNs have also been recognised as an alternative approach to building satellite applications [14], in particular to cope with the intermittent channels typical of LEO constellations systems with inter-satellite links (ISLs) [24].

In general, DTN architecture is designed to support interoperability with other networks (including non-disruptive ones) by translating between different communication protocols (that is, an overlay network) and to tolerate long disruptions and delays between and within those networks by exploiting intermediate storage, as illustrated in Figure 7.1. In providing these functions, DTNs accommodate the mobility and dynamics of time-evolving wireless communication devices operating on Earth, in underwater, aerial, space, and other extreme environments. In this type of disruption-tolerant WSNs, data is routed through network nodes not necessarily

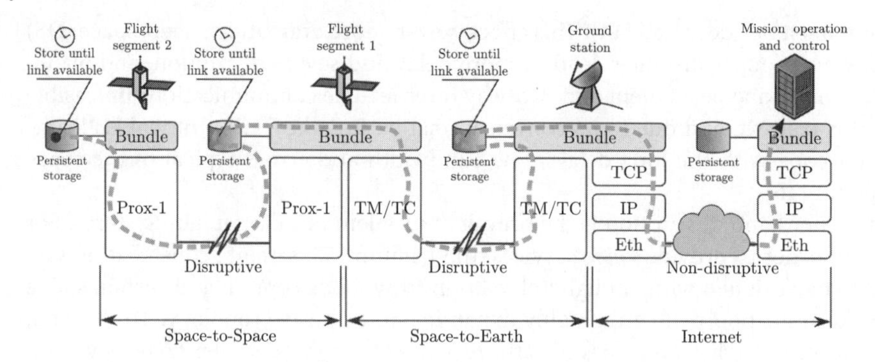

Figure 7.1 Bundle protocol as an overlay layer with storage to tolerate disruptions.

having end-to-end connectivity with the final destination [15]. To achieve this, DTNs overcome the problem of channel delay and disruption using a store-carry-and-forward message scheme, as illustrated by the dashed arrow in Figure 7.1. This is analogous to postal systems, where messages are stored in a given place (node) until they are able to move (forward) to another place (or places), before reaching their final destination. In contrast to to the very short-term storage (measured in milliseconds) provided by traditional Internet routers, DTN nodes require long-term persistent storage to hold messages for considerable periods (hours or even days) until a forwarding opportunity becomes available. In this regard, the Internet is a particular case of a DTN in which node buffer capacity is minimal and the delay between them is nonexistent for practical purposes.

Among the efforts to implement practical DTNs, the definition of a new communication protocol, which does not assume an end-to-end connectivity between source and destination nodes, has been addressed by the specification of the Bundle protocol in RFC 5050 [23]. The key capabilities of the Bundle protocol include:

- custody-based retransmission
- the ability to cope with intermittent links by using continuous, scheduled, predicted and opportunistic connectivity
- late binding of overlay network endpoint identifiers to constituent Internet addresses.

In order to route bundles (Bundle protocol data units), several routing mechanisms have been developed [25–28]. In particular, if topology changes are predictable, as in space environments, contact graph routing (CGR) [29] is appealing because it allows DTN nodes to take advantage of a-priori knowledge of forthcoming communications opportunities generated by the orbit of the spacecraft [30]; in other words, the CP. The overall procedure of CPs is depicted in Figure 7.2. Initially, a centralized mission and operation center determines, in advance, all nodes' forthcoming communications opportunities and imprints them in a contact topology (CT). Next, the CT can be further developed into a CP to satisfy SSN resource requirements by combining one or more optimization criteria. Finally, the CP is distributed to SSN nodes so they can become aware of the forthcoming contacts so as to derive efficient routing paths for the traffic they will generate. Since DTNs are expected to act as an overlay layer over heterogeneous underlying protocols, the Bundle protocol requires collection of protocol-specific convergence layer adapters (CLAs) to provide the necessary functions to carry bundles on

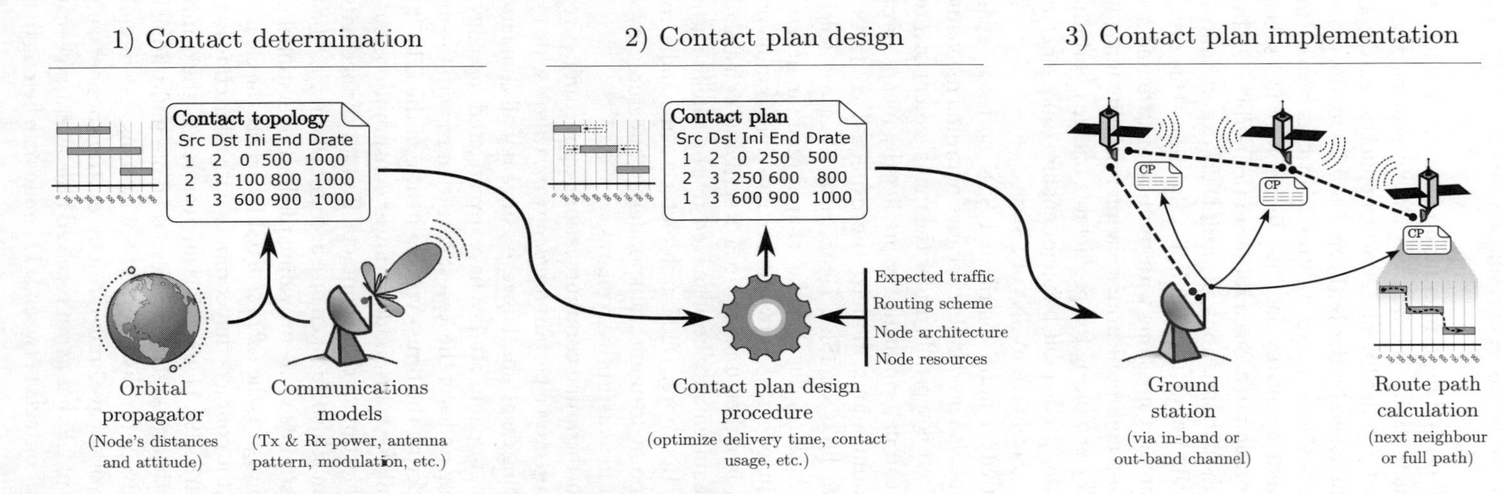

Figure 7.2 Creating, designing, and implementing contact plans for SSNs.

each of the corresponding protocols. As a result, a wide variety of CLAs have been specified and developed [31–33].

The extensive research on and experimentation with the Bundle protocol has led to several software implementations [34–36], NASA's Interplanetary Overlay Network (ION) [37] being the most popular for spaceborne applications. ION is an open-source (BSD-licence) project published by the Jet Propulsion Laboratory, and was one of the first DTN-capable protocol stacks successfully tested in a deep-space mission [38]. Also, the UK-DMC satellite built by Surrey Satellite Technology Ltd successfully validated the DTN2 [34] implementation in a Cisco router in LEO orbit [39]. In general, and due to the cost associated with space missions, in-orbit DTN experiments have mainly been based on single-spacecraft point-to-point link validation. However, and in general, DTNs are designed to support missions with multiple distributed sensors networked by opportunistic multi-hop communications via inter-satellite links, with data transfer performance optimized by exploiting the predictable nature of orbiting objects. As we describe in this chapter, this predictable behavior enables unique network planning and design opportunities for spaceborne WSNs or SSNs.

In particular, connectivity among distributed SSN nodes in space environments (contacts) are sporadic but also predictable because of orbital mechanics. In contrast to the Internet, SSN behaviour is typically under the management of a mission operation and control center that can deterministically predict (by means of orbital mechanics [40] and specific communications models) the expected contacts between nodes, either via Earth–satellite links or ISLs. Henceforth, ISLs are solely thought of as point-to-point communications, disregarding shared medium access schemes, which fail to perform properly in extensive networks because they assume physical adjacency of multiple nodes. This is either unlikely in a free-flying constellation or demands strict flight-formation requirements be managed by the satellite attitude and orbital control system [41]. Furthermore, DTN architecture can handle routing on higher layers, enabling simpler point-to-point transponder architectures, especially if mission requirements can be met in a disruptive scenario.

The set of all feasible forthcoming communications opportunities represents the CT. In general, the CT encompasses a certain *topology interval* on which nodes can rely to take routing and forwarding decisions. However, since not all contacts in the CT might be required to route data, previous studies have investigated topology designs for providing connectivity among nodes at the lowest cost (that is, with the minimum number of contacts) when considering the time-evolving nature of the CT [7, 8]. Furthermore, resource constraints (available transponders, power consumption, and so on) also need to be considered in the selection of these contacts. Besides connectivity, criteria related to capacity and fairness must also be taken into account when selecting the contacts that satisfy a given set of constraints while providing the best operational performance. We refer to this selection or design as the *CP design* (CPD), where the CP is the resulting subset of the CT comprising the contacts that complies with restrictions and maximizes the desired performance metrics [9]. Figure 7.2 is an overview of the complete procedure.

Among the challenges that need to be addressed before multi-hop DTN services can be fully implemented to support distributed SSN systems, the design of CPs is critical, because satellite networks have limited resources (transponders, power, fuel, and so on). However, the design of CPs has as yet received little attention because it is typically assumed that all potential contacts between DTN nodes can belong to the CP. Although the design of CPs with resource constraints has been examined recently [9, 10], the search for a generalised procedure continues, because the problem quickly becomes non-trivial in large-scale systems.

7.2 Contact Plan Design Methodology

In this section, we describe the methodology to efficiently design CPs for predictable disruption-tolerant WSNs. When the traffic to be generated is unknown, CPD procedures such as fair contact plans (FCPs) [9] and route-aware contact plans (RACPs) [10] can be implemented. FCPs aim for an efficient fair link assignation (single-hop) by taking advantage of a well-known matching algorithm [42], while RACPs improve on FCPs by exploiting heuristic techniques (simulated annealing) to optimize multi-hop routing metrics.

In this section we further improve on the efficiency of FCPs and RACPs by focusing on orbiting SSNs with predictable trajectories and scheduled traffic loads, which is the norm for space applications, where spacecraft instrumentation generates data periodically or by operator demand [11]. Then in Section 7.3, we show that the use of traffic information in a TACP allows us to further significantly improve the performance of orbiting disruption-tolerant SSNs. Figure 7.3 compares the expected performance of each CPD methodology according to the SSN information exploited by each of them.

To support our explanation of the TACP model, we introduce an example topology that we will use throughout this chapter to describe the modeling and the problem formulation. This particular SSN will also be used as a performance benchmark in Section 7.3. To this end, we evaluate four disruption-tolerant wireless sensors in polar orbit. These have the orbital parameters shown in Table 7.1, deliberately configured so that one node is ahead of the other on the trajectory vector (by a 5° perigee argument) with a slight variation of the right ascension of the ascending node angle. This scenario is of particular interest for Earth observation missions, where sensors on board the satellites have the maximum distance (coverage) of populated areas while approaching each other at the poles [43]. In these areas, contacts become feasible between adjacent spacecraft, producing a train-like formation, with a separation of 500 km, where two directive point-to-point antennas (placed in front and at the back of each satellite) can optimize the link budget, producing longer contacts. These communications opportunities can be exploited to transfer data between the orbiting sensors, ready for further download to a ground station.

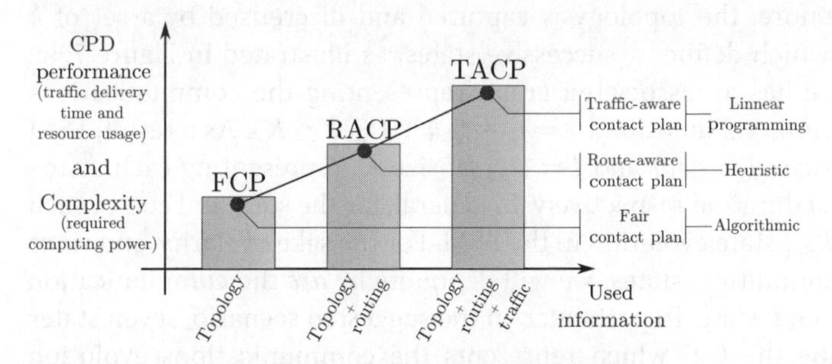

Figure 7.3 Contact plan design methodology performance.

Table 7.1 Case study time interval and orbital parameters.

Topology interval	Date and time
Start	1 Jan 2015, 00:00:00
End	1 Jan 2015, 03:22:36
Parameter	**Value**
Bstar coefficient (/ER)	0
Inclination (deg)	98°
RAAN (deg)	0°, 5°, 10° and 15°
Eccentricity	0
Argument of perigee (deg)	180°, 185°, 190° 195°
Mean anomaly (deg)	0°
Mean motion (rev/day)	14.92
Height (km)	600

Therefore, by using SGP-4 [40] (a well-known satellite propagator), with a maximum communication range of 700 km for the ISL, we can obtain a time-evolving topology suitable for DTN applications, as illustrated in Figure 7.4. In the proposed topology interval of 3 h 22 min 33 s, the constellation orbits four times over the poles, but for the sake of simplicity we only illustrate the first of these opportunities (through the North Pole). The modeling techniques used in Figures 7.4b and 7.4c are detailed in Section 7.2.1.

7.2.1 Delay-tolerant Wireless Sensor Network Model

In order to tackle the CPD problem, a disruption-tolerant SSN modeling technique needs to be specified. Consider the trajectory of the four-satellite network example shown in Figure 7.4a. The time-evolving nature of the contacts between the satellites can be captured by means of graphs, where the vertices and edges symbolise DTN nodes and links respectively. In other words, this representation can be thought as a finite state machine (FSM), where each state is characterised by a graph whose arcs, in turn, represent a communication opportunity during a specific period (in other words, a contact). Therefore, the topology is captured and discretised by a set of k time intervals $[t_k, t_{k+1}]$ which defines k successive states, as illustrated in Figure 7.4a. As illustrated, each state has an associated graph representing the communications opportunities within its interval duration ($i_k = t_k - t_{k-1} : 1 \leq k \leq K$). As a result, FSM can be encoded in matrices $T = \{t_k\}$ and $I = \{i_k\}$ of size K, representing each state's starting time and interval duration respectively. In general, for the start and end of each contact, there is a k_a to k_{a+1} state evolution in the FSM. For the sake of clarity, and since a single contact can span multiple states, we will denonote by *arc* the communication opportunity within a given k state. In particular, in the suggested scenario, seven states are sufficient to describe the CT, which represents the communications evolution during the first half-orbit topology interval. The FSM model of the example network

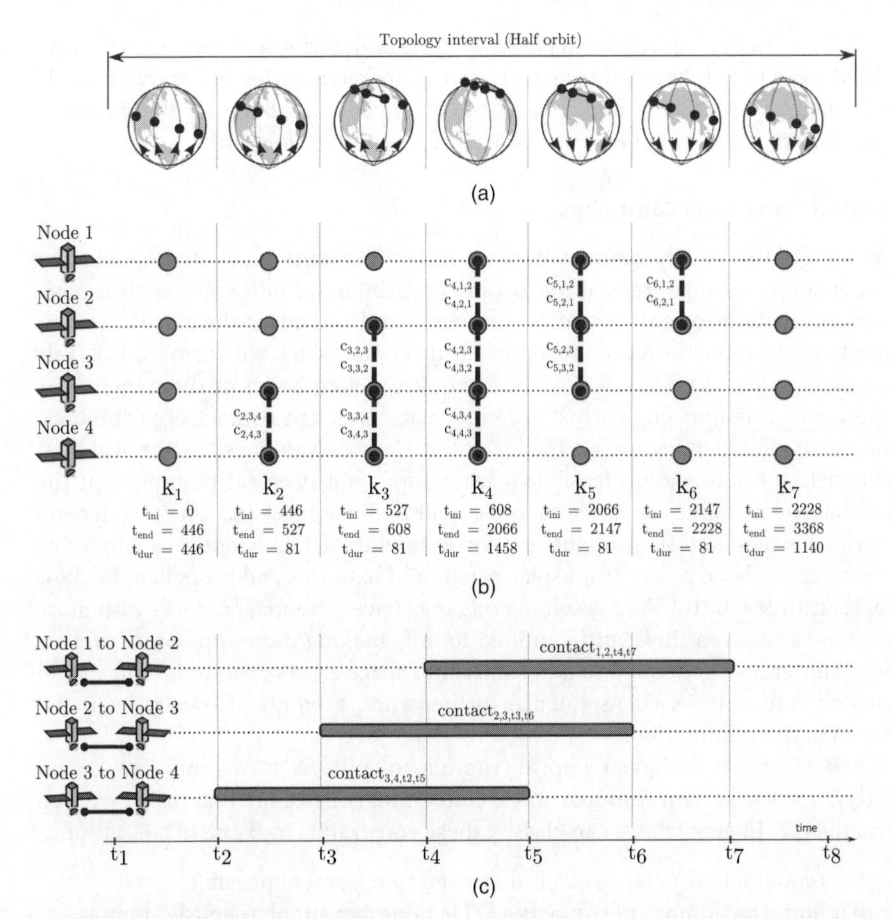

Figure 7.4 Example of a DTN satellite network: (a) trajectories; (b) modeled with finite state machine; (c) modeled with contact list.

is illustrated in Figure 7.4b and the modeling technique used for the formal definition of the CPD process is described in Section 7.2.3. It is worth noting in this figure that the state k_4 represents the train-like formation over the pole (N_1 to N_2, N_2 to N_3 and N_3 to N_4), which has a duration of 1458 seconds.

Alternatively, a topology can be represented by a contact list (CL) where each contact is defined by a source, destination, start time, and stop time (as in $\text{contact}_{1,2,t4,t7}$). Therefore, the first half-orbit of the example network basically consists of three contacts:

- N_1 to N_2 from t_4 to t_7
- N_2 to N_3 from t_3 to t_6
- N_3 to N_4 from t_2 to t_5.

Consequently, the CL model for the example topology, illustrated in Figure 7.4c, is more compact than the FSM since it can be expressed as a contact table instead of a adjacency matrix. Therefore, CL is the format adopted for the ION [37] DTN stack for efficient CP distribution and storage. However, for CPD and engineering, the FSM model granularity might be more convenient to work with, especially when applying MILP optimization techniques [9], as in Section 7.2.3. Finally, as we further describe and discuss in Section 7.4, the FSM model can take advantage of discrete state fractionation in order to provide a more detailed and precise topology description. In general, no matter the modeling technique chosen, translation between FSM and CL is straightforward.

7.2.2 Contact Plan Design Constraints

In the initial CPD phase, communication-subsystem attributes such as transmission power, modulation, bit-error rate, and so on and the orbital dynamics [40], such as position, range and attitude (orientation of the spacecraft and antenna in the inertial system) can be used to determine the feasibility of the future contacts that will form the CT. This technique was applied in Section 7.2 for the example topology and is no different to how single-spacecraft missions currently determine space-to-Earth contact opportunities. Nevertheless, as shown in Figure 7.2, the CT at this stage only defines the physical (that is, RF channel) communication feasibility, which does not necessarily imply that the spacecraft has the resources required to implement it. Since space launch costs depend on the payload weight and size, satellite platforms tend to be highly optimized in terms of architecture, available power from solar panels and batteries and propellant load for station and attitude control. As a result, a node may have potential contacts with more than one node at a given time but be limited to only making use of one of them. Furthermore, interference generated to and from other space assets need to be considered and evaluated in the CP. As a consequence, further work is required to design a CP that considers all of these limitations.

In order to effectively design a CP for disruption-tolerant SSNs, we enumerate, classify and describe the system resource and architectural constraints that might have an impact on the CT. In general, we can classify these constraints into one of two groups:

- those that render a particular contact (in a given timeframe) infeasible
- those that limit the number of contacts a DTN node can simultaneously support.

We refer to the former as *time-zone constraints* (TZ constraints) and to the latter as *concurrent-resources constraints* (CR constraints), where both can relate not only to communications but to general system operational issues.

7.2.2.1 Time-zone Constraints

In general, TZ constraints are those that can forbid communications in a specific geographical area or time to prevent interference or for other agency-specific reasons. As LEO SSN constellation systems orbit over wide geographical areas, complying with international regulations can be challenging. Moreover, as shown in Figure 7.5, since ISLs in LEO constellations are held tangentially with respect to the Earth's surface, GEO satellites can experience interferemce when LEO nodes orbit in polar areas [44]. Some GEO satellites deserve specially consideration becauses they support manned-mission communications, for example for the International Space Station. As a consequence, a proper irradiation policy must be considered so as not to generate (or experience) interference beyond the International Telecommunications Union's recommendations [45].

Furthermore, many other agency-specific reasons might exist regarding irradiating in a particular geographical area. This can be addressed by time and zone constraints that

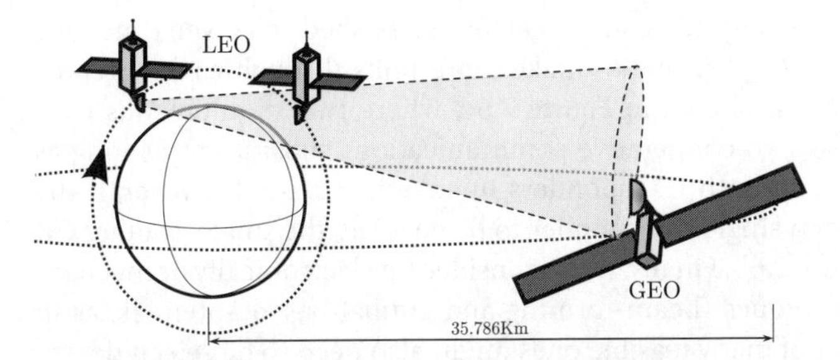

Figure 7.5 Interference generated by ISLs affecting GEO satellites when orbiting over the pole.

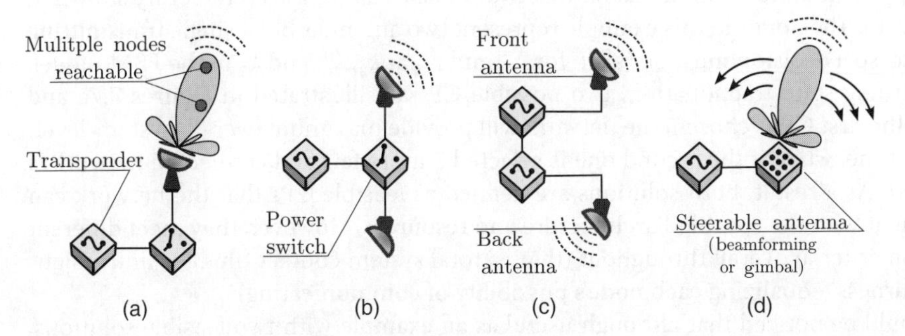

Figure 7.6 Satellite architecture: (a) with multiple target nodes; (b) with a power switch; (c) with two transponders; (d) with a steerable antenna.

prevent the availability of a contact in the corresponding CP in a given topology interval. Consequently, emission in specific countries or regions, among other agency-specific prohibitions, can be avoided by directly disabling the conflicting contacts during the conflict time in the CT. However, and in general, interference with geostationary satellites is not only measured in signal energy, but also in the percentage of time that the signal reaches the interfered node [44, 45]. Therefore, selecting which contact to disable and when to do so is a feasible strategy for CPD. Deriving particular TZ constraints is out of the scope of this chapter.

7.2.2.2 Concurrent-resources Constraints

On the other hand, CR constraints are not as straightforward as TZ constraints, because they usually involve solving a combinatorial problem involving architectural or power limitations in the spacecraft bus. The simplest CR constraint is when a spacecraft antenna radiation pattern reaches two or more neighbors, as shown in Figure 7.6a. In this case, multiple access schemes could automatically negotiate the channel usage, but given the low frequency of occurrence of this scenario in a highly sparse environment like space, and the overhead imposed by traditional negotiation procedures at inter-satellite distances, implementation of such schemes is discouraged in general. Therefore, efficient usage of point-to-point links requires the system operator to decide in advance which communications should be established.

When a satellite platform is expected to accommodate ISL links from different directions, more than one antenna might need to be placed in the structure. For example, consider Figure 7.6b, where a power switch allows a particular transmitter antenna to be selected for use. In this case, only one contact can be established at a given time (one belonging to the CP) even if it is feasible to make more links through each antenna. A more complex architecture is shown in Figure 7.6c, where two simultaneous contacts can be implemented by two cooperative communications subsystems as long as the power supply can maintain both transponders in an active state. However, if the power budget only allows for a single transponder to be enabled, the single-contact CR constraint remains as in Figure 7.6b. Finally, when considering electronically or mechanically steerable antenna techniques (beam-forming and gimbal-based antennas, as in Figure 7.6d), one contact out of many feasible ones might also need to be selected.

In general, CR constraints require a selection process in the CPD phase. To illustrate this, suppose that the example satellite network makes use of the architecture shown in Figure 7.4b. The nodes in this example represent two antennas but a single transmitting resource, so a decision must be taken for N_2 and N_3 at k_3, k_4, and k_5 in the FSM model. Disregarding state fractionation, two possible CPs are illustrated in Figures 7.7a and 7.7b. If the first CP is chosen, the network will provide maximum overall (system-level) contact time, while if the second one is selected a more fair and connected network is obtained. As a result, both solutions are defined as feasible CPs that the network can implement with the specified architecture and resources. However, they meet different selection criteria: overall throughput (that is, total system contact time) or link assignment fairness (equalizing each node's possibility of communicating).

It should be noticed that although useful as an example with two feasible solutions, the longer the topology interval, the more states, and the more nodes, antennas, transponders, and concurrent resources constraints, the more complex the CPD becomes. In general, this results in a non-trivial combinatorial problem with exponentially increasing complexity. This must be solved by the network planner before defining and distributing the final CP. Furthermore, the CPD might be driven by other, more complex selection criteria that take into account not only single-hop considerations as in the present example, but also multiple-hop routing paths (the broken arrow in Figure 7.7). Since several selection methodologies could be required to find the most appropriate CP, the complexity of such procedures might be challenging for the average network operator. Therefore, in Section 7.2.3 we propose a MILP model that is capable of automatically determining suitable CPs, supporting connectivity and data transfers among DTN nodes by exploiting the traffic predictability of SSNs.

7.2.3 MILP Formulation of the Contact Plan Design Problem

In general, the problem of CPD lies in selecting among those contacts that satisfy the communications opportunities represented in the CT, meet TZ and CR constraints, and optimize the topology for a given criterion. In the particular case of space applications, network traffic is generally deterministic, being generated by operator request (instrument or payload acquisition) while the system topology can be predicted by combining high-precision orbital propagators [40] and transponder, antenna and channel models. Under these assumptions, we propose a traffic-aware approach to CPD: a MILP formulation of the CPD problem for these highly deterministic yet complex scenarios. The TACP model assumes a global knowledge of both the CT (including contact capacity) and the forthcoming traffic in the network. As a result, the output of this MILP formulation is an optimal congestion-free routing-and-forwarding assignment that minimizes

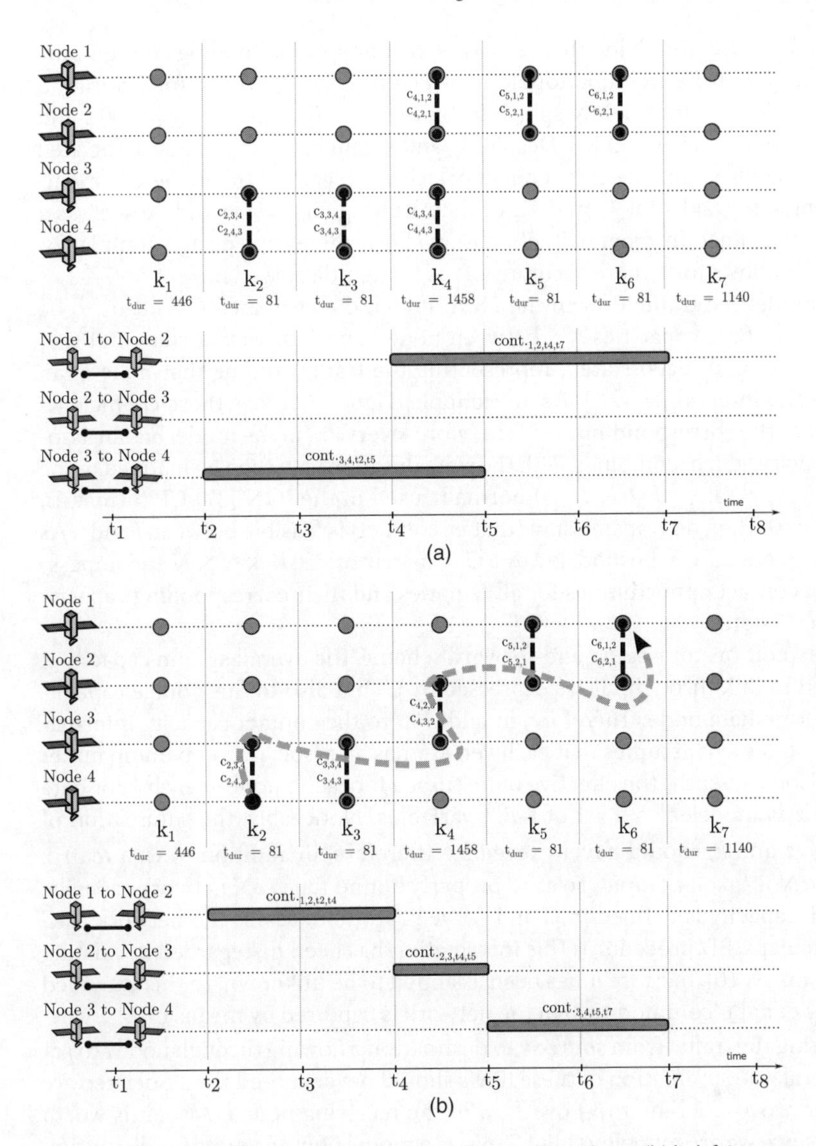

Figure 7.7 Two possible contact plans: (a) maximum throughput; (b) link fairness.

overall delivery delay for the specified traffic while providing the optimal communications interface and resource selection. Indeed, an efficient CP can be directly derived and implemented from this optimal flow assignation by enabling contacts carrying data.

In its core, the TACP formulation represents the topology as a function of time using an FSM representation (as illustrated in Figure 7.4) of the well-known multi-commodity flow problem. Therefore, the MILP formulation assumes the topology is encoded in matrices $T = \{t_k\}$ and $I = \{i_k\}$, as explained in Section 7.2.1. When using FSM for modeling DTNs, state fractionation might give a useful modeling tool before applying the MILP model. Fractionation implies that any k_a state can be deliberately divided into two new states k_b and k_c (implying that $t_{k_a} = t_{k_b} + t_{k_c}$) for specific topology design purposes. For example, in the example topology of Figure 7.4b, k_4 with a total duration of $i_4 = 1458$ s could be conveniently split into three states $i_{4a} = 500$ s, $i_{4b} = 500$ s and $i_{4c} = 458$ s as proposed in Section 7.3. Despite k_b and k_c still having the same associated graph, a CPD procedure can take independent decisions on each of them, therefore generating different sub-graphs for k_b and k_c. In the resulting CP, this would be seen as a contact shortening effect. In general, fractionation allows increasing design granularity, which essentially allows for a more accurate CP, as further discussed in Section 7.4.1.

In order to model the traffic flow in the FSM model, the evaluated CT needs to be expressed as a set of arc capacities $c_{k,i,j}$ between node i and j for each k state. In other words, there is a $c_{k,i,j}$ for each state k representing the traffic volume that node i can transfer to j in the interval $[t_{k-1}, t_k]$. As an example, Figure 7.4b has these coefficients illustrated next to the corresponding arcs. Therefore, every arc in the model has an associated data capacity which, combined with the state duration, can then be mapped to the link data-rate ($c_{k,i,j} = \text{rate}_{i,k} * (t_k - t_{k-1})$) normally used in the ION [37] CP format. In particular, $c_{k,i,j} = 0$ when no transmission (that is, contact) is feasible between i and j. As a result, the complete CT can be modeled by a $C_{k,i,j}$ matrix of size $K \times N \times N$ encompassing the existing contact opportunities for all N nodes and their corresponding capacity discretised into K states.

Since DTN exploits a store-carry-and-forward scheme, the overall system capacity is not only related to link throughput (as expressed by C) but also to the storage capability of each intermediate node. Therefore, in addition to the contact capacity information, the MILP statement assumes that each vertex i has an associated maximum buffer capacity of b_i. Consequently, the effective buffer usage for each i node at each k for data sent from y to z is modeled in a set of a $B^{y,z}_{k,i}$ variables. Noticeably, the summation of all $B^{y,z}_{k,i}$ for all y, z and k, should never exceed b_i. Therefore, in addition to C, a matrix $B = \{b_i\}$ of size N is taken as input, so as to properly bound the DTN network capacity.

Once the CT capacity is defined in C and B, we propose to assess the network traffic in this particular CPD procedure. This information has been disregarded in existing CPD mechanisms [9, 10], because it has been assumed to be unknown. In this extended model, the flow of data (commodities) in the network is captured by means of $X^{y,z}_{k,i,j}$ variables representing the traffic from source y to destination z flowing through node i to j at state k. In general, the summation of these flows should never exceed the associated arc capacity $c_{k,i,j}$ or provoke a buffer (b_i) overflow in the receiving node i. Also, it is worth clarifying that since we are modeling LEO SSNs, the model only accounts for disruption disregarding delays in the network ($X^{y,z}_{k,i,j}$ flow is assumed to arrive instantly at node j). However, the model can easily be generalised to incorporate delay by including a delay parameter in each contact [46].

Each of the resulting flows $X^{y,z}_{k,i,j}$ might be triggered to evacuate the source flow expressed in an input traffic matrix D (where $D = \{d^{i,j}_k\}$), which is known in advance for this type of DTN. In particular, such a traffic plan is formed by a set of $d^{i,j}_k$ traffic volumes representing data to be generated at time t_k at node i and with j as the final destination. Indeed, as we discuss below, the multi-commodity flow-problem constraints will allow the efficient routing of this traffic. It should be noted that the MILP model as formulated allows for multiple data-generation points throughout the topology time, even for the same source and destination tuple (i, j) in different states.

When traffic generation is expected to happen within a given k, fractionation can be implemented to precisely model the creation time of the flow. Finally, the units of C, B, D and $X^{y,z}_{k,i,j}$ should all be the same—typically bits, bytes, or packets—if their size is constant. Also, if node data-rates are equal, traffic volume can be directly mapped to time (channel-access time), which can also be used as the unit.

Once the CT is defined in C, the buffer capacity in B, and the traffic plan in D, the model should be capable of providing an optimal flow assignment to each of the $X^{y,z}_{k,i,j}$ variables. At this stage, the MILP model can be used as a centralised routing scheme, as further discussed in Section 7.3.1. However, resource restrictions (previously detailed in Section 7.2.2) need to be applied before calculating the final CP. In order to model these resource constraints, we use another matrix $P = \{p_i\}$, whose p_i components encode the maximum quantity of communication ports that node i can simultaneously implement at a given time. Such a flexible expression of feasible port usage is a unique property of this MILP model statement, since previous CPD procedures such as FCP [9] assumed the interface limitation was always $p_i = 1 \quad \forall i$. In order to model the combinatorial nature of the interface selection, we include a set of binary variables $Y_{k,i,j}$, which are meant to adopt a binary value of 1 if the port is used and 0 otherwise.

As a result, the TACP MILP model can be used to select the port in each node that will optimize the delivery of the input traffic matrix D, through a CT expressed in C, with storage B, while satisfying a port-usage constraint modeled in P. In consequence, the formulation delivers the expected traffic flows in $X^{y,z}_{k,i,j}$, the buffer allocation required to store the information in $B^{y,z}_{k,i}$, and the optimal port selection in $Y_{k,i,j}$. The stated coefficients required as input for the MILP formulation and the resulting variables are summarised in Table 7.2.

Finally, the overall goal of the TACP model is to use the minimal and earlier arcs to accommodate the flow detailed in D. Therefore, and considering all the coefficients and

Table 7.2 TACP MILP model parameters.

Input coefficients	
N	Nodes quantity
K	Topology states quantity
$T = \{t_k\}$	State k initialization time ($1 \leq k \leq K$)
$I = \{i_k\}$	State k interval duration ($i_k = t_{k+1} - t_k\ :\ 1 \leq k \leq K$)
$C = \{c_{k,i,j}\}$	Capacity of i to j contact at state k ($1 \leq k \leq K$ and $1 \leq i,j \leq N$)
$B = \{b_i\}$	Node i buffer capacity ($1 \leq i \leq N$)
$D = \{d^{i,j}_k\}$	Traffic from i to j originated at k ($1 \leq k \leq K$ and $1 \leq i,j \leq N$)
$P = \{p_i\}$	Number of simultaneous ports in node i ($1 \leq i \leq N$)
Output variables	
$\{X^{y,z}_{k,i,j}\}$	Traffic from y to z at state k in i to j arc ($1 \leq i,j,y,z \leq N$)
$\{B^{y,z}_{k,i}\}$	Node i buffer occupancy by traffic from y to z ($1 \leq i,y,z \leq N$)
$\{Y_{k,i,j}\}$	Interface selection from i to j at state k ($1 \leq k \leq K$ and $1 \leq i,j \leq N$)

variables described, the problem can be formally expressed as follows:

$$\text{minimize:} \quad \sum_{k=1}^{K}\sum_{i=1}^{N}\sum_{j=1}^{N}\sum_{y=1}^{N}\sum_{z=1}^{N} w(t_k) * X_{k,i,j}^{y,z} \tag{7.1}$$

Subject to:

$$\sum_{j=1}^{N} X_{k,j,i}^{y,z} - \sum_{j=1}^{N} X_{k,i,j}^{y,z} = B_{k,i}^{y,z} - (B_{k-1,i}^{y,z} + d_k^{i,z}) \quad \forall k, i, y, z \tag{7.2}$$

$$B_{k,i}^{y,z} <= b_i \quad \forall k, i, y, z \tag{7.3}$$

$$B_{0,i}^{y,z} = 0 \quad \forall i, y, z \tag{7.4}$$

$$\sum_{y=1}^{N}\sum_{z=1}^{N} X_{k,i,j}^{y,z} <= c_{k,i,j} \quad \forall k, i, j \tag{7.5}$$

$$\sum_{k=1}^{K}\sum_{j=1}^{N} X_{k,i,j}^{y,z} = \sum_{k=1}^{K} d_{k,i,z} \quad \forall i = y, z \tag{7.6}$$

$$\sum_{k=1}^{K}\sum_{i=1}^{N} X_{k,i,j}^{y,z} = \sum_{k=1}^{K} d_{k,y,j} \quad \forall y, j = z \tag{7.7}$$

$$\sum_{j=1}^{N} Y_{k,i,j} <= p_i \quad \forall i, k \tag{7.8}$$

$$\sum_{j=1}^{N}\sum_{y=1}^{N}\sum_{z=1}^{N} X_{k,i,j}^{y,z} <= M * Y_{k,i,j} \quad \forall i, k \tag{7.9}$$

The objective function (7.1) aims at minimizing the sum of the product of data units ($X_{k,i,j}^{y,z}$) with the time associated to the k state (t_k) modified by a weighting function $w(t_k)$. The higher the weighting, the more importance for the usage of earlier arcs; the lower $w(t_k)$, the less usage of arcs throughout the topology. For example, consider the simple scenario illustrated in Figure 7.8, where a single flow of $d_1^{4,1} = 10$ is expected to be delivered from N_4 to N_1 with two paths feasible, as shown in (a) and (b). If the $w(t_k)$ coefficient is used with $w(t_k) = t_k$, case (b) minimizes the objective function to 300. However, despite being an efficient solution in terms of arc quantity, it does not provide the earliest delivery time possible. If delivery time is to be strictly optimized, several arcs in k_1 and k_2 ought to be selected instead of a single one in k_3; to achieve this effect, the $w(t_k)$ coefficients might need to be further raised. In general, a $w(t_k) = K * t_k^2$ is enough to guarantee an earliest delivery time; 17000 s and 27000 s being the result of the objective functions for Figures 7.8a and 7.8b respectively. It is worth emphasising that t_k^2 is a coefficient and not a variable operation, so the model remains linear. Also, as noted, the value of the $w(t_k)$ coefficient might increase drastically and must be observed in order to avoid overflows in the MILP solvers. To summarise, the TACP MILP model as stated has a mixed yet customisable objective of optimizing the delivery time of the maximum traffic volume while minimizing the contact usage.

Among the constraints, (7.2) maps the flow imbalance in each node i to its buffer variation for all states and each (y, z) flow. Also, $d_k^{i,z}$ is included in the imbalance modeling the

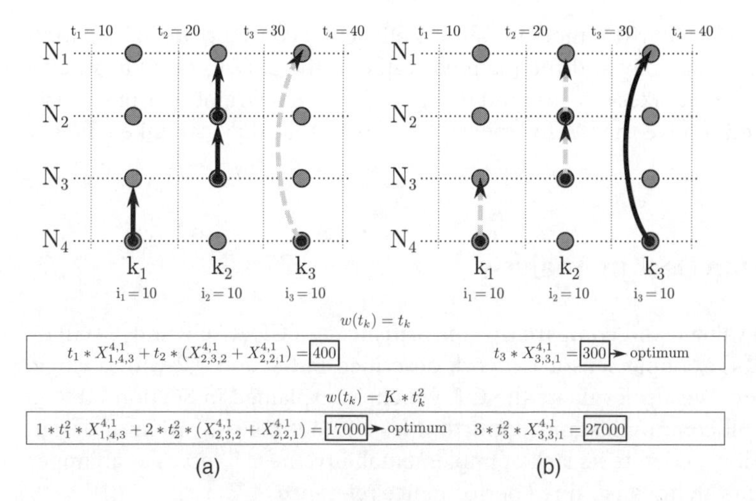

Figure 7.8 MILP $w(t_k)$ coefficient in objective function.

spontaneous generation of traffic in i with destination z. Therefore, $d_k^{i,z}$ is either transmitted (increasing $X_{k,i,j}^{y,z}$) or stored in the local buffer (increasing $B_{k,i}^{y,z}$). Equations 7.3 and 7.4 impose an upper bound b_i and an initial empty condition $B_{0,i}^{y,z} = 0$ on the buffer capacity for each node i and y, z flow respectively. For upcoming states $k > 0$, buffer occupation will increase both by the generation of $d_k^{i,z} > 0$ or the reception of flows from other nodes $X_{k,j,i}^{y,z}$ and will decrease by their transmission. The maximum capacity for each arc is enforced in (7.5). In particular, if all nodes transmit at the same data-rate, all arcs spanning a given contact will have the same $c_{k,i,j}$ value. Next, all source and sink outflow and inflow imbalance is generated by (7.6) and (7.7), respectively. In other words, these equations force the transmitting node to effectively send all generated data and the receiver node to sink it. Also, (7.2) prevents sending a given $d_k^{i,z}$ before the specific k where it is supposed to be generated. As a result, (7.2–7.6) model the CT and buffer capacity limitations and allow the traffic in matrix D to be optimally routed to the final destination.

Equations 7.8 and 7.9 impose the resource constraint of the maximum quantity of simultaneous port use in a given node i. In particular, (7.8) verifies that the summation of the binary variables $Y_{k,i,j}$ meets the p_i bound. On the other hand, (7.9) connects the port selected with the $X_{k,j,i}^{y,z}$ commodities flowing through it. Therefore, if a port is not selected, neither should the corresponding forward or return traffic in the associated arc. To this end, a well-known MILP modeling strategy known as 'big-M' is used in this equation. When a given port is enabled, a sufficiently large coefficient M multiplies the binary $Y_{k,i,j}$ so as to permit the left-hand side of the equation, with the flows associated with that contact, to rise without limit. On the other hand, when a port is disabled, $M * Y_{k,i,j} = 0$, forcing all $X_{k,j,i}^{y,z}$ on the other side of the equation to be 0. To achieve this behaviour, the M coefficient must be bigger than the summation of all the possible traffic flowing through that port ($M > c_{k,i,j}$). Despite (7.9) allowing the model to include a valid mean to select a given port quantity, the big-M approach is known to provoke numerical instability problems, especially for $M >> \sum X_{k,j,i}^{y,z}$. Therefore, M should be carefully chosen to satisfy the requirements of the equation while still being small as possible.

To summarise, the TACP MILP model is able to efficiently combine the information about the predictable topology and the planned traffic of the SSN to provide an efficient flow assignation that takes into account the CPD design constraints. As previously stated, it can be used to give the CP for the SSN nodes so that they can take efficient routing decisions.

7.3 Contact Plan Design Analysis

In this section we evaluate and compare the performance of FCP, RACP and TACP by applying them to the example sensor network described in Section 7.2 with a single usable port per node. We also evaluate the CT, which, as explained in Section 7.1.2, is the aggregation of all communications opportunities without consideration of design or resource restrictions. Despite its lack of implementability, the CT serves as an upper bound and represents an unconstrained performance reference.

7.3.1 Case Study Overview

It should be noticed that the topology interval considered for the example of a disruption-tolerant SSN (3 h 22 min 36 s) allows for four over-the-pole fly-bys (once every 48 min) of the constellation, which essentially makes the pattern illustrated in Figure 7.4 repeat four times. Also, as previously explained in Section 7.2.3, in order to give higher granularity and accuracy in the design, topology states longer than 500 s ($i_k > 500$) are further partitioned into sub-states. In the example topology, the fractionation effect is illustrated in the lower part of Figure 7.9, where state k_4 is subdivided into k_{4a}, k_{4b} and k_{4c} with $i_{4a} = 500$ s, $i_{4b} = 500$ s and $i_{4c} = 458$ s. It is worth noticing that since the complete topology interval accounts for four iterations of this over-the-pole communication scenario, k_{10}, k_{16} k_{22} (seen in the upper part of Figure 7.9) are also fractionated.

In order to evaluate and compare the different CPD procedures, a traffic model needs to be considered as the candidate mechanism to route data through the resulting CP. To this end, we reuse the FSM formulation of the multi-commodity flow problem, as explained in Section 7.2.3. In particular, an optimal flow assignation can be obtained by feeding the obtained CP into the MILP model, but without the interface restriction expressed in (7.8) and (7.9). As further discussed in Section 7.4.3, these routing decisions are optimal and based on a global view of the network that does not necessarily appertain to a real distributed scenario in which forwarding decisions are based on each node's local and limited view of the topology. In general, the lack of a global view of the topology is a DTN routing-specific issue that results in unwanted traffic bouncing and congestion [47], discussion of which is out of the scope of this research and which is therefore disregarded for the SSN CPD procedure.

To complete the description of the case study, we now consider the traffic sources. Data is configured to be equitably generated in all N_2, N_3 and N_4 nodes at the beginning of the topology (k_1). It then flows towards N_1, which is expected to later deliver the data by means of a space-to-Earth high-speed downlink transponder. This behaviour resembles a simultaneous acquisition in the equator area and its consequent accumulation in a single node for later download. In particular, communications systems are constrained

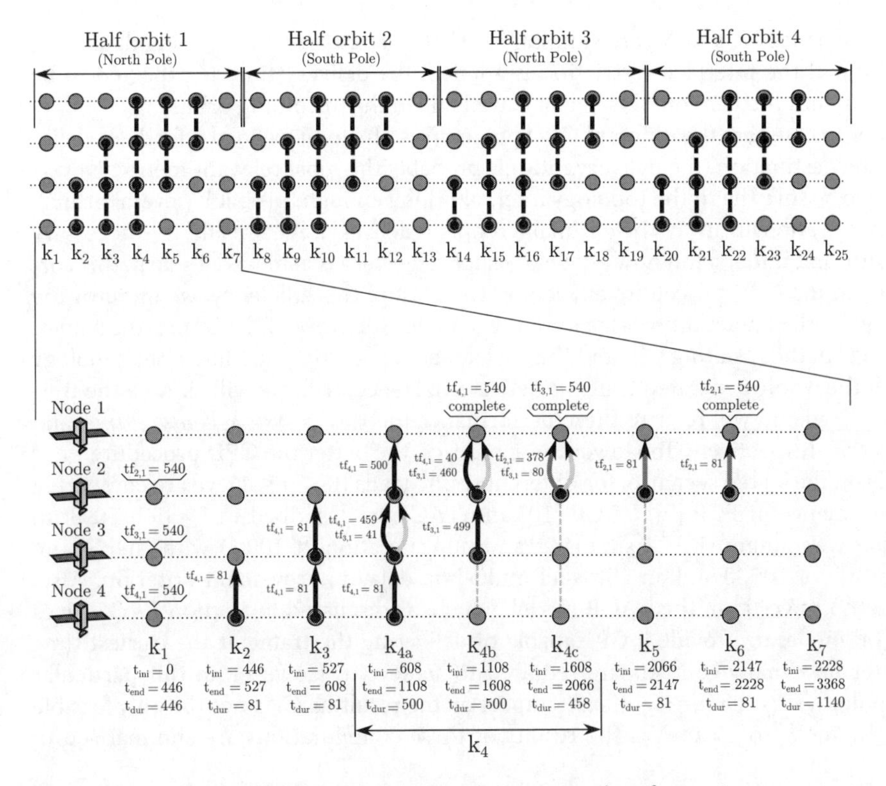

Figure 7.9 Resulting traffic flow in unconstrained contact topology for $\rho = 1$.

to up to one intersatellite link per node, configured with a 1-Mbps full-duplex throughput within the maximum range (700 km). For the sake of simplicity, transmitted bundle (Bundle protocol data units) sizes are set to 1 Mbit or (or 125 KByte), which at 1 Mbps should occupy the channel for 1 s. Such a fixed channel occupation would not be deterministic for a collision-based shared medium access scheme because of their stochastic behavior. Furthermore, since we are modeling a LEO SSN constellation, the time a bundle takes to reach the destination (the propagation delay) is negligible and therefore disregarded in this analysis. Finally, we vary the traffic load from $\rho = 1$ (540 Mbit, 67.5 MByte or 540 packets per node) to $\rho = 0.1$ (54 Mbit, 6.75 MByte or 54 packets per node), where $\rho = 1$ is taken as the traffic that saturates one orbit of the unconstrained CT, as explained below.

The saturation flow for the CT CP is detailed on the first orbit in Figure 7.9, where the optimal flow assignation allows the successful delivery of the three source flows of 540 packets from N_2, N_3 and N_4 to N_1. Within the figure, each traffic flow from N_{src} to N_{dst} is measured in packets (equivalent to seconds in this particular case) and uniquely identified as $tf_{N_{\text{src}},N_{\text{dst}}}$. It should be noted that no further data can flow to N_1 in this orbit because the contact spanning k_4 to k_6 has a duration of 1620 s, which is the limit to accommodate $540 \times 3 = 1620$ source packets of 1 s duration each. Therefore, despite other contacts such as N_4 to N_3 (k_{4b} and k_{4c}) and N_3 to N_2 (k_{4c} and k_5) remaining underutilised, this flow configuration represents network saturation for a single orbit of the proposed SSN. However, such a performance is only achievable when port restrictions are ignored, since N_2 and N_3 are simultaneously transmitting and receiving data at k_3, k_{4a} and k_{4b}. If the interface restriction is applied, the delivery time is expected to be postponed until further contacts become available in the upcoming orbits.

In order to compare the different CPD procedures, the total payload effectively delivered to the destination (the delivery ratio) is probably the most relevant metric. Nevertheless, we be sure that if the topology interval is long enough (probably several orbits), the delivery ratio will always be optimal (complete delivery). As a result, it is necessary to measure and understand how efficiently such a delivery is achieved; in turn, this corresponds to the CPD procedure efficiency. To measure this efficiency, we monitor the overall network contact time usage until the complete delivery of the traffic (the sum of all contacts in the resulting CP) and the payload delivery time (the time in the topology at which the whole generated traffic is delivered). Henceforth, we will assume the traffic is delivered and will refer to these performance metrics as *system contact time*, and *delivery time* respectively. The lower these metrics, the better the CPD procedure.

Finally, in the stated scenario, for different traffic loads ($0 \leq \rho \leq 1$) we are comparing the performance of FCP [9], RACP [10] and TACP, as described in Section 7.2.3. In particular, we configure RACP simulated annealing iterations to 10000, with a maximum temperature of 10000 and an all-to-all multi-hop delay improvement criterion. Also, we set $w(t_k) = K * t_k^2$ in the TACP model, which, as discussed in Section 7.2.3, drives the MILP model to provide a CP capable of delivering the traffic at the earliest time no matter how many contacts are required to achieve it. However, in this particular SSN topology, $w(t_k)$ has no significant impact on the resulting CP since the only feasible path is N_4 to N_3 to N_2 to N_1. The results of these configurations are summarised in Section 7.3.2.

7.3.2 Case Study Results

The delivery time and system contact time metrics for a varying network load ($0 \leq \rho \leq 1$) in the unconstrained CT and each CPD scheme (TACP, RACP, and FCP) are plotted in Figures 7.10a and 7.10b respectively. The delivery time curve in Figure 7.10a has contact opportunity in each of the orbits highlighted. Therefore, all deliveries to N_1 fall within the duration of the contacts with that node. In other words, if the delivery could not be completed in a given over-the-pole fly-over, it might be completed in the next one and so forth.

As originally described in Section 7.3.1, the delivery time of the CT in Figure 7.10a increases up to 2228 s for $\rho = 1$, which is precisely the time when state k_6 and the first orbital period end (Figure 7.9). On the system contact time, the CT plan evidences an increase of this metric each time the traffic flow requires a new state for forwarding the data. It is worth clarifying that this metric only considers all enabled arcs until the full delivery of traffic is complete. For example, for $\rho = 0.3$ the delivery is completed at 1094 s which is within k_4, while for $\rho = 0.35$ it is 1175 s which corresponds to state k_5. In this evolution of ρ, a new arc is required to accommodate the flow and therefore, since the CT has not only the N_2 to N_1 link enabled but also all other arcs (N_4 to N_3 and N_3 to N_2), the overall system contact time increases drastically. In general, routing data using the CT always provides the best delivery time because it accounts for all the network contacts and resources. However, this implies that several arcs remain underutilised, increasing the system contact time. Furthermore, directly implementing the CT can be infeasible due to resource or architectural restrictions, as described in Section 7.2.2.

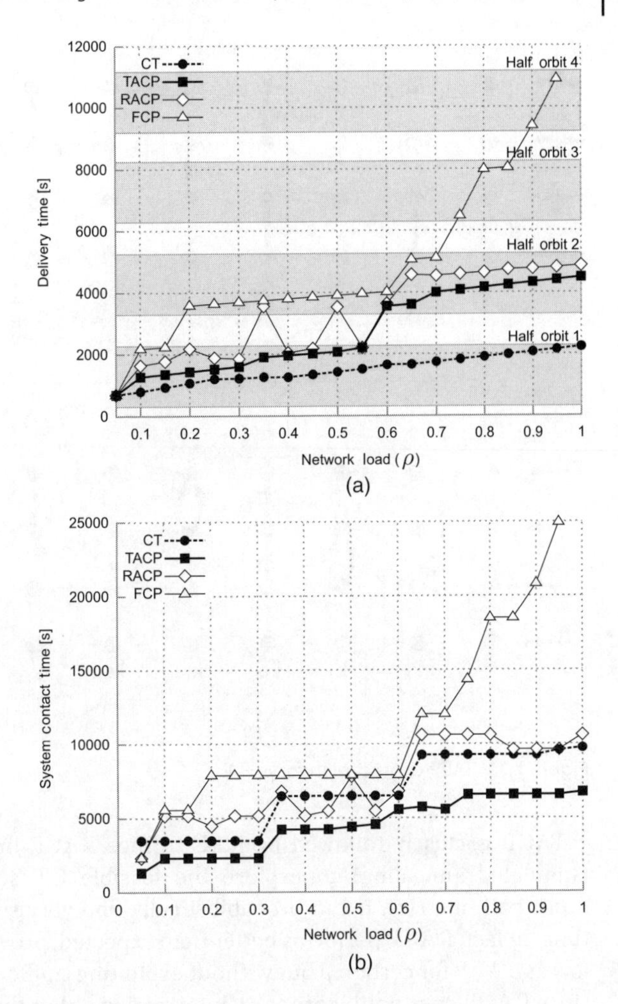

Figure 7.10 Case study: (a) delivery time and (b) system contact time, for varying ρ

Using TACP model, as described in Section 7.2.3, the overall system contact time is optimized because every enabled arc in the final CP is used by the flowing traffic. All remaining arcs are disabled and none of them remains underutilised. This allows to TACP to be the most effective scheme in terms of system resource usage but, as discussed in Section 7.4.1, it leaves no planning error margin in the resulting CP. Regarding delivery time (Figure 7.10a), TACP is not able to match the CT performance due to resource restrictions. This is because CT achieves the optimal delivery time by exploiting simultaneous interface usage, as illustrated in Figure 7.9, while TACP, RACP and FCP are designed and configured to never use two arcs in a given state during the topology interval. Among these resource-aware CPD mechanisms, TACP gives the best delivery time for all ρ. In particular, the optimal CPD for $\rho = 1$ is illustrated in Figure 7.11; the TACP provides a delivery time of 4488 s in state k_{10b} (second half-orbit period). It is interesting to note that in this figure that no node uses more than one arc at all times, and that given this condition, the MILP formulation guarantees that this is the optimal solution in terms of delivery time.

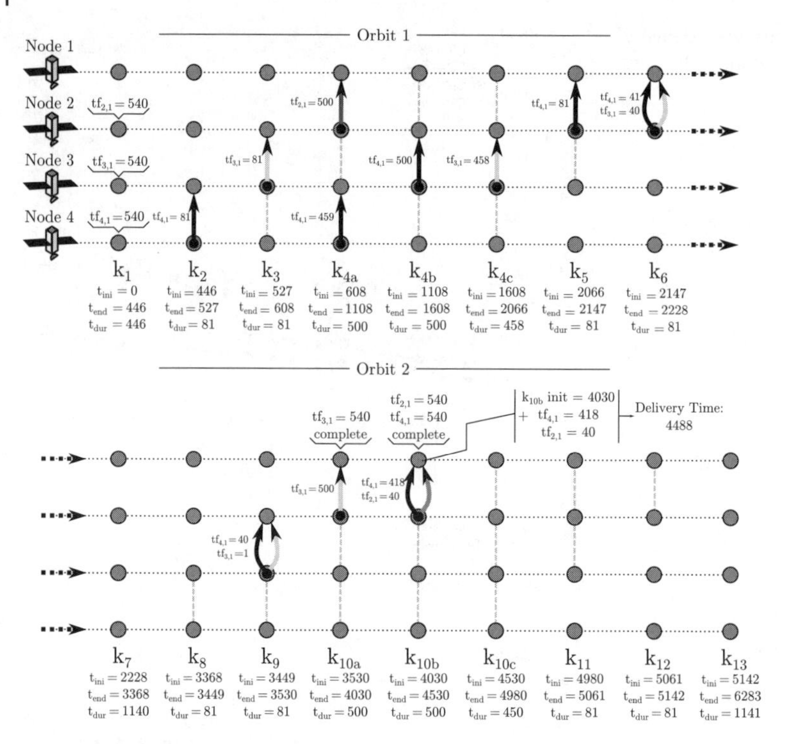

Figure 7.11 TACP contact plan for $\rho = 1$.

TACP is closely followed by RACP. Since RACP implements heuristic algorithms (simulated annealing) to explore the feasible CT solution space while evaluating multi-path metrics, it can probabilistically find very suitable CPs in terms of delivery time. In fact, RACP performs better than expected, providing CPs with delivery times as low as TACP for certain ρ but without exploiting traffic volume predictability. However, since RACP does not incorporate information about the network traffic, it enables arcs with the aim of favouring an all-to-all multi-path flow. Consequently, the RACP has several underutilised arcs, which makes the system contact time considerably higher than in TACP.

Finally, FCP is the most limited CPD mechanism in terms of network information but probably the most efficient in terms of computational complexity. Because of the use of the Blossom algorithm [42], the FCP procedure certainly outperforms RACP and TACP in the processing time required to design a CP. However, since FCP only evaluates single-hop contacts before deciding which arc to disable to maximize a fairness metric [9], the resulting CP provides no particular benefit to the specificity of the traffic flow. As a consequence, FCP-designed CPs require significantly longer to deliver the traffic, which in our test is completely delivered in the fourth half-orbit. Such a delivery time also impacts on the amount of system resources required before the flow reaches its destination (N_1).

To summarise, TACP outperforms the second most efficient scheme (RACP) by 58% in terms of system resource usage while also improving the delivery time by 10% (for $\rho = 1$). Furthermore, this performance improvement becomes more significant for larger topologies. Despite the results metrics being clear and corresponding with the CPD process in Section 7.2, further considerations regarding design complexity and the feasibility of implementing the resulting CP in distributed SSNs will be discussed in Section 7.4.

7.4 Contact Plan Design Discussion

7.4.1 TACP Safeguard Margins and Topology Granularity

As shown in Section 7.3, the TACP scheme gives the best CPs in terms of delivery delay and system contact time. However, this is achieved by exploiting both topology and traffic information, which are assumed to be complete and accurate enough. If unplanned traffic is generated, the consequences could be severe, as the extra information in the network might displace and erroneously utilise resources originally reserved for predicted traffic. This effect is less probable when using traffic-agnostic techniques such as RACP and FCP, which generally enable several arcs throughout the topology interval, allowing for a reallocation of traffic. As a way to mitigate this problem, it is recommended to include safeguard margins in the traffic sources when considering TACP in practical SSN implementations.

On the other hand, as previously stated and described in Section 7.2.3, the fractionation of a state k_a into shorter states k_{a1} and k_{a2} allows CPD schemes to account for a higher granularity in their arc assignations, which in turn enables more efficient CPs. Figure 7.12 illustrates the delivery time metric for TACP with $\rho = 1$ for a varying fractionation parameter. In the example of Section 7.3, the metric was 4488 s for a maximal k duration of 500 s, but it should be clear that it depends on the FSM discretisation factor. As expected, for higher k_{maxDur} durations, the model loses accuracy, resulting in lower quality CPs. However, as the fractionation granularity increases, the model becomes more complex, requiring exponentially increasing computing power to solve it in a reasonable time. In particular, solving this topology with a fractionation of $k_{\text{maxDur}} = 100$ can take about 4 h using a modern processor and a state-of-the-art commercial MILP solver. In other words, solving the topology in these conditions takes longer than the

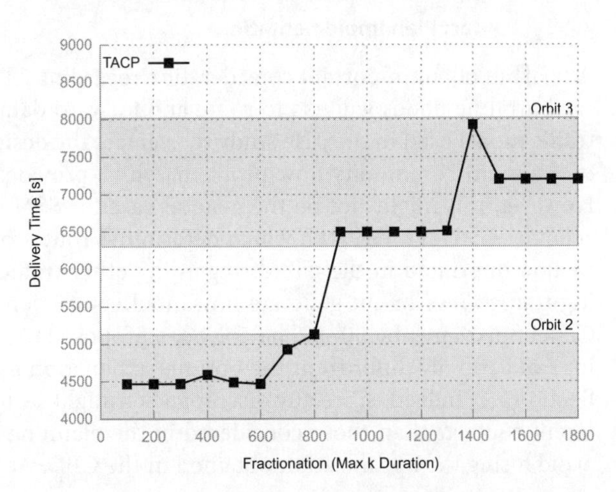

Figure 7.12 TACP contact plan delivery time for $\rho = 1$ and varying fractionation.

actual duration of the topology interval. Furthermore, a highly partitioned plan might bring performance issues due to the increases in communications changes and interruptions making the system more sensitive to synchronism offsets between SSN nodes. Therefore, it is also advisable to adopt a balanced fractionation criterion when modeling CPs with FSM.

7.4.2 Contact Plan Computation and Distribution

There are other operational considerations to be taken into account when designing CPs for SSNs. The first of these is the computational complexity of the CPD schemes. In particular, despite the proposed scenario being composed of a rather limited number of nodes (four) and a short timespan (3 h 22 min 36 s), the analysis required for a successful CPD can be considerable, especially for highly fractionated topologies, as discussed in Section 7.4.1. Furthermore, since several selection methodologies might be needed to find out the most appropriate CP, the overall complexity of such a procedure might be challenging for the average network operator. Therefore, we envision a contact plan computation element (CPCE), which can assist or even automate the design of CPs for future spaceborne WSNs. A CPCE would be capable of determining CPs suitable for supporting connectivity among nodes and data transfers through the network and would typically be part of the mission operations and control center, and would trigger the distribution of the designed CP as discussed below.

Once the design is complete, the problem of CP distribution needs to be solved before the nodes can use it as topological information. Although we have dedicated our effort to the proper and efficient design of CPs for SSNs, there is further work necessary regarding the means by which the designed CPs can safely reach each of the orbiting nodes. It is worth clarifying that these CPs are a key resource, needed by each satellite to take routing and forwarding decisions using contact-graph-based schemes such as CGR [48]. There have been previous studies on the development of protocols to this end, such as the Contact Plan Update Protocol (CPUP) [49], which allows CPs to be distrubuted in-band via Bundle Protocol. Nevertheless, the specific procedure of network commissioning phase remains an implementation and mission-specific issue.

7.4.3 Contact Plan Implementation

A final but not unimportant consideration regarding CPD is the routing scheme that the final orbiting nodes will use to route and forward data. As stated in Section 7.3.1, the traffic model used in the case study to evaluate the designed CPs is a MILP formulation of the multi-commodity flow problem, which provides the optimal flow assignations. However, this might not be the general case for SSN nodes using distributed routing schemes such as CGR [48], which determines routes by relying on limited local information imprinted in the CP (disregarding other traffic sources). Therefore the resulting traffic flows might generate unwanted congestion problems [50]. Despite several CGR improvements [50–52] having been proposed to mitigate congestion, their behavior can certainly differ from the optimal assignation assumed in the CPD, as stated in Section 7.2. Indeed, since any discrepancies might result in traffic flowing through different contacts than those considered in the planning stage, they must be assessed to avoid losing the optimization obtained in the CPD. At the time of writing, the authors are supporting this assessment with detailed system simulations that validate the proper traffic flow in the resulting CP under specific routing algorithms and channel conditions.

7.5 Summary

In this chapter we have discussed the benefits of exploiting spaceborne wireless sensor networks to enhance the functionality and performance of Earth-observation missions. In these extreme environments, where Internet-based protocols fail to perform, DTNs are an appealing communication alternative. We have provided a thorough analysis of the challenges to creating CPDs for effectively implementing DTN-based SSNs. As a result, we have investigated different modeling techniques, including TACPs: a general, unique and novel mechanism that takes advantage of the expected traffic and topology information typically available in SSNs. Finally, by designing the CP for an appealing LEO SSN topology, we have demonstrated TACPs' benefits, showing that it outperforms existing CPD schemes by 58% in terms of system resource usage and by 10% in terms of delivery time.

However, there are still several research opportunities in the area. These relate to the accuracy, scalability and implementability of CPs designed in advance and using a centralised node on Earth. In particular, we have reviewed the necessity for safeguard margins when accounting for traffic in the contact decision, the complexity and limitations of different CPD procedures and the problem of distribution and implementation once the plan is on the corresponding node. This shows that taking advantage of predictability in SSNs can become a complex problem, but one that is worth solving in order to make the best use of valuable orbiting assets.

As a result, an efficient CPD procedures such as TACPs promises to have a significant impact in enhancing the delivery of sensed data from space networks. Indeed, by exploiting the specific characteristics of orbiting assets, such as their predictable nature, such operational and planning solutions are changing the way networked satellite systems are conceived. In particular, since outer space missions have to perform in an inaccessible and extremely harsh environment, they become particularly expensive and their lifetimes are clearly bounded. Therefore, optimizing the delivery time and resource usage in the manner described throughout this chapter will be of great value to the space community and will certainly drive the implementation of future SSN solutions.

References

1 Rashvand, H., Abedi, A., Alcaraz-Calero, J., Mitchell, P., and Mukhopadhyay, S. (2014) Wireless sensor systems for space and extreme environments: A review. *Sensors Journal, IEEE*, **14** (11), 3955–3970, doi:10.1109/JSEN.2014.2357030.

2 Chu, J., Guo, J., and Gill, E. (2013) Fractionated space infrastructure for long-term earth observation missions, in *Aerospace Conference, 2013 IEEE*, pp. 1–9, doi:10.1109/AERO.2013.6496854.

3 Jackson, J. (2005) The interplanetary internet. *IEEE Spectrum*, **42** (8), 31–35.

4 Larson, W. and Wertz, J. (1999) *Space Mission Analysis and Design*, 3rd edn, vol. 8, Microcosm.

5 Consultative Committee for Space Data Systems (CCSDS), Delay Tolerant Networking Working Group (SIS-DTN) webpage, http://cwe.ccsds.org/sis/default.aspx#_SIS-DTN.
6 Internet Engineering Task Force (IETF), Delay Tolerant Networking Working Group (DTNWG) webpage, https://datatracker.ietf.org/wg/dtnwg/charter/.
7 Huang, M., Chen, S., Zhu, Y., and Wang, Y. (2010) Cost-efficient topology design problem in time-evolving delay-tolerant networks, in *Global Telecommunications Conference (GLOBECOM 2010), 2010 IEEE*, pp. 1–5, doi:10.1109/GLOCOM.2010.5684269.
8 Huang, M., Chen, S., Li, F., and Wang, Y. (2012) Topology design in time-evolving delay-tolerant networks with unreliable links, in *Global Communications Conference (GLOBECOM), 2012 IEEE*, pp. 5296–5301, doi:10.1109/GLOCOM.2012.6503962.
9 Fraire, J., Madoery, P., and Finochietto, J. (2014) On the design and analysis of fair contact plans in predictable delay-tolerant networks. *Sensors Journal, IEEE*, **14** (11), 3874–3882, doi:10.1109/JSEN.2014.2348917.
10 Fraire, J. and Finochietto, J. (2015) Routing-aware fair contact plan design for predictable delay tolerant networks. *Ad-Hoc Networks*, **25, Pt B**, 303–313, doi:10.1016/j.adhoc.2014.07.006.
11 Fraire, J., Madoery, P., and Finochietto, J. Traffic-aware contact plan design for disruption-tolerant space sensor networks. *Ad-Hoc Networks*. In Press.
12 Postel, J. (1981), Request for comments: RFC-793, transmission control protocol specification, Network Working Group, IETF.
13 Akyildiz, I., Su, W., Sankarasubramaniam, Y., and Cayirci, E. (2002) A survey on sensor networks. *IEEE Communications Magazine*, **40** (8), 102–114.
14 Caini, C., Cruickshank, H., Farrell, S., and Marchese, M. (2011) Delay- and disruption-tolerant networking (DTN): An alternative solution for future satellite networking applications, pp. 1980–1997, doi:10.1109/JPROC.2011.2158378.
15 Burleigh, S., Hooke, A., Torgerson, L., Fall, K., Cerf, V., Durst, B., Scott, K., and Weiss, H. (2003) Delay-tolerant networking: an approach to interplanetary internet. *Communications Magazine, IEEE*, **41** (6), 128–136, doi:10.1109/MCOM.2003.1204759.
16 Caini, C. and Fiore, V. (2012) Moon to earth DTN communications through lunar relay satellites, in *6th Advanced Satellite Multimedia Systems Conference (ASMS) and 12th Signal Processing for Space Communications Workshop (SPSC)*, Baiona, pp. 89–95.
17 Garrison, T., Ince, M., Pizzicaroli, J., and Swan, P. (1997) Systems engineering trades for the iridium constellation. *Journal of Spacecraft and Rockets*, **34** (5), 675–680.
18 DARPA Tactical Technology Office, System F6 program, http://www.darpa.mil/Our_Work/TTO/Programs/System_F6.aspx.
19 Akyildiz, I.F., Su, W., Sankarasubramaniam, Y., and Cayirci, E. (2002) Wireless sensor networks: a survey. *Elsevier Computer Networks Journal*, **38**, 393–422.
20 Durst, R.C., Miller, G.J., and Travis, E.J. (1997) TCP extensions for space communications. *Wireless Networks*, **3** (5), 389–403, doi:10.1023/A:1019190124953.
21 Fall, K. (2003) A delay-tolerant network architecture for challenged internets, in *Proceedings of the 2003 Conference on Applications, Technologies, Architectures, and Protocols for Computer Communications*, ACM, New York, NY, USA, SIGCOMM '03, pp. 27–34, doi:10.1145/863955.863960.

22 Cerf, V., Burleigh, S., Hooke, A., Torgerson, L., Durst, R., Scott, K., Fall, K., and Weiss, H. (2007) Delay-tolerant networking architecture, *Tech. Rep. RFC-4838*, Network Working Group, IETF.

23 Scott, K. and Burleigh, S. (2007) Bundle protocol specification, *Tech. Rep. RFC-5050*, Network Working Group, IETF.

24 Caini, C. and Firrincieli, R. (2011) *DTN for LEO satellite communications*, Springer, pp. 186–198.

25 Lindgren, A., Doria, A., Davies, E., and Grasic Probabilistic routing protocol for intermittently connected networks, *Tech. Rep. RFC-6693*, Internet Research Task Force.

26 Burgess, J., Gallagher, B., Jensen, D., and Levine, B. (2006) Maxprop: Routing for vehicle-based disruption-tolerant networks, in *25th IEEE International Conference on Computer Communications Proceedings (INFOCOM 2006)*.

27 Balasubramanian, A., Levine, B., and Venkataramani, A. (2007) DTN routing as a resource allocation problem, in *Proceedings of the 2007 Conference on Applications, Technologies, Architectures, and Protocols for Computer Communications (SIGCOMM 2007)*.

28 Spyropoulos, T., Psounis, K., and Cauligi, S. (2005) Spray and wait: an efficient routing scheme for intermittently connected mobile networks, in *Proceedings of the 2005 Conference on Applications, Technologies, Architectures, and Protocols for Computer Communications (SIGCOMM 2005)*.

29 Birrane, E., Burleigh, S., and Kasch, N. (2012) Analysis of the contact graph routing algorithm: Bounding interplanetary paths. *Acta Astronautica*, **75**, 108–119.

30 Caini, C. and Firrincieli, R. (2012) Application of contact graph routing to LEO satellite DTN communications, in *Communications (ICC), 2012 IEEE International Conference on*, pp. 3301–3305, doi:10.1109/ICC.2012.6363686.

31 Demmer, M., Ott, J., and Perreault, S. (2014) Request for comments: RFC-7242, delay-tolerant networking TCP convergence-layer protocol, Internet Research Task Force (IRTF).

32 Kruse, H., Jero, S., and Ostermann, S. (2014) Request for comments: RFC-7122: Datagram convergence layers for the delay and disruption tolerant networking (DTN) bundle protocol and Licklider transmission protocol (ltp), Internet Research Task Force (IRTF).

33 Alfonzo, M., Fraire, J., Kocian, E., and Alvarez, N. (2014) Development of a DTN bundle protocol convergence layer for spacewire, in *Biennial Congress of Argentina (ARGENCON), 2014 IEEE*, pp. 770–775.

34 DTN Research Group, DTN2: a DTN reference implementation. https://sites.google.com/site/dtnresgroup/home/code/dtn2documentation.

35 Viagenie Inc., Postellation: a lean and deployable DTN Implementation. http://postellation.viagenie.ca/.

36 IBR-DTN: A modular and lightweight implementation of the bundle protocol. http://trac.ibr.cs.tu-bs.de/project-cm-2012-ibrdtn.

37 Burleigh, S. (2007) Interplanetary overlay network: An implementation of the DTN bundle protocol, in *Consumer Communications and Networking Conference, 2007. CCNC 2007. 4th IEEE*, pp. 222–226, doi:10.1109/CCNC.2007.51.

38 Wyatt, J., Burleigh, S., Jones, R., Torgerson, L., and Wissler, S. (2009) Disruption tolerant networking flight validation experiment on NASA's EPOXI mission, in

Proceedings of the 2009 First International Conference on Advances in Satellite and Space Communications, IEEE Computer Society, Washington, DC, USA, SPACOMM '09, pp. 187–196, doi:10.1109/.38.

39 Ivancic, W., Eddy, W., Stewart, D., Wood, L., Holliday, P., Jackson, C., and Northam, J. (2010) Experience with delay-tolerant networking from orbit. *International Journal of Satellite Communication and Networking*, **28**, 335–351.

40 Vallado, D.A. (2007) *Fundamentals of Astrodynamics and Applications*, 4th edn, Microcosm.

41 Krishnamurthy, A. and Preis, R. (2005) Satellite formation, a mobile sensor network in space, in *Parallel and Distributed Processing Symposium, 2005. Proceedings. 19th IEEE International*, doi:10.1109/IPDPS.2005.387.

42 Kolmogorov, V. (2009) Blossom V: a new implementation of a minimum cost perfect matching algorithm. *Mathematical Programming Computation*, **1** (1), 43–67, doi:10.1007/s12532-009-0002-8.

43 Fraire, J. and Finochietto, J. (2015) Design challenges in contact plans for disruption-tolerant satellite networks. *Communications Magazine, IEEE*, **53** (5), 163–169, doi:10.1109/MCOM.2015.7105656.

44 Mendoza, H. and Corral-Briones, G. (2013) Interference in medium and low orbit distributed satellite systems, in *XV Reunión de Trabajo Procesamiento de la Información y Control (RPIC)*, San Carlos de Bariloche, Argentina, pp. 1110–1115.

45 ITU-R (2003), *Recommendation ITU-R S.1325-3*, International Telecommunication Union.

46 Alonso, J. and Fall, K. (2003) A linear programming formulation of flows over time with piecewise constant capacity and transit times, *Tech. Rep. IRB-TR-03-007*, Intel.

47 Fraire, J., Madoery, P., and Finochietto, J. (2014) Leveraging routing performance and congestion avoidance in predictable delay tolerant networks, in *Wireless for Space and Extreme Environments (WiSEE), 2014 IEEE International Conference on*, pp. 1–7, doi:10.1109/WiSEE.2014.6973079.

48 Burleigh, S. (2010) Contact graph routing, IETF-Draft.

49 Bezirgiannidis, N., Tsapeli, F., Diamantopoulos, S., and Tsaoussidis, V. (2013) Towards flexibility and accuracy in space DTN communications, in *Proceedings of the 8th ACM MobiCom Workshop on Challenged Networks*, ACM, New York, NY, USA, CHANTS '13, pp. 43–48, doi:10.1145/2505494.2505499.

50 Fraire, J., Madoery, P., Finochietto, J., and Birrane, E. (2015) Congestion modeling and management techniques for predictable disruption tolerant networks, in *40th IEEE Conference on Local Computer Networks (LCN 2015)*, IEEE.

51 Birrane, E. (2013) Congestion modeling in graph-routed delay tolerant networks with predictive capacity consumption, in *Global Communications Conference (GLOBECOM), 2013 IEEE*, pp. 3016–3022, doi:10.1109/GLOCOM.2013.6831534.

52 Yan, H., Zhang, Q., and Sun, Y. (2015) Local information-based congestion control scheme for space delay/disruption tolerant networks. *Wireless Networks*, **21** (6), 2087–2099, doi:10.1007/s11276-015-0911-6.

8

Infrared Wireless Sensor Network Development for the Ariane Launcher

Hendra Kesuma[1], Johannes Sebald[2] and Steffen Paul[1]

[1] *Institute of Electrodynamics and Microelectronics (ITEM), University of Bremen, Germany*
[2] *Airbus Safran Launchers (former EADS Astrium) GmbH, Bremen Germany*

8.1 Introduction

The Ariane Launcher is the heavy-lift launch vehicle of the Ariane rocket family. The rocket family (Ariane 1 to Ariane 5) was developed by Airbus Defence and Space. On 9 July 2013, the European Space Agency (ESA) approved the Ariane 6 final design [1]. The launcher is capable of delivering up to three satellites to geostationary transfer orbit or low-Earth orbit, depending on the size of the satellites.

To give better performance and reliability and lower operational costs in the future, wireless technology has been explored as a replacement for a part of the heavy and complicated wiring in the launcher. This is the main goal of ESA's future launchers preparatory program [2], which maintains Europe's long-term independent capabilities in space technology and increases European competitiveness in the worldwide space-launch market.

The use of wireless technologies using commercial components for space applications is intended to shorten development times and costs, as part of the progressive restructuring of European launcher industries [3]. Wireless technology also increases the flexibility of sensor systems and network configurations. Figure 8.1 shows a possible use of a wireless sensor node on Ariane 5's upper stage. Parts of the upper stage are built by different European countries that specialise in particular functionalities. One of the focuses of Airbus Safran Launchers is the enhancement of wireless sensor nodes for the vehicle equipment bay (VEB) [4].

One of the main constraints in developing wireless sensor nodes for the upper stage is that the electromagnetic emissions should not exceed the values shown in Figure 8.2 [4]. Low electromagnetic emission wireless technologies, such as ultra wide band (UWB), and optical transmissions, such as infrared, are required due, for example, to the low electrical fields generated towards other electronic equipment inside the launcher.

The focus of this chapter is infrared wireless sensor network development for the VEB on Ariane launchers. The prototype of the sensor network was built and tested on the Ariane 5 VEB and will be included in the new Ariane 6 launcher.

Wireless Sensor Systems for Extreme Environments: Space, Underwater, Underground, and Industrial.
First Edition. Edited by Habib F. Rashvand and Ali Abedi.

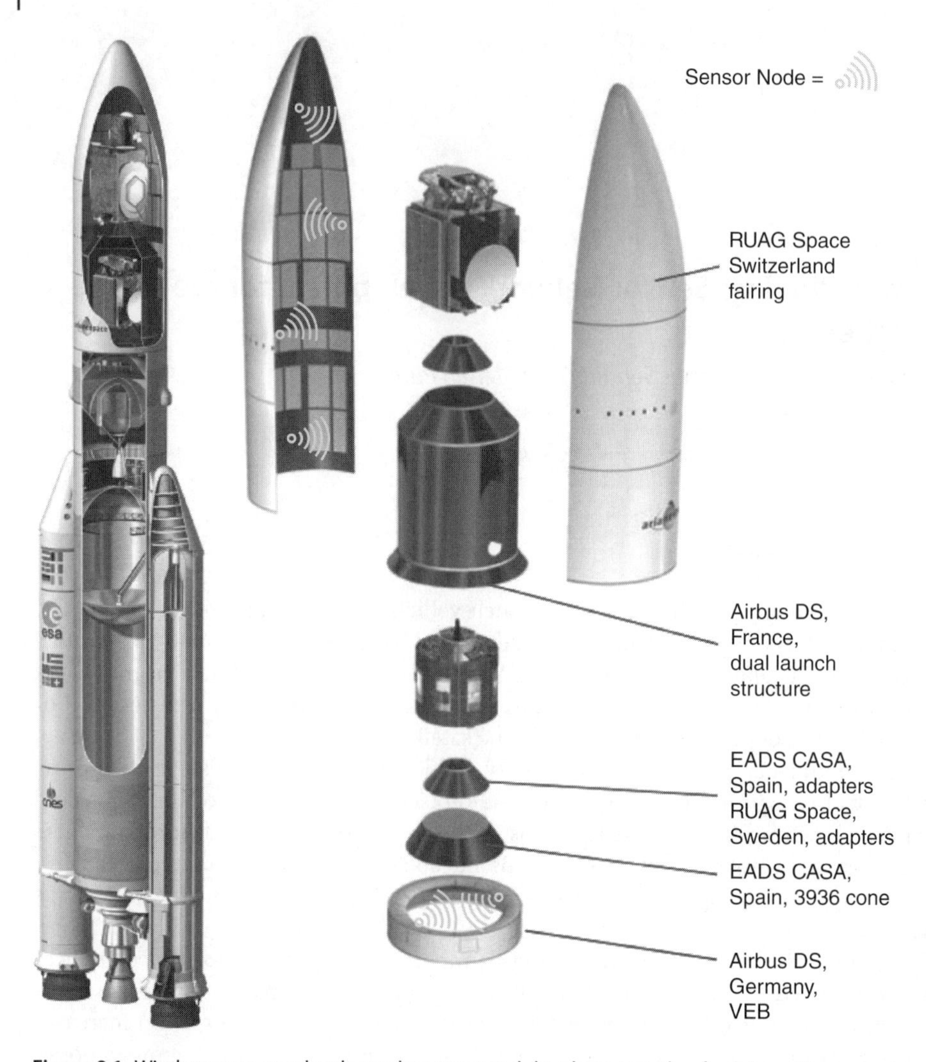

Figure 8.1 Wireless sensor technology placement and development plan for Ariane 5's upper stage.

8.1.1 Objectives

Commercial components, such as smart sensors with microelectromechanical systems are chosen for space sensor nodes for miniaturization purposes. This is also true for the commercial infrared transceivers used in everyday life in home appliances. The application-specific integrated circuit (ASIC) of an infrared transceiver is necessary for some of the technological innovations in the space industries. Apart from its transmission range and bit-error rate, a low-power infrared transceiver design is important to enable longer lifetimes of sensor nodes if they are powered by battery. Other approaches, such as wireless power transmission via visible light, can also be used to increase the operational times of sensor nodes. This method is used to charge the batteries before the mission starts and during the flight.

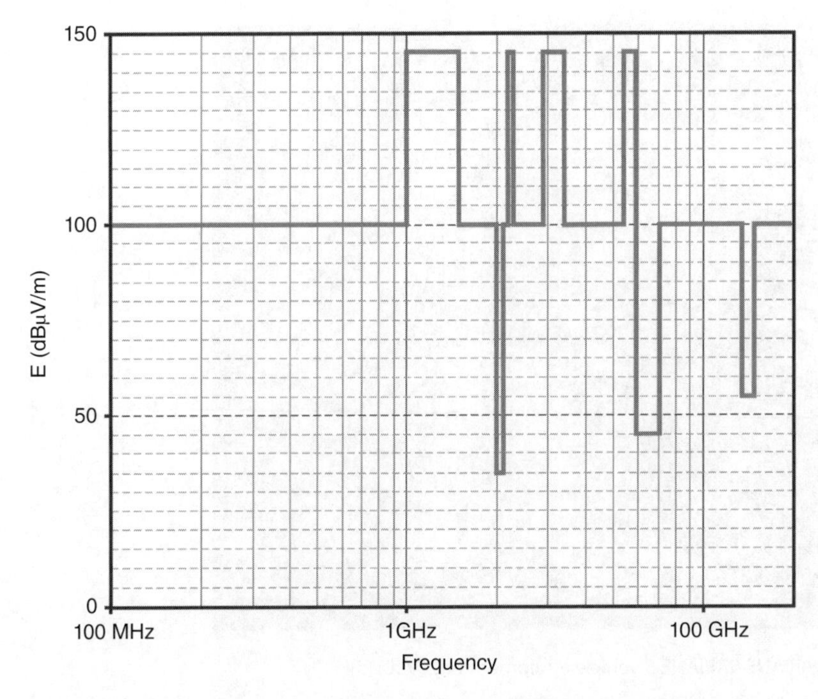

Figure 8.2 The electromagnetic radiation/emissions values permitted in the launch vehicle [4].

Information about the effects of materials and geometry inside the upper stage are utilized for wireless energy transmission and optical communication. The list of the objectives of the infrared wireless sensor network development is then defined as follows:

- Investigate the influence of the material inside the upper stage on infrared communications and wireless power transmission.
- Develop the infrared transceiver ASIC to match the commercial infrared transceiver components.
- Develop the time-synchronization and time-stamping methods for reducing the size of sensor node hardware components and reducing data-packet length.
- Utilize commercial smart sensors to minimize sensor node size in the Ariane 5 telemetry subsystems.

8.1.2 VEB Overview and Internal Surface Material

Multi-layer insulation (MLI) is mainly used as a thermal blanket for the structure and electronics equipment in the VEB. The Ariane launcher's electronics are designed to work between −20 °C and +70 °C and MLI helps maintain the working temperature in that range. Diffuse infrared transmitter LEDs are normally used for home appliances because of their wide coverage transmission area (total transmission angle >30°). This is also advantageous if deployed in a confined space such as the inner compartment of a launcher. Figure 8.3 shows the VEB, of diameter 5.4 m and height 1.56 m. Environmental monitoring in the VEB with the wireless sensor node is important because the VEB

Figure 8.3 The ARIANE 5ARIANE 5 vehicle equipment bay (VEB) [5].

contains the brain of the launch vehicle, which houses most of the electronics for flight control and the on-board computers [5].

The chapter is organized in four sections, covering the points mentioned in the objectives. The development processes and measurement results of each section are explained in the following.

8.2 Development Processes and Measurements of Infrared Transceiver ASIC

The development of the infrared transceiver ASIC requires measurement data for the upper-stage material, which might influence the communication profile. This data helps the design process in relation to determining the ASIC's transmission power and modulation methods. The development process is described as follows.

8.2.1 Influence of Upper-stage Materials on Infrared Communication

The MLI materials used for the launcher are a lightweight film with very high light reflectivity. The layers are made of polyester/polyamide films with 99.99% aluminium vapour deposited on one or both sides. The details of the material are a closely guarded secret of Airbus Defence and Space.

To investigate the material's influence on infrared communications, the equipment in Figure 8.4 was set up. The purpose of this setup was to characterise the effect of the MLI on the bit-error rate (BER) for various measurement angles and distances. This was done by placing the MLI target in front of the infrared transmitter (TX) and the infrared

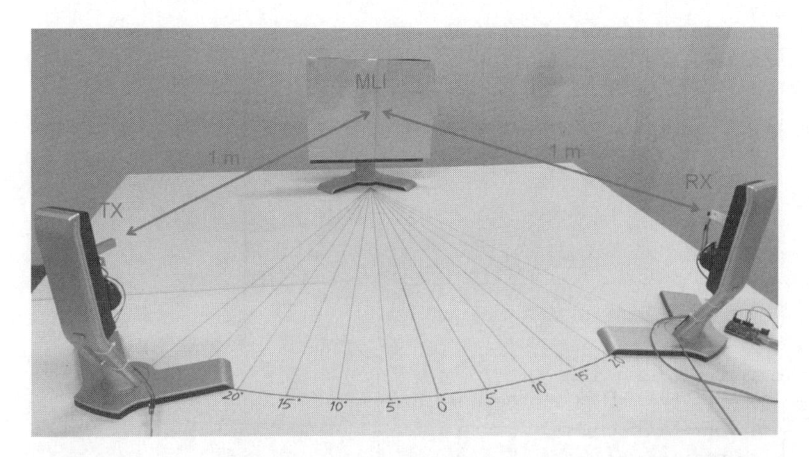

Figure 8.4 Experimental setup of the infrared transceiver and the target MLI.

receiver (RX). The distance between the target holder and the wall behind was arranged to have no significant influence to the measurement results. During the experiment, only a negligible part of the infrared light was reflected from the wall to the receiver. The measurement was performed by introducing noise with an LED DC light, an AC light and in the dark (no light). Equation 8.1 shows the general model of the wireless infrared channel, with $X(t)$ as the output power of the n infrared transmitter LEDs, $Y(t)$ being the output current of the infrared photo detector and $N(t)$ representing visible background light as Gaussian shot noise [6]. The BER of 10^{-9} was measured by placing the receiver and transmitter at 0° in front of the MLI target. This was done in no-light conditions and setting up the optimal transmitter LED current. The increase of the BER has been predicted, and is mainly caused by the Gaussian shot noise from the light sources and the path loss caused by communication angle variation.

$$Y(t) = nX(t) * H(t) + N(t) \tag{8.1}$$

The on-off keying (OOK) uses an IrDa physical layer protocol [7] at 9600 baud and 950 nm (infrared) wavelength for the measurement.

The optimal infrared LED resistance was investigated by performing the BER measurement at 0° at 1 m from the MLI. The measurement results in Figure 8.5 for non line of sight (NLOS) measurement show that varying the value of resistance of the infrared transmitter diode decreases the BER to 10^{-9} if it reaches less than 5 Ω. The short period of current 175 mA is measured through the infrared LED with 3 V forward and at 5 V supply voltage. According to the IrDa standard, the period for turning the LED to '1' is less than 20 μs.

Varying the angles shows that the introduction of the Gaussian shot noise through the LED DC light does not affect the BER much, as compared with the higher BER caused by AC light (see Figure 8.6). For LED DC light and the no-light condition, the BER reaches 10^{-9} at 10°. The higher BER with the communication angle greater than 10° is caused by the weaker power received by the infrared receiver.

Figure 8.7 illustrates the measurement with AC light and the transmission angle between the receiver and the transmitter at 10°.

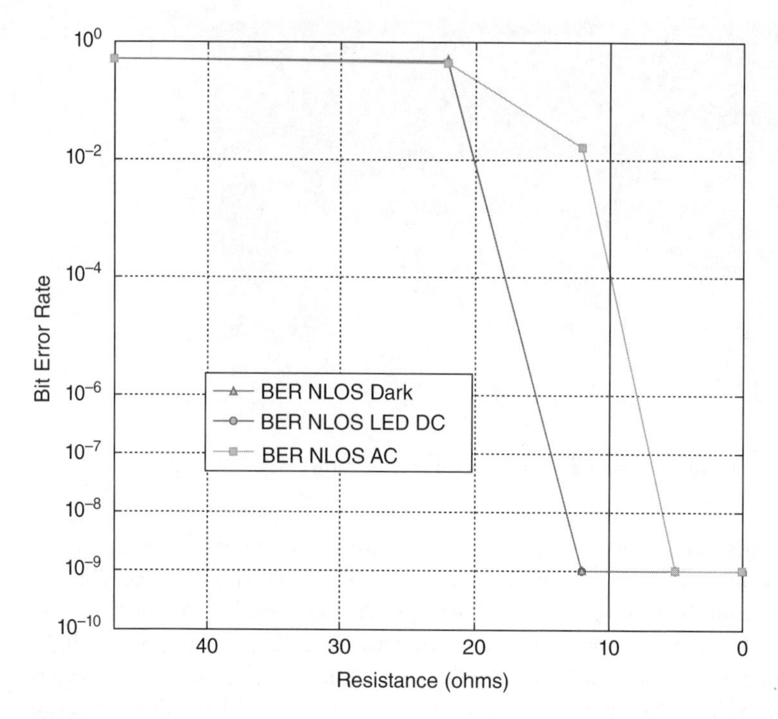

Figure 8.5 Bit-error -rate (BER) measurement versus TX diode resistance in different illumination conditions.

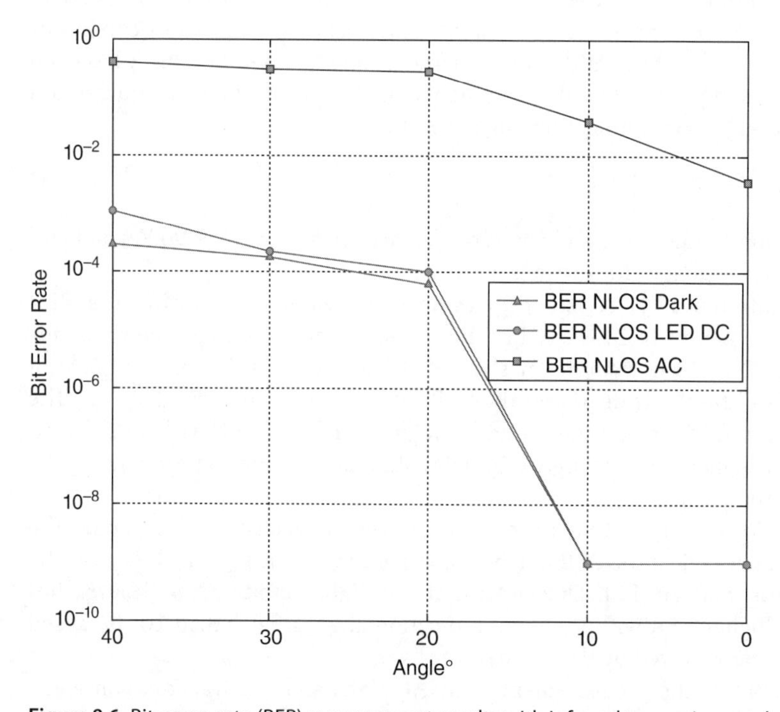

Figure 8.6 Bit-error -rate (BER) measurement results with infrared transceiver angle variation.

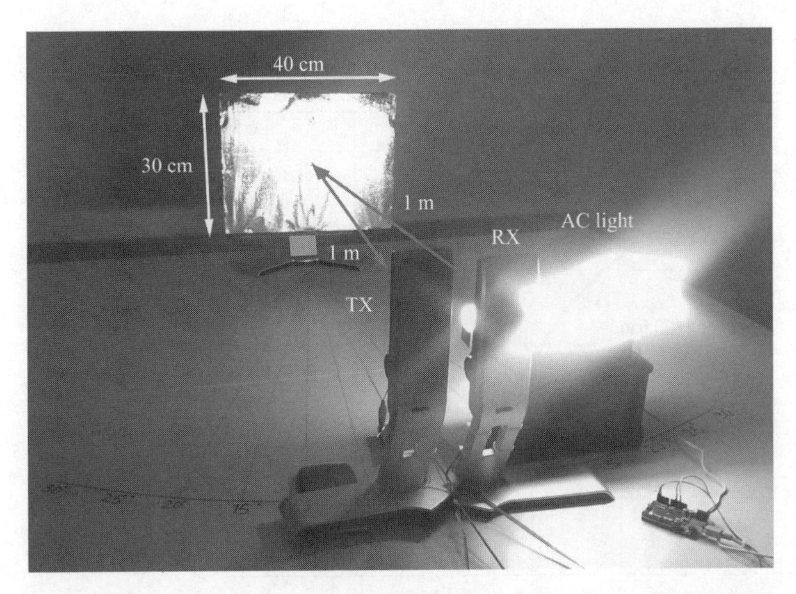

Figure 8.7 Non- line -of- sight experimental setup with fixed angle and fixed distance.

In this section the experimental results show that the use of MLI as a reflector for diffuse infrared communication at 9600 baud inside the VEB of Ariane 5 with a BER of 10^{-9} is possible under some conditions:

- Transmission distances larger than 1 m between the MLI and the infrared transceiver can be achieved with a sufficient infrared transmission supply current (in this experiment larger than 175 mA or a resistor for the LED of less than 5 Ω).
- The MLI materials that were provided by Airbus Safran Launchers show no measurable differences in BER behaviour between DC light and no-light conditions, at various measurement angles. This result prompts the further investigation of DC LEDs for visible-light communication transmitter development.
- The channel model for the infrared receiver does not follow the cosine function reported in the literature [7] and has to be further investigated.

8.2.2 Low-power Infrared Transceiver ASIC development

The ASIC developed in this project is based on the AMS 350-nm technology. This analogue mixed signal 350-nm CMOS technology is currently provided by an Austrian multinational semiconductor manufacturer, Austria Mikro Systeme AG (referred to as AMS).

It is a quite challenging task to design an ASIC for the infrared transceiver in this project. Several steps must be taken carefully to ensure the ASIC meets the requirements of the IrDa standard. These steps are:

- selecting the optimal modulation types for the infrared transceiver through simulation of the power spectral density
- synthesizing and testing the infrared transceiver on the field programmable gate array (FPGA) with the selected modulation types.
- designing the ASIC with AMS 350-nm technology and testing it on the sensor node.

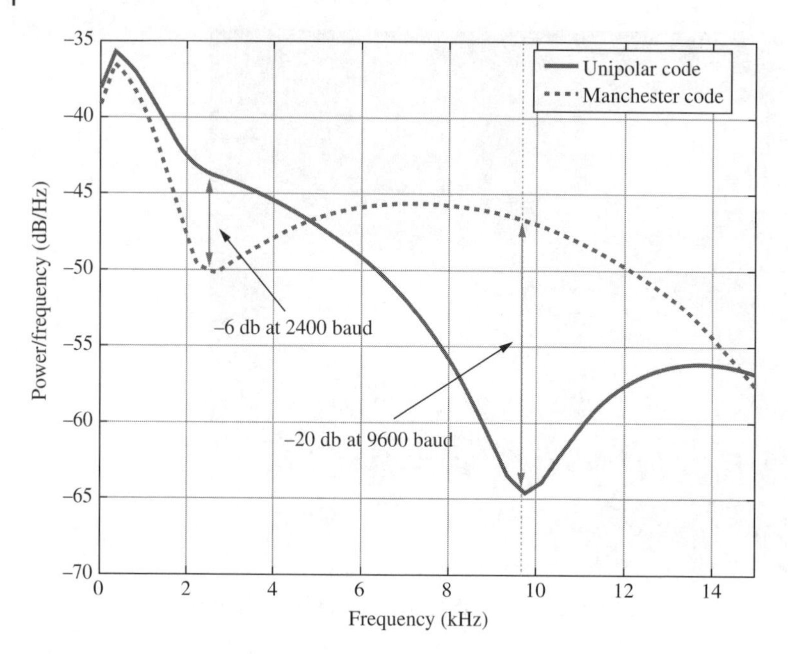

Figure 8.8 The pPower spectral density comparison between thefor Manchester and unipolar coding.

Split-phase Manchester coding and unipolar return zero coding are selected for the power spectral density simulation with Matlab. A random bit stream was generated to simulate the related power spectral density; see Figure 8.8. The simulation results at 9600 baud show that unipolar coding consumes less energy than the Manchester code; by approximately −20 dB. At a lower bit rate of 2400 baud, the Manchester code reduces the power by −6 dB. A transmission speed of 9600 baud is selected because this is sufficient for data transmission from smart sensors, such as those used for temperature, air pressure, air humidity and acceleration.

Next, the hardware block diagrams and signal shapes for both modulation types will be covered in more detail.

The Manchester coding block diagram shown in Figure 8.9 consists of a transmitter that has an UART to Manchester block and a 3/16 pulse-shaping block. The Universal asynchronous receiver transmitter (UART) input signal is converted to Manchester coding, with the '0' expressed as a 0 to 1 transition and the '1' as a 1 to 0 transition [8]. The advantage of this coding is that it does not have a DC-component. This enables the receiver to adjust the decision threshold and generate the baud-rate clock from the incoming signal (edge-detection mechanism) [9]. The receiver part has a 3/16 pulse-recovery block and a Manchester to UART block.

In contrast to the Manchester coding, the unipolar return zero coding in Figure 8.10 has simpler blocks. In the transmitter part, there is only a 3/16 pulse-shaping block required. This block inverts and shapes the UART input signal rx_int to become tx_ext. This signal is then used to drive the infrared transmitter LED.

The receiver part is also quite simple. The signal rx_ext at the output of the infrared receiver circuit enters a 3/16 pulse-recovery block and is processed by extending its period from 20 μs to 104 μs according to the UART 9600 baud specification.

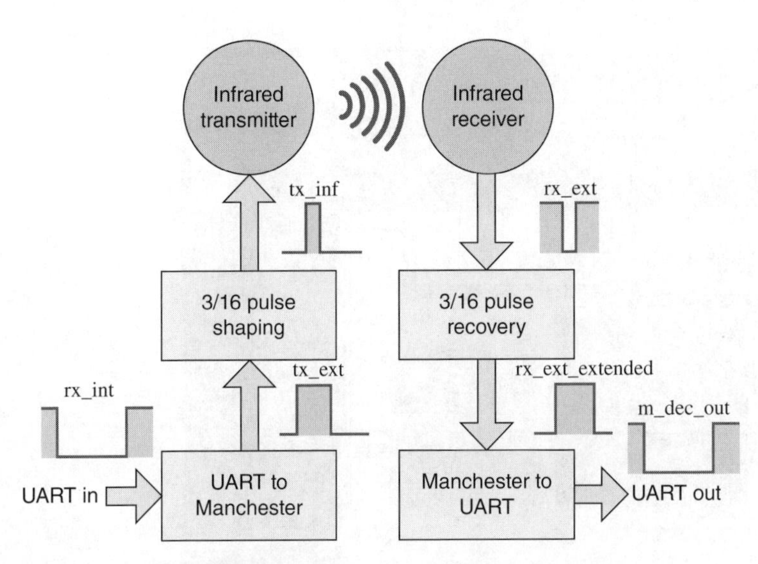

Figure 8.9 Block diagram for split-phase Manchester coding.

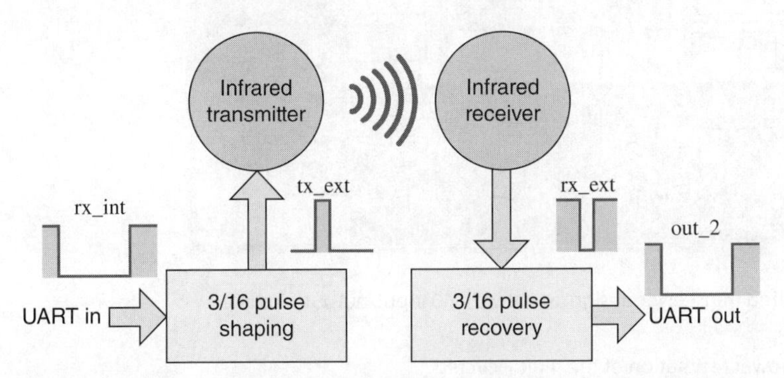

Figure 8.10 Diagram block of the unipolar return zero coding infrared transceiver.

Both of the modulation types were synthesized on the FPGA and tested by measuring the signal shapes with an oscilloscope on various blocks in real time.

The ASIC was synthesized for AMS 350-nm technology. The results of the synthesis are the blocks consisting of primitive logic gates and flip flops connected with each other to perform the functionality of the selected modulation types. Figure 8.11 shows the Manchester_TX/RX (Manchester coding on the ASIC) and Simple_TX/RX (unipolar return zero coding on the ASIC) logic block drawings.

The Manchester_TX/RX has a chip surface of 330 × 330 μm and the Simple_TX/RX is 130 × 130 μm. The total area required for both of the designs, including 40 input/output pads, is 1800 × 1800 μm.

The power consumption estimated for the designs is summarised in Table 8.1. The dynamic power consumption was estimated with 10 ms of simulation using SimVision (from Cadence Design Systems), based on the AMS 350-nm technology parameters. The simulation results were exported as value change dumps and were used to estimate the dynamic power consumption using the register transfer logic compiler.

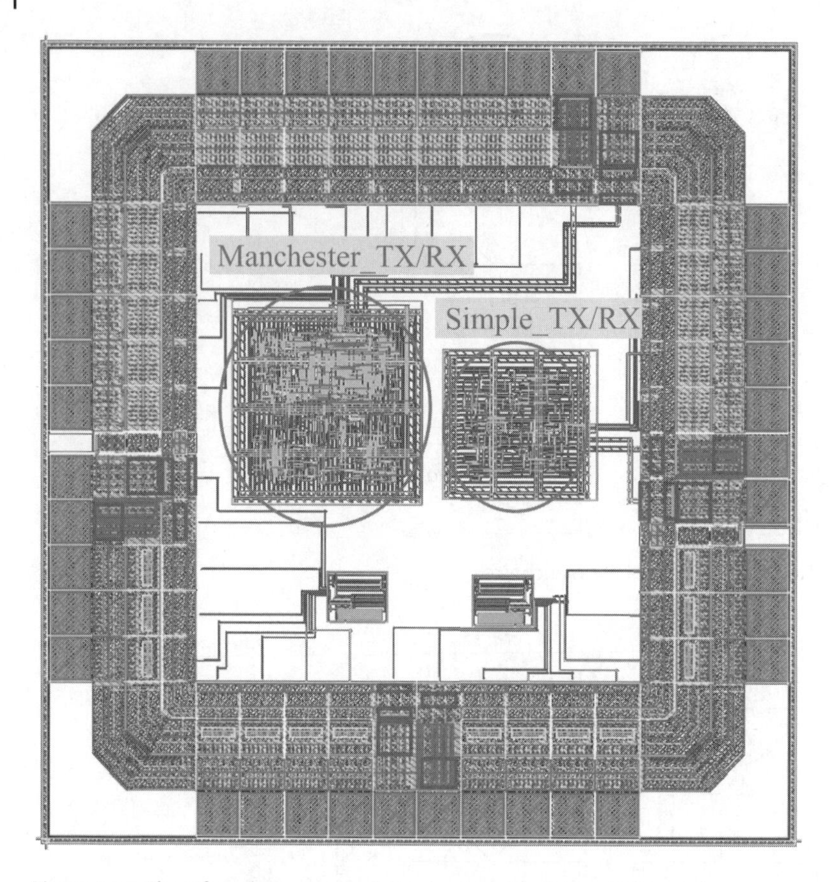

Figure 8.11 The infrared transceiver design layout with 40 input/output pins.

Table 8.1 Dynamic power estimation of the digital circuits.

Module name	Instance	Leakage power (nW)	Dynamic power (mW)	Total power (mW)
Manchester_TX/RX	628	0.598	39.89	39.89
Simple_TX (unipolar)	261	0.248	1.47	1.47

The results show that the split-phase Manchester design, with a layout area 1.4 times larger than the unipolar return zero, consumes almost 26 times more power (39.89 mW) than its counterpart (1.47 mW). The ASIC manufactured with AMS 350-nm technology is shown in Figure 8.12.

Some useful results from the ASIC development are:

- The split-phase Manchester layout area is 1.4 times larger than the unipolar return zero layout and the number of logic cells is more than 2.4 times larger. This leads to an energy consumption for the split-phase Manchester layout that is much larger than that for the unipolar return zero design.

Figure 8.12 The infrared transceiver ASIC.

- The ASIC was tested, and works according to the design. The power consumption is 818 μW for the split-phase Manchester design and 382 μW for the unipolar return zero design.
- The uni-polar return zero coding ASIC function has a much simpler design and lower power consumption than the split-phase Manchester coding ASIC function, and it will be used for future developments.

The ASIC is currently tested for the infrared wireless sensor network inside an Ariane 5 VEB. A radiation hardened by design ASIC is on the way, based on the results from this work. In the next section, the time-synchronization and time-stamping methods will be discussed. These issues have become more important for the Ariane launcher when dealing with wireless sensor networks because it is easier for the on board computer to handle wired sensors with time synchronization/stamping than wireless sensors.

8.2.3 Time-synchronization and Time-stamping Methods

Figure 8.13 shows the reliable sensor network (RSN), a part of the Ariane launcher telemetry system [10]. The time is synchronized from the avionic network gateway down to the sensor node.

In this section, the development of the time-synchronization and time-stamping methods involves minimizing the amount of hardware on the sensor nodes and reducing the workload of the sensor node required to handle this issue.

By using wired communication protocols, the delay $T_A - T_B$ in the RSN becomes deterministic. Wireless communication for the access point and the sensor nodes gives a less deterministic delay T_C and this is the focus of this section. A better method is required for time synchronization between the access point and sensor nodes. This method should be also applicable for UWB sensor networks. Visible light communication (VLC) is used to synchronize the time of all the sensor nodes at the same time.

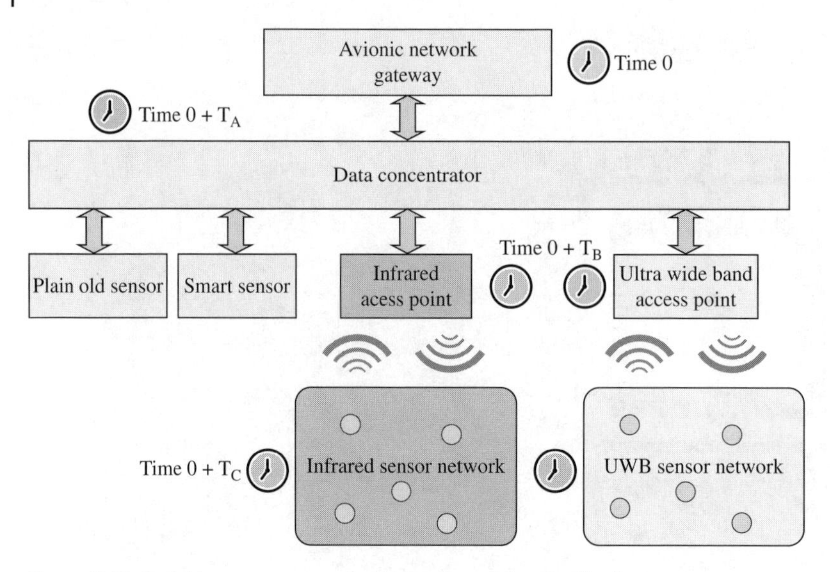

Figure 8.13 Reliable sensor network with its time-delay distribution.

This is possible because each sensor node has a solar cell to receive light from the VLC transmitter.

The sub-system for time synchronization is shown in the block diagram in Figure 8.14. The delay T_1 is caused by the reaction times required by

- the VLC circuit
- the VLC LEDs
- the solar cells of the sensor nodes
- the demodulation circuits of the sensor nodes
- the microcontroller
- the infrared transmitter circuit
- the infrared receiver circuit.

The analogue signal acquisition through an analogue-to-digital converter (ADC) adds delay T_2 and the last delay is T_3 and is caused by the humidity sensor. The delays T_1–T_3 are recorded through repeated measurements on the hardware. These delays are then stored in the access point and later on used to recover the timestamp of each measurement sequence performed by the sensor nodes.

All of the sensor nodes will receive the time information (T_{old}) at the same time, followed by the related commands to perform the sensing sequences. The number of the sensing sequence for ADC0–ADC3 is indicated by n and for the humidity sensor by m. After the access point, the received T_{old}, n, m and the measurement data from the sensor node are used to calculate the time stamp according to:

$$T_{(n,m)} = T_{old} + T_1 + n \times T_2 + m \times T_3 \tag{8.2}$$

For instance, the timestamp for the first measurement of the ADC is $T_{(n=1,\, m=0)} = T_{\mathrm{old}} + T_1 + 1 \times T_2 + 0 \times T_3$. For the second ADC measurement it should be $T_{(n=2,\, m=0)} = T_{\mathrm{old}} + T_1 + 2 \times T_2 + 0 \times T_3$ and for the first humidity measurement it is $T_{(n=0,\, m=1)} = T_{\mathrm{old}} + T_1 + 0 \times T_2 + 1 \times T_3$.

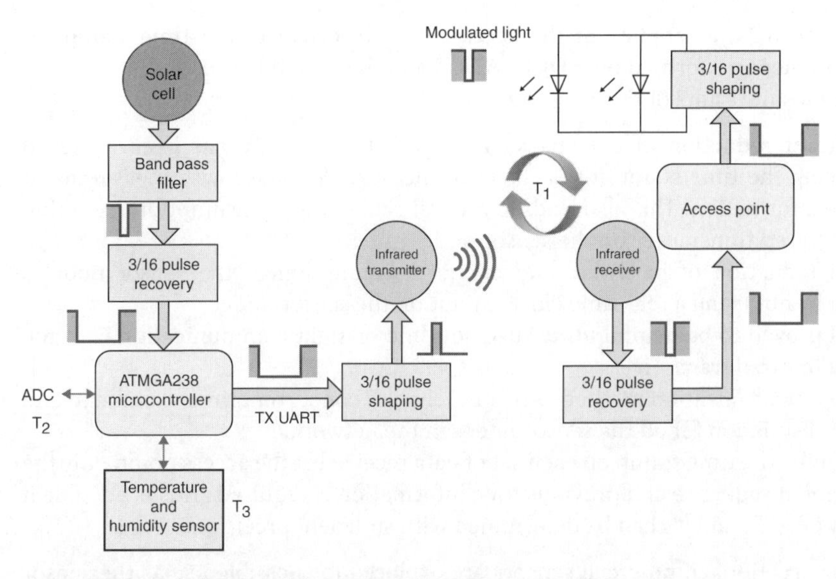

Figure 8.14 Block diagram of the access point and sensor node.

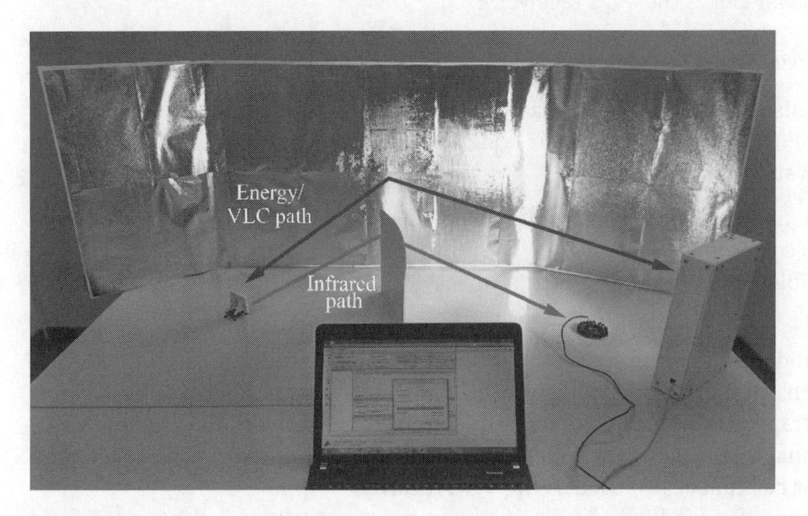

Figure 8.15 Measurement with VLC and energy distribution through the MLI for time synchronization and time stamping.

To test the time-synchronization and time-stamp methods, the equipment in Figure 8.15 was arranged in such a way that there was a no-line-of-sight connection between the sensor node and the access point infrared receiver. In this experiment, diffuse non-line-of-sight visible/infrared propagation via the reflection off the MLI was used [11]. The VLC transmitter consists of six high-power LED lights arranged to focus towards the MLI, with the reflection direct to the sensor node.

The visible and infrared light path in the experiment was 1.5 m long. The measurement results on the oscilloscope show that T_1 is 2.7 ms, T_2 is 1.8 ms and T_3 is

1.4 s. Equation 8.2 was applied at the access point to calculate the time stamp for each measurement performed on ADC0–ADC3 and the humidity sensor.

Some important results of this work are:

- A significant reduction of data package length of up to 40% has been achieved by replacing the time stamp for each set of measurement data with measurement sequence information. This also leads to a significant energy consumption reduction for the infrared transmitter on the sensor node.
- Essential reduction of hardware and energy consumption of the sensor node is achieved by not using a real-time clock circuit on the sensor node.
- The MLI proved to be helpful for diffuse non-line-of-sight communication with the VLC and infrared transceivers.
- The time synchronization method with VLC enables concurrent time, command and energy transmission for all the sensor nodes in the network.
- To determine the time stamp on each set of data received at the access point, only the measurement sequence and previous time information is required. This is possible if the delays T_1, T_2 and T_3 can be determined with sufficient precision.

In the next section, commercial sensors are explored for possible use as the sensor nodes. The sensor node incorporates the infrared transceiver, the infrared transceiver ASIC and the solar cell as the VLC receiver.

8.2.4 Commercial Smart Sensors for Ariane 5 Telemetry Subsystems

Commercial sensors with built in data-acquisition and processing units are selected for the sensor node. The focus of the sensor node is to measure environmental parameters inside the VEB such as:

- air pressure /temperature
- three-axis acceleration
- infrared/visible light intensity
- air humidity,

The sensor node architecture consists of three main parts, as shown in Figure 8.16. The sensing part consist of so called 'smart sensors', connected with the microcontroller with an ADC. The transmission part consists of the infrared transceiver LED and the ASIC. The power management part is supplied by a 3.7 V/150 mAh lithium battery, charged with a 6 V solar cell that is also used as the VLC receiver.

There are four different kinds of smart sensor on the market for the sensor node (Figure 8.17):

- humidity and temperature sensors [12]
- infrared and visible light sensors [13]
- relative air pressure sensors [14]
- three-axis accelerometers [15].

The sensor node architecture, built with commercial components and an infrared transceiver ASIC, is shown in Figure 8.18. On the printed circuit board, the infrared transceiver is placed near the ASIC.

The measurement time for each smart sensor is measured in sequences, as shown in Figure 8.19. The first sequence M_1 is the three-axis acceleration measurement with 100

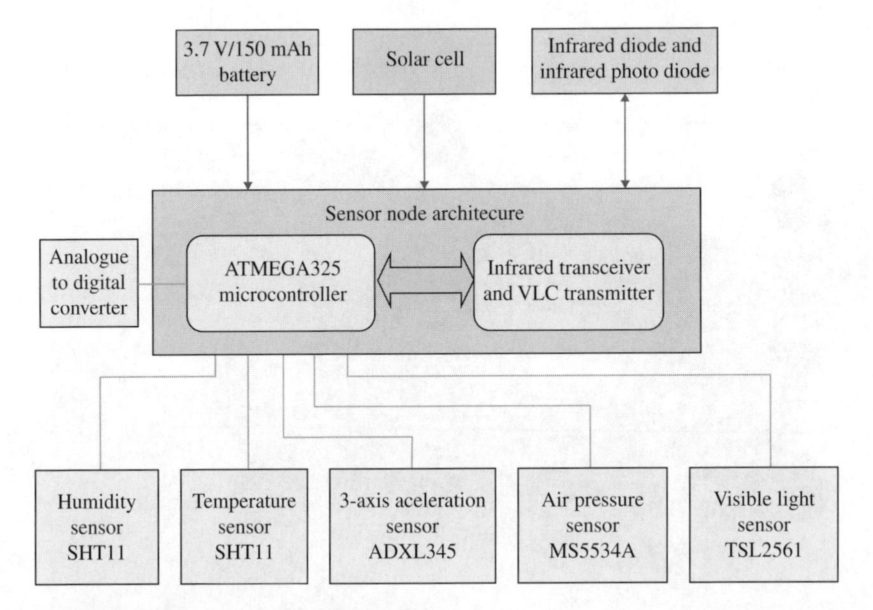

Figure 8.16 Architecture of the sensor node.

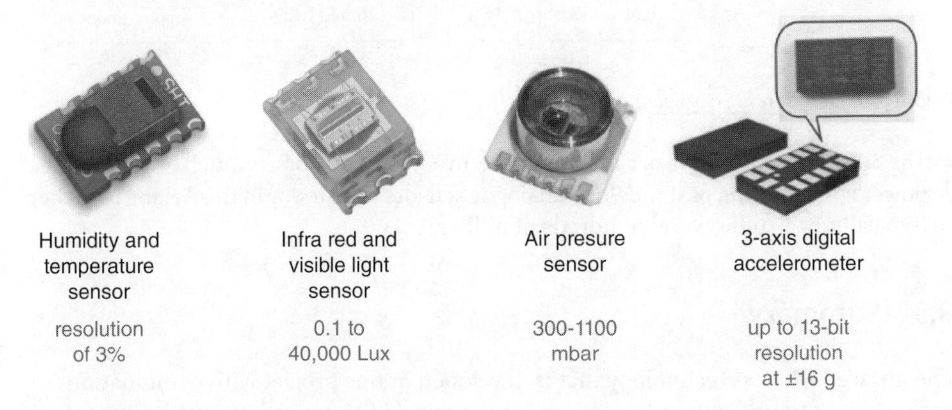

Figure 8.17 The smart sensors for the telemetry subsystems on the wireless sensor node.

data samples. The second sequence M_2 is for measuring the infrared and visible light intensity. The air pressure measurement with temperature compensation is M_3. The last sequence M_4 is the relative air humidity measurement. The total time taken to perform all the measurements in the VEB is 1.684 s.

The subsystem test in the lab, with the access point, three sensor nodes and three VLC transmitters is shown in Figure 8.20. The access point is based on a Xilinx Zynq-7000-FPGA and runs on the Linux operating system to interface with the data concentrator.

Some important results of designing the wireless sensor nodes are:

- Each smart sensor with its required measurement period is verified by the measurement results. The three-axis acceleration sensors require the smallest measurement time, of 1.6 ms.

Figure 8.18 Sensor node hardware.

- The power consumption, of a maximum 3.3 mW for all sensors, is much lower than the power consumption of standalone analogue sensors used today in the Ariane launcher (typically 10 V and several hundreds of milliamps).

8.3 Summary

The infrared wireless technology that is developed in this project with combinations of commercial components and an infrared transceiver ASIC represent a promising future development for the new Ariane launcher.

The MLI that covers most of the inner launcher structures is advantageous for optical communication in the VEB. This is shown in the measurement result when the DC light is introduced to power the solar cell and send information.

The time synchronization and time stamping for the sensor node also take advantage of VLC. This is especially useful if we want to send time information to all sensors at the same time. The working load of the sensor node has been shortened by shifting the calculation of the time stamp to the access point, which has much more computational power.

The combination of the infrared transceiver ASIC and the smart sensors' low power consumption makes the sensor node more durable for space missions. The sensor nodes' ability to perform in many physical measurement environments and resist extremes of temperature, humidity, pressure, acceleration and vibration, covers most of the sensing requirements in the VEB.

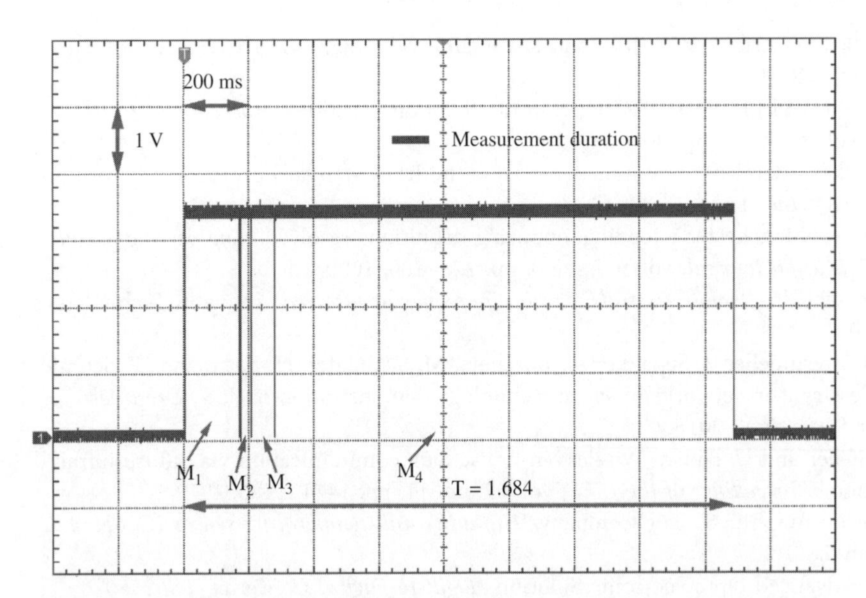

Figure 8.19 The telemetry subsystems measurement sequences.

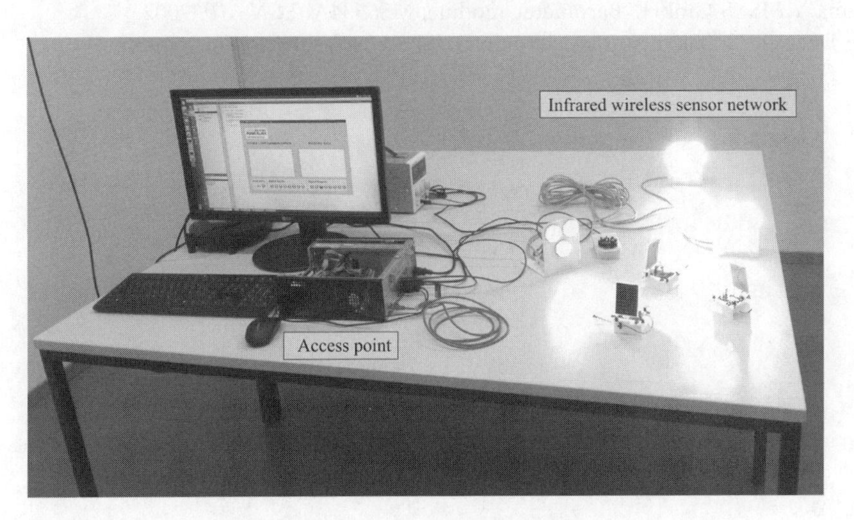

Figure 8.20 The infrared wireless sensor network tested in the lab.

References

1 'Europe okays design for next-generation rocket'. *PhysOrg*, 9 July 2013. Retrieved 9 July 2013.

2 European Space Agency, '*About future launchers preparatory programme (FLPP)*', 4 January 2016.

3 European Space Agency, '*FLPP Preparing for Europe's Next-Generation Launcher*', 9 October 2008.

4 Arianespace, '*Ariane 5 User's Manual Issue 5*', 1 July 2011.

5 'The VEB for Ariane 5G's third and last qualification flight' *ESA/CNES/Arianespace, ID209081*, 2004.
6 J.M. Kahn and J.R. Barry, 'Wireless infrared communication,' *Proceedings of the IEEE*, vol. 85, no. 2, pp. 265–298, 1997.
7 B.C. Baker, 'Wireless communication using the IrDA standard protocol', *Microchip Technology, Inc.*, pp. 1–2, 2003.
8 R. Forster 'Manchester encoding: opposing definitions resolved', *Engineering Science and Education Journal*, vol. 9, issue 6, pp. 278–280, 2000.
9 M. Gotschlich, *'Remote controls – radio frequency or infrared'*. Infineon Technologies AG, 2010.
10 H.J. Besstermueller, J. Sebald, H.-J. Borchers, M. Schneider, H. Luttmann, V. Schmid, 'Wireless-sensor networks in space technology demonstration on ISS', *Dresdner Sensor-Symposium* 2015.
11 F.R. Gfeller and U. Bapst, 'Wireless in-house data communication via diffuse infrared radiation,' *Proceedings of the IEEE*, vol. 67, no. 11, pp. 1474–1486, 1979.
12 Sensorion AG, The Sensor Company, *'Humidity and temperature sensor IC, SHT1x'*, Version 5, 2011.
13 Texas Advanced Optoelectronic Solution, *'Light-to-Digital Converter, TSL2560'*, *TAOS059N*, 2009.
14 Intersema, AMSYS GmbH, 'Barometer module, MS5534A' *ECN 510*, 2002.
15 Analog Devices, '3-axis digital accelerometer, ADXL345', Revision 0, 2009.

9

Multichannel Wireless Sensor Networks for Structural Health Monitoring

Pascale Minet[1], Gerard Chalhoub[2], Erwan Livolant[1], Michel Misson[2], Ridha Soua[1], Rana Diab[2], Badr Rmili[3] and Jean-Francois Perelgritz[4]

[1] *Inria Paris, France*
[2] *Clermont University, Aubière, France*
[3] *CNES Launcher Directorate, Paris, France*
[4] *Airbus Group Innovations, Suresnes, France*

Structural health monitoring has recently been applied to aircraft and space launch vehicles (referred to here as 'launchers'), in which the number of interconnected devices is continuously increasing. Up to now, wired networks have been used, but their high mass leads to increased fuel consumption and high carbon emissions. Wireless sensor networks (WSNs) would certainly reduce the mass and the complexity of the wiring, yet the essential question is whether they able to meet the requirements of non-critical and health monitoring applications in the specific environment of aircraft and launchers.

First, we unify the requirements of non-critical and health monitoring applications in aircraft and launchers, and we show that such requirements require the use of multichannel mesh wireless networks. Multichannel networks bring many advantages in terms of latency, throughput and robustness. However, they do raise a number of challenges, some of which are general, while others are specific to supporting data-gathering applications. Different state-of-the art solutions are described. These solutions, whether designed to take into account the specificities of data gathering or not, range from medium access control to multihop routing. The best performances are obtained when routing and medium access control are tackled together. We establish bounds on the minimum number of time slots needed by a raw data convergecast, taking into account the number of available channels, the number of children of the sink in the routing tree, as well as the number of radio interfaces in the sink. We outline SAHARA, a solution that provides an adaptive multichannel collision-free protocol for data gathering and we present many performance results obtained by simulation.

9.1 Context

Structural health monitoring is an emerging technology in which structures are monitored and inspected for their capacity to serve their intended purpose. Damage and malfunctions are detected and analysed. Structural health monitoring has recently been

Wireless Sensor Systems for Extreme Environments: Space, Underwater, Underground, and Industrial.
First Edition. Edited by Habib F. Rashvand and Ali Abedi.

applied to aircraft. However, the physical systems used in aircraft are becoming more and more complex due to the growing number of interconnected components and the increasing volume of data exchanged. Automated prognostic and health monitoring systems (HMSs) have been proposed to address this issue. Basically, these systems require sensor networks in which different types of sensors are integrated in the aircraft to sense parameters such as temperature, vibration and pressure. Each sensor performs a measurement and then transmits its data to a central entity using wires. The central entity gathers all the data transmitted in the network for further analysis. Nevertheless, intrinsic characteristics of wires, such as overall weight, cost and proneness to breaks and degradation, are hampering the use of wires and the integration of sensors in aircraft. Consequently, there is an urgent need for wireless HMSs that can mitigate these limitations.

Today, WSNs are revolutionising not only home and factory monitoring, but also the aerospace industry. It is argued that transitioning from wired sensor networks to WSNs will radically reduce the amount of wiring [1]. Are WSNs able to meet the requirements of aircraft and launchers?

For the sake of readability, this chapter uses the term 'aircraft' to cover both aircraft and launchers.

9.1.1 Expected Benefits of WSNs in Aircraft

Generally speaking, the benefits of WSNs in aircraft should be a reduction in the complexity and mass of wiring. For instance, cables account for 70% of the 1.5 ton mass of Ariane 5 [2]. The wiring length of a Boeing 747-400 is about 270 km and the most recent and most complex Airbus A380 has over 530 km of cables [3]. In addition, cable routing in aircraft is challenging, because electrical signal cables must be physically separated so as to avoid electrical interference. The developmental flight instrumentation (DFI) for Orion Exploration Flight Test 1 weighs 544 kg for 1200 channels and wires represent 57% of the total weight of the DFI data system.

A NASA study [4] has suggested that there could be a significant potential mass improvement in DFI for future Orion missions. Miniaturisation, optimized architecture and equipment distribution of a 100% wired, 400-channel DFI system might have a mass of 54 kg, but using 100% wireless, it might only weigh 38 kg. A simple computation shows that the mass per channel for a wired approach is 0.45 kg [4]. This mass can be reduced to 0.19 kg using a miniaturised and optimized wired approach, whereas a fully wireless approach has a mass of 0.09 kg per channel. Similar studies conducted on the Ariane launcher have also shown large mass improvements in instrumentation and telemetry subsystems.

As a result, fuel efficiency is improved and carbon emissions are reduced. In addition, replacing wires by wireless connections facilitates the addition or removal of sensors, and introduces the possibility of installing sensors in locations previously inaccessible due to wiring constraints. Thus more sensors can be deployed, helping to ensure system redundancy.

9.1.2 WSN Requirements for Aircraft

To cope with aircraft constraints and specific environments, unified WSN requirements have been defined by aircraft manufacturers and end-users in the SAHARA project, supported by the French aerospace cluster ASTech. This project, led by Airbus Group

Innovations, started at the end of 2011 and comprises academics (CNES, ECE, EPMI, Inria, LIMOS), aircraft manufacturers acting as end-users (Airbus Group Innovations, Airbus Defence and Space, Airbus Helicopters, Safran-ES), as well as small and medium-size businesses (BeanAir, GLOBALSYS, OKTAL-SE, ReFLEX-CES).

In order to shorten development times and to ease the path to technology adoption in aircraft, SAHARA is focused only on non-critical sensors and HMSs and is based only on mature commercial off-the-shelf (COTS) technologies. The goal of this project is to develop existing wireless technologies using COTS components, adapting them into a generic and adaptive WSN for use in aircraft. The scope of the project covers the definition of the unified WSN requirements, the development of WSN technologies and protocols, the development of WSN demonstrators, tests in representative aircraft environments and the development of on-board radio-frequency propagation models.

The unified requirements cover different types of aircraft and launchers. This chapter focuses only on non-critical and HMS measurements. These requirements can be summarised as follows:

- static sensors for temperature, pressure, and so on
- dynamic sensors for vibration, shock, strain and so on
- sampling rates from a few samples per hour to 10 ksps
- sensor networks with various sampling rates
- sensor nodes per network varying from 1 to 50
- expected sensor-node battery capacity from 40 min to 14 h in active mode and 24 months in sleep mode
- sensor nodes integrated in a confined environment leading to propagation issues
- sensor nodes integrated in aircraft fuselage, empennage, wings, engine, landing gear; launcher stage, fuel tank, payload fairing at a distance from the closest sink that is greater than the communication range of the nodes, leading to multihop communication
- latency from a sensor node to the closest sink ranging from 100 to 500 ms
- measurement dating accuracy from 1 to 100 μs.

To meet these strong requirements in terms of latency, energy efficiency and reliability, low-powered wireless mesh networks have been introduced. In addition, to meet determinism and latency constraints, time-slotted and multichannel medium access control is required.

9.1.3 Previous Work

There have been many previous projects dealing with the integration of wireless technologies in aircraft systems. We list several below and outline their objectives.

WISE The Integrated WIreless SEnsing (WISE) European project [5] was aimed at:

- enhancing the aircraft monitoring system by deploying new wireless technologies involving low-power autonomous sensors
- monitoring new parameters where this was not possible using physical links
- continuing the monitoring or improving redundancy when the physical link is impaired;
- improving information segregation.

The final goals were to reduce aircraft operating and installation costs, improve the availability and dispatch rate through a simplification in the maintenance system, reduce the cost of ground and flight-test installations, improve the man–machine interface and reduce accident rates.

SWAN The @MOST SWAN project [6] embraced all engineering issues related to a new aeronautical system, from defining applications and capturing requirements to developing a prototype, and included investigations of aircraft regulation and certification as well as security, safety and reliability issues involved in the design of an aircraft systems architecture integrating WSNs. The main objectives of SWAN were to:

- provide solutions relying on WSN technology for maintenance operations, giving increased efficiency compared to current approaches
- identify potential improvements in different aircraft domains (such as aerodynamics, engine, cabin, systems and structure) where the overall aircraft maintenance could benefit.

AUTOSENS The AUTOnomous SENSing microsystem, (AUTOSENS) project [7] focused on energy harvesting for autonomous wireless sensors embedded in aircraft. The architecture proposed takes into account the specific features of the targeted environment. One module manages energy for sensing and processing while another module manages energy for communication.

SMMART The System for Mobile Maintenance Accessible in Real Time, (SMMART) is a European project [8] investigating a new integrated approach to the maintenance challenges of the transport industry, including air, road and marine transport. Its aims were:

- to reduce the time and cost for scheduled and unscheduled maintenance inspections of increasingly sophisticated and complex products
- to remotely provide adequate up-to-date information for mobile workers in all their tasks, wherever they operate
- to minimize the cost penalties of unscheduled downtime on large transport fleets.

The project was based on radio frequency identification (RFID).

Although these projects have ended, their results have not been made public. It is nonetheless worth noting that all these projects considered wireless technologies to be very promising for preventive maintenance and structural health monitoring in aircraft. However, the autonomy of wireless sensors remains a challenging issue.

9.1.4 Chapter Organization

This chapter is organized as follows. Section 9.2 describes the challenges resulting from multichannel use and presents different solutions. Section 9.3 focuses more particularly on challenges in data-gathering applications and gives examples of solutions. Section 9.4 presents SAHARA, a solution designed to be compliant with the unified WSN requirements for aircraft and using IEEE 802.15.4 COTS technologies. Finally, in Section 9.5, the issues arising are discussed and various perspectives are given. This chapter is an extended version of a paper [9] published at the IEEE WISEE 2015 conference.

9.2 General Multichannel Challenges

In a multichannel network, several channels are used simultaneously by nodes to communicate. Additional challenges arise for communication protocols in multichannel networks. In this section, we will discuss these challenges and present state-of-the-art solutions.

9.2.1 Signal Propagation in an Aircraft Cabin or inside a Launcher

The analysis and behaviour prediction of a 2.4-GHz signal in an enclosed area depends on the physical conditions in which these signals will propagate. At this relatively short wavelength (12.5 cm), reflection from lateral walls, ceiling surfaces, diffraction on seats, headrests and people, and refraction by metallic or hard surfaces are important. The number of parameters renders the signal-analysis task fairly complex. The important parameters for adequate signal characterisation, such as propagation path loss, delay spread and coherence bandwidth, must be validated using both simulation and experiment.

In addition, since the cabin in itself is normally made up of hard surfaces, the possibility of multipath propagation due to successive high-energy reflections on the walls and ceiling must be characterised in order to select suitable signal modulation and symbol speed to avoid excessive intersymbol interference. The noise level from other transmitting devices onboard (such as on-board WiFi or Bluetooth) as well as possible interference of the proposed system must be fully understood prior to the finalisation of the system design.

These considerations have led us to consider a multihop and multichannel adaptive solution. There are publicly available measurements and research dealing with signal propagation for the 2.4-GHz ISM frequency band in enclosed areas such as trains. Rigorous studies have given a working knowledge of the parameters [10, 11]. The results show that because of the many reflections from metallic or hard surfaces, the path loss exponent is smaller than for signal propagation in free space. The positioning of the transmitting/receiving antennas has also been found to be, in many cases, critical for adequate transmission at all times. Hence simulation models have to be modified to correctly predict either the coherence bandwidth or the delay spread. The time-variability of the radio link parameters is probably much more important for an aircraft cabin than for a particular level of a launcher. Experimental investigations are absolutely necessary for such confined and obstructed areas.

9.2.2 Mesh Multichannel Wireless Networks

Even though the size of an airplane or of a launcher is not that large (less than 80×80 m for an airplane) compared to the average communication range in the 2.4-GHz frequency band, the complexity and the time-varying behaviour of the wireless medium led us to consider a multihop topology; this would allow multipath communication for data, thanks to the natural redundancy of such topologies. For reliability reasons, a single point of failure should be avoided. As a consequence, star topologies are not suitable. Mesh topologies are preferred because if link or node failure occurs, another route to the destination can be found.

9.2.3 Network Build-up

During the network build-up phase, nodes try to be part of the network in order to be able to exchange and relay data packets. When a node is activated, there is usually a network discovery phase during which it scans for existing networks. The scanning procedure in a multichannel network is not trivial. A new node should be able to detect activity on a certain channel so that it can try to communicate with the network in order to gain access to and become part of it. Special advertisement frames, usually termed 'beacons', are used for signalling the presence of the network. This is the case for WiFi [12] and ZigBee [13], for example.

The challenge lies in making sure that the new node is able to find the network. In other words, the node should be able to be on the same channel at the same time as neighboring nodes that have already joined the network. Without a fixed and known control channel, this procedure can last a considerable time and, in some cases, drain a lot of the node energy resources.

One solution to making this phase less time- and energy-consuming is to exchange control traffic on a fixed and known channel. This would allow new nodes to scan only one channel. Nodes should periodically switch to this control channel and send beaconing frames in order to be detected by new nodes.

9.2.4 Node Synchronization

When guaranteeing access to the medium and a maximum end-to-end delay in a multi-hop network, node synchronization becomes a must. The fact that nodes are working on different channels makes it even more important to have network scale synchronization in order to manage network discovery and neighborhood updates. Protocols such as TSCH, from IEEE 802.15.4e, are based on one-hop synchronization in order to allocate timeslots on multiple channels.

Extending this synchronization in order to reach nodes that are multiple hops away is a challenge. It could be achieved using an external synchronization device, but the difficult part would be to make sure that this device could be reached by all the nodes of the network. For example, synchronizing nodes using GPS is only possible when all the nodes are able to communicate with satellites. In addition, this has consequences for the weight and energy consumption of the nodes.

Another approach would involve achieving relative synchronization based on an internal reference that was part of the network. For instance, a designated node in the network could broadcast a synchronization beacon that would be propagated by other nodes in order to reach all the nodes of the network. Similar approaches have been reported in the literature [14, 15].

9.2.5 Selection of Channels

Wireless standards based on IEEE 802.15.4 have 16 available channels in the 2.4-GHz frequency band. Other wireless standards, such as WiFi and Bluetooth also use the 2.4-GHz band. This makes IEEE 802.15.4 networks vulnerable to interference from nearby networks, including other IEEE 802.15.4 networks. We call this 'external interference'. External interference happens when nodes of the network receive perturbations coming from sources that are not part of the network.

Internal interference is caused by nodes that are part of the network. Depending on the modulation and the frequencies used in the physical layer, nodes using the same channel are tend to generate interference. In order to avoid this, orthogonal channels should be used.

Internal interference can be avoided using channel assignment techniques, discussed in the Section 9.2.6. To avoid external interference, a scan is usually carried out to identify the energy level on each channel. The scan procedure should result in what is called a 'channel blacklist', which is a list of channels that should not be used in the network. This blacklisting technique can be done in a distributed manner. Every node in the network scans its own environment and builds its own blacklist. This local blacklist is then used locally to choose a convenient channel, or is sent to a controller node that is in charge of distributing channels to all the nodes of the network. This of course generates a significant delay before a channel can be chosen. In addition, if the interference is not stable (which is usually the case), the channel blacklist must be updated frequently. Blacklisting of channels can also be performed for the whole network. This allows the frequency band to be segmented over different networks in order to enhance performance while avoiding external interference. This can be achieved if nearby networks are manageable, but this is often not the case.

9.2.6 Channel Assignment

Channel assignment represents one of the main challenges of multichannel MAC protocols. Constrained by cost and the size considerations, most nodes are equipped with just one radio transceiver, so most protocols propose solutions that allow nodes to only send or receive at any given time. Some assignment schemes allocate a channel for receivers and another for the transmitters. Some protocols combine channel allocation with time-slot allocation.

Some protocols propose a static channel assignment where nodes keep using the same channel until neighborhood or interference conditions change and force them to seek a more suitable one [16]. This approach is often lightweight and does not waste energy on frequent switches of channel. Other protocols use a semi-dynamic channel assignment where nodes switch channels according to the destination [17, 18, 19]. This approach is adaptive and allows more flexibility in the choice of a suitable channel. A more dynamic channel assignment method consists of changing channels at each transmission [20, 21]. This approach is more robust because it avoids bad channels and enables nodes to use all available channels, but at the cost of energy and time-wasting due to frequent switches of channel.

Other protocols propose a multi-interface sink in order to enhance the reception throughput of a particular node [22, 23]. Indeed, assigning different channels to the sink transceivers allows simultaneous receptions.

9.2.7 Network Connectivity

Multichannel assignment solutions may differ in the assumptions made regarding network connectivity, as illustrated in Figure 9.1, where channel *ch*1 is represented by a solid line whereas channel *ch*2 is depicted in dashed line:

- The same topology exists on all channels and connectivity is assumed on each channel like in Figure 9.1a.

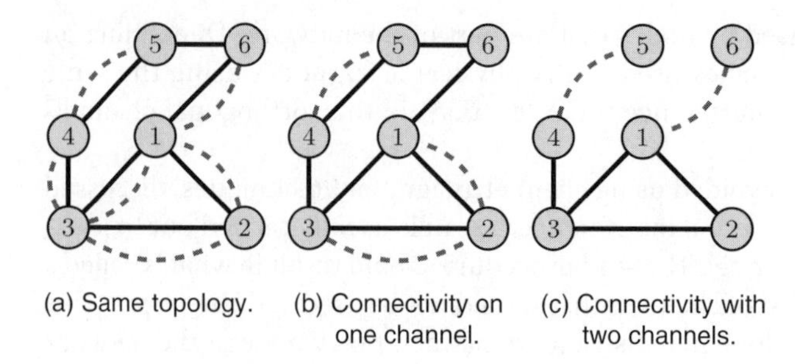

Figure 9.1 Different cases of connectivity.

- The topology may differ from one channel to another but connectivity is assumed at least on one channel, usually the channel used for control messages; this assumption is frequently made. For instance, in Figure 9.1b, there are two disconnected parts on channel *ch*2.
- Several channels are needed to ensure network connectivity. This is the most permissive assumption like in Figure 9.1c.

9.2.8 Neighborhood Discovery

For any node *u*, a node *v* is said to be a one-hop neighbor of *u* on channel *c* if and only if *u* is able to receive messages from *v* on channel *c* and *v* is able to receive messages from *u* on channel *c*: a symmetric link exists between them on channel *c*. Some solutions do not check the symmetry of a link before using it. This will cause useless retransmissions when an acknowledgment is required and the link is not symmetric [24]. Notice that in a real multichannel environment, a node may be a one-hop neighbor on one channel and not on another one [25]. There are different types of solutions according to the assumptions made–assumptions similar to those on network connectivity. The simplest solutions perform neighborhood discovery on only one channel, usually the control channel, whereas the most sophisticated ones perform as many neighborhood discoveries as channels used. Some solutions take advantage of large similarities of links between channels to store this knowledge efficiently.

9.2.9 Medium Access Control

Multichannel medium access control (MAC) protocols that have been proposed in the literature use either time division multiple access (TDMA), or carrier sense multiple access with collision avoidance (CSMA/CA), or a combination of both techniques. In what follows, we briefly describe the best-known multichannel protocols.

9.2.9.1 Contention-based Protocols

Lohier et al. proposed Multichannel Access for Sensor Networks (MASN) [17], a multichannel protocol for hierarchical ZigBee networks with many-to-one transmissions. Assignment of the different channels is centralised by a coordinator and based on the hierarchical address assignment process used in ZigBee. The main advantage of this solution is its simplicity of integration in IEEE 802.15.4 devices with only slight modifications in the MAC layer. However, the authors used a single topology for the simulations, which might be the most convenient topology for MASN.

Zhou et al. proposed Multifrequency Media Access Control for Wireless Sensor Networks (MMSN) [19], which uses a semi-dynamic channel assignment approach for channel allocation based on different strategies. Strategies differ according to the level of overhead and the effectiveness of the channel allocation. At the beginning of each time slot, nodes contend for the medium to broadcast control traffic on a common broadcast channel. When a node transmits a packet, it switches between its own channel and the destination channel during the preamble sending time, which results in increases in the protocol overhead and the number of repetitions due to frequent switches of channel.

9.2.9.2 Contention-free Protocols

Incel et al. proposed the Multichannel Lightweight MAC Protocol (MC-LMAC) [21], which guarantees that the same slot/channel pair is not simultaneously used in the neighborhood up to two hops. MC-LMAC suffers from the overhead of the control messages that are exchanged to discover the channels used in the neighborhood up to two hops. The problem increases with network density.

Pister and Doherty proposed the time synchronized mesh protocol (TSMP) [26], on which industrial standards such as WirelessHART and ISA100.11a are based, as well as IEEE 802.15.4e. It uses a channel-hopping technique to enable nodes to switch channels at each transmission. However, one drawback is its inability to support changes in topology.

9.2.9.3 Hybrid Protocols

The Energy-efficient Multichannel MAC Protocol for Dense Wireless Sensor Networks (Y-MAC) [27], is based on algorithms proposed in LMAC and MMSN. Time slots are assigned to the receivers. It allows new nodes to join the network and assigns time slots in a dynamic way. At the beginning of each time slot, potential senders to the same receiver compete in order to access the medium using a CSMA/CA algorithm. Multiple packets are sent successively on different channels, and the receiver and the sender hop to a new channel according to a predetermined sequence. Y-MAC reduces latency by offering the possibility for nodes that were not able to send their traffic to compete in a consecutive slot. However, the channel allocation method is not detailed in the paper; the authors only insist on not using the same channel in a one-hop neighborhood, which leads to high interference caused by simultaneous transmissions received from nodes that are two hops away.

Priya and Manohar proposed the Energy Efficient Hybrid MAC for WSN, (EE-MAC) [18]. This is a centralised protocol that uses a semi-dynamic channel assignment approach for channel allocation. EE-MAC operates in two phases: a setup phase and a transmission phase. During the first phase, neighbor discovery, slot assignment and global synchronization are achieved. These operations run only during the set-up phase and every time a change in the topology occurs. During the transmission phase, time is divided into time slots. Each slot is divided into schedule subslots and contention subslots. Each cycle starts with scheduled slots followed by contention slots. Nodes use low power listening (LPL) [28] during contention slots and send Hello messages to the base station. EE-MAC suffers from the overhead of the Hello messages that are exchanged and sent to the base station.

Borms et al. proposed the Multichannel MAC protocol (MuChMAC) [20]. It uses a dynamic channel assignment approach. Time is divided into slots. Each node is able to independently choose its receiving channel switching sequence based on its identifier and the current slot number using a pseudo-random generator. A broadcast slot is inserted every n slots. These common broadcast slots also follow a pseudo-random channel-hopping sequence. A sender is thus able to calculate the channel of the receiver. The main drawback of MuChMAC is that channel allocation is based on a random mechanism that does not take into consideration the channel usage in the neighborhood.

Diab et al. proposed the Hybrid MultiChannel MAC protocol (HMC-MAC) [22], based on TDMA for signalling traffic and CSMA/CA combined with FDMA for data exchange. Time is divided into intervals. A TDMA interval is dedicated for neighbor discovery and the channel allocation process. HMC-MAC aims at reducing the control traffic overhead. It allows nodes to share slots on the same channel in order to send data to the same destination. Related results [29] show that the approach enhances network performance in terms of the number of collisions and also packet delivery rates. However, it suffers from high end-to-end delays due to packet accumulation in nodes close to the sink.

9.2.10 Dynamic Multihop Routing

The energy constraints of sensor nodes are challenging issues in the design of routing protocols for WSNs. Proposed protocols aim at load balancing, minimizing the energy consumed by the end-to-end transmission of a packet and avoiding nodes with low residual energy. Initially designed for a single-channel network, they can also be used for multichannel networks. Different families of multihop routing protocols can be distinguished, as described next.

Data-centric protocols send data only to interested nodes in order to avoid useless transmissions. Such protocols make the assumption that data delivery is described by a query-driven model. Two main approaches are proposed for interest dissemination. The first is Sensor Protocols for Information via Negotiation (SPIN) [30], where any node advertises the availability of data and waits for requests from interested nodes. The second is Directed Diffusion (DD) [31], in which the sink broadcasts an interest message to sensors, and only interested nodes reply with a gradient message. Hence both interest and gradient messages establish paths between the sink and interested sensors. Many other proposals have been made, examples being rumour routing and gradient-based routing.

Hierarchical routing determines a hierarchy of nodes to simplify routing and reduce its overhead. The most famous hierarchical routing protocol is the Low-Energy Adaptive Clustering Hierarchy (LEACH) [31]. LEACH organizes sensor nodes into clusters, with one node acting as a cluster head. To balance energy consumption, a randomised rotation of the cluster head is used. Power-Efficient GAthering in Sensor Information Systems (PEGASIS) [31] enhances LEACH by organizing all the nodes into a chain, where the head of the chain varies.

Opportunistic routing takes advantage of the broadcast nature of wireless communications or node mobility. Some techniques maintain multiple forwarding candidates for any given node and select the forwarding candidate taking into account the transmission made. Zeng et al. highlight how these protocols achieve better energy efficiency [32]. Other techniques merge routing and mobility to obtain smaller energy consumption when compared to classical techniques. They use mobile sinks [33, 34], mobile relays [35] or data mules [36, 37] when the connectivity of the network is not permanently assured.

Geographical routing uses the geographical coordinates of the nodes to build routes. Akkaya et al. propose a Geographic and Energy Aware Routing (GEAR) protocol [31], where any message is forwarded first to the target region, and second to the destination within the region. Geographic Adaptive Fidelity (GAF) [31] builds virtual grids based on location information about the nodes. In each cell, a single node is active and routes messages; all the other nodes are sleeping.

Energy based selective routing means avoiding nodes with low residual energy, so as to maximize the network lifetime [38]. These approaches should also save the energy of nodes by selecting routes that minimize the energy consumed by the end-to-end transmission of packets on the routes selected.

9.2.11 Energy Efficiency

The most challenging concern in WSN design is how to save node energy while maintaining the desired network behavior. Any WSN can only fulfil its mission so long as it is considered alive, but not afterwards. As a consequence, the goal of any energy-efficient technique is to maximize network lifetime. This depends markedly on the lifetime of any single node.

To fulfil the mission required by the application, sensor nodes consume energy in sensing, processing, transmitting and receiving data. Hence minimizing the data processed will save the energy of very constrained sensors. In addition, redundancy inherent to WSNs will mean that many similar reports are routed in the network. Communication is a greedy consumer of energy, as confirmed by experimental results.

9.2.11.1 Reasons for Energy Waste

With regard to communication, there is also a great deal of energy wasted in states that are useless from an application point of view:

- *Collision*: when a node simultaneously receives more than one packet, these packets collide. All colliding packets are discarded. If the senders want their packets to be received by their destinations, they retransmit them.
- *Overhearing*: energy is wasted by nodes that are within the transmission range of the sender but are not the intended destination.
- *Control packet overhead*: the number of control packets should be minimized to leave the bandwidth available for data transmissions.
- *Idle listening*: is one of the major sources of energy dissipation in MAC protocols. It happens each time a node listens to an idle channel in order to receive possible traffic.
- *Interference*: when a node receives a packet but cannot decode it.

There are many different techniques aimed at minimizing energy consumption and improving network lifetime.

9.2.11.2 Classification of Energy-efficient Techniques

We can identify five main classes of energy-efficient techniques: data reduction, protocol overhead reduction, energy-efficient routing, duty cycling and topology control.

- *Data reduction*: reduces the amount of data produced, processed and transmitted. Data compression and data aggregation are examples of such techniques.
- *Protocol overhead reduction*: increases protocol efficiency by reducing the overhead. Different techniques exist. Tuning the transmission period of control messages to the stability of the network is one such technique. More generally, optimization based on a cross-layering approach between the application, the network and the MAC layers is another example.
- *Energy-efficient routing*: maximizes network lifetime by minimizing the energy consumed when a packet is transmitted from its source to its final destination. Opportunistic routing, hierarchical routing, data-centric routing, geographical routing and routing selecting routes according to energy criteria are the main examples of energy-efficient routing approaches. Multipath routing protocols use multiple routes to achieve load balancing and robustness against route failures.
- *Duty cycling*: duty cycling means only a fraction of time nodes are active during their lifetime. Nodes' sleep (and therefore active) schedules should be coordinated to meet application requirements. These techniques can be further subdivided into:
  - high granularity techniques that focus on selecting active nodes among all the sensors deployed in the network
  - low granularity techniques that switch off the radio of active nodes when no communication is required and switch it on when a communication involving this node may occur; they are highly related to the medium access protocol.
- *Topology control*: reduces energy consumption by adjusting transmission power while maintaining network connectivity. A new reduced topology is created based on local information.

Table 9.1 shows the impact of each energy-efficient technique on sources of energy waste. The 'M' symbol denotes a main impact, and 'S' a secondary impact.

Table 9.1 Impact of energy-efficient techniques on sources of energy waste.

	Data reduction	Protocol overhead reduction	Energy-efficient routing	Duty cycling	Topology control
Sensing and processing	M	-	-	M	-
Communication	M	M	M	M	M
Collision	S	S	S	M	M
Overhearing	S	S	M	M	M
Control packets	-	M	S	S	-
Idle listening	-	-	-	M	-
Interference	S	S	M	M	M

9.2.12 Robustness and Adaptivity of WSNs

As discussed in Section 9.2.2, signal propagation in confined areas such as aircraft is prone to link failures due to the complex nature of the surroundings. In order to ensure a robust protocol, nodes should be able to adapt to the changing link conditions on the MAC level and the routing level. In many real deployments, the WSN encounters dynamic changes, so it is not sufficient to have a WSN operational. This WSN must also self-adapt to the following factors:

- *Topology changes.* This is usually provided by the routing protocol, which automatically selects a new route when the current one breaks.
- *Traffic changes.* An adaptive time-slot and channel assignment must take into account the traffic changes, in order to assign more slots to nodes that have a higher load.
- *Environment perturbations.* These perturbations can be due to an external source of perturbation, such as radar, or an internal one, such as when there is interference within the WSN itself. The MAC and routing protocols used should be able to select the channels that are not subject to perturbation in order to improve the delivery rate and more generally the quality of service experienced by the users.

9.3 Multichannel Challenges for Data Gathering Support

Data-gathering is the typical applications supported by WSNs in aircraft. Each node senses its environment and generates data that are transferred to the sink in charge of collecting and processing them. Each sensor node plays the role of data source and/or router node to deliver data messages to the sink, without aggregation by intermediate routers. This data collection is called a *raw data convergecast.*

Two key issues for data convergecast are:

- minimizing latencies and guaranteeing packet delivery
- energy saving.

Minimized end-to-end delays ensures the freshness of collected data. In addition, guaranteed packet delivery leads to more accurate monitoring. A limiting factor for fast data collection is interference. To mitigate this problem, researchers resort to multichannel communications. Indeed, multichannel communications are exploited to increase, on the one hand, network capacity with parallel transmissions and, on the other hand, robustness against internal or external perturbations. Hence the data gathering delays can be greatly reduced.

As convergecast involves a large number of sensors that may transmit simultaneously, collisions and retransmissions represent a major challenge for bounded latencies. Collisions lead to data losses and retransmissions that increase packet latency and result in non-deterministic packet delivery times. Unlike contention-based protocols, which suffer from inefficiency due to backoff and collisions, collision-free protocols guarantee bounded latencies. In fact, these protocols, also called *deterministic-access protocols,* ensure that any transmission by a node does not interfere with any other simultaneous transmission. This is achieved by allocating channels and time slots to nodes in such a way that these interferences are avoided, making it easy to control the packet delay needed to reach the final destination. Furthermore, collision-free protocols are more energy efficient than contention-based protocols. They eliminate major sources of energy waste such as idle listening, overhearing and collisions. In addition, a node is active only when it is transmitting to its parent or receiving from its children. If this is not the case, the nodes turn off their radios. Therefore, collision-free protocols are ideal for battery-powered nodes and contribute to energy saving.

For the sake of simplicity, we assume in the following that each data packet can be transmitted in one slot. The slotframe is periodically transmitted, and is made up of a sequence of time slots. In each time slot, transmissions can occur on the number of channels used simultaneously in the network.

9.3.1 High Concentration of Traffic around the Sink

In raw data convergecast, sensor nodes close to the sink forward more packets than more remote sensor nodes; consequently they have a heavier traffic load.

For any node u, let $Gen(u)$ denote the number of data packets generated by u in a slotframe. We can compute $Trans(u)$, the number of data packets transmitted by u in a slotframe (assuming that each node is able to send all the traffic that it generates). We have $Trans(u)$ that is equal to $Gen(u)$ plus the number of data packets received by u from its children and forwarded to its parent. We can write:

$$Trans(u) = \sum_{v \in subtree(u)} Gen(v), \tag{9.1}$$

where $subtree(u)$ denotes the subtree rooted at u in the routing tree.

9.3.2 Time-slot and Channel Assignment

With regard to traffic load, we can distinguish two approaches for time-slot and channel assignment.

- The simplest time slot and channel assignment does not take into account varying traffic loads. All the nodes have the same number of slots assigned, even though nodes close to the sink have a considerably higher traffic load. As a consequence, in the absence of message loss and message aggregation, all the data sent by the sensor nodes in a slotframe may need up to $max_{u \in WSN} Trans(u)$ slotframes to reach the sink. This worst case is when a data message progresses one hop toward the sink at each slotframe.
- Traffic-aware time slot and channel assignment assigns the exact number of slots needed by each node per slotframe. Consequently, in the absence of message loss, a single slotframe is sufficient to enable all the data gathered in this slotframe to reach the sink.

In the following, we consider only the second approach, which ensures the smallest data-gathering delays.

9.3.3 Conflicting Nodes

In the time-slot and channel assignment problem, two conflicting nodes prevent a node from receiving a data or an acknowledgment intended for it when they use the same channel during the same time slot. Assuming that immediate acknowledgment is used at the MAC level, each unicast data packet is acknowledged in the slot it is sent and we determine two types of collision: data–data and data–acknowledgment. Taking into account that data sent by sensor nodes are collected by the sink using a routing tree,

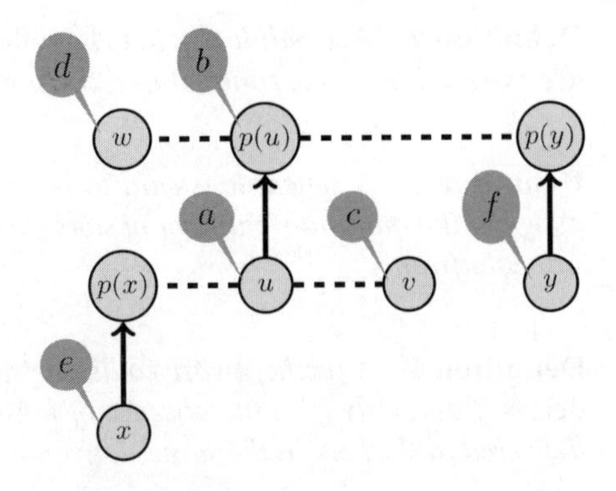

Figure 9.2 Conflicts with immediate acknowledgment.

we can show that the only possible conflicts are those given by Property 1 below, and illustrated in Figure 9.2. In this figure, a solid line between two nodes means a radio link belonging to the routing tree. In contrast, a dashed line represents a radio link that is not used by the routing.

Property 1 *For any node u, its conflicting nodes, when using the immediate acknowledgment, are:*

a) node u itself
b) node Parent(u), denoted p(u) in Figure 9.2
c) one-hop neighbors of u, see node v
d) one-hop neighbors of Parent(u), see node w
e) nodes whose parent is a one-hop neighbor of u, see node x
f) nodes whose parent is a one-hop neighbor of Parent(u), see node y.

9.3.4 Multi-interface Sink

As most of the WSNs deployed support data-gathering applications, the sink node is the destination of all the data generated in the network. Thus in order to enhance the sink reception throughput, a sink with multiple radio interfaces–in short a multi-interface sink–is used. These multiple radio interfaces enable the sink to receive simultaneously on different channels. It is useless to have a sink with a number of radio interfaces strictly higher than its number of children or the number of channels available, as confirmed by theory and experimental results [39].

9.3.5 Optimal Number of Slots in a Collision-free Schedule

We first give some definitions related to schedules.

Definition 1 *A schedule is said to be* valid *if and only if in each time slot there is no node that:*

- *either transmits on the same channel or on the same radio interface more than once,*
- *or receives on the same channel or on the same radio interface more than once,*
- *or transmits and receives on the same channel or on the same radio interface.*

Definition 2 *A schedule is said to be* collision-free *if and only if no two conflicting nodes are assigned the same timeslot and the same channel.*

Definition 3 *A schedule is said to be* traffic-aware *if and only if each sensor node is assigned the minimum number of slots that enables it to transmit all its messages in the same slotframe.*

Definition 4 *A traffic-aware collision-free schedule is said to* minimize data gathering delays *if and only if in the absence of message loss, each message sent in a slotframe is delivered to the sink in the same slotframe.*

Property 2 *The delivery time of data in a collision-free schedule minimizing data gathering delays is bounded by one slotframe plus the duration of slots granted to data gathering.*

In the worst case, the data message is generated at the end of the slot granted to the node considered. Consequently, this message has to wait until the next slotframe. Since, in the absence of message loss, the message is delivered to the sink in the same slotframe, the data message considered reaches the sink in the last slot of the slotframe. Hence the property.

Property 3 *In a raw data convergecast, the minimum number of slots assigned to sensor nodes is lower bounded by* $\max(S_n, S_t)$*, with:*

$$S_n = \left\lceil \sum_{u \in WSN} Gen(u)/g \right\rceil \tag{9.2}$$

$$S_t = Gen(c_1) + 2 \sum_{\substack{v \in subtree(c_1) \\ v \neq c_1}} Gen(v) + \delta \tag{9.3}$$

with $g = \min(ninterf, nchild, nchannel)$*, where* $ninterf$ *denotes the number of radio interfaces of the sink,* $nchild$ *the number of children of the sink,* $nchannel > 1$ *the number of available channels for the convergecast and* $\delta = 1$ *if the* $(g+1)$*th child requests the same number of slots as the first one, the children of the sink being sorted in decreasing order of slot demands, or* $\delta = 0$ *otherwise.*

The proof of this property can be found in the literature [39]. According to this property, we can define two types of topology:

- T_n *topologies,* where the minimum number of slots is given by Equation 9.2. In such topologies, the requests are more uniformly distributed in all subtrees. Such a topology is illustrated in Figure 9.3a.
- T_t *topologies,* where the minimum number of slots is given by Equation 9.3. In such topologies, the number of slots is imposed by the most demanding subtree. An example is depicted in Figure 9.3b.

The minimum number of slots needed by a raw data convergecast is given in Table 9.2, assuming that each node generates one data message per data-gathering cycle.

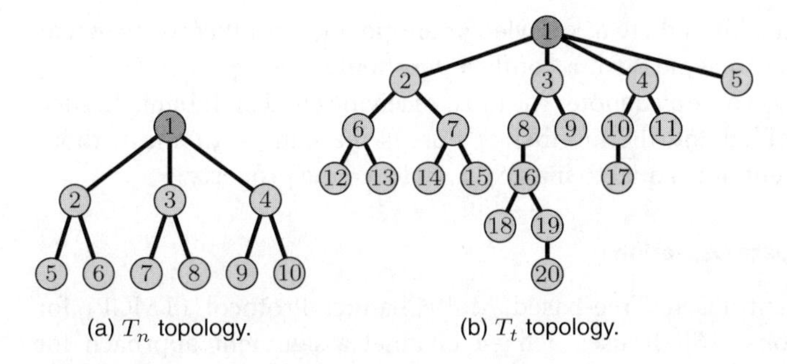

Figure 9.3 Examples of topologies.

Table 9.2 Minimum number of slots needed.

	Sink with 1 interface	Sink with 2 interfaces		Sink with 3 interfaces	
Topology	**2 channels**	**2 channels**	**3 channels**	**2 channels**	**3 channels**
T_n	19	13	13	13	13
T_t	9	6	6	6	5

Table 9.2 shows that with two channels the T_n configuration considered needs only six slots with two radio interfaces, instead of nine slots with a single interface. Increasing the number of radio interfaces to three decreases the number of slots to five only if the number of channels increases to three. The T_t configuration with two channels requires nineteen slots if the sink has a single interface and thirteen slots with a sink equipped with two radio interfaces. Increasing either the number of channels, the number of interfaces or both to three does not bring any decrease in the number of slots. As a consequence, it is useless to equip the sink with a number of radio interfaces strictly higher than the number of available channels or the number of the sink's children.

Property 3 highlights the importance of building a routing tree where no subtree has slot request considerably higher than the average. A routing tree where all subtrees are balanced in terms of slot requests would require fewer slots to ensure data gathering.

Property 4 *If nchannel* $\leq$ *ninterf* $<$ *nchild and* $g \times \max(S_t, S_n) \geq \sum_{u\in\ WSN} Gen(u) +$ *Rcv(ch), the minimum number of slots required for raw data gathering in a multichannel WSN generating heterogeneous traffic is equal to* $\max(S_t, S_n)$*, with and without the immediate acknowledgment, where Rcv(ch) denotes the number of packets received by ch, the child of the sink that receives the most.*

The quantity $g \times \max(S_t, S_n)$ denotes the number of transmission opportunities available for the children of the sink. The number of messages that must be transmitted to the sink is equal to $\sum_{u\in\ WSN} Gen(u)$. Each of these transmissions uses an interface and a channel among the g available interfaces and channels. If *nchannel* $\leq$ *ninterf* $<$ *nchild*, the only possibility for any child of a sink child is to transmit on a channel selected among the

g channels. Hence, with the immediate acknowledgment policy, a conflict would occur with the child of the sink. To avoid such a conflict, we should have: $g \times \max(S_t, S_n) \geq \sum_{u \in WSN} Gen(u) + Rcv(ch)$, where ch denotes the most receiving child of the sink. In such a case, the conflict is avoided and the number of slots is the same with and without immediate acknowledgment; it is equal to $\max(S_n, S_t)$. Hence the property.

9.3.6 MAC dedicated to Data Gathering

Wu et al. describe a centralised Tree-based MultiChannel Protocol (TMCP) for data-collection applications [16]. It uses a fixed channel assignment approach for channel allocation. The whole network is partitioned into multiple subtrees having the base station as a common root, where each subtree is allocated a different channel. TMCP finds available orthogonal channels, partitions the whole network into subtrees and allocates a different channel to each. TMCP improves the throughput with regard to a single-channel solution, while maintaining a high packet-delivery ratio and low latency. However, TMCP blocks direct communications between nodes belonging to different subtrees.

9.3.7 Multichannel Routing for Convergecast

In convergecast scenarios, where all the nodes send their traffic to one destination, namely the sink, it is useless to build routes to all destinations. It is sufficient to build a route from each sensor node to the sink. This routing tree is generally built using a gradient method: the sink broadcasts a message including a cost. A node receiving this message selects the transmitting node as parent if and only if it is the one-hop neighbor that provides the smallest cost. In such a case, the receiving node updates the cost before forwarding the message. The most famous example is given by the IPv6 Routing Protocol for Low-Power and Lossy Networks (RPL) [40]. Other examples exist, such as MOD-ESA, an optimized multichannel slot assignment for raw data convergecast in WSNs that combines neighborhood discovery and route building. A more detailed description of MODESA is given in Section 9.4.

A node may also select several potential parents. For instance, assuming that the cost is simply equal to the depth of a node, any node $u \neq sink$ selects as its potential parents its one-hop neighbors (that is, nodes with which it has a symmetric link) that have a smaller depth than itself. The depth of a node is recursively computed: the sink has a depth 0, its one-hop neighbors have a depth 1, and so on. Notice that the depth of a node in a real multichannel network may differ from one channel to another.

The selection of potential parents may also take into account statistics about link quality. A potential parent with a high link quality is more frequently used to transfer application messages, as can be done in IEEE 802.15.4e TSCH networks [41].

9.3.8 Centralised versus Distributed Collision-free Scheduling Algorithms

Any collision-free schedule consists of a sequence of tuples (sender, receiver, channel, time slot) that is reproduced periodically. Finding a collision-free schedule with the minimum number of slots has been shown to be NP-hard in a single channel network for arbitrary topologies [42]. That is why heuristics are generally used to compute time-slot and channel assignments. This schedule can be computed in a centralised or distributed way. Centralised scheduling algorithms can reach the minimum number of slots, but do not scale. In contrast, distributed algorithms are able to support a large number of nodes but may be far from optimal. Examples of centralised scheduling algorithms are TMCP [16] and MODESA [23]; distributed algorithms include DeTAS [43] and Wave [39].

We now evaluate the number of control messages needed by each of them, assuming a traffic-aware slot and a channel assignment that minimizes the data-gathering delays. We first observe that both the centralised and the distributed assignments use:

- *Neighborhood discovery,* where each node discovers its neighbors and checks the symmetry of the links.
- *Routing tree construction,* where the routing tree used for data gathering is built by exchanging messages including the depth of the sending node. The depth of a node represents its distance to the sink, and this distance is expressed in number of hops.

In a centralised assignment, each node $u \neq sink$, whose depth in the routing tree is $Depth(u)$, transmits the list of its neighbors, including its parent and its children in the routing tree, and its traffic demand $Gen(u)$ to the sink. This message needs $Depth(u)$ hops to reach the sink. We get a total of $\sum_u Depth(u)$ transmissions, which can also be written as $AverageDepth \cdot (N - 1)$. Then, the sink computes the collision-free schedule and broadcasts it to all sensor nodes. Thus, the total number of messages required to establish the collision-free schedule in centralised mode is $AverageDepth \cdot (N - 1)$ transmissions + transmissions to broadcast the *Schedule* to all nodes. Let us assume that the message including the *Schedule* must be fragmented into K fragments to be compliant with the maximum frame size allowed by the standard MAC protocol. Broadcasting the *Schedule* to all nodes requires $K \cdot (N - 1)$ messages in the worst case. Hence the centralised assignment requires $(AverageDepth + K) \cdot (N - 1)$ transmissions.

In a distributed assignment, we assume that any node u uses $Trans(u)$ as its priority for slot and channel assignment. Any node u computes $Trans(u)$ according to Equation 9.1 and transmits it to its parent, denoted $Parent(u)$. This requires $N - 1$ messages to enable all nodes to know their own value of $Trans(u)$, where N is the total number of nodes in the WSN. Any node $u \neq sink$ should notify its priority first, and then its slot assignment, to its conflicting nodes. Hence, u notifies its slot assignment to its one-hop neighbors. This notification is forwarded by nodes that are parents and are one hop away from u or $Parent(u)$. Therefore, the slot assigned to u needs $1 + V + V = 2V + 1$ messages, where V denotes the average number of neighbors per node. Since we have $N - 1$ sensor nodes and each node notifies first its priority and then its slot assignment to its conflicting nodes, we need $2 \cdot (2V + 1) \cdot (N - 1) = (4V + 2) \cdot (N - 1)$ messages to establish the collision-free schedule.

Hence centralised assignment outperforms distributed assignment in terms of the number of required messages if and only if: $(AverageDepth + K) \cdot (N - 1) \leq (N - 1) \cdot (4V + 2)$.

Property 5 *Centralised assignment requires fewer control messages than distributed assignment if and only if $K \leq 4V + 2 - AverageDepth$, where V is the average number of neighbors per node and AverageDepth is the average of the depth of all nodes different from the sink and K is the number of fragments of the Schedule.*

9.4 Sahara: Example of Solution

The work described in this section is part of the larger SAHARA project.

9.4.1 Description of the Solution Proposed

We now present an example of a solution performing joint routing, time-slot and frequency assignment. This ensures multihop synchronization by using cascading beacons. A dedicated channel is used for signalling traffic. However, it should be pointed out that it can also be used by data traffic. Network connectivity is assumed to be ensured on this control channel.

9.4.1.1 A Solution based on the IEEE 802.15.4 Standard

The physical layer of the IEEE 802.15.4 standard has been adopted in many other standards such as ZigBee [13], WirelessHART [44], and ISA100.11a [45]. It is based on a robust modulation that allows symbol redundancy and ensures good resilience against multipath interference, which makes it convenient for indoor deployment. It offers 250 Kbps in the 2.4-GHz band and 16 orthogonal channels, making it a good candidate for multichannel MAC protocols. This physical layer has therefore been adopted in the SAHARA project.

9.4.1.2 Network Deployment

As previously stated, we tackle static deployments where nodes remain static once they have been deployed. The first node to be activated is the node in charge of creating the network and managing the time synchronization. This specific node starts signalling its presence using periodic beacon frames. The sink node can be chosen to play this role. The beacon frame will help other nodes to detect its presence when scanning on the specified control channel. When the other nodes are activated, they will detect the beacon and send a join request to the sink. Once a node is allowed to be part of the network (the network admission process is beyond the scope of this chapter) this node will propagate the beacon that it has received. In order to avoid collisions between beacon frames, the sink indicates in what order the beacons should be propagated. Thus, beacons are sent in a collision-free TDMA manner where each node has its own slot for broadcasting the beacon frame. Nodes that are multiple hops away from the sink will send their join request in a multihop manner to reach the sink. Nodes are informed about the propagation order in the joint response messages. In order for other nodes to update their propagation order, the beacon will include the updates during m consecutive beacon cycles (a beacon cycle is the period separating two successive transmissions of the beacon by the sink). Including updates and not the complete list of nodes makes this mechanism scalable. More details about the beacon propagation mechanism can be found in the literature [46].

9.4.1.3 Slotframe

The slotframe describing the organization of node activities consists of four periods:

- *a synchronization period* that contains the multihop beacon propagation, using the control channel

- *a control period* that allows the transmissions of network control messages in CSMA/CA, using the control channel
- *a data period* that collects the applicative data according to a collision-free schedule; all available channels, including the control channel, are used
- *a sleep period* where all network nodes sleep to save node energy.

9.4.1.4 Multi-interface Sink

To enhance the sink reception throughput, we use a multi-interface sink, so the sink node is able to receive simultaneously on different channels. The number of radio interfaces of the sink is at most equal to the larger of the number of available channels and the number of its children.

9.4.1.5 Neighborhood Discovery

Neighborhood discovery in a multichannel environment is done successively on each channel belonging to the channel list defined by the application. On the channel where the neighborhood is being discovered, each node broadcasts a *Hello* message containing a list of its one-hop neighbors. The sink initiates the *Hello* message cascade. The *Hello* messages are sent, hop by hop, from the sink to the farthest nodes of the network. Each node at a depth d receives *Hello* messages from those of its neighbors that are closer to the sink than itself and sends its own *Hello* message with a random jitter to avoid collisions. After several exchanges of *Hello* messages checking the symmetry of links, and if the neighborhood of the node is stable, each node sends a *Notify* message to the sink. The *Notify* message sent by node $u \neq sink$ contains its depth, its neighborhood and some applicative information (such as the number of slots needed or the number of radio interfaces). The *Notify* messages are processed by the sink acting as the entity in charge of computing the joint time-slot and channel assignment.

9.4.1.6 Collision-free Schedule

Since in our application, the condition of Property 5 is met, we use a centralised algorithm for a joint time-slot and channel assignment. This is called Multichannel Optimized DElay time Slot Assignment (MODESA) [23]. The conflicting nodes and the potential parents of any node $u \neq sink$ are computed from the collected one-hop neighbors, according to Property 1. The routes used for collecting data packets through the whole network are selected jointly with the channel and slot assignment.

During the initial discovery of channels, we want to reduce the latency of the first data gathering. That is why, just after the discovery of the first channel, a first schedule is built, making possible data gathering. The schedule is rebuilt each time a new channel is discovered to take advantage of parallel transmissions. After that, any change in the information contained in the *Notify* messages may cause updates in the current schedule or the creation of a new schedule.

Livolant et al. have compared the cost of installing a centralised schedule into the network using various standards (CoAP, RPL, IPv6, IEEEE 802.15.4e) with the SAHARA solution [47]. This cost is evaluated in terms of the number of packets sent in the network and the latency. This cost is much higher when using the standards than with the SAHARA solution.

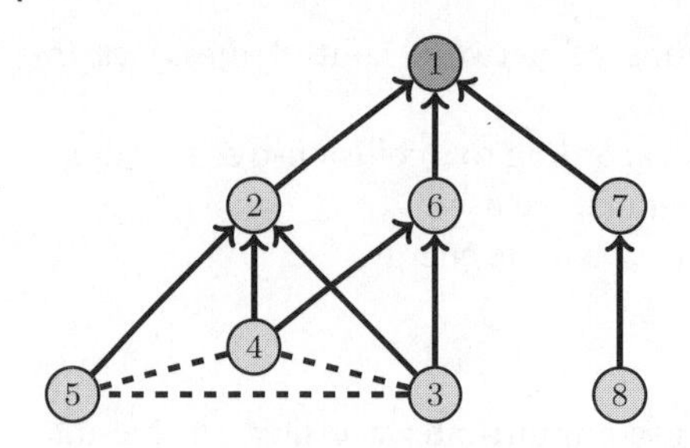

Figure 9.4 An example topology.

Table 9.3 The schedule obtained with MODESA for a sink with three radio interfaces and a network with three channels.

Ch \ S	1	2	3	4	5	6
1	2→1	2→1	6→1	2→1	6→1	2→1
2	6→1	3→6 8→7	7→1	4→6	5→2	
3	7→1		5→2			

9.4.2 Illustrative Example

We consider a network with eight nodes and three channels where, for the sake of simplicity, the topology is the same on all channels and is as depicted in Figure 9.4. Each node $\in \{3, 4, 6, 7, 8\}$ generates one data packet per slotframe, whereas nodes 2 and 5 generate two packets per slotframe. The sink, denoted node 1, has three radio interfaces.

For this network example, where $AverageDepth = 1.57$ and $V = 3$, the centralised assignment of time slots and channels outperforms the distributed assignment as long as $K \leq 12.43$ fragments.

MODESA provides the schedule given in Table 9.3. We observe that the three channels are simultaneously used in slots 1 and 3. The three sink interfaces are active in slot 1: they are receiving from nodes 2, 6 and 7. In slot 2, we notice spatial reuse on channel 2: nodes 3 and 8, which are 4 hops away, transmit simultaneously on the same channel.

9.4.3 Performance Evaluation of the Solution

9.4.3.1 Impact of Multiple Channels and Multiple Radio Interfaces on the Aggregated Throughput

We start by evaluating the benefits of using multiple channels and multiple radio interfaces on the sink node. We conducted simulations using the NS-2 simulator. Each point in the graphs represents an average of 100 repetitions generated with random topologies of 50 nodes. The size of the packets is 50 bytes. We compared the performance in terms of aggregate throughput calculated in terms of the number of packets received by the sink per second for different MAC protocols:

- HMC [29] the multichannel MAC protocol for WSNs presented in Section 9.2.9.3
- random allocation, a multichannel MAC protocol that allocates channels to nodes in a random fashion
- '1 channel', standard CSMA/CA using a single channel.

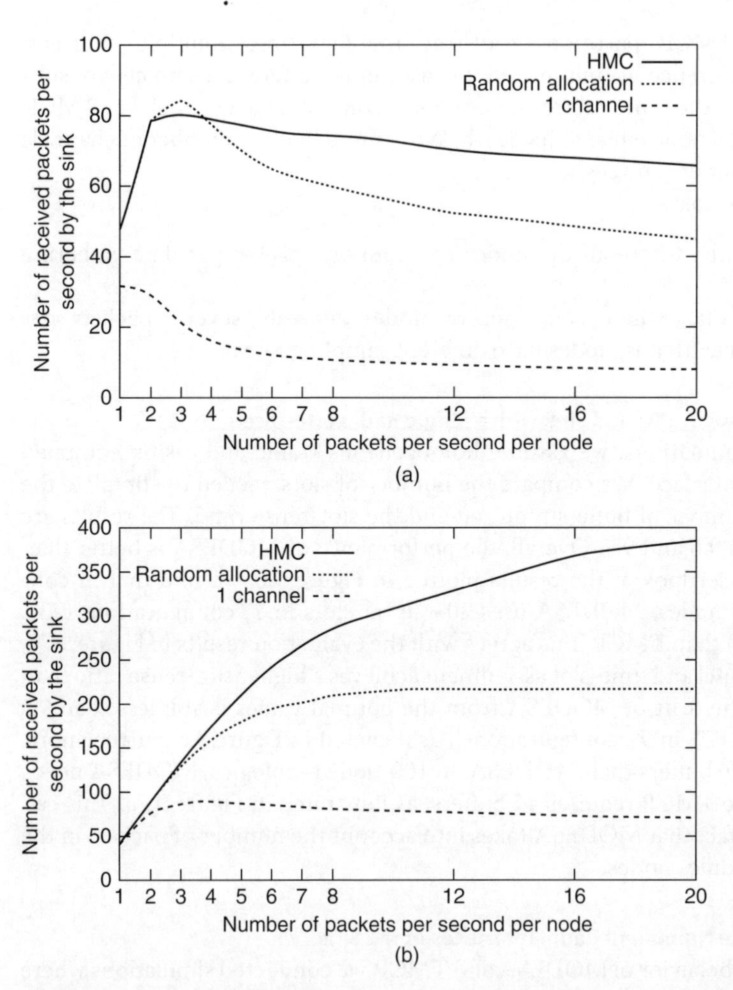

Figure 9.5 Number of packets per second received by the sink. (a) With multiple channels (b) With multiple channels and interfaces.

First, we show in Figure 9.5a the benefits of using multiple channels. Simulation results show that even allocating the channels in a random manner gives much better performance than using a single channel. HMC outperforms random allocation because it takes into account the interference between neighbors. Figure 9.5b shows simulation results of the same protocols, but using three radio interfaces on the sink. Here we show the benefits of using multiple radio interfaces on the sink. The '1 channel' protocol is a MAC protocol that uses one channel per radio interface of the sink. The results clearly show the enhancement in the aggregate throughput when we combine multiple channels with multiple radio interfaces on the sink.

In the following, we evaluate the performance of MODESA against:

- Optimal scheduling, which provides the smallest number of slots for any configuration
- TMCP, a relevant cluster-based multichannel scheduling.

We recall that TMCP partitions the tree topology into multiple subtrees. The inter-tree interference is minimized by assigning different channels to subtrees. All the nodes in the same subtree communicate on the same channel. For TMCP, we set the priority of a node equal to its depth. We assume that the number of channels is equal to the number of subtrees.

We distinguish two cases:

- homogeneous traffic, where all the nodes generate one packet per data-gathering cycle
- heterogeneous traffic, where some sensor nodes generate several packets per data-gathering cycle (that is, nodes have different sampling rates).

9.4.3.2 Homogeneous Traffic and Sink with a Single Radio Interface

In the first set of simulations, we assume homogeneous traffic and a sink equipped with a single radio interface. We compare the number of slots needed to complete the convergecast, the number of buffers required and the slot-reuse ratio. The results are presented in Figures 9.6 and 9.7. Overall, the performance of MODESA is better than TMCP. Taking a closer look at the results plotted in Figure 9.6, we find that in configurations with 100 nodes, MODESA uses 20% fewer slots in T_t configurations (23% in T_n configurations) than TMCP. This agrees with the evaluation results in Figure 9.7b, where our joint channel and time-slot assignment achieves a higher slot-reuse ratio than TMCP. Moreover, the drift of MODESA from the optimal values is still less than 9% in T_s configurations (7% in T_n configurations), as depicted in Figure 9.6. Furthermore, TMCP requires more buffers than MODESA: in 100-node topologies, MODESA needs only 15 buffers while TMCP requires 44 buffers, as illustrated in Figure 9.7a. This can be explained by the fact that MODESA takes into account the number of packets in the buffers when it schedules nodes.

9.4.3.3 Impact of the Number of Radio Interfaces of the Sink

To further study the behavior of MODESA and TMCP, we conducted simulations where the sink is equipped with a number of radio interfaces equal to the number of subtrees.

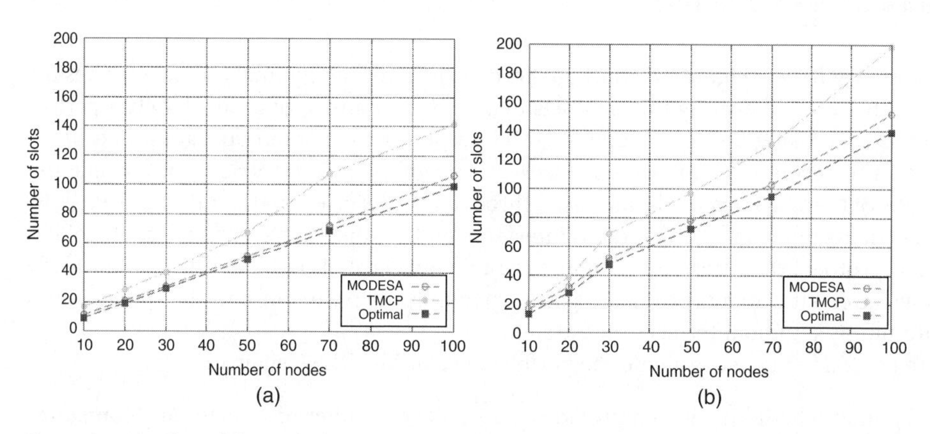

Figure 9.6 Number of slots used by TMCP, MODESA and optimal values. (a) In T_n configurations (b) In T_t configurations.

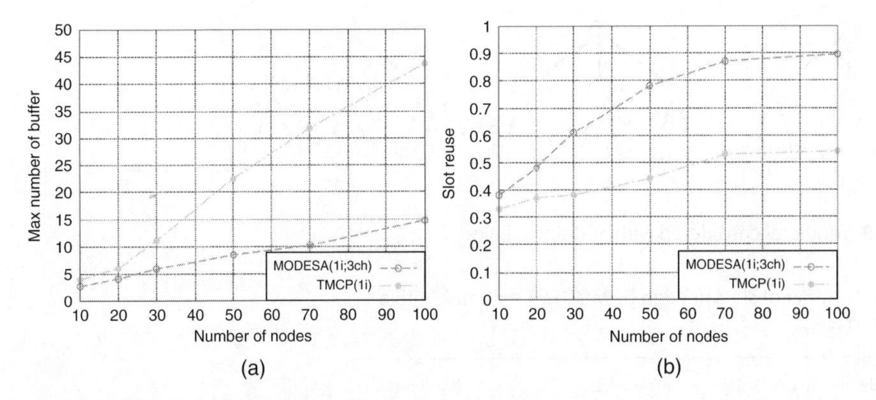

Figure 9.7 Buffers needed and slot reuse by TMCP and MODESA. (a) Buffers for TMCP and MODESA (b) Slot reuse by TMCP and MODESA.

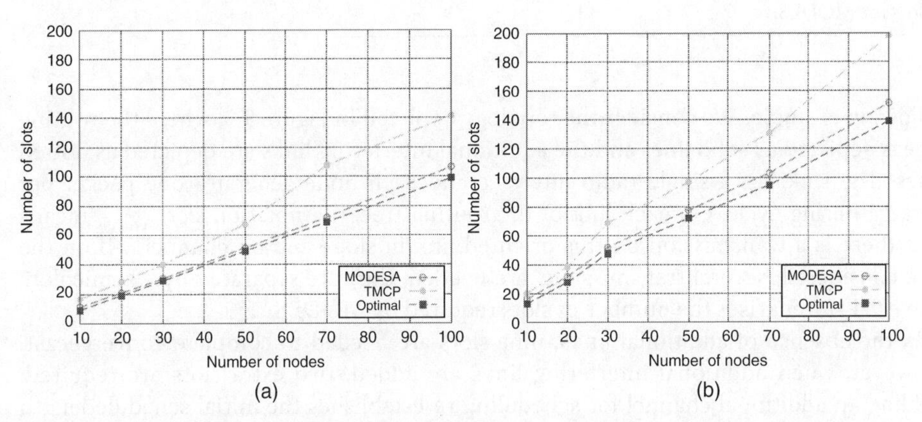

Figure 9.8 Number of slots used by TMCP and MODESA and optimal values, for multiple radio interfaces. (a) In T_n configurations (b) In T_t configurations.

Each radio interface operates on a different channel, so the sink can receive simultaneously from its children. We always assume that each node generates one packet per data-gathering cycle.

Figure 9.8 shows the same behavior of curves as in Figure 9.6. For small topologies (≤ 30), MODESA and TMCP are close. But when the number of nodes increases, the gap between the algorithms becomes huge. These results unambiguously display the excellent performance of MODESA in schedule length.

9.4.3.4 Impact of Additional Links

A frequent assumption in algorithms computing collision-free schedules for data-gathering applications is that all interfering links that do not belong to the routing tree have been eliminated by a receiver-based channel assignment. However, it has been proved [49] that assigning a minimum number of channels to receivers such that all interfering links are removed is NP-complete. It is important to note that MODESA does not require all interfering links to be removed. MODESA easily takes into account the presence of additional interfering links, as illustrated by the

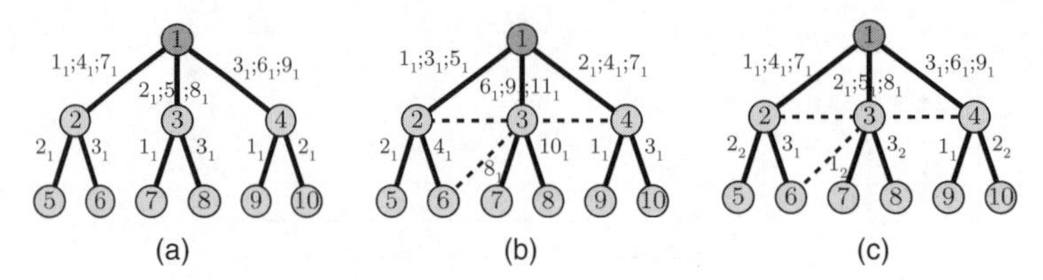

Figure 9.9 Topology considered with additional links.

Table 9.4 Number of slots needed by MODESA for a sink with a single interface.

Channels	1	1	2
Additional link	No	Yes	Yes
Illustrated in	Figure 9.9(a)	Figure 9.9(b)	Figure 9.9(c)
Slots for MODESA	9	11	9

following example. We consider the topology depicted in Figure 9.9, where the routing tree is depicted as solid lines and the additional interfering links are depicted as dotted lines. The sink has a single radio interface and each node generates one packet per data-gathering cycle. On each link of the routing tree, the notation $sloti_{channelj}$ means that there is a transmission on this oriented link in slot i and on channel j. If on the link considered, several transmissions are needed, they are separated by a semicolon. Table 9.4 summarises the number of slots required by MODESA.

In the absence of additional links, nine slots are needed to complete convergecast. However, when additional interfering links are added, two extra slots are required. Adding an additional channel for scheduling re-establishes the initial schedule length because more parallel transmissions are allowed.

To further investigate the impact of interfering links on schedule length, another set of simulations was conducted. In the results presented below, the sink is equipped with a number of radio interfaces equal to its number of children. The number of channels is equal to the number of radio interfaces. Additional links are added: for each node at even depth d in the tree, an additional link is generated with a node at depth $d - 1$ different from its parent. Furthermore, with a probability equal to 0.5, another link is added with a node of depth $d + 1$ different from its children. On average, 60% additional links are added. As can be seen in Figure 9.9, the impact of the additional links depends on the routing tree. The worst routing trees are T_t ones for both MODESA and TMCP. For 100 nodes, MODESA needs thirteen additional slots to complete convergecast in T_t configurations, while only five slots are needed in T_n configurations. It is also worth noting that, for MODESA, the number of additional slots due to the additional links is smaller than for TMCP. This shows the capacity of MODESA to easily incorporate the additional conflicting links.

9.4.3.5 Heterogeneous Traffic

In this second series of simulations, the sink is also equipped with as many radio interfaces as children. We first consider the three topologies depicted in Figure 9.10,

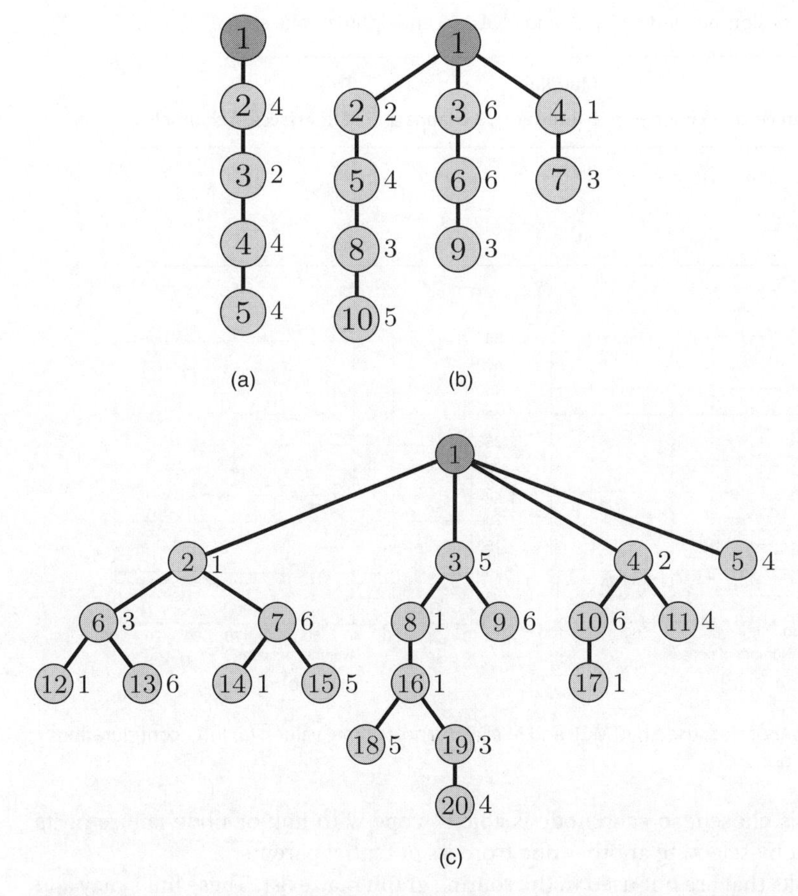

Figure 9.10 Topologies with heterogeneous traffic. (a) Line (b) Multiline (c) Tree.

where the number beside the node represents the number of slots requested by the node considered.

For all these topologies, MODESA is optimal, requiring the minimum number of slots, as illustrated in Table 9.5. However, TMCP needs more slots in all these topologies. In addition, for the multiline and tree topologies, MODESA needs only two radio interfaces and two channels to reach the optimal number of slots to complete convergecast. However, TMCP, even when the sink is equipped with three or four radio interfaces, does not achieve the optimal values. This can be explained by the fact that scheduling all the nodes of the same subtree on a single channel cannot ensure a high spatial-reuse ratio.

The results depicted in Figure 9.11 show again that MODESA is close to the optimal values of slot numbers: the distance is 5% in T_t configurations (3% in T_n configurations). In addition, MODESA obviously outperforms TMCP.

9.4.4 Robustness and Adaptivity of the Solution Proposed

The solution proposed supports any network topology that is connex on the control channel. The topology may vary from one channel to another. For reliability reasons, a

Table 9.5 Number of slots needed by TMCP and MODESA and optimal values.

	Line 1 interface, 1 channel	Multiline 3 interfaces, 3 channels	Tree 4 interfaces, 4 channels
Optimal	24	26	45
MODESA	24	26	45
TMCP	32	34	58

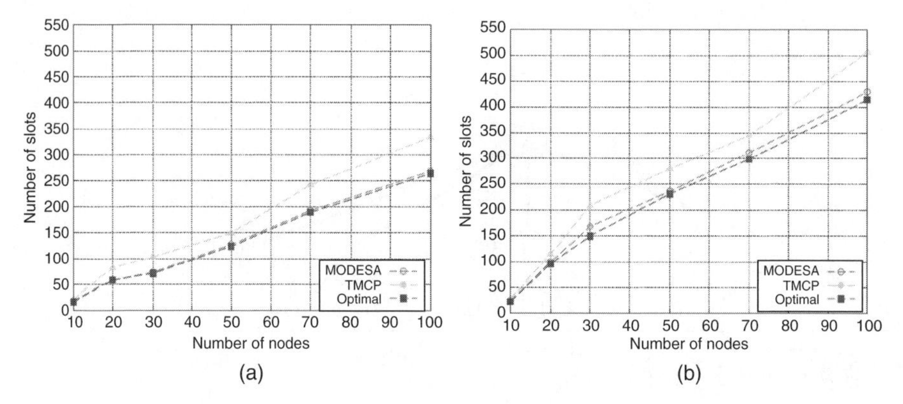

Figure 9.11 Number of slots used by TMCP and MODESA and optimal values. (a) In T_n configurations (b) In T_t configurations.

mesh topology is chosen, so each node is able to cope with link or node failure of its preferred parent by selecting another one from its potential parents.

Additional links that are not used in the routing graph may exist. These links may add interferences that are avoided by our solution. The number of radio interfaces of the sink is a parameter that may vary. To cope with transmission errors, the number of slots granted to each node is higher than the number requested, taking into account the error rate of the link considered at the physical layer.

This solution is able to adapt to topology changes. After joining the network, a node discovers its neighbors and among them its potential parents. It then sends a *Notify* message to the sink that allocates time slots and channels for the transmissions originating from this node. This update of the schedule is then broadcast to all nodes to be applied in the next slotframe. When a node leaves the network, the collision-free schedule is updated to free the slots allocated to the transmissions originating from this node. If a link with a parent is broken, the collision-free schedule is updated to select another parent among the potential parents. If a new link is created, the conflicting nodes are updated. If the sink detects no conflict in the current schedule, this schedule is kept. Otherwise, it is updated to avoid conflicts. Furthermore, statistics about link quality may be used by the sink to select the preferred parent from all the potential ones.

In addition, changes in the application needs (such as generation of more traffic) are taken into account [50], leading to the existing schedule being updated accordingly.

To summarise this section, MODESA relies on an efficient heuristic that provides a schedule length that is close to optimal values. It is significantly better than the state-of-the art TMCP solution. The gain can be up to 20%. It also incurs less buffer consumption. Another advantage of MODESA is that it is flexible enough to take into account additional interfering links.

9.5 Summary

Aircraft have demanding requirements–regarding throughput, robustness and latency–for wireless gathering of non-critical data. Wired networks meet such requirements, but at a high cost due to their wiring mass. For instance, the Airbus A380 has 530 km of wires. Similarly, the wiring in Ariane 5 represents 70% of its avionic mass. Wireless networking is able to dramatically reduce this mass. However, radio propagation in confined environments such as aircraft cabins is versatile and difficult to predict. Despite the low size of the deployment area, a multihop and multichannel solution is required to cope with interference and obstructed paths. Multichannel WSNs are able to meet such requirements. First, multichannel WSNs increase the parallelism in transmissions and hence improve the throughput. In addition, they mitigate both internal and external interference, taking advantage of channel diversity. As a consequence, a higher delivery rate and better throughput are achieved. The gains brought by multichannel WSNs on the one hand, and a sink equipped with multiple radio interfaces on the other hand, have been evaluated through intensive simulations.

Based on the multichannel paradigm, the SAHARA solution has been proposed. At the MAC level, it provides collision-free scheduled medium access, combining time slots and multichannel access. Theoretical results for the minimum number of slots required by data gathering have been established. The impact of the data-gathering tree has been highlighted. The centralised schedule built by MODESA is close to optimal. The SAHARA solution is also self-adaptive to changes in both topology and traffic. The feasibility of this solution has been proved through the implementation of simple testbeds.

As a further work, a performance evaluation of real demonstrators placed in an aircraft will be conducted, in order to obtain performance results in an environment very close to the real one. Some issues are still open. Distributed scheduling algorithms are still being investigated. They provide better scalability than centralised scheduling algorithms, but may be far from optimal. Self-adaptivity of WSNs could be improved. For instance, WSNs should be provided the means of finding the best tradeoff, taking into account both the real environment in which they are as well as the application requirements in terms of quality of service. Furthermore, a software defined radio is attractive because of its auto-adaptivity to the various conditions of radio propagation and various traffic profiles. However, technical issues (weight, size and energy consumption) do not allow its immediate integration in aircraft. Another open issue concerns the development of a field programmable gate array and/or smart antennas that could implement this solution and would be compliant with aeronautics standards (such as DO-160).

Acknowledgments

We would like to thank the members of the SAHARA project and Richard James for his valuable help.

References

1 L.M. Miller, C. Guidi, T. Krabach, *Space sensors for human investigation of planetary surfaces (SpaceSHIPS),* In Proceedings of the 2nd International Conference on Micro/Nanotechnology for Space Applications, NASA/JPL, Pasadena, CA, April 1999.

2 J.M. Collignon, B. Rmili, *An ultra low power RFID sensor platform for launchers applications,* WISEE 2013.

3 T. van den Berg, G. La Rocca and M.J.L. van Tooren, *Automatic Flattening Of Three-Dimensional Wiring Harnesses For Manufacturing,* ICAS 2012.

4 E.R. Martinez, J.A. Santos, R. David, M. Mojarradi, L. del Castillo, S.P. Jackson, *Challenge of developmental flight instrumentation for Orion exploration flight test 1: Potential benefit of wireless technology for future Orion missions,* IEEE WISEE, October 2014.

5 WISE partners, *D6.3.3 WISE Project, Publishable Final Activity Report,* http://cordis.europa.eu/docs/publications/1270/127030191-6_en.pdf, November 2008.

6 SWAN partners, *Wireless Sensor Networks for Aircraft Maintenance Operations,* http://triagnosys.com/swan

7 AUTOSENS partners, *AUTOnomous SENSing microsystem,* http://www.fnrae.org/1-39841-Detail-projet.php?id_theme=2&id_projet=7

8 SMMART partners, *System for Mobile Maintenance Accessible in Real Time,* http://www.lintar.disco.unimib.it/space/Progetti/SMMART/SMMART-6220-Project_Flyer.pdf

9 P. Minet, G. Chalhoub, E. Livolant, M. Misson, B. Rmili, J.-F. Perelgritz, *Adaptive wireless sensor networks for aircraft,* IEEE WISEE 2015, Orlando, FL, December 2015.

10 A. Liccardo, A. Mariscotti, A. Marrese, N. Pasquino, R. Schiano Lo Moriello, *Statistical characterization of the 2.45 GHz propagation channel aboard trains, ACTA IMEKO,* **4** (1): 44–52, 2015.

11 B. Nkakanou, G.Y. Delisle, N. Hakem, Y. Coulibaly, *UHF Propagation parameters to support wireless sensor networks for onboard trains Journal of Communication and Computers,* August 2013, **10**: 1120–1130.

12 *IEEE standard for local and metropolitan area networks Part 11: Wireless LAN medium access control (MAC) and physical layer (PHY) specifications,* IEEE Std 802.11-2012, Institute of Electrical and Electronics Engineers (IEEE), March 2012.

13 *Zigbee Specification,* Document 053474r17, ZigBee Alliance, January 2008.

14 G. Chalhoub, A Guitton, F. Jacquet, A. Freitas, M. Misson, *Medium access control for a tree-based wireless sensor network: Synchronization management,* IFIP Wireless Days, November 2008.

15 G. Huang, A.Y. Zomaya, F.C. Delicato, P.F. Pires, *Long term and large scale time synchronization in wireless sensor networks, Computer Communications,* **37**: 77–91, 2014.

16 Y. Wu, J.A. Stankovic, T. He, Sh. Lin, *Realistic and efficient multi-channel communications in wireless sensor networks,* IEEE Infocom, April 2008.

17 S. Lohier, A. Rachedi, I. Salhi, E. Livolant, *Multichannel access for bandwidth improvement in IEEE 802.15.4 Wireless Sensor Network,* IFIP/IEEE Wireless Days, November 2012.

18 B. Priya, S.S. Manohar, *EE-MAC: Energy efficient hybrid MAC for WSN*, International Journal of Distributed Sensor Networks 2013. Article ID 526383.

19 G. Zhou, Ch. Huang, T. Yan, T. He, J.A. Stankovic, *MMSN: Multi-frequency media access control for wireless sensor networks*, IEEE Infocom, 2006.

20 J. Borms, K. Steenhaut, B. Lemmens, *Low-overhead dynamic multi-channel MAC for wireless sensor networks*, EWSN, February 2010.

21 O.D. Incel, P. Jansen, S. Mullender, *MC-LMAC: A multi-channel MAC protocol for wireless sensor networks*, Technical Report TR-CTIT-08-61, Centre for Telematics and Information Technology, 2008.

22 R. Diab, G. Chalhoub, M. Misson, *Evaluation of a hybrid multi-channel MAC protocol for periodic and burst traffic*, IEEE LCN, September 2014.

23 R. Soua, P. Minet, E. Livolant, *MODESA: an optimized multichannel slot assignment for raw data convergecast in wireless sensor networks*, IPCCC 2012, December 2012.

24 P. Minet, S. Mahfoudh, G. Chalhoub, A. Guitton, *Node coloring in a Wireless sensor network with unidirectional links and topology changes*, IEEE WCNC, April 2010.

25 T. Watteyne, A. Mehta, K. Pister, *Reliability through frequency diversity: why channel hopping makes sense*, ACM PE-WASUN, October 2009.

26 K. Pister, L. Doherty, *TSMP: Time synchronised mesh protocol*, in Parallel and Distributed Computing and Systems (PDCS), Orlando, Florida, November 2008.

27 Y. Kim, H. Shin, H. Cha, *Y-MAC: An energy-efficient multi-channel MAC protocol for dense wireless sensor networks*, International Conference on Information Processing in Sensor Networks, 2008

28 J. Polastre, J. Hill, D. Culler, *Versatile low power media access for wireless sensor networks*, ACM Sensys, November 2004.

29 G. Chalhoub, R. Diab, M. Misson, *HMC-MAC Protocol for high data rate wireless sensor networks*, *Electronics*, **4** (2): 359–379, 2015.

30 E.-O. Blass, J. Horneber, M. Zitterbart, *Analyzing data prediction in wireless sensor networks*, in Proc. IEEE Vehicular Technology Conference, VTC Spring 2008, 2008.

31 K. Akkaya, M. Younis, *A survey on routing protocols for wireless sensor networks. Ad Hoc Networks*, **3** (3): 325–349, 2005.

32 K. Zeng, W. Lou, J. Yang, D.R. Brown III, *On geographic collaborative forwarding in wireless ad hoc and sensor networks*, International Conference on Wireless Algorithms, Systems and Applications (WASA), 2007.

33 Jun Luo; Hubaux, J.-P., *Joint mobility and routing for lifetime elongation in wireless sensor networks*, 24th Annual Joint Conference of the IEEE Computer and Communications Societies (INFOCOM 2005), 2005.

34 I. Papadimitriou, L. Georgiadis, *Energy-aware routing to maximize lifetime in wireless sensor networks with mobile sink. Journal of Communications Software and Systems*, **2** (2), 141–151, 2006.

35 L. Pelusi, A. Passarella, M. Conti, *Opportunistic networking: data forwarding in disconnected mobile ad hoc networks, IEEE Communications Magazine*, **44** (11): 134–141, 2006.

36 R.C. Shah, S. Roy, S. Jain, W. Brunette, *Data MULEs: modeling a three-tier architecture for sparse sensor networks*, IEEE International Workshop on Sensor Network Protocols and Applications (SNPA 2003), 11 May 2003, pp. 30–41.

37 C.H. Ou, K.F. Ssu, *Routing with mobile relays in opportunistic sensor networks,* in 18th Annual IEEE International Symposium on Personal, Indoor and Mobile Radio Communicatioons (PIRMC'07), 2007.

38 S. Mahfoudh, P. Minet, I. Amdouni, *Energy-efficient routing and node activity scheduling in the OCARI wireless sensor network, Future Internet* 2010, **2** (3): 308–340, 2010.

39 R. Soua, P. Minet, E. Livolant, *Wave: a distributed scheduling algorithm for convergecast in IEEE 802.15.4e TSCH networks, Transactions on Emerging Telecommunications Technologies,* **27** (4): 557–575, 2015.

40 T. Winter, P. Thubert, A. Brandt, J. Hui, R. Kelsey, P. Levis, K. Pister, R. Struik, J. Vasseur, R. Alexander, *RPL: IPv6 Routing Protocol for Low Power and Lossy Networks,* RFC 6550, March 2012.

41 *IEEE Standard for Local and Metropolitan Area Networks-Part 15.4: Low Rate Wireless Personal Area Networks (LR-WPANs) Amendment 1: MAC Sublayer,* IEEE Std 802.15.4e-2012, Institute of Electrical and Electronics Engineers (IEEE), April 2012.

42 H. Choi, J. Wang, E.A. Hughes, *Scheduling on sensor hybrid network,* IEEE ICCCN, October 2005.

43 N. Accettura, M.R. Palattela, G. Boggia, L.A. Grieco, M. Dohler, *Decentralized traffic aware scheduling for multi-hop low power lossy networks in the Internet of Things,* WoWMoM'13, Madrid, Spain, 2013.

44 HART Communication Foundation, *HART field communication protocol specifications,* Tech. Rep., 2008.

45 International Society of Automation, *ISA100.11a : 2009 wireless systems for industrial utomation: Process control and related applications,* Draft standard, 2009.

46 G. Chalhoub, M. Misson, *Cluster-tree based energy efficient protocol for wireless sensor networks,* IEEE ICNSC, April 2010.

47 E. Livolant, P. Minet, T. Watteyne, *The cost of installing a new communication schedule in a 6TiSCH low-power wireless network using CoAP (Extended version,* Research Report RR-8817, Inria, November 2015.

48 Z. Shelby, K. Hartke, C. Bormann, *The constrained application protocol (CoAP),* Technical report RFC 7252, Internet Engineering Task Force, June 2014.

49 A. Ghosh, O.D. Incel, V.S.A Kumar, B. Krishnamachari, *Multichannel scheduling algorithms for fast aggregated convergecast in sensor networks,* 6th IEEE International Conference on Mobile Adhoc and Sensor Systems, MASS 2009, October 2009.

50 R. Soua, E. Livolant, P. Minet, *Adaptive strategy for an optimized collision-free slot assignment in multichannel wireless sensor networks, Journal of Sensor and Actuator Networks, Special Issue on Advances in Sensor Network Operating Systems,* **2** (3): 449–485, 2013.

10

Wireless Piezoelectric Sensor Systems for Defect Detection and Localization

Xuewu Dai[1], Shang Gao[2], Kewen Pan[3], Jiwen Zhu[1] and Habib F. Rashvand[4]

[1] Northumbria University, Newcastle upon Tyne, UK
[2] Nanjing University of Aeronautical and Astronautics, China
[3] School of Electrical and Electronic Engineering, University of Manchester, UK
[4] Advanced Communication Systems, University of Warwick, UK

10.1 Introduction

Structural health monitoring (SHM) is a system to continuously and/or periodically monitor and assesses the integrity of structures, such as bridges, trains, offshore oilrigs and aircraft wings, so that the structure's performance and safety can be maintained at an appropriate level. Defect detection is an emerging field in SHM, which not only reduces costs by minimizing maintenance and inspection cycles, but also prevents catastrophic failures at an earlier stage. This is particularly useful for developing self-monitoring structures, into which 'smart' materials are integrated.

As a non-destructive evaluation method, the well-known Lamb wave-based defect detection and localization approach has been widely used in SHM [1, 2]. It utilises guided ultrasonic waves for identifying degradation and defects (such as cracks) in the structure. The fundamental of Lamb-wave condition monitoring is the interactions between Lamb wave propagation and defects in structures. For example, a crack in a thin structure introduces new boundaries in the Lamb wave propagation path, which leads to reflecting, scattering and mode conversion. These interactions can be observed through the changes of the Lamb wave propagation characteristics, such as amplitude, mode, frequency and propagation delays. As a result, one can determine if a path contains defects and even the defect's location and size. Numerous approaches have been developed to improve the defect-localization accuracy, decrease the size and reduce the deployment and maintenance costs. In the last decade, there has been intensive interest in the use of piezoelectric lead zirconate titanate (PZT) sensors in Lamb wave-based condition monitoring. A comprehensive study of PZT-based Lamb-wave condition-monitoring and defect-detection technologies is found in the literature [1]. Due to its unique features of simplicity, durability, small size and low-cost, PZT-based Lamb-wave monitoring technology has shown great promise for defect detection in SHM [3].

Typical applications of PZT networks in SHM can be found in condition monitoring of bridges, pipes, aircraft wings and unmanned vehicles. Yapar et al. [4] used acoustic emission piezoelectric sensors in monitoring bridges. The effectiveness and

Wireless Sensor Systems for Extreme Environments: Space, Underwater, Underground, and Industrial.
First Edition. Edited by Habib F. Rashvand and Ali Abedi.

applicability of acoustic PZT monitoring were evaluated in both experimental and numerical investigations for three types of representative bridge: steel-girder, reinforced concrete and pre-stressed concrete. Gu et al. [5] introduced a piezoelectric-based strength-monitoring technique using embedding PZT sensors in a concrete specimen at the stage of casting. The recent development of embedded PZT sensors for SHM of a concrete structure has been reported [6]. Surface-bonded PZT sensors have been used in monitoring surface acoustic wave propagation in beam elements [7]. PZT sensors with nonlinear Lamb wave techniques have been developed and validated for fatigue-crack detection and online defect monitoring in high-speed train bogies [8].

Another emerging application of PZT Lamb wave technology is the health and event monitoring of unmanned aerial vehicles (UAVs). Compared to manned air systems, online SHM of UAVs is more demanding, as UAVs do not have pilots to be a sensing system for observing unexpected events, as they do in manned aircraft. Furthermore, most UAV components are made of multilayered composite materials and are more prone to internal defects, which is harder to detect. Qiu and Yuan [2] designed wired multichannel PZT arrays for a UAV wing box for defect detection (Figure 10.1a). Oliver et al. used PZT sensors, either embedded or surface mounted, for detecting defects in composite UAV wings [9]. As a part of the KASHMOS project sponsored by the Agency for Defence Development in Korea, PZT sensors for defect detection were tested [10, 11]. A novel PZT-based impact event sensing system with up to 64 PZT channel parallel data acquisition was proposed to detect the location of impact events for unmanned vehicles [12]. A damage detection method, which used a compact PZT sensor array for localizing single or multiple defects in composite aircraft structures has been described [13, 14]. An innovative protocol that integrates UAVs and image processing and data acquisition procedures for crack detection and assessment of surface degradation has also been described [15]

The maturity of wireless communication techniques and wireless sensor networks (WSNs) means that wireless SHM systems are a promising solution for rapid, accurate and low-cost structural monitoring [17]; see Figure 10.1b. The latest development in the field is low-cost wireless PZT sensor networks [18]. As illustrated in Figure 10.1c, these next-generation SHMs will integrate the PZT-based Lamb wave technologies with wireless transmission, allowing for large-area SHM with easy and cable-free deployment. In this kind of online PZT-based SHM and event-monitoring system, a set of PZT sensors is either attached to the surface of the structure or embedded inside it. One or more PZT sensors works as exciters to induce Lamb waves into the structure. Since the propagation of Lamb waves is affected by degradation of and defects in the structure, the characteristics of the Lamb waves propagating from the exciter to these receiving PZT sensors can be closely monitored and so that defects within the structure can be identified.

Although there have been some studies on using Lamb waves for defect localization [16], the majority are wired systems requiring intensive cabling and therefore incurring high deployment and maintenance costs, which makes them unsuitable for distributed sensing in large-scale SHM. Martens et al. [19] investigated a wired PZT sensor platform using a TI TMS320F28335 digital signal processor with high-resolution pulse width modulation and a multichannel analogue-to-digital (ADC) converter at a 4 MHz sampling rate. However, this platform is wired and, without a wireless module, is not suitable for PZT networks for wide-area monitoring.

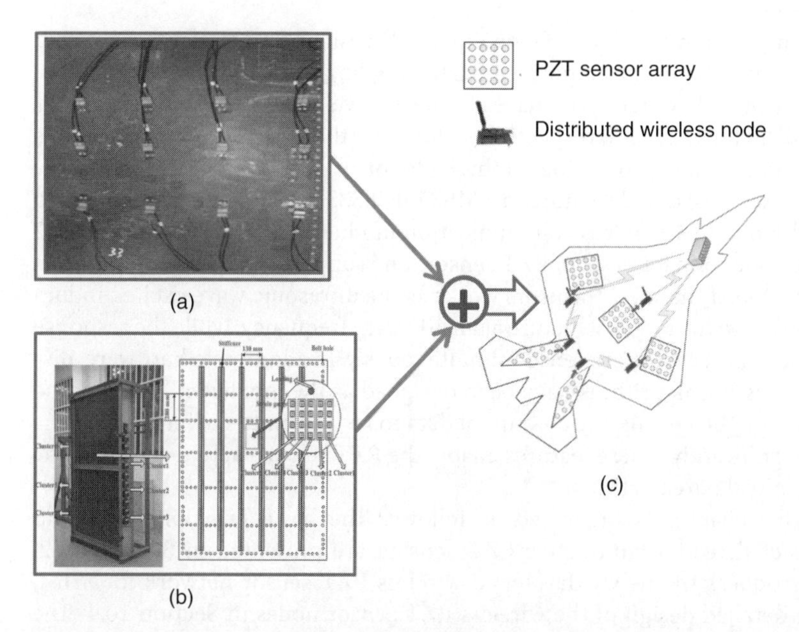

Figure 10.1 Application examples and evolution of PZT sensor networks for SHM: (a) wired PZT sensor network in UAV wing box [16]; (b) high-throughput wireless data acquisition system [17]; (c) next-generation wireless active PZT sensor system.

Beside their advantages, wireless PZT-based SHMs have brought new challenges to the structural engineering community because conventional designs of WSN nodes are not suitable for active sensing in SHM. Most existing applications (for example, smart homes and environmental monitoring) of WSNs do not requires high sampling rates, so conventional WSN nodes only support rates between a few Hertz and a few hundred Hertz. However, in Lamb wave-based SHM, the waves to be sensed are high-frequency (of the order of hundreds of kilohertz), and require a high sampling rate. A high sampling rate is also preferred because it has direct impacts on the resolution of defect localization. In addition, due to the complexity of Lamb-wave propagation, defect detection algorithms usually are computationally intensive and require considerable data-processing capabilities. However, existing WSN hardware motes (such as Mica2, MicaZ and TelosB) are designed for low data rates and applications requiring relatively low amounts of computation. This makes it difficult to process the Lamb-wave data locally at existing WSN hardware motes. The low-data-rate wireless link (up to 125 kbps in IEEE 802.15.4) will take quite a long time to send raw Lamb-wave data to a remote central server for data processing.

To address the issues of the large quantities of Lamb-wave data and low wireless data transmission rates, new compressive sensing methods have been proposed for Lamb wave-based SHM [20]; it has been verified that it is possible to reconstruct Lamb waves after compressive sensing. However, it is still challenging to embed the compressive sensing method into a wireless node, due to the intensive computation required for distributed Lamb-wave data processing. A wireless node using a field programmable gate array for active PZT sensing has been proposed [21]. Another solution is the introduction of digital signal processor hardware into the PZT sensor node to enhance the on-board computation capability. Dong et al. [22] designed a Martlet node with a TMS320F28069 chip, which can support a 3-MHz sampling rate MEMS accelerometer. Wireless PZT sensor and exciter nodes have also been developed, using TMS320C2811 and TMS320F2812 chips [23, 24], respectively. However, there is a lack of compressive sensing and distributed data processing on these platforms.

A newly designed wireless node features a TMS320F28335 digital signal processor and an improved IEEE 802.15.4 wireless communication module with a data rate of up to 2 Mbps [25]. Each node connects a set of PZT sensors and supports system sampling rates up to 12.5 MHz. One of these PZT sensors works as the ultrasonic wave exciter, inducing the Lamb wave into the target structure at an arbitrary frequency, while the response vibrations at other PZT sensors are sensed simultaneously. In addition to hardware, distributed data processing algorithms have been designed as the intelligent 'brain' of the wireless PZT-sensor. As a result, the amount of data to be transmitted over the wireless link is reduced significantly. These features enable the PZT sensor–actuator node to be deployed easily in wide-area SHM.

The rest of this chapter is organized as follows. The basic principles of Lamb wave-based defect detection and active PZT sensing are presented in Section 10.2. Section 10.3 introduces the newly developed wireless PZT sensor network for SHM, followed by the detailed design of the wireless PZT sensor nodes in Section 10.4. The procedure for distributed signal processing for SHM is discussed in Section 10.5 and a summary and an outline of future trends are given in Section 10.6.

10.2 Lamb Wave-based Defect Detection

Lamb waves, also known as long range ultrasonic waves, are guided elastic waves that can propagate over relatively long distances in thin plate-like structures with little attenuation, thus having a lot of potential for non-destructive SHM. PZT sensors are robust, low-cost, small, lightweight and energy-efficient, all favoured features for WSNs. Thus the combination of Lamb-wave monitoring and emerging wireless PZT sensors is considered the next-generation approach for real-time SHM of large structures. This section will present the principles of active PZT sensing and Lamb-wave analysis in SHM.

10.2.1 Active Piezoelectric Sensing Technology

PZT sensors make use of the electromechanical interaction between the mechanical and the electrical state in PZT materials to measure mechanical strain by translating it to electrical charge, or vice versa, to generate a mechanical strain from an electrical field applied to the PZT material. The latter is referred to as the *inverse piezoelectric effect* and is widely used in the production of ultrasonic waves. The PZT sensor can be surface-mounted on existing structures or embedded inside composite materials.

In recent years, active sensing of Lamb-waves with PZT sensors has been explored for SHM. The PZT sensors work actively in an ultrasonic excitation mode and a passive detection mode alternately. In active sensing, a set of PZT sensors is deployed at the surface of the structure to be monitored, and one or more PZT sensors works as exciters to induce Lamb waves into the structures. Other PZT sensors work in passive mode to detect the arrival of Lamb waves as they propagate through the structure. A switch circuit is also needed to change the role of these sensors, so that Lamb waves are excited from and monitored at various locations to improve defect-detection performance.

Because of their durability, low weight, low cost, dual role (either ultrasonic wave excitation or detection) and energy efficiency, PZT sensors are the most widely used exciters and sensors in ultrasonic acoustic applications. For example, a 25 mm round PZT sensor mounted on the surface of a metal plate (see the inserted pictures in Figure 10.5) weight around 1 g and costs a few pounds. PZT sensors deliver excellent performance in Lamb wave generation and detection and these features are greatly favoured in Lamb wave-based embedded wireless SHM.

10.2.2 Lamb Wave-based Defect Detection

Named after Horace Lamb [2] who analysed a particular type of acoustic wave in solids, Lamb waves are a kind of mechanical elastic deformation wave that propagates in thin plate-like thin-walled structures, guided by the structure's two parallel surfaces. Lamb waves are dispersive and consist of two types of wave mode: symmetric modes (represented by S_n) and anti-symmetric modes (represented by A_n). The frequency response of the Lamb wave can be expressed as

$$H(l,\omega) = \sum_{A_n} a_n(l,\omega)\cdot e^{-j\cdot K_{An}\cdot l} + \sum_{S_n} b_n(l,\omega)\cdot e^{-j\cdot K_{Sn}\cdot l} \tag{10.1}$$

where l is the propagation distance, ω denotes the angular frequency and the superscript A_n and S_n represent the nth anti-symmetric mode and nth symmetric mode, respectively. a_n and $K_{An} = \omega/C_{An}$ are the amplitude and wavenumber of the A_n mode, respectively, where C_{An} denotes the phase velocity of the A_n mode. b_n and $K_{Sn} = \omega/C_{Sn}$ are the amplitude and wavenumber of the S_n mode, respectively, where C_{Sn} denotes the phase velocity of the S_n mode. The same notation applies to the nth symmetric mode S_n. The number of modes for a Lamb wave propagating in a structure and their dispersive characteristics are strongly linked to the Lamb wave's frequency and the structure itself. In addition, different modes have different phase velocities, varying with the frequency f and the thickness of the material d. For any given frequency, at least two modes (S_0 and A_0) are present. At lower frequency (usually below 100 kHz), A_0 and S_0 are the main modes of the Lamb wave signal (as shown in Figure 10.3). Therefore, (10.1) can be approximated as

$$H(l,\omega) = a_0(l,\omega)\cdot e^{-j\cdot K_{A0}\cdot l} + b_n(l,\omega)\cdot e^{-j\cdot K_{Sn}\cdot l} \tag{10.2}$$

The simultaneous presence of two or more Lamb wave modes complicates the use of Lamb waves for detecting structural defects.

As shown in the literature, the unique feature of Lamb waves is their interference with defects in structures, manifesting as scattering and/or mode conversion. For example, a crack in a thin structure introduces new boundaries in the Lamb wave propagation path. When a wave passes through the crack, reflecting and scattering occur, which leads to possible changes in Lamb wave propagation characteristics: the amplitude, mode, frequency, propagation delays (referred to as *time of flight*) and direction. By looking at the changes of the propagation characteristics, one can determine if a path contains defects and even an estimate of the type and size of the defect. Many algorithms can be used to measure these characteristics, such as the cross-correlation for time of flight, baseline test for amplitude, fast Fourier transform and short-time frequency transform for spectrum. From these characteristics, scattering in the Lamb wave's pathway can be identified and the defect can be localized.

When the Lamb waves are measured synchronously by a set of sensors distributed at several locations in the structure, it is possible to find their propagation characteristics. Changes in these Lamb wave signals caused by defect in the structure can be measured and processed to extract information such as the location of the defect. However, certain kinds of defect are more detectable with particular Lamb wave modes. In particular, the S_0 wave mode is better for the detection of through-the-thickness cracks and the A_0 mode is better for finding corrosion and disbonding.

The challenge in processing Lamb wave signals into a defect image is the waves' dispersive and multimode nature, which complicates the interpretation of experimental data and reduces the wave spatial resolution. It is still a challenging task to decompose Lamb wave modes into symmetric and anti-symmetric modes for SHM applications. Various data techniques have been developed to address the challenge for the purpose of better interpretation and improved defect detection performance, and these are described in the following subsections.

Hanning window modulated Lamb waves: One technique is to transmit a modulated Lamb wave as a pulse (referred to as tone burst) with a narrow spectrum. Short durations and narrow spectra make it easier to separate different Lamb wave modes in the time and frequency domains, respectively. An example is Hanning window modulated 5-cycle sine waves, see Figure 10.2c. Due to their different phase velocities, different modes have different TOFs and will arrive at the sensor at different times. A short pulse makes different modes of the Lamb waves arrive at different times in the time domain. The Hanning window also features reduced-frequency sidelobes caused by the abrupt beginning and end of the wave forms (Figure 10.2c and 10.2d), so that the energy of the induced Lamb wave concentrates in a narrow frequency band around the excitation frequency. With reduced dispersion effects and less bandwidth, the Hanning window modulated Lamb wave helps in Lamb wave data interpretation and improves defect detection.

Frequency tuning: Since the mode depends on frequency and the dispersion varies with the frequency f and the thickness of the material d, it is better to find an optimal frequency for Lamb waves that contains low-dispersion modes with less overlapping of different modes. The process of selecting an appropriate frequency is referred to as *tuning*. It is possible to find an ideal frequency at which only one mode is excited or at which the multiple modes can be easily separated, as shown in Figure 10.3. Here, the excitation frequency of 100 kHz is purposely selected to avoid overlap between the two modes, which have different group velocities. As a result, the A_0 mode wave apparently takes a longer time than the S_0 mode wave to arrive at the PZT sensor.Although Lamb wave tuning requires a lot of time and effort in analysis and tests, it brings considerable advantages and allows the Lamb waves that are most appropriate for the particular application to be selected.

Data processing and imaging for defect detection: Once the optimal frequency is selected and modulated, Lamb waves are excited and sensed by the PZT sensors, and the various modes should be decomposed from the detected signals to extract the propagation characteristics. After this, the defect can be estimated from these propagation characteristics. A common defect detection method is a comparative analysis

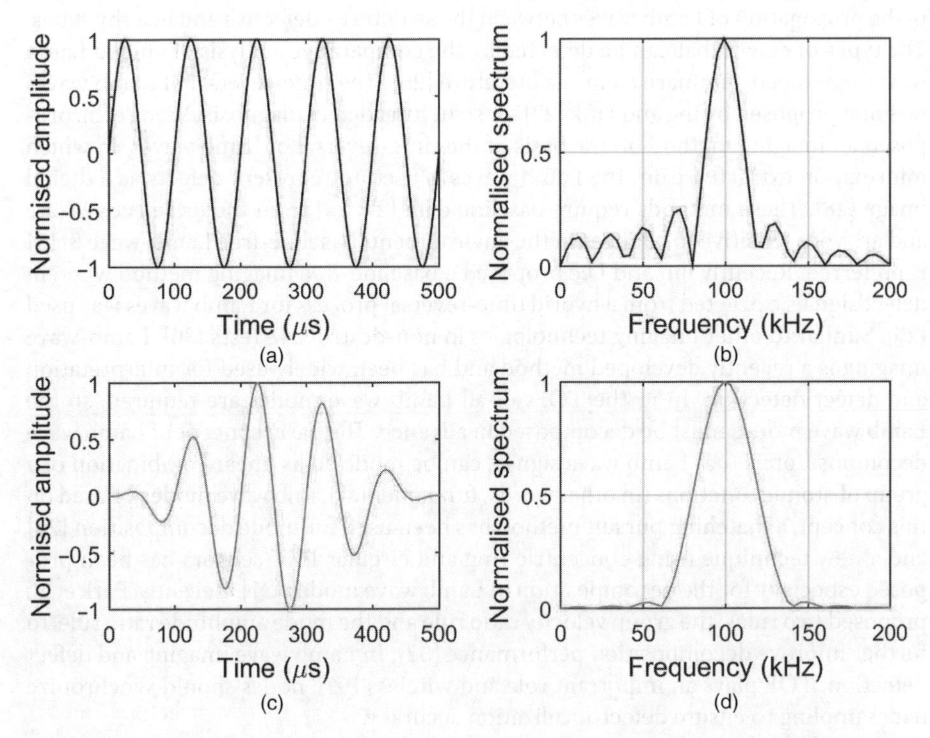

Figure 10.2 Hanning window modulation: (a) original five cycle 100 kHz sine wave (b) spectrum of five cycle sine wave; (c) Lamb wave after Hanning window modulation; (d) the spectrum after Hanning window modulation.

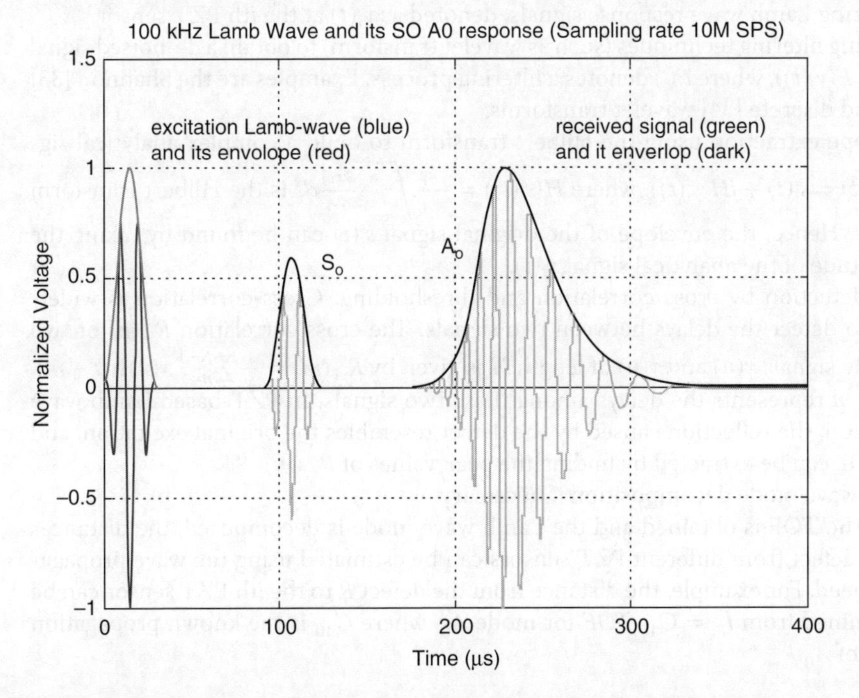

Figure 10.3 100-kHz Lamb wave and its responses. The signals are normalised for the purposes of illustration.

of the propagation of Lamb waves between the structure's defective and healthy states. The types of defect that can be detected by the comparative analysis of guided Lamb waves have been summarised in the literature [26]. The time-reversal of Lamb waves was first proposed by Ing and Fink [27] for structural defect diagnosis. Wang et al. proposed an imaging method on the basis of the time-reversal of Lamb waves, in which information extracted from the Lamb waves is used to represent defects as a digital image [28]. These methods require baseline data [27, 28] from the undefective state and are very sensitive to changes in the environment. Baseline-free Lamb-wave SHM is preferred. Recently Jun and Lee proposed a baseline-free imaging method wherein defect signals extracted from a hybrid time-reversal process for Lamb waves was used [29]. Similar to other imaging technologies in non-destructive tests [30], Lamb-wave imaging is a recently developed method and has been widely used for interpretation and defect detection. In it, the TOFs of all Lamb wave modes are required, so the Lamb wave modes must be decomposed in advance. The basic concept of Lamb wave decomposition is that Lamb wave signals can be modeled as linear combination of a group of atomic functions (in other words, fundamental Lamb wave modes). Based on this concept, a matching pursuit method has been used for mode decomposition [23] and a new technique using concentric ring and circular PZT sensors has been proposed especially for the decomposition of Lamb wave modes [31]. Recently, Park et al. proposed two rules: the group velocity ratio rule and the mode amplitude ratio rule, to further improve decomposition performance [32]. In Lamb wave imaging and defect detection, TOF plays an important role and wireless PZT nodes should synchronize data sampling to ensure defect-localization accuracy.

The principle of PZT-based Lamb wave imaging for SHM is illustrated in Figure 10.4. The implementation process of the TOF-based defect imaging method is as follows:

1. Acquiring Lamb wave response signals, denoted as $v_i(t)$ at the ith PZT sensor.
2. Applying filtering techniques (such as wavelet transform) to obtain a de-noised signal $s_i(t) = F(v_i(t))$, where $F(\cdot)$ denotes a filtering process. Examples are the Shannon [33] and and discrete [34] wavelet transforms.
3. Envelope extraction using the Hilbert transform to build a complex analytical signal $x_i(t) = s_i(t) + jH(s_i(t))$, where $H(s_i(t)) = -\frac{1}{\pi}\int_{-\infty}^{+\infty} \frac{S_i(\tau)}{t-\tau}dt$ is the Hilbert transform of $s_i(t)$. Hence, the envelope of the original signal $s_i(t)$ can be found by taking the magnitude of the analytical signal $x_i(t)$.
4. TOF detection by cross correlation and thresholding. Cross-correlation is widely used to detect the delays between two signals. The cross correlation $R_{xy}(n)$ of two discrete signals $x(n)$ and $y(n)$ of length N is given by $R_{xy}(m) = \frac{1}{N}\sum_{n=0}^{N-1} x(n)y(n-m)$, where m represents the delays among these two signals. In PZT-based Lamb wave detection, the reflection caused by the defect resembles the original excitation, and the TOF can be extracted by finding the peak values of $R_{xy}(m)$.
5. Lamb wave mode decomposition if necessary.
6. Once the TOF is obtained and the Lamb wave mode is decomposed, the distances of the defect from different PZT sensors can be estimated using the wave propagation speed. For example, the distance from the defect R to the ith PZT sensor can be determined from $l_{Ri} = C_{A0}\cdot TOF$ for mode A_0, where C_{A0} is the known propagation speed of A_0.

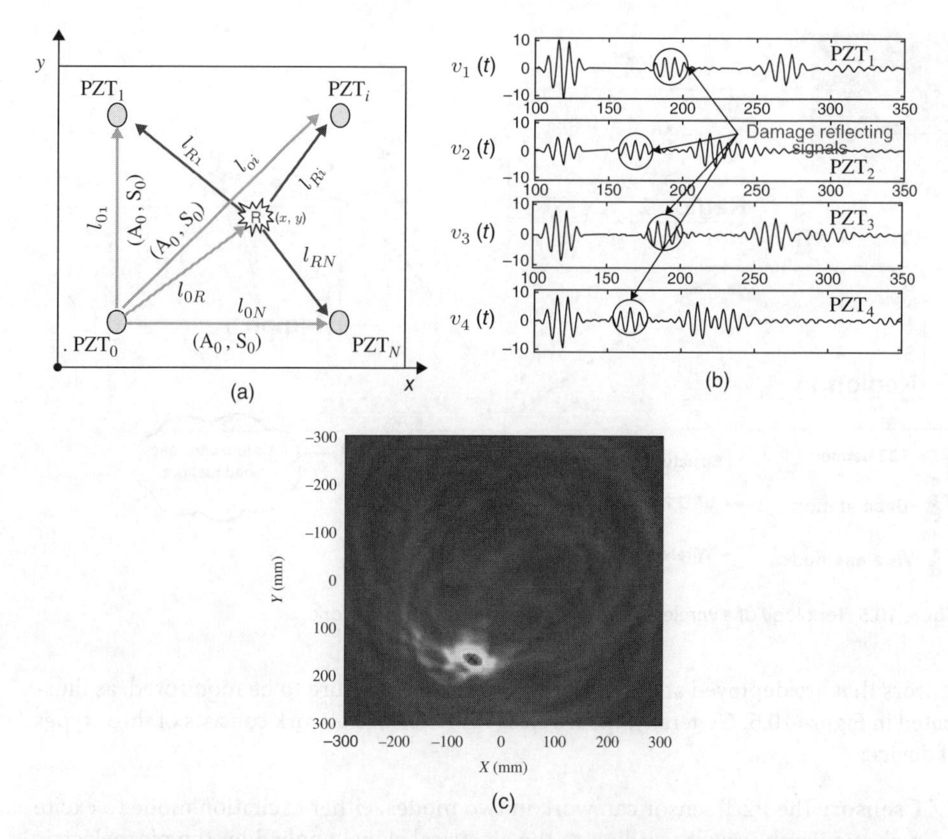

Figure 10.4 Principles and data processing for PZT-based Lamb wave imaging for SHM: (a) layout of PZT sensors; (b) detected Lamb wave signals; (c) Imaging of defect detection.

7. A geometric process can be used to determine the defect location. Multiple ellipses that share the same foci between two PZT sensors can be drawn from the estimated distance. The possible location of the defect is at the intersection points of the two ellipses. The number of the intersection points at a location (x, y) represents the possibility of a defect, so an image can be acquired representing the location of the defect.

Another widely used defect-localization method is time-reversal focusing [35]. The time-reversal focusing signal will have high amplitude at or near the position of the actual defect, so the defective area can be represented as an amplitude distribution image of the synthesised time-reversal signal.

10.3 Wireless PZT Sensor Networks

Similar to most PZT-based SHM systems, a wireless piezoelectric sensor network consists of a set of distributed wireless nodes, and each node connects a group of PZT

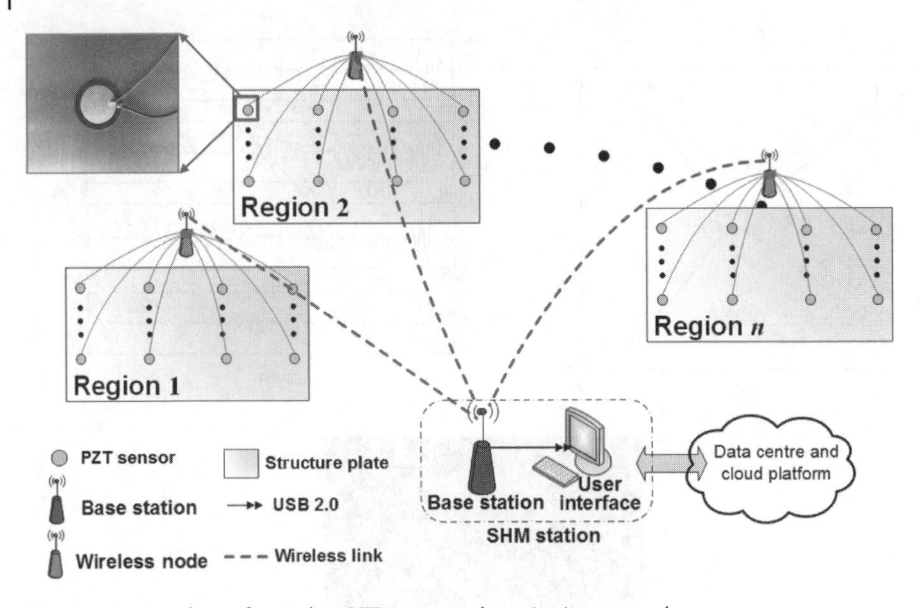

Figure 10.5 Topology of a wireless PZT sensor and monitoring network.

sensors that are deployed at different regions of the structure to be monitored, as illustrated in Figure 10.5. Generally, the wireless PZT sensor network consists of three types of device:

PZT sensor: The PZT sensor can work in two modes, either excitation mode to excite an elastic Lamb wave according to the electrical signal applied on the piezoelectric crystal, or detection mode to transform the responding elastic Lamb waves into an electrical signal.

Wireless sensor node: Each node connects to several PZT sensors (up to 16 PZT channels in the proposed wireless node) and decides the working mode of the PZT sensor by either generating an excitation signal and driving the PZT sensor or by reading the electrical signals from PZT sensors. The wireless node has computational resources at an appropriate level for processing Lamb wave signals. In particular, considering the number of PZT sensors in a wireless network, the large amounts of Lamb wave data collected and the low data transmission rates in low-powered wireless communications (such as IEEE 802.15.4), it is preferable that the data is first processed locally at the wireless node and only the useful information (signal envelopes, TOFs, modes) is sent to the base station for further processing.

Base station and user interface: This is the data sink node where the Lamb wave propagation characteristics from the wireless nodes are aggregated to finalise the defect detection and localization. The base station may also work as a coordinator to optimize the wireless data transmission and provide time-synchronization services for Lamb wave generation and detection among multiple wireless sensor nodes.

Depending on the application requirements and the size of the area to be monitored, the topology of the network can be a single cluster or multicluster. Figure 10.5 illustrates a single-cluster wireless PZT network, which makes synchronized data acquisition easier.

10.4 Wireless PZT Sensor Node

The function diagram of the proposed wireless PZT node is shown in Figure 10.6. The wireless PZT node consists of three components: the conditioning unit, the signal processing unit and the wireless communication unit.

The conditioning unit is an analogue signal processing circuit that has two tasks: Lamb wave excitation and Lamb wave detection. Lamb wave excitation amplifies the low-voltage (say, 3.3 V) Lamb waveforms generated by the signal processing unit to an appropriate higher voltage, so that the desired Lamb wave can be induced into the structure by the PZT sensor. Lamb-wave detection amplifies and filters the weak and noisy signals picked up by the PZT sensor to an appropriate level (say, 3.3 V) for analogue-to-digital conversion in the signal processing unit. Typical excitation and switching circuits are reported in the literature [16, 36]. In our recent work [25], the Lamb wave detection circuit consists of a group of charge amplifiers implemented using AD8608 operational amplifiers. Unlike voltage preamplifiers, which suffer from distance/attenuation effects, a charge amplifier can maintain the signal sensitivity regardless of distance from the PZT sensor to the preamplifier. The AD8608 is chosen because of its low bias current (1 pA maximum), low noise (12 nV maximum) and low offset voltage (65 μV maximum). The output of the charge amplifier is set to 0.1–3.2 V, which matches the input range of the ADC (0–3.2 V) with a 100 mV protection margin to maintain linearity.

The data processing unit in the proposed system is based on TI's TMS320F28335 digital signal processor, which is capable of synchronized high-speed data acquisition. The built-in ADC is 2-bit, with a 80 ns conversion time and two independent sample-and-hold units to support simultaneous conversion. The direct memory access module allows the wireless PZT node to collect sensor data at a sampling rate of up to 12.5 MHz without increasing the computation burden of the CPU. Another core function of the signal processing unit is distributed data processing by carrying out 'intelligent' Lamb wave processing algorithms, such as mode decomposition and cross-correlation checks, wavelet and/or Hilbert transforms, TOF estimation and compressive sensing.

The wireless communication unit is based on a 32-bit ARM Cortex-M0+ processor (ATSAMR21 by Atmel) and a built-in ultra-low-power IEEE 802.15.4 transceiver (RF233). It supports a data rate of 2 Mbps, much higher than the standard rate 125 kbps for IEEE 802.15.4. Once the signal processing unit has finished data processing, the

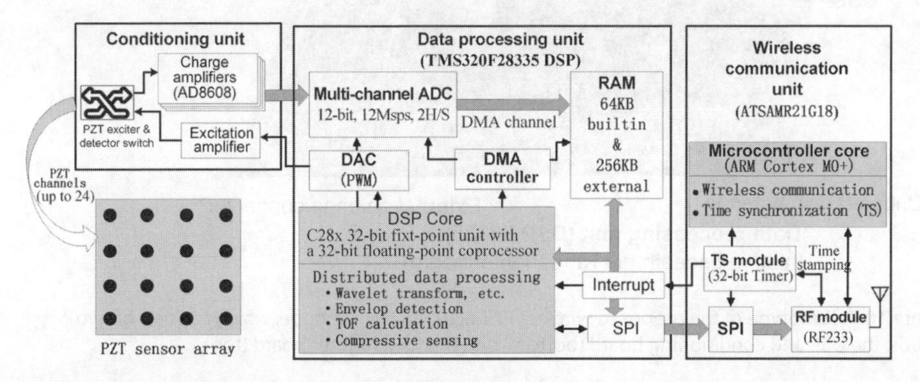

Figure 10.6 Block diagram of the wireless PZT node.

results are sent to the wireless communication unit via an SPI interface. Another task of the wireless communication unit is time synchronization. It is well known that time synchronization plays a key role in all TOF-based SHM systems. The accuracy of arrival time calculations and the resolution of defect localization are reduced by errors in time synchronization. Recognising the importance of time synchronization, a dual-processor architecture is adopted in the proposed hardware system, with the second processor (the ATSAMR21G18 in the radio unit) separated from the computation-intensive data processing processor. The ATSAMR21G18 is reserved for wireless communication and time synchronization. Therefore, the time accuracy of interrupt-based time synchronization and time-stamping is not compromised by other tasks (such as ADC interrupts and delays for intensive data processing).

The proof-of-concept prototype of the proposed wireless PZT sensor node is shown in Figure 10.7. It consists of two PCB boards connected through two 20-way connectors. The bottom board contains the data processing unit (a TI F28335 DSP), located just under the RF daughter board, and the signal conditioning unit (an array of amplifiers). The daughter board is at the top and contains the wireless communication unit (the ATSAMR21). Meanwhile, the 32-kHz crystal at the wireless unit provides a real-time clock, which is reserved for precise time synchronization and synchronized data acquisition.

10.5 Distributed Data Processing

10.5.1 Operation Overview

Figure 10.8 is an operational overview of the distributed data processing and communication. The diagnosis process is initiated from the user interface at the SHM station

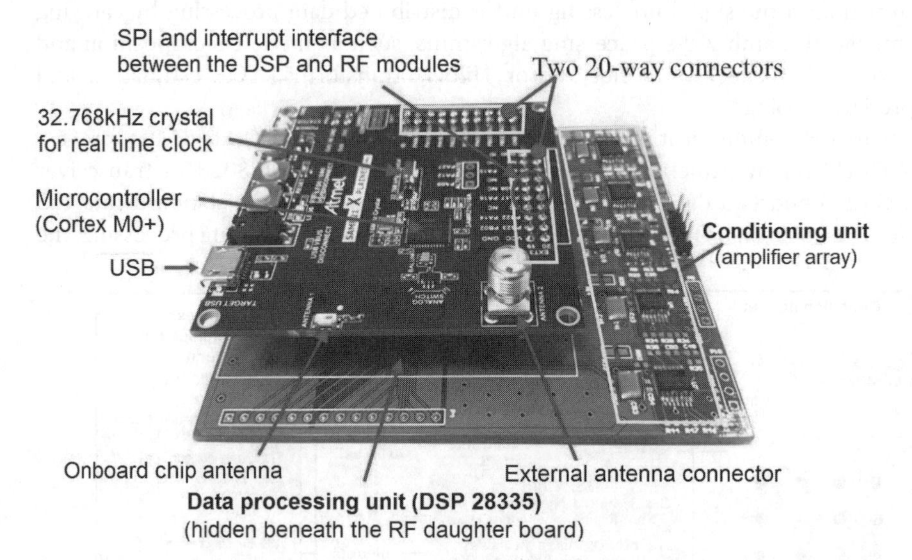

Figure 10.7 Prototype of the proposed wireless PZT actuator/sensor node, which consists of two boards: the DSP and conditioning board (bottom) and the RF daughter board (top).

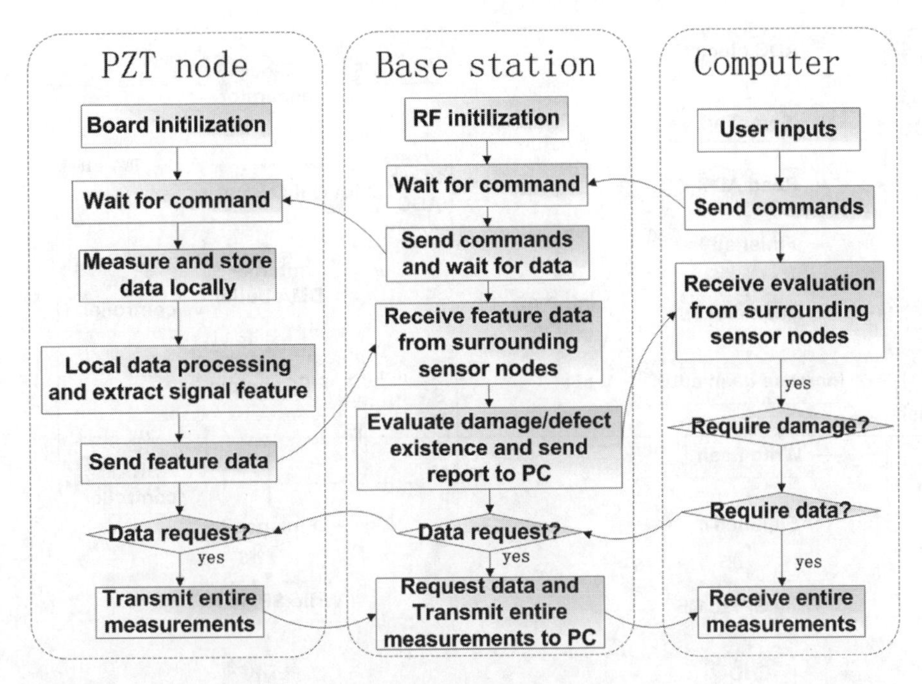

Figure 10.8 Operational overview of the wireless PZT network.

and broadcast to child wireless nodes through the base station (see Figure 10.5). After the handshaking process is completed, the base station broadcasts an initial command to the wireless PZT node. Once the initial command is received, the selected PZT node triggers the signal excitation and other PZT nodes immediately start data acquisition.

Conventional wireless sensor nodes normally use a centralised architecture. In this architecture, ADCs and memory chips are directly connected to the I/O ports of the microcontroller. The microcontroller must access the peripheral chips sequentially and multiple clock cycles are required to complete an operation that involves several peripheral chips. Most peripheral chips should be active while waiting for signals from the microcontroller. For such an architecture in a conventional design, a high sampling rate is almost impossible.

10.5.2 Synchronized High Sampling-rate Sensing and Data Processing

In Figure 10.9a, the flowchart demonstrates the typical data-sampling cycle of a conventional design, such as the Telosb, Imote2 and MICA motes widely used in WSN prototypes. These conventional motes have an internal 12-bit ADC in the microcontroller. The microcontroller sends a clock signal and control signals to the ADC to trigger the sampling cycle conversion. The internal buffer is set up to get data from the ADC. The reading takes 12 periodic operations and an ADC clock signal is generated in each period for filling the buffer bit by bit. After the data have been read out entirely, the microcontroller provides a clock signal and control signals to the on-board flash memory. Before writing the data into the flash bit by bit, the microcontroller takes some instructions to send an address to the flash. The flash saving operation also takes 12 periodic operations, each of which involves several instructions and a few clock cycles. Then

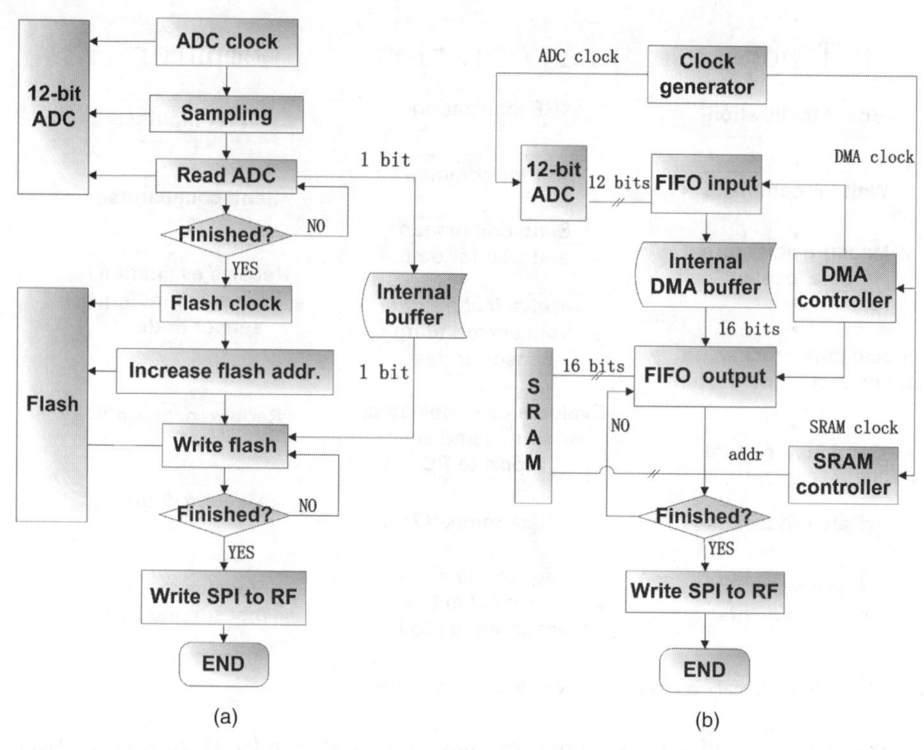

Figure 10.9 Data sampling and collection process: (a) conventional wireless sensor node with low sampling rate; (b) the proposed wireless PZT sensor/actuator node.

the sampling cycle ends and another sampling cycle can begin. This architecture clearly reveals the inefficiency of a wireless sensor node using an ordinary microcontroller.

Figure 10.9b is an improved sampling design using a TMS320F28335 chip. The chip internally has some controllers, including a first-in, first-out (FIFO) DMA controller for sampling data input and output, a SRAM controller and a clock generator. The DMA algorithm has been widely used to allow hardware to access memory independently. To enable a wireless sensor node for high-speed applications, algorithms such as semi-DMA are applied to extend the traditional architecture of wireless sensor nodes. In our design, with the DMA controller, the main microcontroller TMS320F28335 chip can be released from the task of data transfer. The sampling data transfer and data acquisition can be more efficient when DMA is adopted. The 12-bit sampling data is acquired by the ADC and saved into the internal DMA buffer, in FIFO input mode, by the DMA controller. Meanwhile, the DMA controller and address controller control the data to access the SRAM in FIFO output mode. The sampling data is moved from the internal DMA buffer to SRAM at 16-bit length, which is the I/O width of the SRAM. Each clock cycle has eight continuous sampling operations and one writing operation for writing data to the SRAM. As a result, the sampling data is encapsulated according to the bit alignment in the internal DMA buffer and the DMA, ADC, and the SRAM can be fully utilised without compromise.

10.6 Summary

This chapter reviews the state-of-art of PZT-based active acoustic sensing technologies for SHM and presents a next-generation, low-cost, distributed wireless PZT sensor network for large-area online defect detection and event monitoring. The basic principles of PZT sensors and guided Lamb wave-based signal processing for defect detection are reviewed and the design of a high performance wireless PZT sensor network is discussed. The designed wireless node is a powerful wireless platform for performing precise high-frequency data acquisition and distributed local data processing. The benefits of an online distributed SHM include failure prevention at an earlier stage, cost reduction through cable-free deployment and condition-based maintenance with fewer inspection routines. It is a key enabling technology for developing self-monitoring smart structures.

By combining the advantages of PZT sensors (low cost, small size, simplicity and durability), Lamb wave defect detection (non-destructive and online operation) and wireless data transmission (cable-free, easy and fast development), wireless PZT sensor systems have shown their potential for online large-scale SHM, in areas such as defect detection and event monitoring of aircraft, UAVs, trains and bridges. However, as an emerging cross-disciplinary technology, it also raises new challenges to the existing sensor, wireless-communications and signal-processing communities. Future research and technical trends in this area are likely to be in the areas of synchronization, scalability and reliability.

Synchronization: In SHM, precise timing in data acquisition is a key enabling technology for achieving higher resolution for defect localization, which depends on the phase and amplitude relationship of distributed sensor measurements. The crystal oscillator in a low-cost wireless node is not a sufficiently accurate clock source and various techniques have been proposed to synchronize drifting clocks using time-stamped packet exchanges. Examples include RBS [37], TSPN [38] and FTSP [39]. The recent trend is to adapt IEEE 1588 PTP into WSN [40]; advanced data processing is needed to address the temporal jitter and spatial jitter, which is larger and fluctuates more in wireless communications than it does in wired networks.

Scalability: As SHM requires a large number of PZT sensors and a high sampling rate to measure the state of a structure, a great quantity of raw data is generated. It is important to take the scalability of the system into account. If the number of nodes increases, the system has to send large amounts of data over the band-limited wireless link. Distributed signal processing and compressed sensing have shown the potential to address scalability issues by extracting useful information from big data collected locally, so that the data transmitted over the wireless link is reduced. On the other hand, bandwidth-efficient high-throughput communication scheduling is preferable in wireless PZT sensor networks. Packet collision is the main cause of low throughput in CSMA-based wireless communications. TDMA and synchronized duty-cycle management can prevent packet collisions to a great extent, thus improving the utilisation of wireless bandwidth.

Reliability: A wireless node could be unreliable and the power supply of a wireless node is constrained. Condition monitoring of PZT sensors and sensor nodes

[41] is an attractive technique for building fault-tolerant PZT sensor networks. Energy-harvesting techniques are a promising potential solution to the power-supply issue.

Conflict of Interests

The authors declare that there is no conflict of interests regarding the publications of this paper.

Acknowledgment

This work is supported by European Commission project 'Health Monitoring of Offshore Wind Farms' (HEMOW) under grant FP7-PEOPLE-2010-IRSES-GA-269202 and the EPSRC project 'Novel Sensing Network for Intelligent Monitoring' (NEWTON) under grant EP/J012343/1.

References

1 Z. Su and L. Ye., *Identification of Damage Using Lamb Waves: From Fundamentals to Applications*. Springer-Verlag, 2009.

2 L. Qiu and S. Yuan, 'On development of a multi-channel PZT array scanning system and its evaluating application on UAV wing box,' *Sensors and Actuators A: Physical*, vol. **151**, no. 2, pp. 220–230, 2009.

3 V. Giurgiutiu, 'Lamb wave generation with piezoelectric wafer active sensors for structural health monitoring,' in *SPIE's 10th Annual International Symposium on Smart Structures and Materials and 8th Annual International Symposium on NDE for Health Monitoring and Diagnostics*, San Diego, CA, 2002.

4 O. Yapar, P. Basu, P. Volgyesi, and A. Ledeczi, 'Structural health monitoring of bridges with piezoelectric AE sensors,' *Engineering Failure Analysis*, vol. **56**, pp. 150–169, 2015.

5 H. Gu, G. Song, H. Dhonde, Y. L. Mo, and S. Yan, 'Concrete early-age strength monitoring using embedded piezoelectric transducers,' *Smart Materials and Structures*, vol. **15**, no. 6, p. 1837, 2006.

6 D. Ai, H. Zhu, and H. Luo, 'Sensitivity of embedded active PZT sensor for concrete structural impact damage detection,' *Construction and Building Materials*, vol. **111**, pp. 348–357, 2016.

7 F. Song, G. L. Huang, J. H. Kim, and S. Haran, 'On the study of surface wave propagation in concrete structures using a piezoelectric actuator/sensor system,' *Smart Materials and Structures*, vol. **17**, no. 5, p. 055024, 2008.

8 Q. Wang, Z. Su, and M. Hong, 'Online damage monitoring for high-speed train bogie using guided waves: Development and validation,' in *EWSHM 7th European Workshop on Structural Health Monitoring*, 2014.

9 J. Oliver, J. Kosmatka, C. Farrar, and G. Park, 'Development of a composite uav wing test-bed for structural health monitoring research,' in *Proceedings of 14th SPIE Conference on Smart Structures and Nondestructive Evaluation*, San Diego, CA., March 2007.

10 Y.-K. An, M. K. Kim, and H. Sohn, 'Airplane hot spot monitoring using integrated impedance and guided wave measurements,' *Structural Control and Health Monitoring*, vol. **19**, no. 7, pp. 592–604, 2012.

11 C. Y. Park, J. H. Kim, and S.-M. Jun, 'A structural health monitoring project for a composite unmanned aerial vehicle wing: overview and evaluation tests,' *Structural Control and Health Monitoring*, vol. **19**, pp. 567–579, 2012.

12 X. P. Qing, S. J. Beard, R. Ikegami, F.-K. Chang, and C. Boller, 'Aerospace applications of smart layer technology,' in *Encyclopedia of Structural Health Monitoring*, 2009.

13 Y. Zhong, S. Yuan, and L. Qiu, 'Multiple damage detection on aircraft composite structures using near-field MUSIC algorithm,' *Sensors and Actuators A: Physical*, vol. **214**, pp. 234–244, 2014.

14 L. Qiu, B. Liu, S. Yuan, and Z. Su, 'Impact imaging of aircraft composite structure based on a model-independent spatial-wavenumber filter,' *Ultrasonics*, vol. **64**, pp. 10–24, 2016.

15 S. Sankarasrinivasan, E. Balasubramanian, K. Karthik, U. Chandrasekar, and R. Gupta, 'Health monitoring of civil structures with integrated UAV and image processing system,' *Procedia Computer Science*, vol. **54**, pp. 508–515, 2015.

16 L. Qiu, S. Yuan, and Q. Wu, 'Design and experiment of PZT network-based structural health monitoring scanning system,' *Chinese Journal of Aeronautics*, vol. **22**, no. 5, pp. 505–512, 2009.

17 S. Gao, S. Yuan, L. Qiu, B. Ling, and Y. Ren, 'A high-throughput multi-hop WSN for structural health monitoring.' *Journal of Vibroengineering*, vol. **18**, no. 2, 2016.

18 X. Liu, J. Cao, and W. Z. Song, 'Distributed sensing for high-quality structural health monitoring using wsns,' *IEEE Transactions on Parallel and Distributed Systems*, vol. .**26**, no. 3, pp. 738–747, 2015.

19 O. Martens, T. Saar, and M. Reidla, 'TMS320F28335-based piezosensor monitor-node,' in *4th European Education and Research Conference (EDERC)*, 2010, Dec 2010, pp. 62–65.

20 X. Zhao, H. Gao, G. Zhang, B. Ayhan, F. Yan, C. Kwan, and J. L. Rose, 'Active health monitoring of an aircraft wing with embedded piezoelectric sensor/actuator network: I. defect detection, localization and growth monitoring,' *Smart Materials and Structures*, vol. **16**, no. 4, p. 1208, 2007.

21 L. Liu and F. Yuan, 'Active damage localization for plate-like structures using wireless sensors and a distributed algorithm,' *Smart Materials and Structures*, vol. **17**, no. 6, Article ID 055022, 2008.

22 X. Dong, D. Zhu, and Y. Wang, 'Design and validation of acceleration measurement using the martlet wireless sensing system,' in *ASME 2014 Conference on Smart Materials, Adaptive Structures and Intelligent Systems*, 2014.

23 A. Perelli, T. D. Ianni, A. Marzani, L. D. Marchi, and G. Masetti, 'Model-based compressive sensing for damage localization in lamb wave inspection,' *IEEE Transactions*

on Ultrasonics, Ferroelectrics, and Frequency Control, vol. **60**, no. 10, pp. 2089–2097, 2013.

24 A. Perelli, L. De Marchi, and A. Marzani, 'Acoustic emission localization in plates with dispersion and reverberations using sparse pzt sensors in passive mode,' *Smart Materials and Structures*, vol. **21**, no. 2, 2012.

25 S. Gao, X. Dai, Z. Liu, G. Tian, and S. Yuan, 'A wireless piezoelectric sensor network for distributed structural health monitoring,' in *Wireless for Space and Extreme Environments (WiSEE), 2015 IEEE International Conference on*, Dec 2015, pp. 1–6.

26 V. Giurgiutiu, A. Zagrai, and J. Bao, 'Damage identification in aging aircraft structures with piezoelectric wafer active sensors,' *Journal of Intelligent Material Systems and Structures*, vol. **15**, no. 9–10, pp. 673–687, 2004

27 R.K. Ing and M. Fink, 'Time-reversed lamb waves,' *IEEE Transactions on Ultrasonics, Ferroelectrics, and Frequency Control*, vol. **45**, no. 4, pp. 1032–1043, 1998.

28 C. Wang, J. Rose, and F. Chang, 'A synthetic time-reversal imaging method for structural health monitoring,' *Smart Materials and Structures*, vol. **13**, pp. 415–423, 2004.

29 Y. Jun and U. Lee, 'Computer-aided hybrid time reversal process for structural health monitoring,' *Journal of Mechanical Science and Technology*, vol. **26**, no. 1, pp. 53–61., 2012.

30 B. Gao, W. L. Woo, and G. Y. Tian, 'Electromagnetic thermography nondestructive evaluation: Physics-based modeling and pattern mining,' *Scientific Reports*, vol. **6**, 2016.

31 C. M. Yeum, H. Sohn, and J. B. Ihn, 'Lamb wave mode decomposition using concentric ring and circular piezoelectric transducers,' *Wave Motion*, vol. **48**, no. 4, pp. 358–370, 2011.

32 I. Park, Y. Jun, and U. Lee, 'Lamb wave mode decomposition for structural health monitoring,' *Wave Motion*, vol. **51**, no. 2, pp. 335–347, 2014.

33 S. Gao, X. Dai, Z. Liu, and G. Tian, 'High-performance wireless piezoelectric sensor network for distributed structural health monitoring,' *International Journal of Distributed Sensor Networks, vol.* 2016, **2016**.

34 L. Yu and V. Giurgiutiu, 'In-situ optimized PWAS phased arrays for Lamb wave structural health monitoring,' *Journal of Mechanics of Materials and Structures*, vol. **2**, no. 3, pp. 459–487, 2007.

35 L. Qiu, S. Yuan, X. Zhang, and Y. Wang, 'A time reversal focusing based impact imaging method and its evaluation on complex composite structures,' *Smart Materials and Structures*, vol. **20**, no. 10, p. 105014, 2011.

36 D. Musiani, '*Design of an active sensing platform for wireless structural health monitoring*,' Ph.D. dissertation, University of Bologna, 2006.

37 J. Elson, L. Girod, and D. Estrin, 'Fine-grained network time synchronization using reference broadcasts,' *ACM SIGOPS Operating Systems Review*, vol. **36**, no. SI, pp. 147–163, 2002.

38 S. Ganeriwal, R. Kumar, and M. B. Srivastava, 'Timing-sync protocol for sensor networks,' in *Proceedings of the 1st International Conference on Embedded Networked Sensor Systems*. ACM, 2003, pp. 138–149.

39 M. Maróti, B. Kusy, G. Simon, and Á. Lédeczi, 'The flooding time synchronization protocol,' in *Proceedings of the 2nd International Conference on Embedded networked Sensor Systems*. ACM, 2004, pp. 39–49.

40 Y. Huang, T. Li, X. Dai, H. Wang, and Y. Yang, 'TS2: a realistic IEEE1588 time-synchronization simulator for mobile wireless sensor networks,' *Simulation*, vol. **91**, no. 2, pp. 164–180, 2015.

41 X. Dai, F. Qin, Z. Gao, K. Pan, and K. Busawon, 'Model-based on-line sensor fault detection in wireless sensor actuator networks,' in *IEEE 13th International Conference on Industrial Informatics (INDIN), 2015*, Cambridge, United Kingdom, 2015, pp. 556–561.

11

Navigation and Remote Sensing using Near-space Satellite Platforms

Wen-Qin Wang[1] *and Dingde Jiang*[2]

[1] *School of Communication and Information Engineering, University of Electronic Science and Technology of China, Chengdu, China*
[2] *College of Information Science and Engineering, Northeastern University, Shengyang, China*

11.1 Background and Motivation

Advances in wireless sensor technology promise to give effective, widely applicable, low-cost, and secure information exchange between airborne vehicles. For example, if in-flight sensors in commercial airlines can allow information about adverse weather conditions and emergency situations to be shared, especially when the flights are in the areas outside the reach of ground control stations, aviation accidents will be significantly reduced. Unmanned airborne vehicles rely on such sensor networks for safe manoeuvring. It is anticipated that airborne networks will be able to provide information exchange among airborne vehicles and connect them with space and ground networks to create future multiple-domain communication networks [1]. The development of reliable routing protocols that minimize the number of packets lost due to link and path failures has been reported [2–4].

Although satellites and aircraft are well-established platforms for wireless sensors, they cannot provide sensing over an areas of interest for periods of days, weeks, or months over a selected area of interest. Even if we can launch a satellite on demand for a particular mission, it would only be in view for very short periods. Table 11.1 shows the pass time of typical satellites in low-Earth orbit (LEO) for different angles above the horizon [5]. It shows just how short these times are. Satellites usually operate in orbits higher than 200 km and aircraft routinely operate at altitudes lower than 18 km.

11.1.1 What is Near-space?

Near-space is defined as the atmospheric region from about 20–100 km altitude above the Earth's surface, as shown in Figure 11.1. Note that the lower limit is not determined from operational considerations, but from the international controlled airspace altitude.

Very few sensors are currently operating in near-space, because the atmosphere is too thin to support flying for most aircraft and yet too thick to sustain orbit for

Wireless Sensor Systems for Extreme Environments: Space, Underwater, Underground, and Industrial.
First Edition. Edited by Habib F. Rashvand and Ali Abedi.

Table 11.1 Typical LEO satellite pass times for different angles above horizon.

Pass time (min:sec)		Angle above horizon				
		0°	5°	10°	30°	45°
	200	7:49	5:37	4:08	1:40	1:00
Altitude (km)	300	9:35	7:16	5:34	2:24	1:27
	400	11:10	8:44	6:54	3:08	1:54

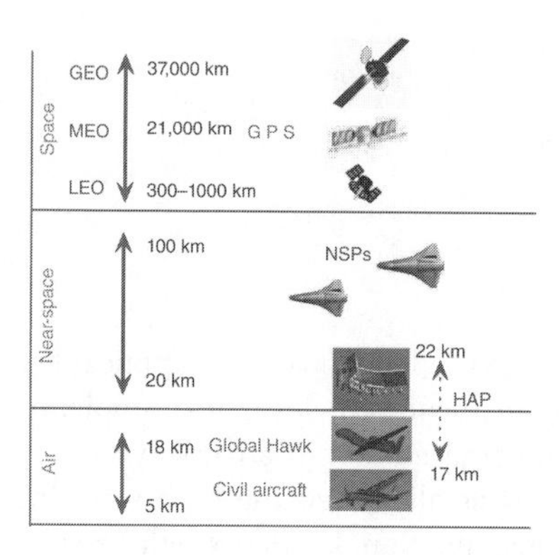

Figure 11.1 Near-space definition.

satellites. However, advances in several technologies have led to a revolutionary advance in capability:

- power supplies, including thin, lightweight solar cells, small, efficient fuel cells, and high-energy-density batteries
- the miniaturisation of electronics and an exponential increase in computing power, enabling extremely capable sensors in very small and lightweight packages
- very lightweight, strong, and flexible materials that can resist degradation under strong ultraviolet illumination and that are relatively impermeable to helium or hydrogen [6].

Another two emerging technologies that show promise for enabling an operational foothold to be established in near-space are high-altitude buoyant lift and plasma thrust. As altitude increases, electrodynamic power transfer into the air can be used for propulsion, for cooling or for control. This allows many novel electromagnetic circuits to be used. Plasma technology developments have prompted both high-altitude and global observing research in many institutes or laboratories.

Near-space thus has become an area of exceptional interest in recent years. Possible real-world uses in communication radar and navigation applications have come to light.

11.1.2 Advantages of Near-space for Sensor Platforms

Compared to current satellite and aircraft platforms, near-space platforms have many advantages for microwave remote sensing applications.

11.1.2.1 Inherent Survivability

There are defensive options available to help near-space vehicles, such as deceptive 'chaff', similar to that carried on modern fighter aircraft. This is small and light enough to make it feasible. Chaff could be dispensed in the hope of confusing radar guidance systems. Also, using decoy vehicles is an option. The vehicles themselves are relatively inexpensive and a simple fake payload could be attached. They might have electronic and infrared signatures that mimic those of the real vehicle, further making it difficult to discriminate between them.

11.1.2.2 Persistent Monitoring or Fast Revisiting Frequency

The most useful and unique aspect of near-space vehicles is their ability to provide persistent regional coverage or fast revisiting frequencies. Space technologies have revolutionised modern battlefields and remote sensing, but persistent coverage, which is highly desired, is still unavailable through satellites or aircraft. Table 11.1 shows the observation times for selected LEO orbits. Lower down in the atmosphere, the longest persistence that we can currently expect from an air-breathing aircraft is about a day or so. However, persistent coverage can be achieved by using near-space vehicles. Near-space is above the troposphere, the atmospheric region in which most weather occurs; there are no clouds, thunderstorms or precipitation. Moreover, modern propulsion techniques can be used to counter any light winds in near-space.

11.1.2.3 High Sensitivity and Large Footprint

Near-space vehicles are much closer to their targets than their orbital cousins. Distance is critical when receiving low-power signals. From the radar equation, we know that the received signal power attenuates as the square of the distance from the transmitter to the target, while that of an active antenna attenuates as the fourth power of the transmitter distance. Considering a point at nadir, near-space vehicles are 10–20 times closer to their targets than a typical LEO satellite at 400 km. This distance differential implies that near-space vehicles could detect signals that are 10–13 dB weaker.

On the other hand, near-space vehicles will have impressive ground coverage from such high altitudes. Figure 11.2 shows the ground coverage area as a function of

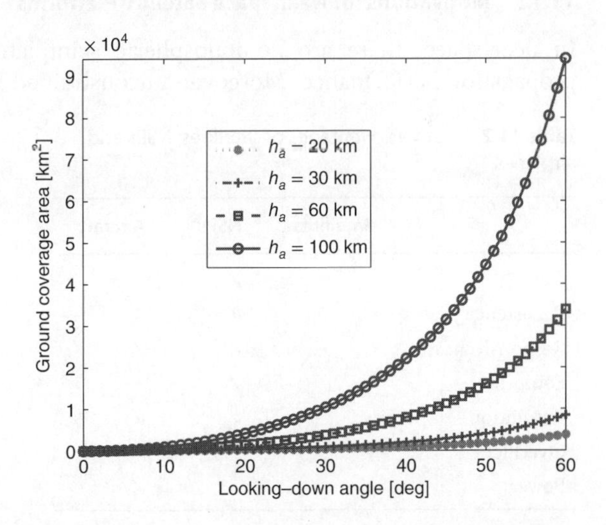

Figure 11.2 Ground coverage area as a function of looking-down angle for different flying altitudes h_a.

looking-down angle. Although orbiting satellites traditionally have a much larger footprintthan near-space vehicles, they do so at the cost of persistence and signal strength. Additionally, being lower than satellites also brings another advantage over near-space vehicles: they fly below the ionosphere. Ionospheric scintillation is very difficult to predict, but it can disrupt satellite communication and navigation significantly. Fortunately, ionospheric scintillation has no impact on near-space vehicles.

11.1.2.4 Low Cost

The inherent simplicity, recoverability, and relatively less complex infrastructure of near-space vehicles all contribute to its cost advantage. The price of a high-altitude vehicle will be of the order of millions of dollars. These costs can be compared to current unmanned aerial vehicles (UAVs) such as the Predator and the Global Hawk. The cost of satellites is enormous, typically from $60–300 million each. In addition to this cost is the expense of launching them into orbit, which adds another $10–40 million. Additionally, not being exposed to the electronic radiation that is found in space, payloads in near-space vehicles do not require costly radiation-hardening.

Table 11.2 summarises the relative advantages of satellites, near-space platforms (NSPs) and airplanes. Satellites have advantages in footprint and overflight. Aircraft, both manned and unmanned, are extremely responsive; they can be launched in minutes or at least hours, and once on station they can be redirected at will. NSPs are also extremely responsive compared to satellites and almost as responsive as aircraft. The cost of development of satellites is much greater, and thus it is more efficient to cover a large area with several NSPs rather than using multiple terrestrial stations or a satellite. In addition, due to their long development periods, satellites always carry the risk of becoming obsolete once they are in orbit. NSPs also enjoy favourable path-loss characteristics compared with both terrestrial and satellite sensors. They can frequently take off and land for maintenance and upgrading. It is thus necessary to construct a mixed infrastructure of NSPs and terrestrial and satellite systems. In doing so, a powerful integrated network infrastructure can be constructed, each part of which makes up for the weaknesses of the others.

11.1.3 Motivations for Near-space Satellite Platforms

In near-space, there are no ionospheric scintillations to degrade wireless signal propagation performance. Moreover, unconstrained by orbital mechanics and having

Table 11.2 Relative advantages of satellites, NSPs and airplanes.

	Satellites	NSPs	Aircraft
Cost		✓	
Persistence		✓	
Responsiveness		✓	✓
Footprint	✓	✓	
Resolution		✓	✓
Overflight	✓		
Power	✓		✓

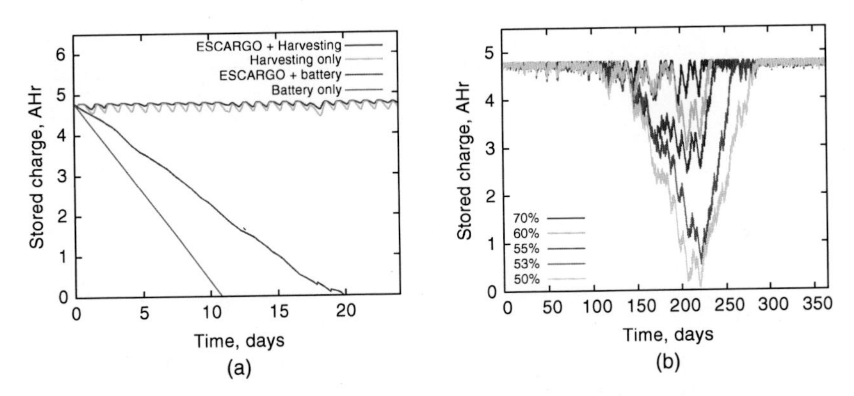

Figure 5.3 Stored charge. (a) Mean stored charge during first month of operation; (b) Mean stored charge over time by charging efficiency.

Wireless Sensor Systems for Extreme Environments: Space, Underwater, Underground, and Industrial.
First Edition. Edited by Habib F. Rashvand and Ali Abedi.

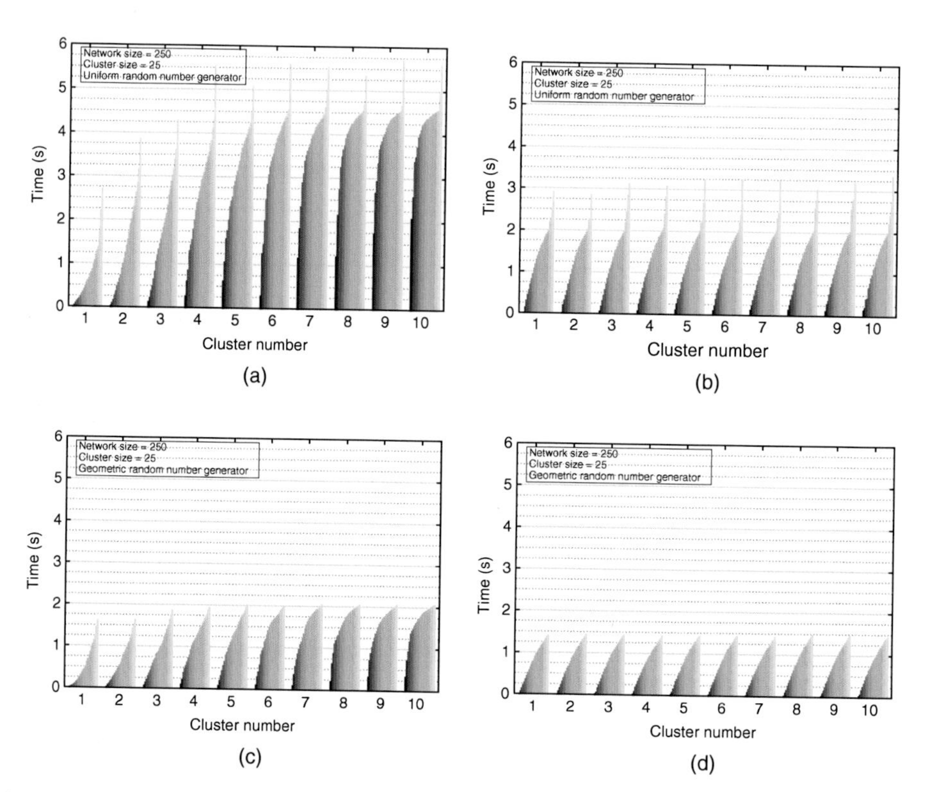

Figure 5.16 Packet arrival time ($N = 250, c = 25$). (a) IEEE802.15.4 with uniform random backoff slot selection; (b) Cluster-centric MAC with uniform random backoff slot selection; (c) IEEE802.15.4 with geometric random backoff slot selection; (d) Cluster-centric MAC with geometric random backoff slot selection.

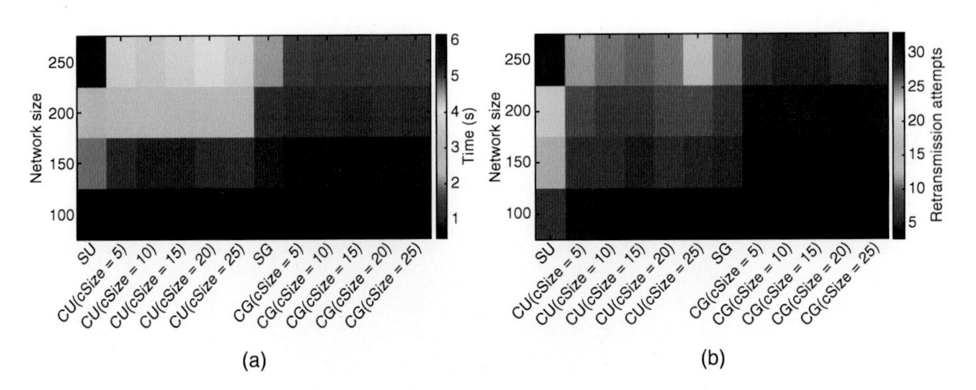

Figure 5.17 Performance for different network and cluster sizes to transmit all packets: S, standard IEEE 802.15.4 MAC; C, cluster-centric MAC; U, G, uniform/geometric random number generator; cSize, cluster size. (a) Average packet transmission time; (b) Average number of retransmissions.

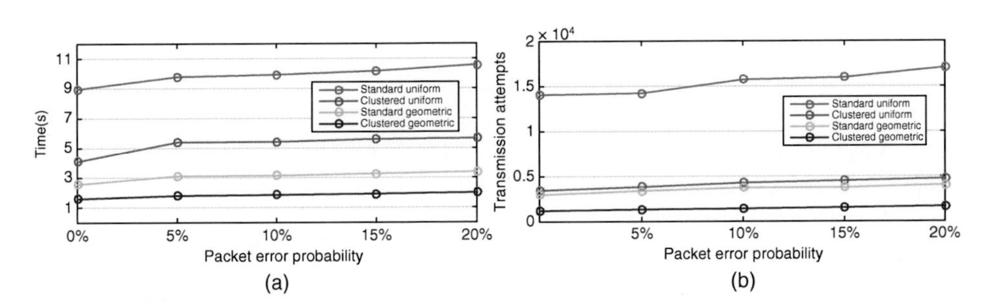

Figure 5.18 Performance in lossy environment ($N = 250, C = 10$). (a) Total time to transmit all packets; (b) Total retransmission attempts.

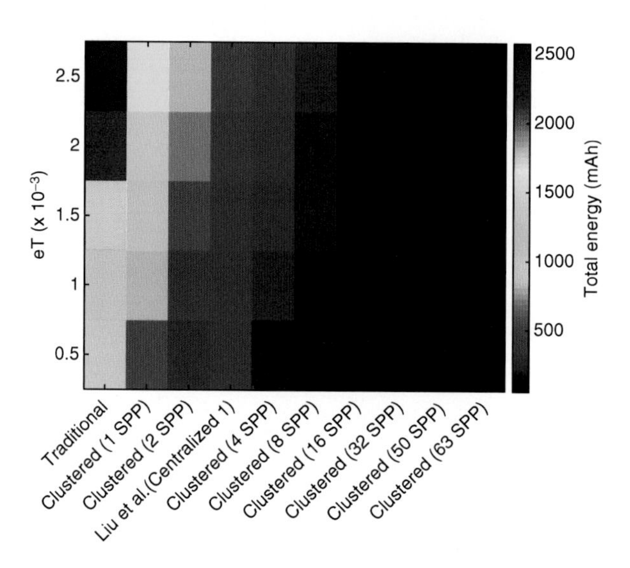

Figure 5.19 Energy consumption comparison with WSN for SHM proposed by Liu et al. [49], where e^T (*y*-axis) is the transmission power (mAh).

the possibility of having control over the flight path mean that NSPs can remain almost stationary over the service region [7, 8]. They can be moved on demand in order to serve different regions, and are able to take off and land for payload maintenance and upgrading. Therefore, NSPs can combine the large sensor footprints and long-durations that are commonly associated with satellites and the responsiveness that is associated with UAVs. These features make NSPs attractive for a large class of applications, among which telecommunications is one of the most promising in terms of the commercial revenues that might be possible. Other interesting applications worth mentioning are remote sensing [9], environmental monitoring, and agriculture support [10].

The applications of NSPs in broadband communications have been surveyed by Karapantazis and Pavlidou [11]. The survey begins with an introduction to NSPs, followed by discussions about suitable platforms, possible architectures, and some points on channel modeling, antennas, and transmission and coding techniques. Avagnina et al. [12] suggested high-altitude platforms (HAPs) for joint provision of cellular communication services and support services for navigation satellite systems. It was shown that NSPs are suitable for implementing macrocells with large radii. The role of NSPs in providing global connectivity for future communication services was discussed by Mohammed et al. [13]. Additional work about the role of NSPs in various communications systems and networks can be found in the literature [14–18].

This chapter includes an overview of applications of integrated wireless sensor systems on near-space and satellite platforms, but without discussing actual implentation details [19]. We also call for more research. The application focus is placed on communications relaying and passive radar for regional remote sensing. Although NSPs have a much smaller coverage area than satellites due to their lower altitude, they can offer regional coverage, with an effective observation area hundreds of kilometers in radius, as well as providing cost-effective communication, navigation and localization services, particularly for areas with limited or no access to spaceborne and terrestrial sensors.

11.2 Near-space Platforms in Wireless Sensor Systems

11.2.1 Near-space Platforms

Traditionally, very few vehicles could operate in near-space because the atmosphere is too thin to support flight for aircraft and too thick to sustain satellites in orbit. Long-term persistent NSPs have now become possible because of

- advances in ultraviolet-resistant hull materials and designs for super-pressure airships [20]
- computer-aided design models
- high-altitude aircraft technologies and better buoyancy/ballonet management technologies
- advances in knowledge of the stratosphere.

Currently, aircraft and airship on-station loitering times in the stratosphere are limited to a few days, limited by the amount of fuel that can be carried; their payload

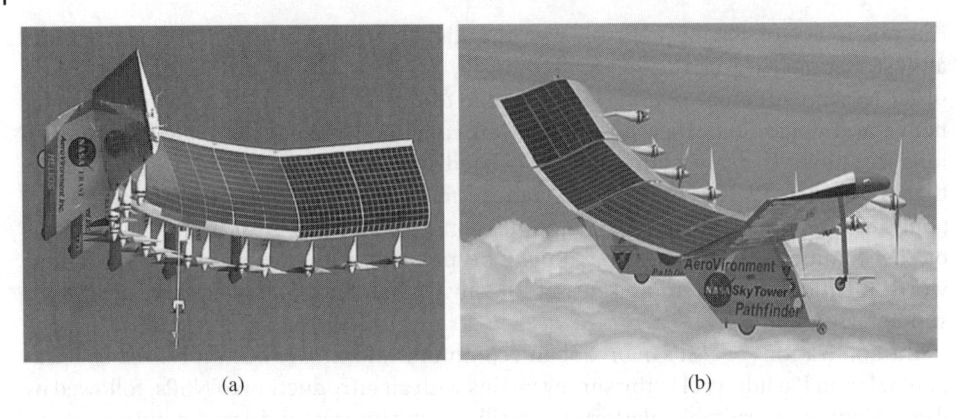

(a) (b)

Figure 11.3 Typical NSPs designed by NASA. (a) HELIOS (b) Pathfinder

power-supply capabilities are limited to a few kilowatts. However, there is increasing interest in delivering multiple kilowatts or even megawatts of power by wireless, so as to provide for continuous operations in the stratosphere and higher-powered payloads [21].

Some NSPs already exist and more are currently at the prototype stage [11, 22, 23]. Figure 11.3 shows two typical NSPs designed by NASA. HAPs can be classified into three major categories:

- free-floaters
- steered free-floaters
- manoeuvring vehicles.

A free-floater is essentially a large balloon floating with the wind. They are normally classifed into two types: zero-pressure or super-pressure. Limited steering is possible by using variable ballast; this enables them to float at different altitudes to take advantage of different wind directions and speeds. However, free-floaters have no station-keeping ability because no active steering or propulsion systems are used. Free-floaters have already demonstrated their commercial viability for communication platforms. They can lift payloads of tens to thousands of kilograms to altitudes higher than 30 km, depending on the volume.

Steered free-floaters also drift in the winds, but they can exploit the winds in the way that a sailing ship does and they can be navigated with high precision. With limited steering, a steered free-floater can maintain a position for a short time. Its sensors are more complex than those on a free-floater. Steered free-floaters are commercially mature and military deployments are imminent.

Manoeuvring vehicles have propulsion and control capabilities. They can manoeuvre and thus fly to and stay at an area for a long time. Manoeuvring vehicles provide large footprint and long mission duration that are associated with satellite and responsiveness that is commonly associated with UAV. Manoeuvring vehicles are the most useful type of NSPs for applications that require a fast revisiting frequency. They are possible substitutes for the satellites that support wireless sensor systems [15].

11.2.2 Why Near-space Platforms should be used in Wireless Sensor Systems

Current spaceborne and airborne communication and navigation techniques have two disadvantages [24]:

- they cannot provide persistent regional monitoring
- there are few sensors deployed at altitudes between satellites and aircraft.

Moreover, the requirements for stealth and robust survivability in military applications has led to a call for new platforms other than satellites and aircraft. These objectives can be simultaneously implemented by using NSPs.

NSPs are 10–20 times closer to the targets than a typical LEO satellite at 400 km altitude. This distance difference implies that NSPs can detect much weaker signals. Although an orbiting satellite has a much larger footprint than an NSP, the latter has an impressive footprint, with the possibility of persistent coverage and good signal strength. NSPs are more flexible than satellites because the former can be repositioned and repaired on the ground in case of failure. NSPs also provide a platform for the 'last mile' broadband problem because of the large coverage they can give without deploying multiple terrestrial base stations [25].

Another feature for which NSP-based wireless sensor systems are highly favoured is their line-of-sight path-loss characteristics. NSPs can be directly linked to a satellite communication system to provide fixed wireless access services. Multiple NSPs can also be linked together. Due to their lower latency and more favourable link budgets, NSPs are more cost-effective than satellites for the provision of services to mobile users. Furthermore, frequency reuse can be achieved by spot beam antennas, thus avoiding the need to instal huge antennas to achieve high spectral efficiency [26].

Additionally, smart sensors [27] that are capable of manipulation and computation of the sensor-derived data can be installed on NSPs. Much work has been done on developing common communication protocols for smart sensor communications. Examples include the inter IC bus (I^2C), the serial peripheral interface, and internet-based communication. Some authors have investigated frameworks for smart sensor and intelligent systems. For example, Schmalzel et al. [28] developed a new architecture for intelligent systems based on smart sensors. We think smart sensors will enable NSP systems to support implementation of highly autonomous methodologies, with appropriate communication protocols for timely and high-quality interactions with the system elements being developed.

In summary, NSPs can become a novel stratospheric segment in the wireless sensor marketplace. They are able to overcome the main drawbacks of satellite technology because they are deployed closer to the ground, and because of their quasi-stationary characteristics in the sky. Additionally, their costs for construction, deployment, launch and maintenance are much lower than for satellites. Certainly, it is evident that NSPs can replace neither satellites nor terrestrial radio links, for reasons of coverage, reliability, safety and cost. In fact, satellites, NSPs and terrestrial systems have different but complementary characteristics. While satellites are more suited to covering very large areas and providing broadcast applications, NSPs are able to cover remote or sparsely populated areas at reduced cost, and bringin broadband services to mobile users when fixed directive antennas cannot be used [29].

11.3 Overview of NSPs in Wireless Sensor Systems

In recent years, the applications of NSPs in wireless sensors have received much attention. In the following, we give an overview of such developments.

11.3.1 NSP Enabling Sensor Communications

Establishment of NSP wireless communication services was first described by Djuknic et al. in 1997 [10]. The suggested platforms include high-altitude aeronautical vehicles (HAAVs) such as airships, planes and helicopters, which would operate at stratospheric altitude for long periods and carry multipurpose communications payloads. The authors concluded that HAAVs would have many advantages over their terrestrial and satellite counterparts because they brought opportunities for providing wireless communication services and developing innovative communication concepts such as cell scanning and stratospheric radio relaying. Through communications established between the platforms and user terminals on the ground, NSPs can serve more users with less infrastructure than terrestrial networks. They can therefore provide high-speed broadband wireless services, a complement to terrestrial and satellite systems.

Many research projects have been dedicated to NSP-based communication services. Comprehensive overviews are available in the literature [22, 30, 31]. The first NSP program was the Stationary High Altitude Relay Platform in Canada, but the first commercial video telephony and internet service via NSPs was initiated by the Sky Station Inc. in the USA [32]. The SkyTower aims to market NSPs as ways to deliver fixed broadband, narrowband and broadcast communications [33]. The Advanced Concept Technology Demonstration was initiated to design, build, and test a high-altitude aerostat prototype that could maintain a geostationary position at over 21 km for up to six months, generate its own power and carry multiple payloads. It was designed for military and civilian activities including:

- weather and environmental monitoring
- short- and long-range missile early warning
- surveillance
- target acquisition.

In Europe, two organizations have funded research activity into NSPs: the European Space Agency and European Commission. The projects include [34, 35]:

- HeliNet, a network of stratospheric platforms for traffic monitoring, environmental surveillance and broadband services
- CAPANINA, a broadband communications technology
- UAVNET, a network of UAVs
- CAPECON, civil applications of UAVs and the economic effects of potential configuration solutions
- USICO, UAV safety issues for civil operations.

The CAPANINA project is to investigate possible broadband applications and services, and the most appropriate integration options for aerial platforms, including deploying multiple platforms to serve the same coverage area, along with the most appropriate backhaul and network infrastructure. Such multi-platform configurations can be used for both fixed and mobile users [36]. Skynet was a Japanese project for the development of a balloon to be deployed at an altitude of over 20 km for communications, broadcasting and environmental observation [37]. In Korea, research activities on NSPs have been conducted by Electronics and Telecommunications Research Institute [38]. Their main objective is to develop a full-scale 200-m-long airship to carry telecommunications and remote-sensing payloads weighing up to 1000 kg. Additionally, the Chinese National Natural Science Foundation and 863 Programs have funded several research projects on NSPs [39–42].

As the demand for high-capacity wireless communication services has brought increasing challenges, especially relating to the 'last mile' delivery, the use of NSPs for delivering wireless broadband has been suggested [43]. The role of NSPs in beyond-3G networks has also been extensively investigated [44], with different hybrid system architectures considered, and with an emphasis on the merits of NSPs and integrated terrestrial–NSP–satellite systems. A method of significantly improving the capacity of NSP communication networks operating in the millimeter-wave band has been described [45], showing how NSP constellations can share a common frequency allocation by exploiting the antenna directionality. Several good review papers have discussed the role of NSPs in broadband communications [11, 13] and global wireless connectivity [29]. The potentialities and challenges have been presented from the perspective of integrated terrestrial–NSP–satellite communication infrastructure [29].

Several sensor systems for global connectivity are presented by Mohammed et al. [13]. The problem of deploying NSP sensor networks to provide wireless communications for terrestrial users has also been investigated, by Zong et al. [46]. The aim was to provide communication services to ground users with quality of service guaranteed. The channel estimation for a long-term evolution downlink that uses orthogonal frequency division multiplexing for NSP systems was investigated by Kahar and Iskandar [47]. The performance of propagation models for efficient handoff in NSP systems to sustain quality of service was analysed in two other papers [48, 49]. Radio resource allocation for multicast transmissions using NSPs was investigated by Ibrahim and Alfa [50], who formulated an optimization problem for a scenario in which different sessions are multicast to user terminals across an NSP service area. They then solved it to find the best allocation of NSP resources, such as radio power, subchannels and time slots. The optimization problem turns out to be a mixed integer non-linear program, which is solved using the Lagrangian relaxation algorithm [51]. Additionally, data and optical communications from NSPs have been considered in the literature [52–55].

11.3.2 Using NSPs for Radar and Navigating Sensors

NSPs are also a promising platform for radar and navigation sensors. Since NSPs fill the gap between satellites and aircraft [24], the applications of NSPs in microwave remote sensing have been extensively investigated [56]. NSPs are especially useful for synthetic aperture radar (SAR) imaging, which obtains its high resolution by using the transmitted wide-band waveform and high azimuth resolution, exploiting the relative motion between the target and the radar platform. The wide swath of SAR sensing is of great utility, but it cannot be efficiently achieved by current spaceborne and airborne systems due to the minimum antenna area constraint [57]. Spaceborne SAR systems have an imaging capability with a wide coverage but with limited azimuth resolution. In contrast, airborne systems have an imaging capability with a high resolution but with limited coverage. Moreover, neither spaceborne nor airborne SARs provide persistent imaging. Typical solutions are based on the displaced phase center processing technique, but they may bring a non-uniform spatial sampling problem. Another approach is to use multiple orthogonal waveforms [58]. This approach significantly reduces the ambiguous signal peaks, but the ambiguity energy is unchanged, and it is not suitable for distributed targets. Suppressing range ambiguity through azimuth modulation is also feasible [59], but a key problem is how to design practical orthogonal waveforms [60, 61]. Therefore, there is an incentive to increase the coverage width and azimuth resolution simultaneously by using NSPs as the radar platforms [62].

In the following discussion, we introduce two potential applications of NSP-based SAR systems; a more comprehensive overview is available in the literature [63].

One potential application is in homeland security [64]. It is necessary to protect mass transit systems, civil aviation and critical infrastructure from terrorist attacks, without impacting normal activities. Radar has been used in various military and civilian applications. Many countries have military radar systems, which are specifically designed to detect threats. However, attacks may not be handled well by current radar systems. It appears that passive radar using an NSP receiver and an opportunistic illuminator can provide a solution to these problems [5]. Rather than emitting signals, passive radar relies on opportunistic transmitters and passive receivers to detect threats. This is particularly attractive for homeland security applications, because the sensor can also serve other ends, such as traffic monitoring and weather prediction.

Another potential application is in disaster monitoring. The frequency of natural disasters has shown a rapid increase in recent years. Taking tsunami detection as an example, tsunamis can originate in earthquakes, submarine landslides, volcanic eruptions, meteorite impacts or a combination of these factors. Tsunamis have sufficient energy to cross an entire ocean. In the deep ocean they are low, and therefore hard to detect, but they become higher, and therefore more detectable, near to shore [65]. It was reported that the Sumatra tsunami had a wave height of 60–80 cm in the deep ocean, but a maximum wave height of 15 m near to Banda Aceh [66]. It is has been proved [67] that there is a link between a tsunami's wave amplitude and its radar cross-section. Significant variations of a few decibels in the radar cross-section, synchronous with sea-level anomaly, have been found in both C and Ku band records from the sea-level altimeter satellite, Jason-1.

Tsunami detection by NSP radar has been investigated by Galletti et al. [68], who concluded that tsunamis can be detected by measuring:

- their wave height
- their orbital velocity
- the induced radar cross-section modulations.

11.3.3 Integrated Communication and Navigation Sensors

The increasing number of space missions has led to the development of a network able to support services such as navigation and communications [69]. Provision of telecommunication services by means of NSPs is thus becoming a relevant topic of interest for next-generation systems. Mobility on demand, large coverage, and sensor re-configurability are only some of the expected benefits in wireless sensors based on NSPs [12].

High-performance communication and navigation sensors for interplanetary networks have been investigated by de Cola and Marchese [70]. The authors proposed transmission strategies relying upon a packet-layer coding approach to improve overall performance. An integrated mobile satellite broadcast, paging, communications and navigation system was proposed by Noreen [71], whose designed networks support a mix of mobile satellite services optimized for wireless market applications and, in particular, for the needs of the rural populace and the travelling public.

Dreher et al. [72] designed an antenna and receiver with digital beamforming for satellite navigation and communications. An integrated communication, navigation and surveillance satellite system for air traffic management was studied by Galati et al. [73], while Chen et al. presented a multimodal wireless network with communications and surveillance on the same infrastructure [74]. However, as noted by Avagnina et al. [12], research regarding the use of NSPs for communications usually neglects the fact that it is also an important infrastructure for the provision of navigation and positioning services. The authors believe that NSPs can act as a supplementary infrastructure for global navigation satellite systems (GNSSs) to perform direction of arrival estimation and broadcast position information.

11.4 Integrated Wireless Sensor Systems

Both past and on-going research activities have shown that NSP-based sensors can either be integrated into current major terrestrial and satellite wireless networks (such as WiMAX or 3G/4G) [75–77], or provide services as standalone systems [78]. Certainly the trend is not to replace existing communication technologies, but instead to co-exist with them in a complementary and integrated manner [79]. Although NSPs provide the flexibility to accommodate a wide range of applications, ranging from communications to navigation and remote sensing [80], this chapter focuses on integrated wireless sensor systems for communication and navigation.

Figure 11.4 shows an extensive system architecture, which can be categorised as an integrated terrestrial–NSP sensor system (see Figure 11.5). and an integrated terrestrial–NSP–satellite sensor system. In the terrestrial–NSP sensor system, connections to other communication networks via a gateway stations (GS) are allowed [81]. It can be further divided into two topologies taking into account where the switching takes place [82]:

- *With on-ground switching*: the path between two users encompasses an uplink from user to NSP, a feeder downlink to the GS where the switching is performed, a feeder uplink from the GS to the NSP and a downlink to the users.
- *With on-board switching*: the path between two users takes only the uplink from the user to the NSP, where switching is performed, and a downlink from the NSP to the users.

The terrestrial–NSP–satellite sensor system is particularly useful when the NSPs are placed above an environment with deficient or non-existent terrestrial infrastructure (rural and remote areas). In this case, the NSP sensor can be used as a relay for satellite communications [83]. Additionally, we can use only a standalone NSP sensor system [84], as shown in Figure 11.6. This NSP system can be deployed economically and

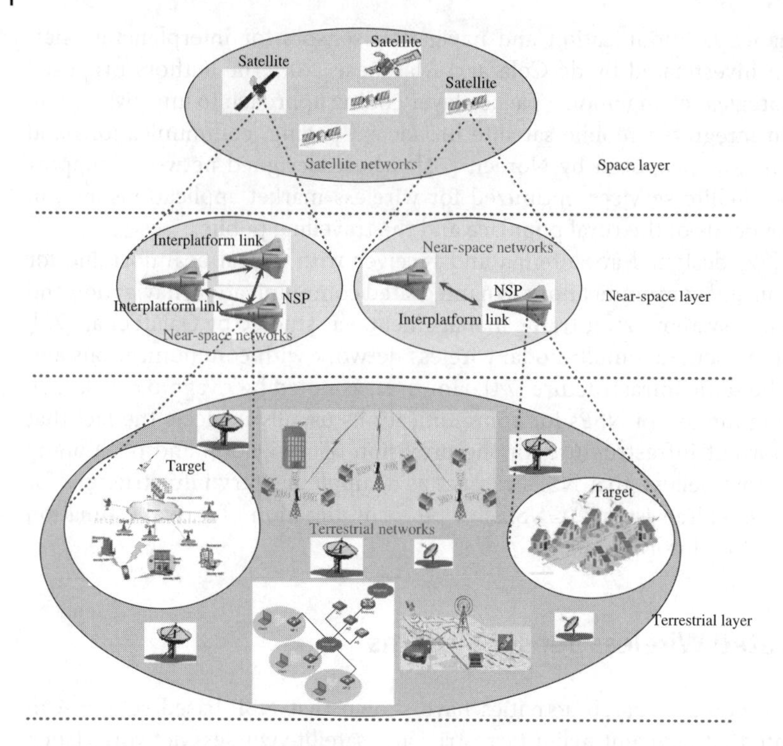

Figure 11.4 Integrated wireless sensor systems with NSP and satellite platforms.

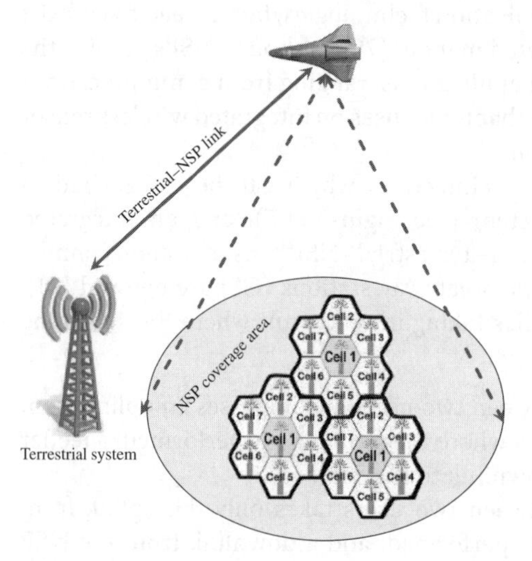

Figure 11.5 Illustration of NSP-terrestrial system.

efficiently in a rural or remote area where it is rather expensive and inefficient to deploy terrestrial sensor systems.

Multiple NSPs can be interconnected via ground stations or inter-platform links to provide a sensor network in the sky [85, 86], especially making use of a sensor network in the sky. It has been demonstrated [87] that 16 NSPs can serve Japan at a minimum elevation angle of 10°, and a network of 18 NSPs can cover the whole of Greece

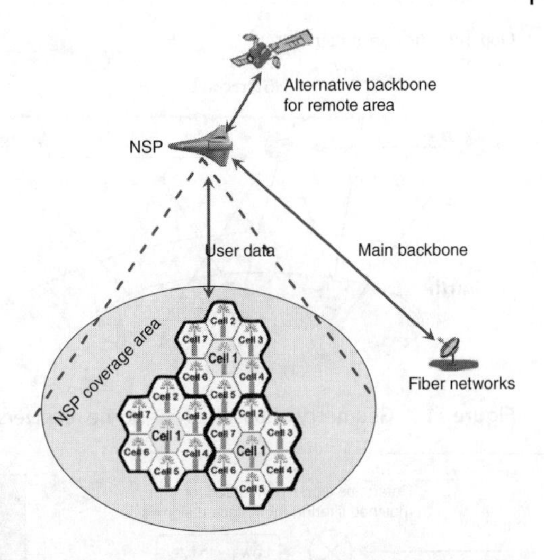

Figure 11.6 Illustration of standalone NSP system.

including all the islands [88]. Multiple NSPs can also be deployed in some planar or vertical arrangement so as to cover the same single coverage area. By exploiting spatial discrimination, such configurations provide incremental rollout and the systems can be expanded on demand to give higher capacity [45]. In response to increasing traffic load and service demands there is a possibility for 3–6 NSPs to be deployed to cover the same cell area [13]. If the NSPs are interconnected via ground stations, they can be enabled only above the area where the ground station is placed. In contrast, when interplatform links are employed, the NSP operation can be independent of terrestrial sensors and thus highly flexible system coverage and lower signal delays can be obtained.

Localization services can also be integrated into the sensor network because exploiting the user-position information allows improvement in the system capacity. GNSSs are widely used to provide localization services, but they require at least four satellites in view, a situation seldom met in an urban environment [12]. On the other hand, current terrestrial networks cannot provide effective localization services for mobile users, particularly in rural environments. In contrast, the integrated NSP–satellite system can be complementary to existing localization systems. For instance, the localization accuracy provided by GNSSs for critical applications and harsh environments can be improve by the NSP networks. Long-endurance flight and several other features make NSPs good calibrators for GNSS localization services. Moreover, they provide an additional ranging signal that can be exploited by users, who can choose the best set of measurements according to the specific scenario [12].

The capability to detect passive targets with opportunistic spaceborne illuminators can also be integrated into the sensor systems. As shown in Figure 11.7, the passive NSP receiver consists of two channels. One heterodyne channel pointing directly towards the satellite is used to extract the reference signal for matched filter and synchronization compensation. The signal is sampled in a delayed window that can be predicted using knowledge of the NSP position information. The other antenna, directed at the ground area of interest, is used as the radar channel to receive the reflected signals. Taking GPS satellite as an example, Figure 11.8 illustrates the functional blocks for extracting the reference signal from the direct-path channel for matched filtering of the reflected signals. Note that the configuration using a single-channel receiver is also feasible. In this

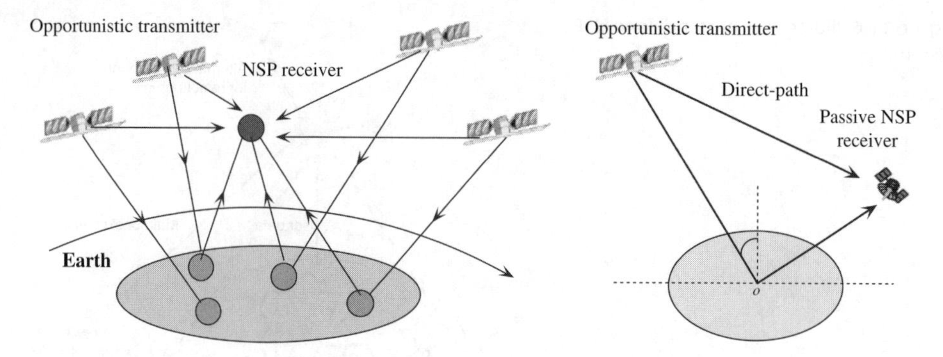

Figure 11.7 Geometry of passive NSP-borne receivers and opportunistic spaceborne transmitter.

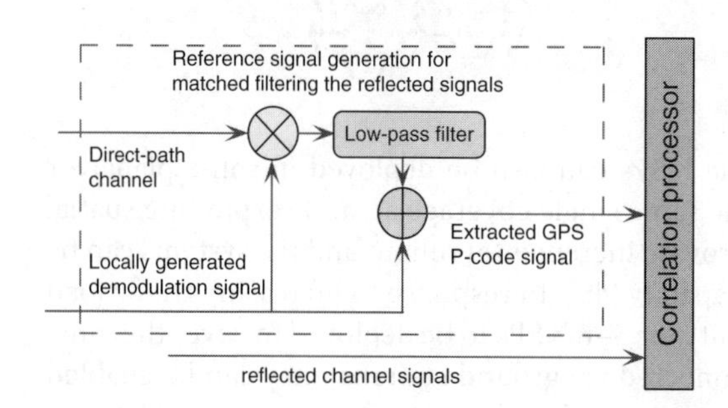

Figure 11.8 Functional blocks for extracting the reference signal from the direct-path channel for matched filtering the reflected signals.

case, decoding the navigation message is required, which needs the synchronization information.

11.5 Arrangement of Near-space Platforms

To arrange multiple NSPs to monitor a given region, we consider the formation geometry shown in Figure 11.9, where θ_{in} is the incidence angle, θ_e is the geocentric angle, R_e is Earth's radius, and h_s is the NSP altitude.

The projection distance from the user to a NSP can be expressed as

$$D_e = R_e\theta_e \tag{11.1}$$

where

$$\theta_e = 90^\circ - \theta_{\text{in}} - \sin^{-1}\left[\frac{R_e \sin(90^\circ + \theta_{\text{in}})}{R_e + h_s}\right]. \tag{11.2}$$

Supposing the distance between two NSPs is L_h and its projection distance on the ground is L_e, we then have the following geometry relation

$$\frac{L_h}{L_e} = \frac{R_e + h_s}{R_e} \Rightarrow L_h = \frac{(R_e + h_s)L_e}{R_e}. \tag{11.3}$$

If the triangle geometry shown in Figure 11.9b is employed, L_e should be

$$L_e \leq D_e. \tag{11.4}$$

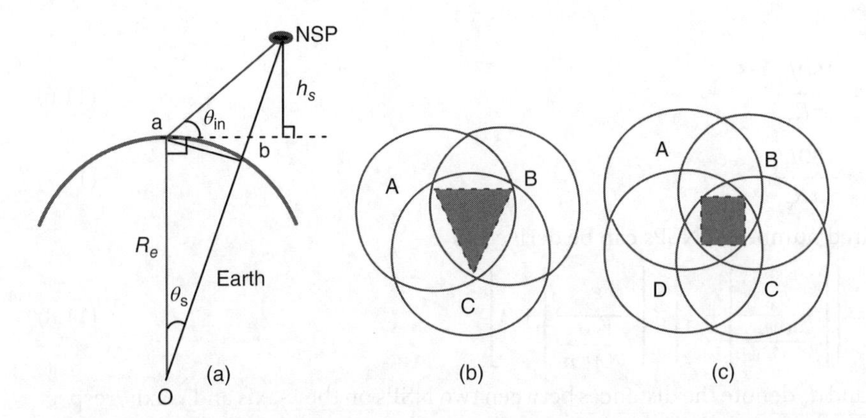

Figure 11.9 Illustration of NSP observation geometry and coverage region: (a) geometry; (b) triangle coverage; (c) quadrate coverage.

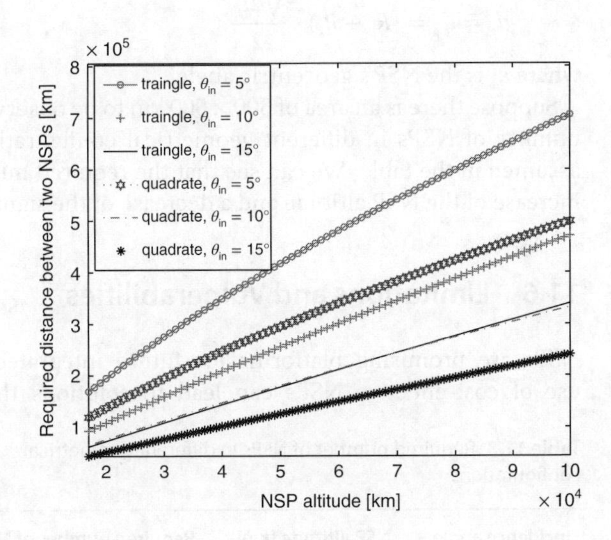

Figure 11.10 Required distance between two NSPs.

Otherwise, if the quadrate geometry shown in Figure 11.9c is employed, L_e should be

$$L_e \leq \frac{\sqrt{2}}{2} D_e. \tag{11.5}$$

As an example, suppose Earth's radius is $R_e = 6378$ km. Figure 11.10 gives the required NSP arrangement parameters for different combinations of the incidence angle θ_{in} and platform altitude h_s. We can see that the required NSP distance decreases with the increase of incidence angle θ_{in}, but it increases with an increase of the platform altitude h_s. The received signal-to-noise ratio will deteriorate with a decrease of the incidence angle [89].

To determine the required number of NSPs to monitor a given region, we consider the formation geometry illustrated in Figure 11.9c. Suppose the length and width of the region to be monitored are L_l and L_w, respectively, and the corresponding geocentric angles are

$$\varepsilon_x \approx \frac{180L_l}{\pi R_e} \tag{11.6}$$

$$\varepsilon_y \approx \frac{180L_w}{\pi R_e} \tag{11.7}$$

The required number of NSPs can be derived as

$$N = \left\lceil \left\lceil \frac{\varepsilon_x}{\frac{180d_x}{(R_e+h_s)\pi}} \right\rceil + 2 \right\rceil \times \left\lceil \left\lceil \frac{\varepsilon_y}{\frac{180d_y}{(R_e+h_s)\pi}} \right\rceil + 2 \right\rceil \tag{11.8}$$

where d_x and d_y denote the distances between two NSPs on the x-axis and y-axis, respectively, and $\lceil x \rceil$ denotes the nearest integer greater than or equal to x. According to the geometry relation, d_x and d_y can be derived as

$$d_x = d_y = (R_e + h_s) \cdot \frac{\sqrt{2}\varepsilon_s}{2} \tag{11.9}$$

where ε_s is the NSP's geocentric angle.

Suppose there is an area of 500 × 500 km to be observed. Table 11.3 gives the required number of NSPs in different geometrical configurations. Note that $R_e = 6378$ km is assumed in the table. We can see that the required number of NSPs decreases with an increase of the NSP altitude and a decrease of the signal incidence angle θ_{in}.

11.6 Limitations and Vulnerabilities

NSPs are promising platforms for future integrated wireless sensor systems. The use of cost-effective NSPs can lead to solutions that were previously thought to

Table 11.3 Required number of NSPs in different geometrical configurations.

Incidence angle	NSP altitude [*km*]	Required number of NSPs
5°	20	5 × 5
	30	5 × 4
	60	4 × 3
	100	3 × 3
15°	20	10 × 8
	30	8 × 6
	60	5 × 5
	100	4 × 4
20°	20	13 × 10
	30	10 × 8
	60	6 × 5
	100	5 × 4

be out of reach for wireless customers. However, there are several limitations and vulnerabilities:

11.6.1 Launch Constraints

Weather will be a risk factor that could be significant if the NSP does not have reliable sensing of meteorological data around it to allow a vehicle controller to predict turbulence, icing and violent wind gusts that might jeopardise the craft. The experience with high-altitude tropospheric operations from around-the-world balloon teams and their weather teams should be collected and codified to aid computer predictions. Since NSPs can operate in the troposphere for over 5 h, the weather conditions must be within allowable parameters before a launch can commence. Note that a satellite faces similar launch constraints but these only have to be met during the launch. Manned and unmanned aircraft are also subject to similar launch and recovery constraints, although their limitations are less stringent than those for NSPs.

Currently it is difficult to make NSPs that will operate at altitudes higher than 20 km. Increasing the altitude greatly increases the NSP's size, quickly making it difficult to manage on the ground and during launch. A rule of thumb is that for every extra 30 km of altitude, the NSP's volume will double.

11.6.2 Survivability Constraints

As NSP technology develops, it should be assumed that target-detection technology will also develop. Near-space was not historically a region in which vehicles could operate, and so missiles were not designed to reach these altitudes. However, modern surface-to-air missiles can reach as high as jet aircraft fly and consequently NSPs are not beyond the range of current conventional weapons. Therefore, NSPs need to be flexible, replaceable and cost-effective; otherwise their use in military applications may be limited. On the other hand, once an NSP becomes autonomous, even high-performance radar may have difficulty in detecting and tracking it due to its small radar cross section.

11.6.3 Legal Constraints

Freedom of overflight is another limitation. The legal status of the near-space regime is a grey area that is not directly addressed by any treaty or policy. Near-space is not a new legal regime; the question is whether it falls under air law, where nations claim sovereignty, or space law where overflight rights exist. Due to a lack of clear legal precedent governing the near-space regime, there is considerable disagreement over whether overflight rights exist or not [90].

11.6.4 System Implementation Issues

For integrated wireless sensor networks via near-space and satellite platforms, synchronization compensation is a technical challenge. In the case of indirect phase synchronization using identical local oscillators in the transmitter and receiver, phase stability is required over the whole coherent integration time. Even if the toleration of low-frequency phase errors can be relaxed to 45° [91], the requirement for phase stability is only achievable with ultra-high-quality oscillators [92]. Moreover, as aggravating circumstances often accompany NSPs, the phase stability will be further degraded. Several potential synchronization techniques or algorithms, such as using ultra-high-quality oscillators [91] and an appropriate bidirectional link [93], have been proposed. However, in most cases, we cannot alter the transmitter, so it is necessary to develop a practical synchronization technique that does not do this. One potential solution is to follow the direct-path signal-based synchronization approach [94].

Problems may also arise due to the presence of atmospheric turbulence, which introduces NSP trajectory deviations from the normal position, as well as the altitude (roll, pitch and yaw angles). As a consequence, motion compensation is often required. In current communication and navigation systems, GPS and inertial navigation systems are usually employed for this task. However, for NSP sensors, motion-compensation facilities may be impossible, because NSPs have very limited load capabilities. Therefore, efficient motion compensation techniques will need to be developed.

Antenna design and implementation is also a key technique in NSP systems. Dielectric lens antennas are widely used in NSP systems due to their superior performance [95, 96]. A dielectric lens antenna ground plane is discussed in the literature [96]. The designed antenna has relatively high aperture efficiency and multiple scanned beams over a wide angle. However, if a lower carrier frequency is used, the feed-waveguide volume and weight will increase greatly and consequently the system is difficult to install on an NSP. To resolve this problem, a novel multi-beam dielectric lens antenna array operating from 1.77–2.44 GHz and fed by a Yagi-Uda unit has been designed [97]. In fact, there is a trade-off between operating frequency and cost in the deployment of directional antennas on NSPs, where the speed, the aerodynamics and the drift caused by winds introduces many challenges in establishing an effective link between NSPs [98]. A few studies on overcoming such challenges have been reported. For example, Alshbatat et al. have proposed an adaptive MAC protocol for UAV communication networks [99]. In their model, the request to send/clear to send (RTS/CTS) exchange is conducted omnidirectionally and data transfer is conducted directionally, with one of four beams. They assume four antennas on a UAV where two of them are located beneath the wing and the others above. In this model, it is assumed that the UAV has the capability to plan and tilt to the desired transmission direction, which actually is not a realistic assumption.

11.7 Summary

In this chapter we have demonstrated the NSPs' technological advantages over satellites and other airborne networks in a wide range of monitoring, earth-observation, and other sensing applications. The relatively large coverage area and some navigation support capabilities of NSPs make them part of a promising infrastructure for the co-existence of navigation and communication stations. These systems will provide integrated services by connecting terrestrial stations with the satellites. In many persistent surveillance applications, the superiority of NSP stationary monitoring over periodic snapshots from moving satellite or airplane platforms enables the integration of NSPs, satellite and airplane platforms, for applications such as infrared, electro-optical, hyper-spectral imagery and smart sensors and other traditional and upcoming sensors. We have also explained how NSPs can play a significant role as an additional, always-present layer between satellite and terrestrial networks. We urge the engineering community to support the research and development of NSPs into applications that will serve humanity.

References

1 Sampigethaya, K., Poovendran, R., Shetty, S., Davis, T., and Royalty, C. (2011) Future e-enabled aircraft communications and security:The next 20 years and beyond. *Proceedings of the IEEE*, **99** (11), 2040–2055.

2 Fu, B. and DaSilva, L.A. (2007) A mesh in the sky: A routing protocol for airborne networks, in *Proceedings of the IEEE Military Communication Conference*, Orlando, FL, pp. 1–7.

3 Kuiper, E. and Nadjm-Tehrani, S. (2011) Geographical rounting with location service in intermittertly connected manets. *IEEE Transactions on Vehicular Technology*, **60** (2), 592–604.

4 Rohrer, J., Jabbar, A., Cetinkaya, E., Perrins, E., and Sterbenz, J. (2011) Highly-dynamic cross-layered aeronautical network architecture. *IEEE Transactions on Aerospace and Electronic Systems*, **47** (4), 2742–2765.

5 W.-Q. Wang, Cai, J.Y., and Peng, Q.C. (2010) Near-space microwave radar remote sensing: potentials and challenge analysis. *Remote Sensing*, **2** (3), 717–739.

6 Tomme, E.B. (2005) Balloons in today's military: an introduction to near-space concept. *Air Space Journal*, **19** (1), 39–50.

7 Zhao, J., Jiang, B., He, Z., and Mao, Z.H. (2014) Modelling and fault tolerant control for near-space vehicles with vertical tail loss. *IET Control Theory & Applications*, **8** (9), 718–727.

8 Shen, Q., Jiang, B., and Cocquempot, V. (2013) Fuzzy logic system-based adaptive fault-tolerant control for near-space vehicle attitude dynamics with actuator faults. *IEEE Transactions on Fuzzy Systems*, **21** (2), 289–300.

9 Wang, W.Q. (2012) Regional remote sensing by near-space vehicle-borne passive radar system. *ISPRS Journal of Photogrammetry and Remote Sensing*, **69** (2), 29–36.

10 M. Djuknic, G., Freidenfelds, J., and Okunev, Y. (1997) Establishing wireless communications services via high-altitude aeronautical platforms: A concept whose time has come? *IEEE Communications Magazine*, **35** (9), 128–135.

11 Karapantazis, S. and Pavlidou, F. (2005) Broadband communications via high-altitude platform: a survey. *IEEE Communications Surveys & Tutorials*, **7** (1), 2–31.

12 Avagnina, D., Dovis, F., Ghilione, A., Mulassano, P., and di Torino, P. (2002) Wireless networks based on high-altitude platforms for the provision of integrated navigation/communication services. *IEEE Coomunication Magazine*, **40** (2), 119–125.

13 Mohammed, A., Mehmood, A., Pavlidou, F., and Mohorcic, M. (2011) The role of high-altitude platforms (haps) in the global wireless connectivity. *Proceedings of the IEEE*, **99** (11), 1939–1493.

14 Chaumette, S., Laplace, R., Mazel, C., Mirault, R., Dunand, A., Lecoute, Y., and Perbet, J.N. (2011) CARUS, an operational retasking application for a swarm of autonomous UAVs: First return on experience, in *Proceedings of the IEEE Military Communication Conference*, Baltimore, MD, pp. 2003–2010.

15 Liu, Y., Grace, D., and Mitchell, P.D. (2009) Exploiting platform diversity for QoS improvement for users with different high altitude platform availability. *IEEE Transactions on Wireless Communications*, **8** (1), 196–203.

16 Holis, J. and Pechac, P. (2008) Elevation dependent shadowing model for mobile communications via high altitude platforms in built-up areas. *IEEE Transactions on Antennas and Propagation*, **56** (4), 1078–1084.

17 Likitthanasate, P., Grace, D., and Mitchell, P.D. (2008) Spectrum etiquettes for terrestrial and high-altitude platform-based cognitive radio systems. *IET Communications*, **2** (6), 846–855.

18 Panagopoulos, A.D., Georgiadou, E.M., and Kanellopoulos, J.D. (2007) Selection combining site diversity performance in high altitude platform networks. *IEEE Communication Letters*, **11** (10), 787–789.

19 Wang, W.Q. and Jiang, D.D. (2014) Integrated wireless sensor systems via near-space and satellite platforms—A review. *IEEE Sensors Journal*, **14** (11), 3903–3914.

20 Onda, M. (2001), Super-pressured high-altitude airship, aist. 6 305 641.

21 Dickinson, R.M. (2013) Power in the Sky: requirements for microwave wireless power beamers for powering high-altitude platforms. *IEEE Microwave Magazine*, **14** (2), 36–47.

22 David, G. and Mihael, M. (2011) *Broadband Communications via High Altitude Platforms*, John Wiley & Sons, Hoboken, NJ.

23 Alejandro, A.Z., Lius, C.R.J., and Antonio, D.P.J. (2008) *High-Altitude Platforms for Wireless Communications*, John Wiley, New York, USA.

24 Wang, W.Q. (2011) Near-space vehicles: supply a gap between satellites and airplanes. *IEEE Aerospace and Electronic Systems Magazine*, **25** (4), 4–9.

25 Cianca, E., Prasad, R., Sanctis, M.D., Luise, A.D., Antonini, M., Teotino, D., and Ruggieri, M. (2005) Integrated satellite-HAP systems. *IEEE Communications Magazine*, **43** (12), 33–39.

26 Bayhan, S., Gur, G., and Alagoz, F. (2007) High altitude platform (HAP) driven smart radios: A novel concept, in *Proceedings of the International Workshop on Satellite and Space Communications*, Salzburg, pp. 201–205.

27 Sveda, M. and Vrba, R. (2003) Integrated smart sensor networking framework for sensor-based appliances. *IEEE Sensors Journal*, **3** (5), 579–586.

28 Schmalzel, J., Figueroa, F., Morris, J., Mandayam, S., and Polikar, R. (2005) An architecture for intelligent systems based on smart sensors. *IEEE Transactions on Instrumentation and Measurement*, **54** (4), 1612–1616.

29 Falletti, E., Laddomada, M., Mondin, M., and Sellone, F. (2006) Integrated services from high-altitude platforms: A flexible communication system. *IEEE Communications Magazine*, **44** (2), 85–94.

30 Aragón-Zavala, A., Cuevas-Ruíz, J.L., and Delgado-Penín, A. (2008) *High-Altitude Platform for Wireless Communications*, Wiley.

31 Cook, E.C. (2013) *Broad Area Wireless Networking Via High Altitude Platforms*, Pennyhill Press.

32 Falletti, E., Mondin, M., Dovis, F., and Grace, D. (2003) Integration HAP with terrestrial UMTS network: Interference analysis and cell dimensioning. *Wireless Personal Communications*, **25** (2), 291–325.

33 Wierzbanowski, T. (2006) Unmanned aircraft systems will provide access to the statosphere. *RF Design*, pp. 12–16.

34 Pent, M., Tozer, T.C., and Penin, J.A.D. (2002) HAPs for telecommunication and surveillance applications, in *Proceedings of the 32nd European Microwave Conference*, Milan, Italy, pp. 1–4.

35 Lopresti, L., Mondin, M., Orsi, S., and Pent, M. (1999) Heliplat as a GSM base station: a feasibility study, in *Proceedings of the Data Systems in Aerospace Conference*, Lisbon, Portugal, pp. 1–4.

36 Grace, D., Capstick, M.H., Mohorcic, M., Horwath, J., and Pallaricini, M.B. (2005) Integrating users into the wider broadband network via high-altitude platforms. *IEEE Wireless Communications*, **12** (5), 98–105.

37 Yokomaku, Y. (2000) Overview of stratospheric platform airship R&D program in Janpan, in *Proceedings of the 2nd Stratospheric Platform Systems Workshop*, Tokyo, Japan.

38 Lee, Y.G., Kim, D.M., and Yeom, C.H. (2005) Development of Korean high altitude platform systems. *International Journal of Wireless Information Network*, **13** (1), 31–41.

39 Jiang, B., Gao, Z.F., Shi, P., and Xu, Y.F. (2010) Adaptive fault-tolerant tracking control of near-space vehicle using Takagi-Sugeno fuzzy models. *IEEE Transactions on Fuzzy Systems*, **18** (5), 1000–1007.

40 Hu, S.G., Fang, Y.W., Xiao, B.S., Wu, Y.L., and Mou, D. (2010) Near-space hypersonic vehicle longitudinal motion control based on Markov jump system theory, in *Proceedings of the 8th World Congress Intelligent Control Automation*, Jian, China, pp. 7067–7072.

41 Ji, Y.H., Zong, Q., Dou, L.Q., and Zhao, Z.S. (2010) High-oder sliding-mode observer for state estimation in a near-space hypersonic vehicle, in *Proceedings of the 8th World Congress Intelligent Control and Automation*, Jinan, China, pp. 2415–2418.

42 He, N.B., Jiang, C.S., and Gong, C.L. (2010) Terminal sliding mode control for near-space vehicle, in *Proceedings of the 29th Chinese Conference*, Beijing, China, pp. 2281–2283.

43 Tozer, T.C. and Grace, D. (2001) High-altitude platforms for wireless communications. *Electronics & Communication Engineering Journal*, **13** (3), 127–137.

44 Karapantazis, S. and Pavlidou, F.N. (2005) The role of high altitude platforms in beyond 3G networks. *IEEE Wireless Communications*, **12** (6), 33–41.

45 Grace, D., Thornton, J., Chen, G.H., White, G.P., and Tozer, T.C. (2005) Improving the system capacity of broadband services using multiple high-altitude platforms. *IEEE Transactions on Wireless Communications*, **4** (2), 700–709.

46 Zong, R., Gao, X.B., Wang, X.Y., and Lv, Z.T. (2012) Deployment of high altitude platforms network a game theoretic approach, in *Proceedings of the International Conference on Computing Networking and Communications*, Maui, HI, pp. 304–308.

47 Kahar, M.R. and Iskandar, A. (2013) Channel estimation for LTE downlink in high altitude platforms (HAPs) systems, in *Proceedings of the International Conference on Information and Communication Technology*, Bandung, pp. 182–186.

48 Alsamhi, S.H. and Rajput, N.S. (2014) Performance and analysis of propgation models for efficient handoff in high altitude platform system to sustain QoS, in *Proceedings of IEEE Students' Conference on Electrical, Electronics and Computer Science*, Bhopal, pp. 1–6.

49 Hasirci, Z. and Cavdar, I.H. (2012) Propagation modeling dependent on frequency and distance for mobile communications via high altitude platforms (haps), in *Proceedings of the 35th International Conference on Telecommunications and Signal Processing*, Prague, pp. 287–291.

50 Ibrahim, A. and Alfa, A.S. (2013) Radio recource allocation for multicast transmissions over high altitude platforms, in *Proceedings of the IEEE Globecom Workshops*, Atlanta, GA, pp. 281–287.

51 Ibrahim, A. and Alfa, A.S. (2014) Solving binary and continuous knapsack problems for radio resource allocation over high altitude platforms, in *Proceedings of the Wireless Telecommunications Symposium*, Washington, DC, pp. 1–7.

52 White, G.P. and Zakharov, Y.V. (2007) Data communications to trains from high-altitude platforms. *IEEE Transactions on Vehicular Technology*, **56** (4), 2253–2266.

53 Fidler, F., Knapek, M., Horwath, J., and Leeb, W.R. (2010) Optical communications for high-altitude platforms. *IEEE Journal of Selected Topics in Quantum Electronics*, **16** (5), 1058–1070.

54 Wang, X.Y. (2013) Development of high altitude platforms in heterogeneous wireless sensor network via MRF-MAP and potential games, in *Proceedings of the IEEE Wireless Communications and Networking Conference*, Shanghai, pp. 1446–1451.

55 Raafat, W.M., Fattah, S.A., and El-motaafy, H.A. (2012) On the capacity of multicell coverage MIMO systems in high altitude platform channels, in *Proceedings of the International Conference on Future Generation Communication Technology*, London, pp. 6–11.

56 Wang, W.Q. (2013) Large-area remote sensing in high-altitude high-speed platform using MIMO SAR. *IEEE Journal of Selected Topics in Applied Earth Observation and Remote Sensing*, **6** (5), 2146–2158.

57 W.-Q. Wang (May 2013) *Multi-Antenna Synthetic Aperture Radar*, CRC Press, New York.

58 Wang, W.Q. (2013) Mitigating range ambiguities in high PRF SAR with OFDM waveform diversity. *IEEE Geoscience and Remote Sensing Letters*, **10** (1), 101–105.

59 Bordoni, F., Younis, M., and Krieger, G. (2012) Ambiguity suppression by azimuth phase coding in multichannel SAR systems. *IEEE Transactions on Geoscience and Remote Sensing*, **50** (2), 617–629.

60 Wang, W.Q. (2013) MIMO SAR imaging: Potential and challenges. *IEEE Aerospace and Electronic Systems Magazine*, **27** (8), 18–23.

61 Wang, W.Q. (2014) MIMO SAR chirp modulation diversity waveform design. *IEEE Geoscience and Remote Sensing Letters*, **11** (9), 1644–1648.

62 W.-Q. Wang and Shao, H.Z. (2014) High altitude platform multichannel SAR for wide-swath and staring imaging. *IEEE Aerospace and Electronic Systems Magazine*, **29** (5), 12–17.

63 Wang, W.Q. (2011) *Near-Space Remote Sensing: Potential and Challenges*, Springer, New York.

64 Wang, W.Q. (2007) Application of near-space passive radar for homeland security. *Sensing and Imaging: An International Journal*, **8** (1), 39–52.

65 Meyers, R.G., Draim, C.J.E., Cefola, P.J., and Raizer, V.Y. (2008) A new tsunami detection concept using space-based microwave radiometry, in *Proceedings of the IEEE Geoscience and Remote Sensing Symposium*, Boston, MA, pp. 958–961.

66 Borrero, J.C. (2005) Field data and satellite imagery of tsunami effects in Banda Aceh. *Science*, **308** (5728), 1596–1596.

67 Kouchi, K. and Yamazaki, F. (2007) Characteristics of tsunami-affected areas in moderate-resolution satellite images. *IEEE Transactions on Geoscience and Remote Sensing*, **45** (6), 1650–1657.

68 Galletti, M., Krieger, G., Thomas, B., Marquart, M., and Johanness, S.S. (2007) Concept design of a near-space radar for tsunami detection, in *Proceedings of the IEEE Geoscience and Remote Sensing Symposium*, Barcelona, pp. 34–37.

69 Camana, P. (1988) Integrated communications, navigation, identification avionics (ICNIA)-the next generation. *IEEE Aerospace and Electronics Systems Magazine*, **3** (8), 23–26.

70 de Cola, T. and Marchese, M. (2008) High performance communication and navigation systems for interplanetary networks. *IEEE Systems Journal*, **2** (1), 104–113.

71 Noreen, G.K. (1990) An integrated mobile satellite broadcast, paging, communications and navigation system. *IEEE Transactions on Broadcasting*, **36** (4), 270–274.

72 Dreher, A., N. Niklash, Klefenz, F., and Schroth, A. (2003) Antenna and receiver system with digital beamforming for satellite navigation and communications. *IEEE Transactions on Microwave Theory and Techniques*, **51** (7), 1815–1821.

73 Galati, G., Giorgio, P., Girolamo, S.D., Dellago, R., Gentile, S., and Lanari, F. (1996) Study of an integrated communication, navigation and surveillance satellite system for air traffic management, in *Proceedings of the CIE International Radar Conference*, Beijing, China, pp. 238–241.

74 Chen, J.J., Safar, Z., and Sorensen, J.A. (2007) Multimodal wireless networks: communication and surveillance on the same infrastucture. *IEEE Transactions on Information Forensics and Security*, **2** (3), 468–484.

75 Wang, T., de Lamare, R.C., and Mitchell, P.D. (2011) Low-complexity set-membership channel estimation for cooperative wireless sensor networks. *IEEE Transactions on Vehicular Technology*, **60** (6), 2594–2607.

76 Razi, A., Afghah, F., and Abedi, A. (2011) Binary source estimation using a two-tiered wireless sensor network. *IEEE Communications Letters*, **15** (4), 449–451.

77 Suryadevara, N.K. and Mukhopadhyay, S.C. (2012) Wireless sensor network based home monitoring system for wellness determintion of elderly. *IEEE Sensors Journal*, **12** (6), 1965–1972.

78 Avdikos, G. and Papadakis, G. (2008) Overview of the application of high altitude platform (hap) systems in future telecommunication networks, in *Proceedings of the 10th International Workshop on Signal Processing for Space Communications*, Rhodes Island, pp. 1–6.

79 Yang, Z. and Mohammed, A. (2008) Business model design for capacity-driven services from high altitude platforms, in *Proceedings of the 3rd IEEE/IFIP International Worksjop on Business-driven IT Management*, pp. 118–119.

80 Elabdin, Z., Elshaikh, O., Islam, R., Ismail, A.P., and Khalifa, O.O. (2006) High altitude platform for wireless communications and other services, in *Proceedings of the International Conference on Electrical and Computer Engineering*, Dhaka, pp. 432–438.

81 Hatime, H., Namuduri, K., and Watkins, J.M. (2011) OCTOPUS: An on-demand communication topology updating strategy for mobile sensor networks. *IEEE Sensor Journal*, **11** (4), 1004–1012.

82 Kandus, G., Svigelj, A., and Mohorcic, M. (2005) Telecommunication network over high altitude platforms, in *Proceedings of the 7th International Conference on Telecommunications in Modern Satellite, Cable and Broadcasting Service*, pp. 344–347.

83 Yao, H., McLamb, J., Mustafa, M., Narula-Tam, A., and Yazdani, N. (2009) Dynamic resource allocation DAMA alternatives study for satellite communication systems, in *Proceedings of the IEEE Military Communication Conference*, Boston, MA, pp. 1–7.

84 Wicaksono, B.I. (2012) On the evaluation of techno-economic high altitude platforms communication, in *Proceedings of the 7th International Conference on Telecommunication Systems, Services, and Applications*, Bali, pp. 255–260.

85 Anastaspoulos, M.P. and Cottis, P.G. (2009) High altitude platform networks: a feedback suppression algorithm for reliable multicast/broadcast services. *IEEE Transactions on Wireless Communications*, **8** (4), 1639–1643.

86 Celcer, T., Javornik, T., Mohorcic, M., and Kandus, G. (2009) Virtual multiple input multiple output in multiple high-altitude platform constellations. *IET Communications*, **3** (11), 1704–1715.

87 Miura, R. and Oodo, M. (2001) Wireless communications system using stratospheric platforms–R& D program on tecommunication and broadcasting system using high altitude platforms. *Journal of the Communications Research Laboratory*, **48** (4), 33–48.

88 Milas, V., Koletta, M., and Constantinou, P. (2003) Interference and compatibility studies between satellite systems and systems using high altitude platform stations, in *Proceedings of the 1st International Conference on Advances on Satellite Mobile Systems*, Frascati, Italy, pp. 1–4.

89 Thornton, J., Grace, D., Capstick, M.H., and Tozer, T.C. (2003) Optimizing an array of antennas for cellular coverage from a high altitude platform. *IEEE Transactions on Wireless Communication*, **2** (3), 484–492.

90 E. B. Tomme, The paradigm shift to effects-based space: Near-space as a combat space effects enabler. http://www.airpower.au.af.mil.

91 Gierull, C. (2006) Mitigation of phase noise in bistatic SAR systems with extremely large synthetic apertures, in *Proceedings of the European Synthetic Aperture Radsr Symposium*, Dresden, Germany, pp. 1–4.

92 Weiss, M. (2004) Time and frequency synchronization aspects for bistatic SAR systems, in *Proceedings of the European Synthetic Aperture Radar Symposium*, Ulm, Germany, pp. 395–398.

93 Younis, M., Metzig, R., and Krieger, G. (2006) Performance predication of a phase synchronization link for bistatic SAR. *IEEE Geoscience and Remote Sensing Letters*, **3** (3), 429–433.

94 W.-Q. Wang, Ding, C.B., and Liang, X.D. (2008) Time and phase synchronization via direct-path signal for bistatic synthetic aperture radar systems. *IET Radar Sonar and Navigation*, **2** (1), 1–11.

95 Thornton, J. (2004) A low sidelobe asymmetric beam antenna for high altitude platform communications. *IEEE Microwave and Wireless Components Letters*, **14** (2), 59–61.

96 Thornton, J. (2006) Wide-scanning multi-layer hemisphere lens antenna for Ka-band. *IEE Proc.-Microwave, Antennas and Propagation*, **153** (6), 573–578.

97 Cai, R.N., Yang, M.C., X.-Q. Zhang, and Li, M. (2012) A novel multi-beam lens antenna for high altitude platform communication, in *Proceedings of the IEEE 75th Vehicular Technology Conference*, Yokohama, pp. 1–5.
98 Temel, S. and Bekmezci, I. (2013) On the performance of flying Ad Hoc networks FANETs utilizing near space high altitude platforms HAPs, in *Proceedings of the 6th International Conference on Recent Advances in Space Technologies*, Istanbul, pp. 461–465.
99 Alshbatat, A.I. and Dong, L. (2010) Adaptive MAC protocol for UAV communication networks using directional antennas, in *Proceedings of the International Conference on Networking, Sensing and Control*, Chicago, IL, pp. 598–603.

Part III

Underwater and Submerged WSS Solutions

12

Underwater Acoustic Sensing: An Introduction

Habib F. Rashvand[1], Lloyd Emokpae[2] and James Agajo[3]

[1] *Advanced Communication Systems, University of Warwick, UK*
[2] *U.S. Naval Research Laboratory, Code 7160 – Acoustics Division, USA*
[3] *Federal University of Technology Minna, Ihiagwa, Nigeria*

12.1 Introduction

Nearly three quarters of the surface of the earth is covered in water, from shallow waters and seaside to deep waters in places that only a handful of people have glimpsed. There is currently a new surge of interest in this unique, easily reachable environment.

New smart sensors could enable us to penetrate the waters more, to tame, to inhibit, and even in many cases to feel at home in it. This could be the start of a new beginning for advanced and civilised and industrial societies, who will begin to consider its potential and true hidden treasures. It is certain that its unlimited resources will enable us to enhance our quality of life. However, lack of visibility, poor access, and lack of control over what goes on in the deep water could make us regard this as wishful thinking rather than a new potential, long-term humanitarian development. It is a fact that new trends begin when new and traditional industries and new and old generations become enthusiastic about them. We need only look at the literature generated by industrial programs and forward-looking researchers to see the many foreseeable paths and trends in the field.

One prospective view on the importance of oceans and a better understanding of the underwater environment comes from Toma et al. [1], who claim that, as an integral source of life, oceans regulate the Earth's climate. This work urges us to study the underwater environment and to use advanced smart sensors to see, interact and learn more about the biological, geological, and chemical processes of the oceans. This knowledge could help us with the sustainability of the Earth and life on it. The authors emphasise the need for global-scale in-situ observation programs. They also argue that due to the very high water coverage of the planet and poor use of our existing technological capabilities, underwater projects are and will remain costly, slow, and very time consuming. We have noticed that most of recent development programs for the underwater environment are for monitoring of climate and scientific discoveries rather than intense industrial applications which we need to initiate a new underwater industrial paradigm.

Wireless Sensor Systems for Extreme Environments: Space, Underwater, Underground, and Industrial.
First Edition. Edited by Habib F. Rashvand and Ali Abedi.

However, Toma et al identify some points of concern in EU member states that are inhibiting this:

- Environmental monitoring requires dealing with an extensive amount of data: collection, analysis, and processing, where integration of single-purpose devices and programs is essential.
- Environmental monitoring is fragmented, with geographically scattered activities at the national and regional levels.
- There is a lack of commonality, quality control and standards.
- There is limited support for deployment of economically viable applications and limited flexibility in available data for further potential manipulations.
- There are different general policies, different intellectual property rights and different costs.
- There is a shortage of experts and skilled workers.

They recommend future investigations of the following nature:

- generic geographical and biological monitoring for assessing the underwater environment and climate change
- RFID style classification and information collection (as in temperature and pressure measurements)
- prediction of natural events through seismic signals and hurricane measurements
- underwater object and bomb detection and associated robotic removals
- underwater cultivation and food-resource projects and planning (marine science and technologies, fisheries research and so on)
- underwater pipe leak detection and associated robotic repair and maintenance
- use of advanced buoys for communicating and issuing warnings to ships and underwater vehicles
- underwater treasure hunts and leisure activities
- underwater manned vehicle guidance and rescue
- underwater mineral, oil and gas resources projects and planning.

On the technical side, communications through water remains hugely under-developed. Our main underwater data communication technology is acoustics, for example, underwater acoustic sensor networks (UASNs), and we have many good practical solutions for sensor applications. With available acoustic transmission technologies, long-distance underwater communications are somewhat limited in bandwidth and there is often excessive background noise, as shown in Figure 12.1, so quality video and other wideband applications are difficult. For most urgently needed low-speed applications – monitoring and communication of sensitive information – current technologies are more than adequate.

This chapter has been structured as follows. Section 12.2 looks at the underwater environment and examines the communication capabilities of acoustic and non-acoustic systems. Section 12.3 looks at the underwater components of an underwater sensor network and the development of its nodes and links in particular. Here, for the node, we provide mathematical models for the acoustic antenna and the acoustic links. Section 12.4 looks at networking aspects of underwater acoustic sensor networks, covering the issues and requirements in one subsection and the solutions in another. Section 12.5

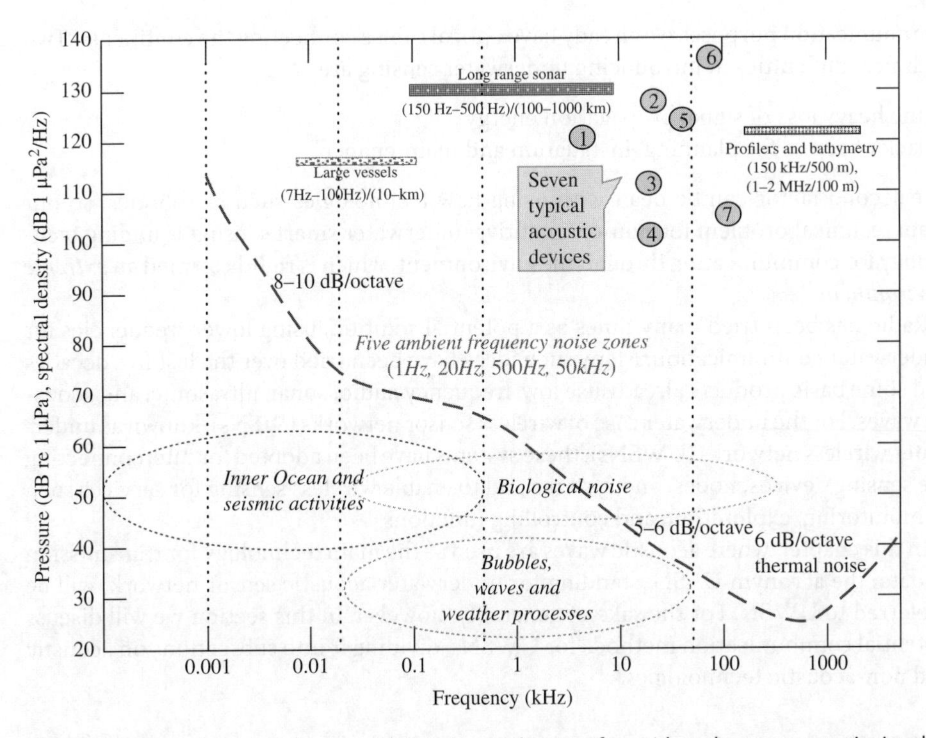

Figure 12.1 Typical sources of noise and disturbance that interfere with underwater acoustic signals depending on water depth of the oceans. Seven acoustic devices are indicated: (1) sub-bottom profiler (5 kHz/1 km); (2) modem (20 kHz/1 km); (3) long-baseline positioning (20 kHz/10 km); (4) modem (20 kHz/10 km); (5) depth sounder (40 kHz/1 km); (6) ADCP (75 kHz/500 m); (7) modem (80 kHz/1 km). Source: Toma et al. [1].

provides a brief review of the applications of underwater networks for mobile and fixed platforms. Section 12.6 is the chapter summary.

12.2 Underwater Wireless Smart Sensing

The idea of sending and receiving information underwater can be traced back to the time of Leonardo da Vinci, who discovered the possibility of detecting a distant ship by listening on a long tube submerged under the sea [2]. This was the first record of using the water environment for communications. How much better are our solutions today? Have they really improved along with the abundant and superior technologies we have available today? Are they comparable to recent terrestrial wireless systems? If the answers are negative, then it is logical to look for three policy areas:

- non-technical
- demand
- technical.

Prior to the advent of smart sensors and concerns about the climate and the environment, the development of technologies in this area was sporadic and minimal. For communication purposes we already have optical cables connecting the continents. Two technical difficulties in introducing underwater sensing are

- the heavy loss of signal propagation energy
- lack of access for planning, installation and maintenance.

The second factor can be dealt with using new technologies such as robotics, so our main technical problem for non-competitive underwater smart sensing is finding technology for communicating through the environment, which is rightly termed an *extreme environment*.

Radio has been tried many times as a potential solution, using lower frequencies for underwater communication. Many such ideas have been tried over the last five decades and some basic products already use low-frequency audio, sonar, ultrasonic, and acoustic waves. For the underwater case of wireless sensor networks (WSNs), known as underwater wireless networks (UWSNs), these systems have been adopted for interconnecting the sensing devices, nodes, and systems and to enable wireless sensing for services such as monitoring, exploration, and controlling functions.

In this chapter, where acoustic waves are used as the main technology for transmission of data, the acronym UASN, standing for underwater acoustic sensor network, will be preferred to UWSN. For the sake of generality, however, in this section we will discuss potential communication methods for UWSNs, dividing it into subsections on acoustic and non-acoustic technologies.

12.2.1 Non-Acoustic Sensors

Acoustics are the proven approach to wireless underwater communications. But let us survey the three most promising technologies for future underwater communications.

12.2.1.1 Radio Systems

Well-established terrestrial wireless sensing systems bring ever-increasing capabilities and very low costs and are widely available. It has therefore been tempting to use them for underwater sensing applications too. Unfortunately, due to their very high absorbent energy losses and highly dispersive propagation rays when travelling through water, this is not currently possible.

Unlike terrestrial waves, where higher frequency results in more throughput, in underwater environments radio waves follow the 'microwave oven' effect: the higher the frequency carrier signal, the higher the loss of energy, resulting in poorer throughput. In other words, the propagation loss factor increases more rapidly with either distance or frequency. Thus in water we must use moderate-frequency radio waves for communication over short distances or lower frequencies – down to 500 Hz – for longer distances.

Radio frequency communication can help us with applications that require greater data rates but only short distances are involved. As an example, for a specific underwater scenario, de Freitas evaluated the performance of a specially configured IEEE 802.11 system using the 700-MHz, 2.4-GHz and 5-GHz wireless bands [3]. A newly designed antenna operating at 768 MHz shows a 50% increase in distance over a terrestrial antenna using the same signal frequency. De Freitas's system has receiving data rates around 400 kbit/s and 11 Mbit/s at distances of 2 and 1.6 m. His measurements for the 2.4- and 5-GHz bands show them working at distances of 32 and 10 cm, respectively. Similarly, at 20 cm distance and 2.462 GHz his system gives a throughput of 100 Mbit/s; at 10 cm and 5 GHz the figure is 10 Mbit/s.

12.2.1.2 Optical Systems

Fully unguided underwater optical communication systems, also called free-space optics (FSO), can be used for short distances. De Freitas explains that although optical solutions can achieve higher data rates over short distances, due to their higher costs and other severe limitations, such as the need for line-of-sight (LOS) communications and the need to compensate for the strong disturbances caused by sunlight, makes this technology rather unattractive.

Another interesting optical underwater example is from Arnon and Kedar [4] who propose a non-line-of-sight (NLOS) optical solution. They claim that optical wireless technology is a promising alternative for underwater links for sensor networks. They have demonstrated a 1-Gb/s data transmission link in the laboratory equipped with a simulated aquatic medium of oceanic waters. They achieved long-range underwater communication for FSO technology, providing 5 GHz bandwidth over a distance of 64 m of clear ocean water. A drop of 1 GHz for 8-m communication in turbid harbour water was also examined. Further experiments on oceanic FSO over 100 m, using a LED transmitter in a hybrid acoustic–optical wireless link, showed how affordable underwater sensor networks with higher data rates might be built. However, in practice, when sensing devices are deployed, either on the seabed or fixed somewhere near the depth of the sensors, LOS communication is not always possible. This led the authors to experiment with multi-hop communication for long distances. They tested this approach in a prototype terrestrial NLOS optical wireless sensor network, showing that although the back-scattering lights in the atmosphere function for communication in a similar way to the deployment of numerous tiny reflecting mirrors, the main reflections come from the ocean–air surface, when applicable.

12.2.1.3 Magnetic Induction Systems

As an alternative to underwater acoustic and optical techniques for deep waters, Allen et al. proposed the use of magnetic induction (MI) technologies, particularly for shallow and cloudy waters [5]. This idea was based on the historical use of magnetic induction for naval sensing. It was a leading technology in underwater environments for well over a century, used for target detection and tracking. However, as magnetic signals are poor for long-distance transmission, Allen et al. suggest that it would only be applicable in small autonomous underwater vehicles (AUVs).

Due to their specific features, MI technologies have been used for underground as well as underwater environments, and this has led to some interesting developments. One is the use of meta-material-enhanced magnetic induction (M^2I) in underwater communication systems, an approach that shows great improvements over MI technologies. Guo et al. used special meta-material-enclosed devices and explained how, for a noise power of −100 dBm (well above the usual −140 dBm), they achieved data rates of the order of several kilobits per second for distances over 30 m. This is nearly twice the rate in classic MI systems lacking the meta-material. Therefore, all underwater networked-sensor applications that make use of short-distance M2I connectivity can now provide much higher bandwidths. Due to demand for higher numbers of nodes, the complexity of the network increases rapidly as the size of the coverage area grows. However, the new low-cost and competitive M2I networking system may introduce a new and important trend that will be important in the future of underwater sensor applications [6].

12.2.2 Acoustic Sensors

In nature, dolphins and bats use acoustic techniques, an approach which we have used (at first unrecorded but more recently with recording) for detection of objects under water. One significant record is from Leonardo da Vinci, who in 1490 inserted a tube into water to detect objects under the surface, indicating the possibility of using audible acoustic waves underwater. Nowadays, acoustic transmission technology has been developing further, using various forms and frequencies. Sonar, for example, is a form of scan-to-display technology for deep waters. One major step towards the development of UWSN is finding a suitable common, practically proven technology for underwater communication. For example, acoustics have been experimenting with for a long time: first used for voice communication since about 1945, they have become widely available since then and there is now a vast family of products, for example underwater telephones to communicate with submarines. This is known as an underwater acoustic channel. Underwater acoustic technology, however, comes with some natural limiting characteristics:

- due to the low carrier frequency it has limited bandwidth
- it is weak against narrowband jamming
- there is a low attainable range (typically a maximum of 40 km-kbps).

Today acoustic waves are a widespread method for underwater communication with ranges of hundreds of kilometres. Due to the severe frequency-dependent attenuation and surface-induced spread reflections, the data rates normally achieved are around 20 kbit/s. The distance can be extended to 100 km in shallow waters and 200 km in deep oceans, but with lower data rates of 0.5 kbit/s. Due to the low speed of acoustic waves in water, 1500 m/s, long delays inject high latencies that pose a problem for synchronization. Normally, acoustic links operate in half-duplex mode [4].

The features of a pure acoustic underwater technology for netted sensors are listed below [7, 8]:

- The nodes have 3D mobility.
- The nodes move at 1–3 m/s due to currents.
- Stronger self-configuration is needed to handle non-uniform deployments.
- Drifting data sinks due to water currents make predefined paths unstable.
- Much lower data rates than terrestrial: 40 kbit/s at 1 km.
- Higher loss and longer distances means much higher energy consumption.
- Routing optimization is more challenging, mostly due to moving nodes.
- Communication speed decreases from speed of light to speed of sound, so much higher propagation delays
- Ad-hoc connectivity is commonly used for monitoring applications, the number of hops depending on the depth; normally 4 to 7.
- Due to drifting, fouling or corrosion the nodes are more error prone and can die at much faster rates than in terrestrial networks, so stronger self-recovering routing capabilities are required

- Battery power is costly so alternative sources of energy are recommended.
- Only sparse deployments are possible due to cost of underwater equipment and large areas involved.

12.2.3 Received Signal Model

A ray propagation model is a widely accepted method for modeling signal propagation in shallow water. There are typically four basic types of eigenrays that are of interest:

- refracted-surface-reflected (RSR)
- refracted-bottom-reflected (RBR)
- refracted-surface-reflected-bottom-reflected (RSRBR)
- direct-path (DP).

The length of each eigenray, which is reflected at the surface first before being reflected i times during the entire propagation is:

$$r_i = \sqrt{(D_{\mathrm{TX}} + a_i D_w + b_i D_{\mathrm{RX}})^2 + d^2} \tag{12.1}$$

The geometrical descriptions of the transmitter and the receiver are illustrated in Figure 12.2. The coefficients a_i and b_i are defined in (12.2)–(12.4).

$$a_1 = 0 \tag{12.2}$$

$$a_{i+1} = a_i + (1 + (-1)^{i+1}) \tag{12.3}$$

$$b_i = (-1)^{i+1} \tag{12.4}$$

Given an approximation for each eigenrays r_i, the received signal $r(t)$ is simply a convolution of the transmitted signal $e(t)$ with the acoustic channel $h(t)$ with an additive

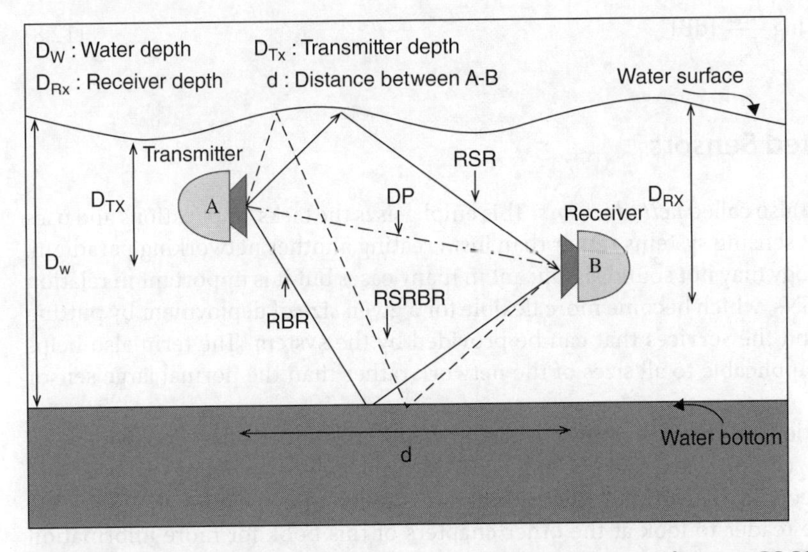

Figure 12.2 Illustration of multipath channel with the four eigenrays of interest: DP, RSR, RBR and RSRBR. We propose using RSR and RBR eigenrays for directional communication for NLOS links [7].

white noise $w(t)$:

$$r(t) = e(t) * h(t) + w(t)$$

$$r(t) = s(t) + w(t) = \sum_{k=1}^{K} \beta_k e(t - \tau_k) + w(t) \quad (12.5)$$

where * is the convolution operator and τ_k and β_k correspond to the time delay and attenuation factor for the kth path respectively. The time delay can be determined by the geometry of the transmitter and receiver depths $\{D_{TX}, D_{RX}\}$, the water depth D_w and the LOS distance d:

$$\tau_k = \frac{r_k}{\bar{c}_{ij}} = \frac{\sqrt{(D_{TX} + a_k D_w + b_k D_{RX})^2 + d^2}}{\bar{c}_{ij}} \quad (12.6)$$

For an eigenray (travelling wave) of length r_k, that is dependent on the average sound speed $\bar{c}_{ij}$ between the two connected nodes $\{i,j\}$. The attenuation factor β_k can be modeled for the kth path as:

$$\beta_k = 10^{\frac{TL_{\mathrm{TOTAL}}}{10}} = 10^{\frac{TL_{\mathrm{MULT}}(k)+TL_{\mathrm{LOS}}}{10}} \quad (12.7)$$

where TL_{TOTAL} is the total attenuation in the shallow-water environment. In shallow-water environments, most of the total transmission loss will be due to multipath effects. In underwater acoustics, the surface-area (that is, the water surface or the bottom) and volume (that is, plane-wave) scattering strength S is typically used as a parameter of reverberation. Reverberation is the total sum of all scattering contributions in the medium due to the inhomogeneity in the sea boundaries. The total scattering index is a function of the ratio of I_{scat}, the intensity of the sound scattered by a unit surface area or volume, referenced to a unit distance, and I_{inc}, the incident plane-wave intensity:

$$S = 10 \log \frac{I_{\mathrm{scat}}}{I_{\mathrm{inc}}} \text{ [dB]} \quad (12.8)$$

12.3 Netted Sensors

The WSNs are also called *netted sensors*. This emphasises the tasks, applications and uses of these smart sensing systems rather than just creating another networking paradigm. This terminology may not sound significant in many cases but it is important in relation to smaller WSNs, which become more flexible for a given size of deployment by putting more weight on the services that can be provided by the system. The term also helps with matters applicable to all sizes of the network rather than the normal large sensor networks.

In this section we briefly address smart sensing and underwater deployment of WSNs – the components, networking, and technologies – avoiding generic and common aspects of networking where these are similar to conventional WSNs. We encourage the reader to look at the other chapters of this book for more information about these. For ease of flow we first look at the importance of the node. We survey recent technical research papers and then there is a brief discussion of a typical array-style antenna model. Then we look at the underwater links to extend our understanding of the underwater environment, and propagation and long-distance communications through it, including sound-speed modeling of the link.

A smart sensing system is generally characterised by the few basic but predominant WSN functionalities that influence its dynamics, behaviour, performance, and capabilities. For example, for applications of WSN in underwater environment we normally gauge a project with respect to five feasibility factors:

- ease of communication
- lifetime and survivability
- cost
- maintenance and serviceability
- security.

Most underwater applications are usually used over long distances, ignoring the costs. We therefore normally consider the first two factors as the most important.

12.3.1 Nodes

In underwater netted sensors (UWSNs), the nodes of the network can vary greatly in both size and functionality. A node could be a single sensing device, a multiple/complex sensing device, a buoy, an AUV, a platform, a data sink, or simply a remote-control system located either under or above the water. Let us look at some recent developments in the area.

One common question for deployment of a new application is minimizing the overall cost versus the number of devices used in the network. This can be reduced by making use of common, low-cost sensors. Unfortunately, this does not apply to underwater systems, for obvious reasons such as:

- premature technology
- bulky acoustic modems
- low-volume use due to lack of applications viability.

Low-volume production means that progress in underwater technology and development of new services is difficult. Therefore, many designers are considering enhancing the design process. One interesting idea comes from Jurdak et al., who suggest customising standard software to suit short-range mote-based underwater environmental monitoring applications [9]. They redesign the modem software, often called the 'soft-modem', to use built-in microphones and speakers. Upon their Tmote Invent platforms they adopt a modular design to accommodate their built-in hardware modules. Experimental acoustic hardware profiles of the channel favour frequencies below 3 kHz for the FSK soft modem and there is a capability for transferring data at several tens of bits per second at a range of 10 m.

In order to increase network lifetimes, Yang et al. redesign underwater sensor nodes' hardware and use a special sleeping procedure to reduce energy consumption at the medium access control (MAC) level of the protocol [10]. That is, on the top of selecting low power devices and a new architecture, they adopt an autonomous 'sleep–wake' operation mode in the hardware. There are five parts in the pre-channel module of their system: a transducer, a pre-amplifier, band-pass filters, automatic gain control (AGC) and a CymaScope, a scientific instrument that makes sound visible but here is used to wake the modem up when required.

We can extend many characteristics of an underwater node through strategic power allocation. A good method, fully explained in Chapter 3, comes from Alirezaei et al. [11, 12], who enhanced the performance of the nodes regarding optimal power and energy allocation, prolonging the lifetime of the network, increasing the overall sensing performance, making the network robust against node failures, specifying the reliability of each sensor node and automating the on-off time of the nodes.

A hydrophone is a microphone that can be used as an acoustic antenna for communicating underwater; it has a piezoelectric transducer with an acoustic impedance matched to water. A single-element hydrophone is omnidirectional in nature. For directional underwater acoustic communication, the antenna is composed of an array of hydrophones that can be summed at various phases and amplitudes, turning it into a beamformer. A directional beamformer can be expressed by averaging the SNR to the output of the hydrophone array as follows:

$$\frac{S^2}{N^2} = \frac{\overline{\left[\sum_{i=1}^{m} s_i(t)\right]^2}}{\overline{\left[\sum_{i=1}^{m} n_i(t)\right]^2}} = \frac{\sum_{\mathrm{i}=1}^{\mathrm{m}} \sum_{j=1}^{m} s_{ij}}{\sum_{\mathrm{i}=1}^{\mathrm{m}} \sum_{j=1}^{m} n_{ij}} \tag{12.9}$$

where the bar represents the average over time of the m signals $s_i(t)$ and noise $n_i(t)$. The array signals and noise for the ith and jth hydrophones can be expressed as $s_{ij} = s^2 s_{ij}$ and $n_{ij} = n^2 n_{ij}$. Thus, the beamformer can be further simplified to:

$$\frac{S^2}{N^2} = \frac{s^2}{n^2} \frac{\sum_{\mathrm{i}=1}^{\mathrm{m}} \sum_{j=1}^{m} s_{ij}}{\sum_{\mathrm{i}=1}^{\mathrm{m}} \sum_{j=1}^{m} n_{ij}}$$

The gain of the hydrophone array (AG) can then be expressed as:

$$\mathrm{AG} = 10 \log \left(\frac{\frac{S^2}{N^2}}{\frac{s^2}{n^2}} \right) = 10 \log \left(\frac{\sum_{\mathrm{i}=1}^{\mathrm{m}} \sum_{j=1}^{m} s_{ij}}{\sum_{\mathrm{i}=1}^{\mathrm{m}} \sum_{j=1}^{m} n_{ij}} \right)$$

We can further define the antenna gain $G(\theta)$ at the beam steering angle $\theta = \theta_s$ and for an array gain of AG as:

$$G(\theta) \cong \begin{cases} AG, & \text{Combined gain} \\ DI(\theta = \theta_s), & \text{Directional gain} \end{cases} \tag{12.10}$$

A vector hydrophone array has been developed for underwater acoustic communication [13]. Although the design used a hydrophone array that could receive acoustic signals but not transmit them, the same concept can be extended to a hydrophone array with both sensing and emitting properties [14, 15]. Figure 12.3 shows an illustration of the seven-element 3D vector sensor array with sensors oriented along both the positive and negative axes $\pm (x, y, z)$ as well as at the origin (0,0,0). When combined, the seven elements can be used as an omni-directional antenna with the beam pattern shown on the right-hand side of Figure 12.3 (the dipole is the omni-directional beam oriented along the x-axis).

A piezoelectric directional transducer was designed and evaluated by Butler et al. [16]; this generated a directional pattern by combining the fundamental vibration modes of

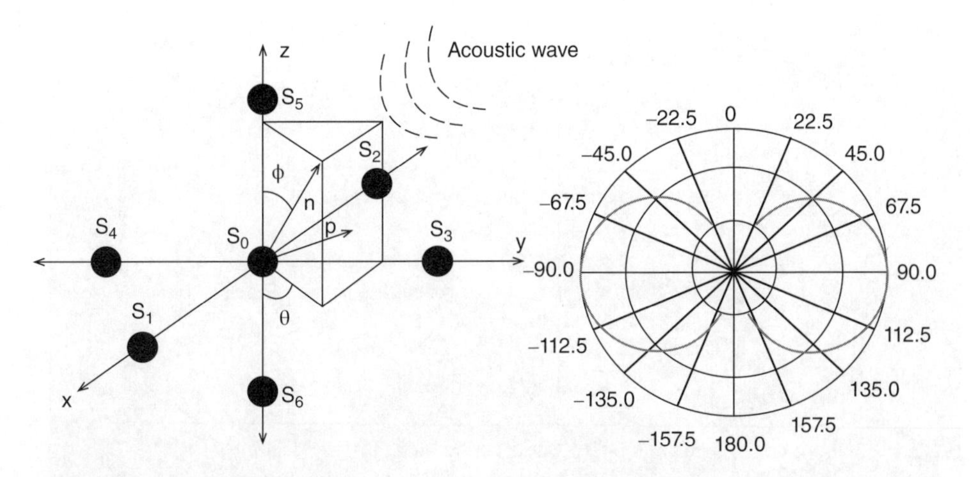

Figure 12.3 A seven-element antenna array and the corresponding beamformer pattern, combining the signals of all elements in one axis

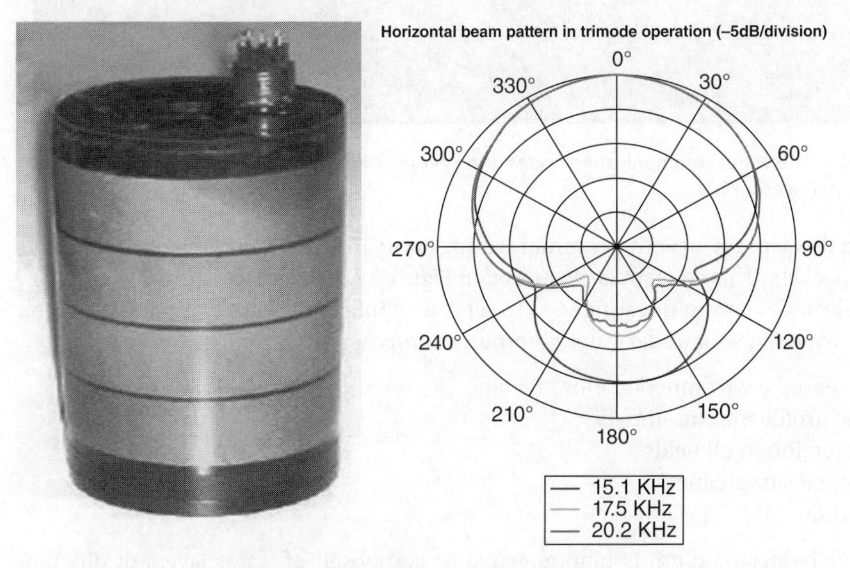

Figure 12.4 A hydrophone of stackable multimode piezoelectric directional transducers. Each cylinder is about 50 mm in height with an outer diameter of 108 mm. The horizontal beam pattern is shown; the vertical beam pattern depends on the number of stacks [7].

a cylindrical acoustic radiator. The transducer and its vertical beam pattern are shown in Figure 12.4. The beam steering angle can be controlled electrically by changing the voltage amplitude rather than the phase, which makes it ideal for our proposed system model.

12.3.2 Links

An underwater acoustic network (UWAN), a UASN for multimedia communications, consists of both stationary and mobile nodes that cooperate to form a network [17].

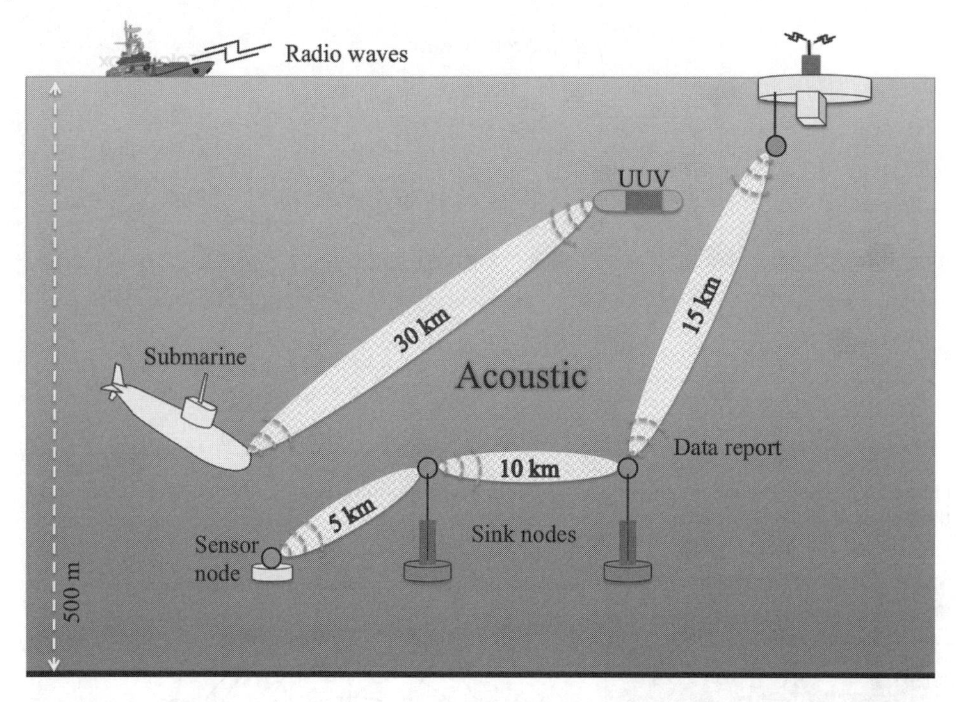

Figure 12.5 Underwater communicating nodes using acoustic links because of their favourable propagation properties [7].

A node in this sense is an active communicating entity in that it can transmit and receive network packets. This concept is illustrated in Figure 12.5, where we see both stationary and mobile nodes communicating so as to serve an application. Applications of UWANs include a wide range of underwater communications as well as:

- sensor-based environmental monitoring
- oceanic profile measurements
- leak detection in oil fields
- distributed surveillance
- navigation.

The underwater medium is inhomogeneous, composed of water layers of different temperatures, densities, and salinities. These varying parameters affect the sound-speed profile [18–20], which can be modeled as a function of the ocean depth. The underwater medium can be split into two different layers: the surface and the thermocline, as depicted in Figure 12.6. The surface layer is mainly affected by seasonal temperature changes. On the other hand, the thermocline layer is affected by the increase in pressure and the decrease in temperature for an increase in depth [21]. To simplify our analysis, we focus on shallow-water communication at depths less than 200 m. This means that we will only be concerned about the surface layer. In shallow water, acoustic waves will be greatly affected by multipath propagation, which ultimately leads to multiple attenuated copies of the same signal arriving at different times [22–25]. A ray-trace simulation of the multipath effect can be seen in Figure 12.7, which shows typical sound propagation with starting angles ranging from 5–15°. Directional communication is often used to mitigate

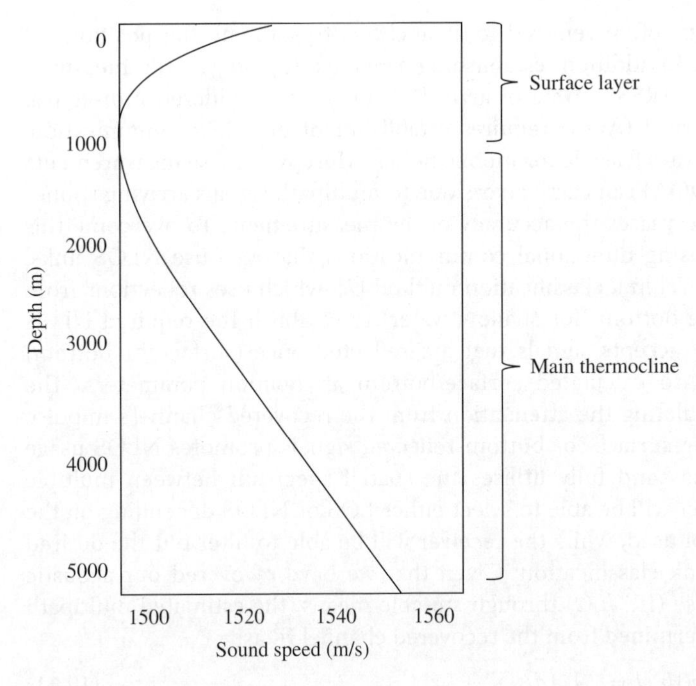

Figure 12.6 Statistical variation of sound speed in the surface layer and the main thermocline [7].

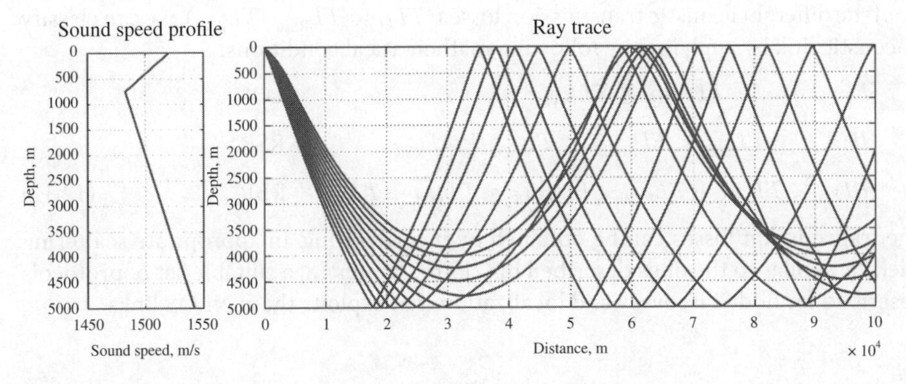

Figure 12.7 Result of a mathematical simulation results demonstrating how average speed of sound being produced [7].

the multipath effects [16, 26, 27]. More importantly, directional communication takes advantage of the spatial spectrum by allowing multiple transmissions to occur simultaneously within the same transmission range. However, most directional communication schemes rely on LOS links, which might not be available underwater. Furthermore, for ad-hoc formation, node localization, medium arbitration and data routing are required for effective network operations.

Node mobility becomes a challenge for localization when there are directional communication constraints. Even non-mobile nodes tend to change position over time due to ocean currents. This poses a challenge for localization techniques that depend on fixed reference nodes, often referred to as anchors, to estimate the positions of the other sensor nodes. In addition, establishing a relative topology using measured received-signal-strength (RSS), time-of-arrival (TOA), time-difference-of-arrival (TDOA), or angle-of-arrival (AOA) requires establishment of LOS communication, something that is not always feasible for mobile nodes. Moreover, these measurements (especially TOA and TDOA) can carry errors due to multipath signals arriving sooner than expected, which decreases the accuracy of the measurement. To overcome this challenge we propose using directional communications that also use NLOS links. We have developed a novel link classification method [7], which uses reflections from the water surface or the bottom (for shallow water) to establish the required NLOS links. The receiver only accepts signals that are reflected once (surface or bottom) by comparing the RSS to calculated surface/bottom attenuation parameters. The RSS is obtained by calculating the attenuation from the recovered channel's impulse response. Only accepting surface- or bottom-reflected signals promotes NLOS usage with directional antennas and fully utilises the spatial spectrum between multiple nodes. Thus a transmitter will be able to select either LOS or NLOS depending on the communication protocol used, while the receiver will be able to filter out the desired link by incorporating link classification. Given that we have recovered our acoustic channel impulse response (IR) $h_r(t)$ through suitable means, the estimated multipath loss (EL_{MULT}) can be determined from the recovered channel IR as:

$$EL_{\mathrm{MULT}} = -10\ log(|h_r(t=\tau)|)\ [dB] \tag{12.11}$$

where τ is the delay–spread of the first multipath signal. Assuming that we have knowledge of the different acoustic transmission losses (TL_{LOS}, TL_{RSR}, TL_{RBR}), we can classify the acoustic link by applying the following mathematical conditions:

$$\begin{array}{lll} (I) & EL_{\mathrm{MULT}} \leq TL_{\mathrm{LOS}} & \mathrm{LOS} \\ (II) & TL_{\mathrm{LOS}} < EL_{\mathrm{MULT}} \leq TL_{\mathrm{LOS}} + TL_{\mathrm{RSR}} & \mathrm{RSR} \\ (III) & TL_{\mathrm{LOS}} + TL_{\mathrm{RSR}} < EL_{\mathrm{MULT}} \leq TL_{\mathrm{LOS}} + TL_{\mathrm{RBR}} & \mathrm{RBR} \end{array} \tag{12.12}$$

The transmission losses can be obtained by incorporating an appropriate scattering model, i.e. similar to the ones described in [7]. In Chapter 13, a suitable set of protocols is described for node discovery and localization that exploits these NLOS links.

12.4 Networking

As mentioned earlier, high-frequency radio signals attenuate under water very quickly, and optical signals generally show high scattering components leaving acoustic signals in a better position compared with radio or optical signals. Therefore, so far they position themselves as the preferred technology for most underwater sensing applications. However, acoustic communications come with several handicaps. (1) The acoustic channels provide a much lower data rate, normally several tens of kilobits per second for distances below 1 km. For above 1 km their data rates drop much faster due to multipath and interference problems. For this reason, practical underwater acoustic modems are the designs to use for short-distance node deployment. Consequently, we need a much larger number of nodes and a huge mass of them to cover larger underwater regions. (2) Another drawback of acoustic communication channels is their propagation quality such as multipath signal propagation and highly variable layers of the medium affected by temperature, conductivity, varying reflection and refraction properties. (3) Also, surface waves increase the time variability of the acoustic channel. (4) Doppler effects, due to low speed of sound in water, may become more significant. (5) Other common underwater physical challenges are energy, maintenance, and protection. Despite all these problems, in practice we have been able to develop a wide range of applications including GPS assisted localization, surveillance, oil platform monitoring, earthquake and tsunami forewarning, climate and ocean observation, and water pollution tracking, etc [28].

12.4.1 Environment

Underwater is an unconventional and extreme environment, where smart sensors expected to penetrate into it using existing technologies to enable us to expand our understanding of it and to control it for future capabilities. We, therefore, consider the distinct features of underwater acoustic sensor networking to new challenges towards an upcoming ubiquitous access era of interacting with remote places of the selected underwater areas and oceans through our Internet friendly protocols in every level. The netted sensor penetrated areas through acoustic technologies are mentioned earlier in this Chapter the UASNs. An UASN normally consists of a large number of sensors and vehicles that are deployed to perform collaborative monitoring tasks over a special target area. To achieve our sensing objectives at the node level, our devices and vehicles are required to work in such a harsh and unconventional environment, to be autonomous, and to be intelligent enough to organize themselves at all levels of operation. At the network level, the network should be able to handle basic functions using the available channels, which face the following characteristics: (a) high error rates due to high noise; (b) high multi-path losses; (c) distorted by Doppler effects; (d) limited bandwidth; (e) high energy consumption; (f) long and highly variable delays; (g) low and variable propagation speeds; (h) highly variable media for pressure, temperature, and salinity; (i) surface waves; surface reflections; and (j) oceanic noises.

In order to make UASNs communicate with the terrestrial Internet the P2P protocol compatibility is essential. For example, the use of IPv6 encapsulation format can help the cost of larger packet headers overhead. When added to the other inefficiencies and incompatibilities the nodes could suffer greatly creating many unpredictable practical issues. Figure 12.8 illustrates the architecture of our P2P based UASN application scenario showing various acoustics enabled sensor nodes, relay nodes, sink nodes and complementary terrestrial nodes [29].

Further internetworking prospects for future industrial applications of UASN are for compatibility with telecommunication systems. LTE interworking would be required for most practical applications such as: climate monitoring, ocean observation, water pollution, oil platform monitoring, naval surveillance, object tracking, and earthquake forewarning [30].

12.4.2 Solutions

Terrestrial communication networks are normally designed and operated to meet 2D requirements. In underwater applications, depth plays a significant role, so we need 3D

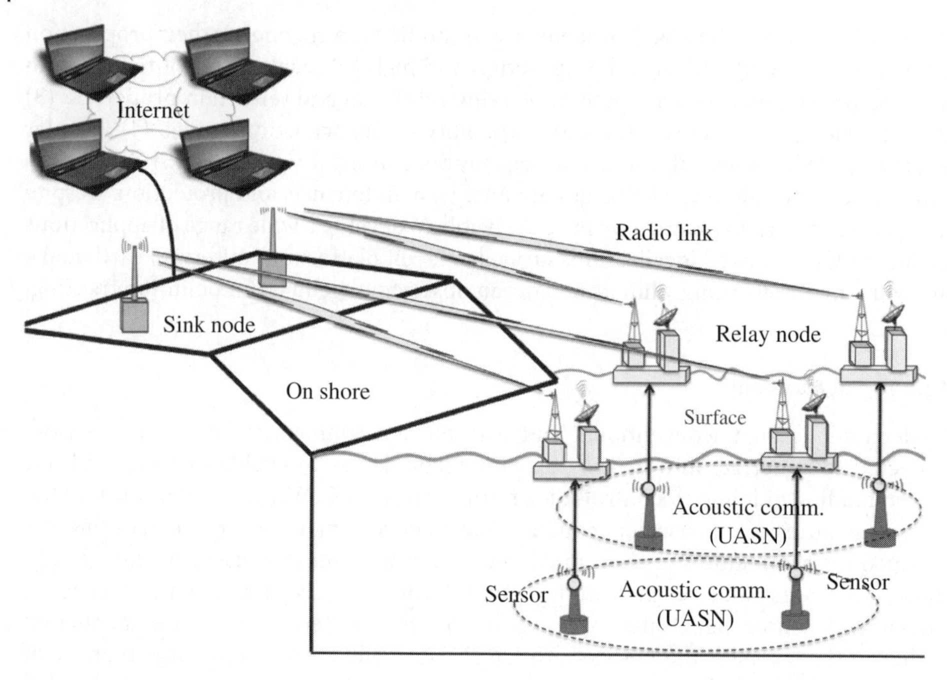

Figure 12.8 Architecture of a typical application scenario illustrating P2P based UASN. {Drawn based on [30]}

networking. 3D networking has been studied for other extreme environments such as in space and underground. Zheng et al. came up with a new way of connecting deep waters right up to satellites using surface buoys and other devices [31]. The proposed architecture included some surface stations and flexible loading platforms. The flexible loading function is claimed to play an important role in the overall operation and the collection of monitoring data.

Like other sensor networking solutions, apart from extremely large WSNs, which are usually used as a model for research and further development, all small- and medium-scale netted sensors require application- or often scenario-specific network designs. A designer can select a suitable set of partial solutions for three main aspects of the deployment:

- hardware
- communication
- protocols.

In the case of UASNs based upon existing solutions, we need to consider the following requirements:

- self-configuration
- acoustic communication (short-range, limited bandwidth)
- time synchronization (often harsh and unpredictable)
- mostly-off operation
- data catching and forwarding
- energy-aware system design

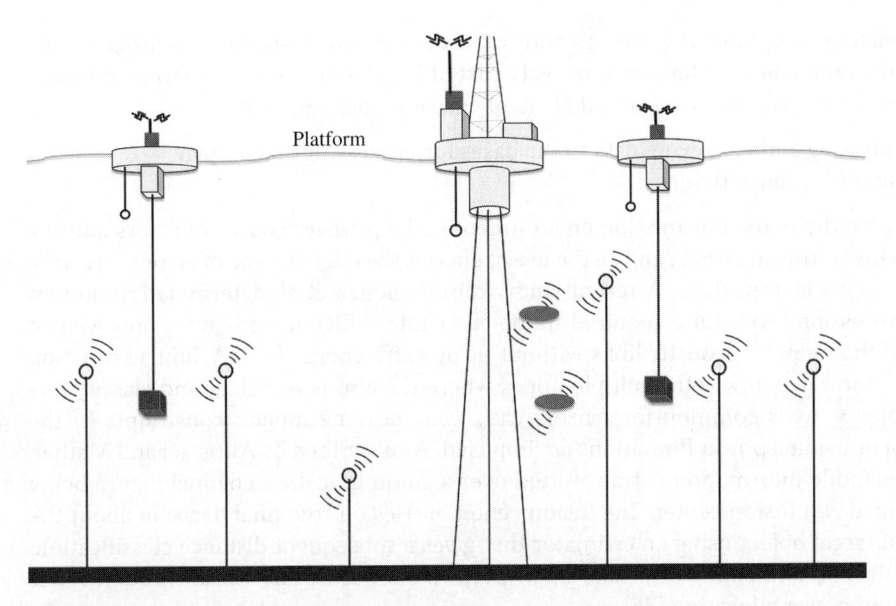

Figure 12.9 A basic core scenario for designing practical applications [32].

- ultra-low duty cycle operations
- low-cost
- MAC protocol
- delay-tolerant protocols.

However, all UWSNs start with the locations of the sensors. One proposal comes from Heidemann et al. [32], as shown in Figure 12.9. The authors propose four different types of nodes for this monitoring solution. At the lowest layer of the network there are a large number of sensor nodes to be deployed on or near the sea floor, usually standing upright near the seabed and equipped with moderate computing power, storage capacity, and possiblywith some sort of energy-harvesting capability. At the upper layers – the platform – there are various ways to connect this network to global networks and the Internet in order to deliver remote control or data storage for further analysis and use. Mobile nodes, shown as oval shapes, either floating or connected to submersible robots and AUVs, can be integrated in the system if required by the application.

In some cases of underwater deployment, visual netted sensors require much greater bandwidth than most single acoustic channel links can offer. For such cases Khan et al. propose a using AUVs to visit and offload the data, images and video files from the sensor nodes [33]. If demand for higher data rates persists then *greedy path planning* may offer a solution.

In order to enhance the throughput and delay performance of acoustic channels we may adopt network coding (NC) in the netted sensors. Manville et al. suggest using NC underwater to create more effective and reliable communication links for UASNs [34]. Considering the extra complexities involved with NC and the proposed forward error correction (FEC) systems, the idea is more applicable to large complex networks.

Supplying energy for the sensors and other devices and systems is regarded as the most difficult, labour-intensive and costly part of UASNs operations. Energy-efficiency is critical, because over 75% of UASN energy use is associated with

- monitoring and conveying data to the bases for processing and storage
- communication activities.

Because the poor transmission environment is the primary source of errors and the associated retransmissions and so the use of energy, the effective use of error correcting codes seems is important. A recent study from de Souza et al. determined minimum retransmission levels for a required frame error rate (FER) demonstrates how we can extend the range of acoustic links without using extra energy [35]. A similar situation applies for a network with multiple nodes where the aim is to detect and classify target objects. As is common for sensors with weak power supplies, constraints by the sum of individual power limitations are imposed. As described by Alirezaei and Mathar, this per-node information is transmitted over a communication channel before being combined at a fusion center. The fusion center carries out the final decision about the type of target object using an estimator that gives a subsequent distance classification. Such object classification can enable high-performance sensor networks through better optimum power allocation [36].

There is still much to do before we can say our netted-sensor-based technological capabilities represent a new UWSN paradigm. Heidemann et al. assess some of requirements that still have not been met or where there is little chance of doing so in the near future. The underwater sensing operations are quite limited. This is because remote controlled UASNs have been deployed mostly in small platforms and submersibles, and active and managed devices, so they are inherently temporary. In terrestrial WSNs, the nodes are cheap, short-range, multi-hop, self-configuring, and have access to utilities for energy; by comparison, underwater acoustic nodes are still expensive, sparsely deployed, and typically communicate directly to a base-station over long distances. For example, one of the potential underwater applications is using permanent seismic sensors for generating real-time images of undersea oilfields, a task currently carried out by a ship towing a large array of hydrophones [37].

Regarding future developments:

- 4-D seismic monitoring is required for judging field performance and motivating intervention.
- Equipment monitoring and control for most underwater establishments is required to be long-term.
- Robotics are maturing rapidly so we should not hesitate to employ flocks of underwater robots.
- Acoustic channels are very poor due to the immaturity of existing technologies.
- Acoustic channels put many constraints on networks of underwater sensor nodes; current protocols cannot handle long-distance communications.

12.5 Typical Underwater Sensing Applications

To show the trends and future potential of underwater systems and services, we now briefly look at a current research in the area.

One interesting development of UWSN technology is to help with maintenance, reducing the cost of resources. Research will increase the quality of services provided by underwater systems. An interesting study from Delauney [38], on systems for cleaning environmental monitoring networks, suggests solutions that will not only improve the performance of UWSNs, but can be applied to all underwater equipment subject to biofouling. Biofouling, short for 'biological fouling', is the undesirable accumulation of microorganisms or algae on wetted surfaces. It happens all the time and very seriously in hot climates. Without proper measures it is a serious problem for long-term underwater monitoring systems and devices.

Node mobility and density are two parameters that vary for different UWSNs. UWSNs are usually static, with the nodes attached to docks, to anchored buoys or fixed to the seafloor. Semi-mobile UWSNs are suspended from buoys on a temporary basis or for long durations, their network topology allowing for efficient connectivity. Mobile UWSNs, however, require dynamic networking and there are technical and performance challenges for localization and maintaining a connected network [39].

The next two subsections cover mobile monitoring vehicles and fixed systems, also known as *developing platforms*.

12.5.1 Monitoring Vehicles Approach

One practical to inspect and monitor underwater is making use of sensor-rich smart and autonomous vehicles. Unmanned underwater vehicles (UUVs), also known as underwater drones, are vehicles that can operate underwater without being driven by a human inside them. They are divided into two categories: remotely operated underwater vehicles (ROVs) and AUVs, which operate independently and are also called 'underwater robots'. These two groups come with some major differences in the logistics of deployment and minor differences in the technologies used, the costs and the maintenance.

One of the key components of navigation for these vehicles is the PZT sensor. These come in different forms for different applications. Inspired by the blindfish imaging mechanism, Asadnia et al. developed thin-film piezoelectric pressure sensor arrays for passive fish-style underwater sensing [40]. AUVs normally make use of such sensor arrays to monitor their movements, the surrounding flows, and other objects. The individual sensing elements of the array respond to the relative motion of the body of the vehicle and the surrounding water, acting as flow sensors. As the vehicle glides by an underwater object, flow-pressure variations occur on the body due to the presence of the object, and these depend on the shape and position of the object. The presence of surrounding objects is therefore detected from analysis of the water disturbances generated by moving objects, so the process is also called 'touch at a distance' sensing.

Underwater gliders are AUVs that use small changes in buoyancy in conjunction with wings to convert vertical motion to horizontal. This approach makes them less expensive and much more energy efficient. They have therefore been very useful in underwater operations. One example is using their passive acoustic capabilities to identify sea traffic [41]. This is done using a small array of 3D acoustic antennas with beamforming and multipath beam cross-correlation. Such systems can detect ships precisely up to 200 m away, and give good estimates up to 850 m.

It may be worth mentioning that sonar can be integrated with communication systems. Sonar (sound navigation and ranging) is an acoustic version of radar (radio navigation and ranging) that is used for underwater sensing applications such as detection and tracking of moving objects. It can be also used for submarine navigation and communication with or detection of objects. It comes in two forms: passive, for listening only, and active, for analysing echoes of transmitted signals. Sonar technology is used in most AUVs and gliders. For example, for detecting specific small underwater objects, Crosby and Cobb [42] claim that underwater object detection is tedious and time-consuming, and if the objects are mines, they are dangerous to the systems and people involved in surveillance, investigation and exploration. AUVs therefore should carry short-range but very high-resolution sonar, the operation of which can be improved through signal processing: pattern classification for artifacts, platform motion and sonar beam patterns. These enhancements would remove litter from the image of the ocean bottom prior to sensing and identifying the objects.

Further advances of sonar technology for AUV path planning and mine-detection comes from Paull et al., who make use of side-looking sensors (SLSs). Many underwater mine countermeasure operations can be conducted with an SLS. There are two possible approaches: synthetic aperture sonar (SAS) or side-scan sonar (SSS). SAS is similar to analogue radar, but adaptive, and therefore requires further signal processing. SSS has been adopted in this work as it returns emitted high-frequency sounds to generate an image of the seabed. The main detection comes from analysis of the image of an object sitting on the seabed, casting a sonar-style shadow. This can be processed and compared to a selected database of objects, such as, for example, mines. The on-board SSS sensors gather sufficient image data as the AUV moves along its path [43].

12.5.2 Developing Platforms Approach

One of the developments in underwater technologies is the use of underwater platforms. The idea is to employ a relatively small unit, rich with sensing technologies as a smart base in a limited area. An example is a technological enabling facilitator, composed of one or a few cooperating platforms. These platforms enable ad hoc networking at various service level connections, so that they can provide shared common observation and monitoring services to users: industrial, scientific and so on. Most applications we examine here are directional and make use of dynamic smart wireless sensors.

Platforms can be used in pilot studies and normally they perform the initial investigations for large projects, such as habituation and acclimatisation of a new environment for penetrating specific developments becomes an important point for trying and experimenting new technologies and services. Platforms can be designed specifically for supporting a single task/services or can be multifunctional, used for many tasks or services. Multifunctional platforms may:

- act as common bases for networking information collection and maintenance
- act as access points for attachable systems and devices (gliders and AUVs)
- facilitate extendable applications and services
- provide shared computing and sensing resources.

Carevic describes a small single-service platform for detecting and tracking multiple underwater targets. Called an observation platform, it uses 13 directional sensors. The layout is as shown in Figure 12.10, with eight sensors shown [44].

In order to use the common measurement method of TDOA for the passive transient signals, the responses of the sensors had directional patterns in the relative position of

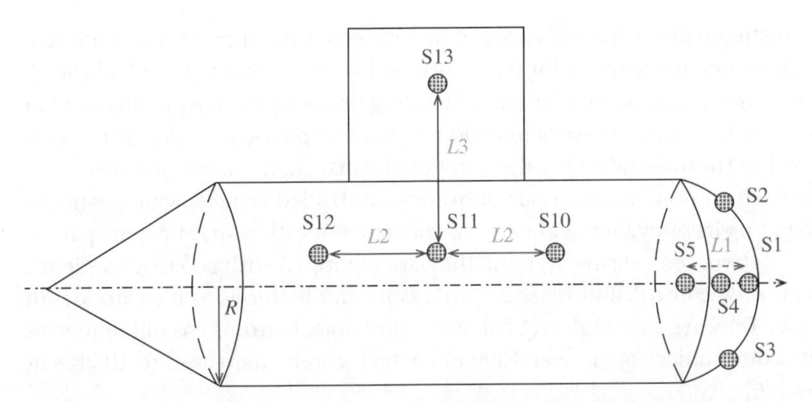

Figure 12.10 A geometric 3D observation model. Thirteen directional sensors are in use (S1–S13): S1 and four surrounding sensors (S2–S5) are on the head. The four remaining sensors (S6–S9) are similar to (S10–S13) but located on the other side of the ship [44].

the source receiving its own radiated signal. The task of automatic estimation of the positions and velocities of moving targets in noisy and harsh underwater environments is quite tedious. A sophisticated special likelihood method is used. The measured data is regarded as a single target and a special modeling process identifies the presence and absence of all targets by analysing the clutter.

It is also interesting to look at practical problems indicated by Zhang et al. whilst carrying out some simple experiments [45]:

- Correct positioning of the nodes to keep them within transmission range is quite difficult; self-deployment and self-configuration would help.
- Reflection of underwater acoustic signals is very severe, causing many systems not to perform when there are substantial errors caused by a reflected arrival measurement signal.
- Online debugging and logging are important for platforms that are under development or that are evolutionary in nature or that deal with real-time information, examining the usefulness of monitoring and collected data.
- Energy and low battery issues are not just annoying but could affect the performance of sensing, and making communication and signal processing tasks unreliable.

Complex platform deployment would br a better and more sustainable way of deploying long-term underwater projects for applications such as monitoring, pollution control, climate recording, prediction of natural disturbances, search and survey missions, and study of marine life. That is, smart underwater sensing could become focal point of a promising futuristic ocean-observation complex. Heidemann et al. analyse the progress of the field and the associated deployment issues of underwater sensor networks on static platforms. Basically, all sensing platforms require underwater sensors, which are mostly common and therefore easy to work with and inexpensive: pressure sensors for depths, photodiodes and thermistors for ambient light and temperature and so on. However, many can become difficult and expensive, including those for measuring carbon dioxide concentrations and turbidity in water, and sonar sensors to detect underwater objects [39].

Another interesting work is from Kastner et al., on sensor platforms for multimodal monitoring underwater, particularly for observation of whales, sharks, and plankton. A fixed or mobile platform that is used for understanding underwater ecosystems, needs a multimodal underwater sensor to work in various cyber and physical modes. These sensor platforms include both aerial and underwater platforms. Their underwater platforms include a drifter, which is an autonomous buoyancy-controlled vehicle, which controls its depth by changing its buoyancy against the ocean currents. They may become part of a true collective networked sensing system. They are equipped with acoustic modems, a piston for buoyancy control, and sensors, processors and batteries to measure depth and temperature. They are able to detect fish and other objects from tens of kilometres away, with data communication at over 1000 b/s, which can be increased to 10 kb/s at short distances [46].

Another interesting passive acoustic platform technology is from Toma et al. [1]. Their multi-platform passive underwater acoustics instrument is used for more cost-efficient assessment of ocean ecosystems. Developing next-generation low-cost multifunctional web enabled ocean sensor systems empowers marine, maritime and fisheries management. The passive acoustic platform is feasible for observation and monitoring applications. These applications include detection of fish reproduction areas, detection of greenhouse gases, rainfall estimation, seismic events detection, ice cracking, thermometry and tomography. Figure 12.11 is futuristic view of passive acoustics applications.

At the core of this work they present their 'smart electronic interface for a sensors and instruments system' with the data collection system.

Figure 12.11 Seascape acoustic environment. Courtesy of NeXOS, a collaborative project funded by the European Commission [47].

12.6 Summary

The vastness of the waters covering the earth has become the center of attention for many scientists and engineers, who are starting a new systematic exploration of the possibilities of smart sensor technologies. This chapter is one of many recent attempts to look at the water with the positive aim of understanding its potential through new technological developments, bringing a new paradigm of applications for civilised use. For this we have examined recent developments in wireless underwater sensing with a specific interest in UASNs. The complex task of surveying such a vast and complex environment through smart sensing is now dependent on our transmission technologies, which due to the extreme underwater environment still have far to go.

The conquest of the underwater environment through smart sensing cannot be achieved in isolation but only in association with other technologies, some of which are described in other chapters in this book. Nodes for sensing and accumulation of information with minimum energy consumption, link capabilities for communications and data transmission in UASNs, speed of signals, delay, energy, and lifetimes, can all be enhanced. However, we need to find technological shortcuts, such as more advanced coding, new modulation methods, and multilink communication techniques. At the application level, have looked at typical underwater sensing systems for monitoring, such as UUVs, AUVs and ROVs, and have given an analysis of a new platform-based approach.

References

1 D.M. Toma, J. d. Río, N. Carreras, L. Corradino, P. Braulte, E. Delory, A. Castro and P. Ruiz, 'Multi-platform underwater passive acoustics instrument for a more cost-efficient assessment of ocean ecosystems,' in *2015 IEEE International Instrumentation and Measurement Technology Conference (I2MTC)*, Pisa, 2015.

2 M. Stojanovic, *The Wiley Encyclopedia of Electrical and Electronics Engineering*, John Wiley, 1999.

3 P.C. de Freitas, '*Evaluation of Wi-Fi underwater networks in freshwater*,' MSc Thesis, Universidade do Porto, 2014.

4 D.K. Shlomi Arnon, 'Non-line-of-sight underwater optical wireless communication network,' *Optical Society of America A*, vol. 26, pp. 530–539, 2009.

5 G.I. Allen, R. Matthews and M. Wynn, 'Mitigation of Platform generated magnetic noise impressed on a magnetic sensor mounted in an autonomous underwater vehicle,' in *OCEANS, 2001. MTS/IEEE Conference and Exhibition*, Honolulu, 2001.

6 H. Guo, J.S. Sun and N.M. Litchinitser, 'M2I: Channel modeling for metamaterial-enhanced magnetic induction communications,' *IEEE Transactions on Antennas and Propagation*, vol. 63, no. 11, pp. 5072–5087, 2015.

7 L. Emokpae, 'Design and analysis of underwater acoustic networks with reflected links,' PhD Thesis, University of Maryland, Baltimore, 2013.

8 N.A.B. Idrus, 'The performance of directional flooding routing protocol for underwater sensor networks,' PhD Thesis, Universiti Tun Hussein Onn Malaysia, 2015.

9 R. Jurdak, C.V. Lopes and P. Baldi, 'Software acoustic modems for short range mote-based underwater sensor networks,' in *OCEANS 2006–Asia Pacific*, 2006.

10 Y. Yang, Z. Xiaomin, P. Bo and F. Yujing, 'Design of sensor nodes in underwater sensor networks,' in *4th IEEE Conference on Industrial Electronics and Applications*, 2009.
11 G. Alirezaei, O. Taghizadeh and R. Mathar, 'Optimum power allocation with sensitivity analysis for passive radar applications,' *IEEE Sensors Journal*, vol. 14, no. 11, pp. 3800–3809, 2014.
12 G. Alirezaei, O. Taghizadeh and R. Mathar, 'Comparing several power allocation strategies for sensor networks,' in *The 20th International ITG Workshop on Smart Antennas (WSA'16)*, Munich, 2016.
13 N. Zou, C.C. Swee and B.A. L. Chew, 'Vector hydrophone array development and its associated DOA estimation algorithms,' in *OCEANS 2006–Asia Pacific*, 2006.
14 D. Billon and B. Quellec, 'Performance of high data rate acoustic underwater communication systems using adaptive beamforming and equalization,' in *Proceedings of OCEANS '94, Oceans Engineering for Today's Technology and Tomorrow's Preservation*, Brest, 1994.
15 D. Chizhik, A.P. Rosenberg and Q. Zhang, 'Coherent and differential acoustic communication in shallow water using transmitter and receiver arrays,' in *OCEANS 2010*, Sydney, 2010.
16 A.L. Butler, J.L. Butler, J.A. Rice, W. Dalton, J. Baker and P. Pietryka, 'A tri-modal directional modem transducer,' in *OCEANS 2003 Proceedings*, San Diego, 2003.
17 I.F. Akyildiz, D. Pompili and T. Melodia, 'Underwater acoustic sensor networks: research challenges,' *Ad Hoc Networks*, vol. 3, no. 3, pp. 257-279, 2005.
18 M.A. Pedersen, 'Normal-mode and ray theory applied to underwater acoustic conditions of extreme downward refraction,' *The Journal of the Acoustical Society of America*, pp. 323–368, 1972.
19 M.B. Porter, 'Acoustic models and sonar systems,' *IEEE Journal of Oceanic Engineering*, vol. 18, no. 4, pp. 425–437, 1993.
20 L. Emokpae and M. Younis, 'Surface based underwater communications,' in *Global Telecommunications Conference (GLOBECOM 2010)*, Miami, 2010.
21 F.B. Jensen, W.A. Kuperman, M.B. Porter, H. Schmidt, *Computational Ocean Acoustics*, Springer, 2011.
22 C.T. Tindle and M.J. Murphy, 'Microseisms and ocean wave measurements,' *IEEE Journal of Oceanic Engineering*, vol. 24, no. 1, pp. 112–115, 1999.
23 F. Shulz, R. Weber, A. Waldhorst and J. Bohme, 'Performance enhancement of blind adaptive equalizers using environmental knowledge,' in *Proc. of the IEEE/OES OCEANS Conference*, San Diego, 2003.
24 A. Jarrot, C. Ioana and A. Quinquis, 'Denoising underwater signals propagating through multi-path channels,' in *Europe Oceans 2005*, 2005.
25 G. Zhang, J.M. Hovem, H. Dong and L. Liu, 'Experimental studies of underwater acoustic communications over multipath channels,' in *Fourth International Conference on Sensor Technologies and Applications (SENSORCOMM)*, 2010.
26 A. Essebar, G. Loubet and F. Vial, 'Underwater acoustic channel simulations for communication,' in *Proc. of the IEEE/OES OCEANS'94 Conference*, 1994.
27 G.S. Sineiro, '*Underwater multimode directional transducer evaluation*,' MSc Thesis, Naval Postgraduate School, Monterey, 2003.

28 M. Erol-Kantarci, H.T. Mouftah and E. Oktug, 'Localization techniques for underwater acoustic sensor networks,' *IEEE Communications Magazine*, vol. 48, no. 12, pp. 152–158, 2010.

29 M. Xu and G. Liu, 'Design of a P2P based collaboration platform for underwater acoustic sensor network,' in *The 11th International Symposium on Communications & Information Technologies (ISCIT 2011)*, 2011.

30 F. Xu, R. Li, C. Zhao, H. Yao and J. Zhang, 'Congestion-aware signaling aggregation scheme for cellular based underwater acoustic sensor network,' in *IEEE ICC 2015–Workshop on Radar and Sonar Networks*, 2015.

31 J. Zheng, S. Zhou, Z. Liu, S. Ye, L. Liu and L. Yin, 'A New underwater sensor networks architecture,' in *IEEE International Conference on Information Theory and Information Security (ICITIS)*, 2010.

32 J. Heidemann, Y. Li, A. Syed, J. Wills and W. Ye, 'Underwater sensor networking: research challenges and potential applications,' USC/ISI, 2005.

33 F.A. Khan, S.A. Khan, D. Turgut and L. Boloni, 'Greedy path planning for maximizing value of information in underwater sensor networks,' in *IEEE 39th Conference on Local Computer Networks Workshops (LCN Workshops)*, Edmonton, 2014.

34 C. Manville, A. Miyajan, A. Alharbi, H. Mo, M. Zuba and J.-H. Cui, 'Network coding in underwater sensor networks,' in *OCEANS 2013*, Bergen, 2013.

35 F.A. d. Souza, B.S. Chang, G. Brante, R.D. Souza, M.E. Pellenz and F. Rosas, 'Optimizing the number of hops and retransmissions for energy efficient multi-hop underwater acoustic communications,' *IEEE Sensors Journal*, vol. 16, no. 10, pp. 3927–3938, 2016.

36 G. Alirezaei and R. Mathar, 'Optimum power allocation for sensor networks that perform object classification,' *IEEE Sensors Journal*, vol. 14, pp. 3862–3873, Nov 2014.

37 Heidemann, W. Ye, J. Wills, A. Syed and Y. Li, 'Research challenges and applications for underwater sensor networking,' in *IEEE Wireless Communications and Networking Conference, 2006. WCNC 2006*, 2006.

38 L. Delauney, 'Biofouling protection for marine underwater observatories sensors,' in *OCEANS 2009–EUROPE*, 2009.

39 J. Heidemann, M. Stojanovic and M. Zorzi, 'Underwater sensor networks: applications, advances and challenges,' *Phil. Trans. R. Soc. A*, vol. 370, pp. 158–175, 2012.

40 M. Asadnia, A.G. P. Kottapalli, Z. Shen, J. Miao and M. Triantafyllou, 'Flexible and surface-mountable piezoelectric sensor arrays for underwater sensing in marine vehicles,' *IEEE Sensors Journal*, vol. 13, no. 10, pp. 3918–3925, 2013.

41 A. Tesei, R. Been, D. Williams, B. Cardeira, D. Galletti, D. Cecchi, B. Garau and A. Maguer, 'Passive acoustic surveillance of surface vessels using tridimensional array on an underwater glider,' in *OCEANS 2015*, Genova, 2015.

42 F. Crosby and J.T. Cobb, 'Sonar Processing for short range, very-high resolution autono- mous underwater vehicle sensors,' in *Proceedings of OCEANS 2005*, 2005.

43 L. Paull, S. Saeedi, M. Seto and H. Li, 'Sensor-driven online coverage planning for autonomous underwater vehicles,' *IEEE/ASME Transactions on Mechatronics*, vol. 18, no. 6, pp. 1827–1838, 2013.

44 D. Carevic, 'Detection and tracking of underwater targets using directional sensors,' in *Proceedings of the Intelligent Sensors, Sensor Networks and Information (ISSNIP'07)*, Melbourne, 2007.

45 K. Zhang, S. Climent, N. Meratnia and P.J. M. Havinga, 'Practical problems of experimenting with an underwater wireless sensor node platform,' in *Seventh International Conference on Intelligent Sensors, Sensor Networks and Information Processing (ISSNIP)*, 2011.

46 R. Kastner, A. Lin, C. Schurgers, J. Jaffe, P. Franks and B.S. Stewart, 'Sensor Platforms for multimodal underwater monitoring,' in *Green Computing Conference (IGCC), 2012 International*, San Jose, 2012.

47 E.Q. Gutiérrez, 'http://www.nexosproject.eu/sites/default/files/Factsheet_1st_update,' 27 July 2016. URL: http://www.nexosproject.eu.

13

Underwater Anchor Localization Using Surface-reflected Beams

Lloyd Emokpae

U.S. Naval Research Laboratory, Code 7160 – Acoustics Division, USA

13.1 Introduction

In this chapter, we focus on using reference nodes that have already been located using an appropriate anchor-free localization scheme to locate a lost node that has drifted away from the network. We consider an underwater acoustic sensor network (ASN) that operates in shallow water, as depicted in Figure 13.1. The ASN consists of a base station (BS) node, geographically positioned (GP) nodes and lost-drifted (LD) nodes. The goal of the UREAL[1] algorithm [1] is to locate all LD nodes; it is assumed that we will only locate one LD node at a time. The BS node resides close to the water's surface, using its antenna array to periodically measure the water surface function, as illustrated in Figure 13.1. Both the BS and GP nodes are equipped with directional piezoelectric underwater transducers, similar to the ones mentioned in Section 12.3.1, which can be configured to allow for both elevation and azimuth angle-of-arrival (AOA) angle measurements [2, 3]. LD nodes are only equipped with an omni-directional transducer.

The reflection points on the water's surface (and the BS) can be used as reference points to solve a standard triangulation problem. So, given our ASN model for the UREAL algorithm in Figure 13.1, the position of the ith GP node (GP_{xi}, GP_{yi}, GP_{zi}) is determined by evaluating the multilateration expression:

$$\begin{bmatrix} (X_1 - GP_{xi})^2 + (Y_1 - GP_{yi})^2 + (Z_1 - GP_{zi})^2 \\ (X_2 - GP_{xi})^2 + (Y_2 - GP_{yi})^2 + (Z_2 - GP_{zi})^2 \\ \dots \\ (X_n - GP_{xi})^2 + (Y_n - GP_{yi})^2 + (Z_n - GP_{zi})^2 \end{bmatrix} = \begin{bmatrix} d_1^2 \\ d_2^2 \\ \vdots \\ d_n^2 \end{bmatrix} \tag{13.1}$$

where, (X_i, Y_i, Z_i) is the ith reflection point on the water's surface, which is used as a temporary reference point. Equation (13.1) also requires the ranging information $d_i = c * \tau_i/2$ from the GP node to each reflection point, which is known since each GP node maintains its own water surface function estimate.

Thus, after initial anchor-free localization, the BS and GP nodes will be used as reference points to locate the LD node that has drifted away because of the water current.

1 Underwater reflection-enabled acoustic-based localization

Wireless Sensor Systems for Extreme Environments: Space, Underwater, Underground, and Industrial.
First Edition. Edited by Habib F. Rashvand and Ali Abedi.

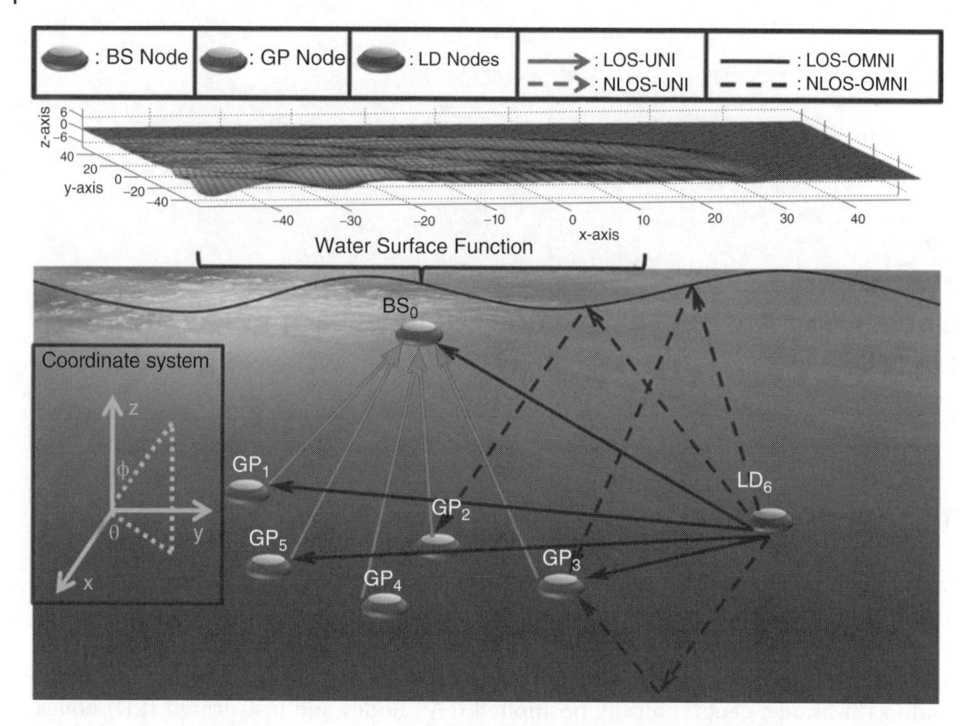

Figure 13.1 Network model of the proposed scheme. UREAL uses both line-of-sight (LOS) and non-line-of-sight (NLOS) AOA range information to locate an LD node that has drifted away from the ASN. The water surface function is used for NLOS position estimation.

In this case, we now only rely on the surface function from the BS node. The lost or LD node will send broadcast omni-directional ping messages with a reference packet number, and this will reach a subset of reference nodes (BS and GPs). Each GP node then will classify the link as line-of-sight (LOS) or surface-reflected non-line-of-sight (NLOS). In this phase, the GP nodes will not need to maintain a water surface function since the BS node will assume that responsibility. Each GP node will also determine the azimuthal (θ) and elevation (ϕ) AOAs of the received ping messages from the lost node. Depending on θ and ϕ, the GP will classify the link to the LD node as LOS or NLOS; in other words, pointing towards the surface. A collection of the classified links (LOS and NLOS) and AOAs will then be sent to the BS node to be used to centrally locate the drifted node. The calculated node position will then be broadcast throughout the network until it reaches the LD node. This proposed underwater reflection-enabled acoustic-based localization scheme (UREAL) is summarised in Figure 13.1. For the sake of clarity, we make the following assumptions:

- The proposed UREAL scheme will only locate one LD node at a time.
- In the case that multiple LD nodes want to join the ASN, the BS and GP nodes can differentiate them using the packet header information.
- The positions of the BS and GP nodes are determined a priori at network setup by applying a suitable anchor-free location scheme [4], as illustrated in Figure 13.2. The position information is maintained throughout the life of the ASN.

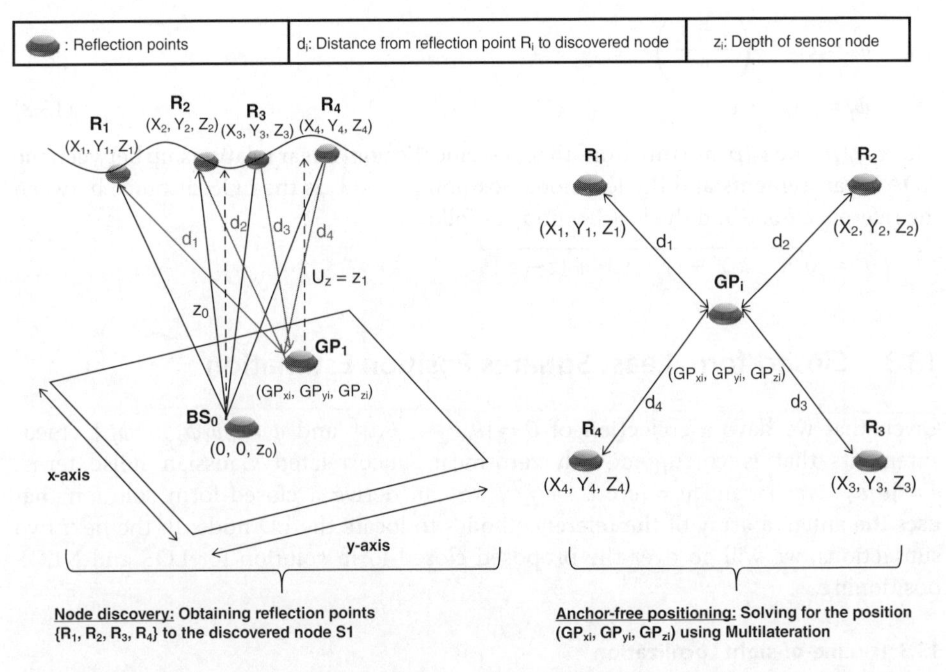

Figure 13.2 An anchor-free localization scheme used to locate the GP nodes. The located BS and GP nodes will then be used as reference nodes to locate an LD node that has drifted away from the ASN.

In the three Sections, we will go over the UREAL method in detail.

13.2 UREAL Angle of Arrival Measurements

As mentioned earlier, traditional relative localization algorithms work by using LOS range measurements. To locate the LD node that has drifted away, we propose an AOA-based closed-form solution that aims to use the antenna array of the reference GP and BS nodes in the localization process.

According to Zamora et al. [5], AOA ranging restricts the source location along the line of bearing (LOB). Hence the location of the source node is obtained from the intersection of multiple LOBs to each reference node. Due to the 3D nature of the underwater environment, we assume that azimuth (θ) and elevation (ϕ) AOA measurements are obtained from up to N reference BS and GP nodes. This information, along with the classified link type, is sent to the BS. We also assume that the AOAs $\{\theta_i, \phi_i\}$ measured at each reference node will be subject to errors denoted by $\{\varepsilon_i, e_i\}$. Thus the measured AOAs between each ith reference node at known position $\boldsymbol{p_i} = [x_i y_i z_i]^T$ and lost node at unknown position $\boldsymbol{p} = [x\ y\ z]^T$ can be expressed as:

$$\theta_i = \tan^{-1}\left(\frac{y - y_i}{x - x_i}\right) + \varepsilon_i$$

$$\theta_i = f_i(\boldsymbol{p}) + \varepsilon_i \tag{13.2}$$

$$\phi_i = \cos^{-1}\left(\frac{z - z_i}{r_i}\right) + e_i$$
$$\phi_i = g_i(\boldsymbol{p}) + e_i \tag{13.3}$$

where, $f_i(\boldsymbol{p})$ and $g_i(\boldsymbol{p})$ are functions that describe the non-linear relationship between the AOA measurements and the lost node position $\boldsymbol{p}$, and r_i is the LOS distance between the reference node and the lost LD node as follows:

$$r_i = \sqrt{(x - x_i)^2 + (y - y_i)^2 + (z - z_i)^2}$$

13.3 Closed-form Least Squares Position Estimation

Given that we have a collection of $\boldsymbol{\theta} = [\theta_1\theta_2 \ldots \theta_N]^T$ and $\phi = [\phi_1\phi_2 \ldots \phi_N]^T$ measurements that is corrupted with zero-mean uncorrelated Gaussian noise terms $\varepsilon = [\varepsilon_1\varepsilon_2 \ldots \varepsilon_N]^T$ and $\boldsymbol{e} = [e_1 e_2 \ldots e_N]^T$, we can derive a closed-form solution that uses the antenna array of the reference nodes to locate the LD node. In the next two subsections, we will go over the proposed closed-form solution for LOS and NLOS positioning.

13.3.1 Line-of-sight Localization

As mentioned above, we assume that azimuth and elevation AOA measurements are obtained from all N reference nodes. This information, along with the classified link type, is sent to the BS. Assuming LOS link classification for up to N reference nodes, as in Figure 13.3, we arrange the AOA measurements into two vectors, which results in:

$$\boldsymbol{\theta} = \boldsymbol{f}(\boldsymbol{p}) + \varepsilon \tag{13.4}$$
$$\phi = \boldsymbol{g}(\boldsymbol{p}) + \boldsymbol{e} \tag{13.5}$$

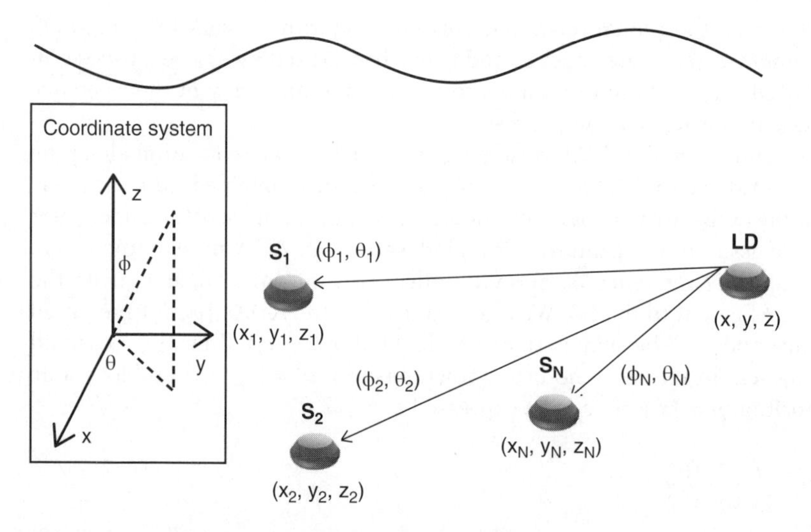

Figure 13.3 LOS position estimation, showing the collection of angles from each sensor node.

such that$\boldsymbol{\theta} = [\theta_1\theta_2 \dots \theta_N]^T$, $\varepsilon = [\varepsilon_1\varepsilon_2 \dots \varepsilon_N]^T$ and $\phi = [\phi_1\phi_2 \dots \phi_N]^T$, and $\boldsymbol{e} = [e_1e_2 \dots e_N]^T$. Thus, estimating both $\boldsymbol{f}(\boldsymbol{p})$ and $\boldsymbol{g}(\boldsymbol{p})$ will allow us to solve for the position of the lost node. To determine $\boldsymbol{f}(\boldsymbol{p})$, we linearise the non-linear AOA function by performing a Jacobian linear map described by $\boldsymbol{J_F}(\boldsymbol{p})$, which can be used to give us the best linear approximation for $\boldsymbol{f}(\boldsymbol{p})$ near the reference point $\boldsymbol{p}_0$ as:

$$\boldsymbol{f}(\boldsymbol{p}) \approx \boldsymbol{f}(\boldsymbol{p}_0) + \boldsymbol{J_F}(\boldsymbol{p}_0)(\boldsymbol{p} - \boldsymbol{p}_0) \tag{13.6}$$

The Jacobian matrix of a 3D spherical coordinate system with:

$$x = r\sin\phi\cos\theta \qquad y = r\sin\phi\sin\theta \qquad z = r\cos\phi$$

is defined as:

$$\boldsymbol{J_F}(\boldsymbol{p}) = \frac{\partial x, y, z}{\partial r, \phi, \theta} = \begin{bmatrix} \frac{\partial x}{\partial r} & \frac{\partial x}{\partial \phi} & \frac{\partial x}{\partial \theta} \\ \frac{\partial y}{\partial r} & \frac{\partial y}{\partial \phi} & \frac{\partial y}{\partial \theta} \\ \frac{\partial z}{\partial r} & \frac{\partial z}{\partial \phi} & \frac{\partial z}{\partial \theta} \end{bmatrix}$$

$$\boldsymbol{J_F}(\boldsymbol{p}) = \begin{bmatrix} \sin\phi\cos\theta & r\cos\phi\cos\theta & -r\sin\phi\sin\theta \\ \sin\phi\sin\theta & r\cos\phi\sin\theta & r\sin\phi\cos\theta \\ \cos\phi & -r\sin\phi & 0 \end{bmatrix} \tag{13.7}$$

with unit vectors:

$$v_1 = [\cos\phi\cos\theta\cos\phi\sin\theta\ -\sin\phi]^T \tag{13.8}$$

$$v_2 = [-\sin\phi\sin\theta\sin\phi\cos\theta\ 0]^T \tag{13.9}$$

If we rearrange the system of equations corresponding to the linear estimation and select the second unit vector needed to approximate $\boldsymbol{f}(\boldsymbol{p})$, we get the following linear system:

$$\boldsymbol{b}(\boldsymbol{\phi}, \boldsymbol{\theta}) = \boldsymbol{H}(\boldsymbol{\phi}, \boldsymbol{\theta})\cdot\boldsymbol{p} \tag{13.10}$$

where,

$$\boldsymbol{H}(\boldsymbol{\phi}, \boldsymbol{\theta}) = \begin{bmatrix} -\sin\phi_1\sin\theta_1 & \sin\phi_1\cos\theta_1 & 0 \\ \dots & \dots & \dots \\ -\sin\phi_N\sin\theta_N & \sin\phi_N\cos\theta_N & 0 \end{bmatrix}$$

and

$$\boldsymbol{b}(\phi, \boldsymbol{\theta}) = \begin{bmatrix} -x_1\sin\phi_1\sin\theta_1 + y_1\ \cos\phi_1\sin\theta_1 - z_1 0 \\ \dots \\ -x_N\sin\phi_N\sin\theta_N + y_N\sin\phi_N\cos\theta_N - z_N 0 \end{bmatrix}$$

Hence, we estimate the (x, y) position of the lost node $\boldsymbol{p} \approx \tilde{\boldsymbol{p}} = [x\ y\ ?]^T$ by computing the least squared solution to (13.10) as:

$$\tilde{\boldsymbol{p}}_{\text{LOS}} = (\boldsymbol{H}(\boldsymbol{\phi\theta})^T\boldsymbol{H}(\boldsymbol{\phi}, \boldsymbol{\theta}))^{-1}\boldsymbol{H}(\boldsymbol{\phi}, \boldsymbol{\theta})^T\boldsymbol{b}(\boldsymbol{\phi}, \boldsymbol{\theta}) \tag{13.11}$$

The z-coordinate of the lost node can be solved by substituting the solution to (13.11) into r_i and into (13.5), which gives us an approximation for $\mathbf{g}(\mathbf{p})$. We then solve the resulting quadratic equation for the unknown z-coordinate resulting in:

$$z = \frac{-2z_i \pm \sqrt{(2z_i)^2 - 4(z_i^2 - \beta_i)}}{2} \tag{13.12}$$

for:

$$\beta_i = \frac{((x - x_i)^2 + (y - y_i)^2)\cos^2\phi_i}{1 - \cos^2\phi_i} \tag{13.13}$$

13.3.2 Non-line-of-sight Localization

Due to the lower transmission loss of water surface links compared to bottom links and the variability of the water surface, we only focus on NLOS positioning for surface-reflected links. Assuming that the links for a set of BS and GP nodes have been classified as surface-reflected NLOS, the goal of the location algorithm will be to estimate the position of the LD node by using only the AOA measurements.

Since the LD node is equipped with an omni-directional transducer, we will have multiple reflection points that correspond to one broadcast signal. Hence, we will need to determine the transmitted vector for each reflection point, as illustrated in Figure 13.4. This will be used in our closed-form expression to solve for the position of the LD node. We can express the spherical unit vector $(\vec{v_i})$ of the received reflected signal for the ith reference node by relation to the Cartesian coordinate system as shown:

$$\vec{v_i} = \begin{bmatrix} \hat{r} \\ \hat{\phi} \\ \hat{\theta} \end{bmatrix} = \begin{bmatrix} \sin\phi_i \cos\theta_i & \sin\phi_i \sin\theta_i & \cos\phi_i \\ \cos\phi_i \cos\theta_i & \cos\phi_i \sin\theta_i & -\sin\phi_i \\ -\sin\phi_i & \cos\theta_i & 0 \end{bmatrix} \begin{bmatrix} \hat{x} \\ \hat{y} \\ \hat{z} \end{bmatrix} \tag{13.14}$$

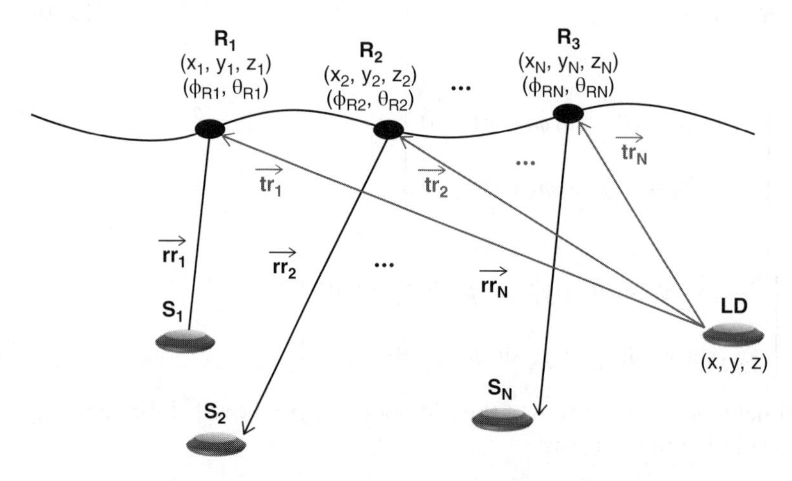

Figure 13.4 4 NLOS position estimation uses the reflection points as reference to solve for LD position

The reflected ray ($\overrightarrow{rr_i}$) for each reference node can then be determined by the following expression:

$$\overrightarrow{rr_i} = p_i + t\overrightarrow{v_i} \tag{13.15}$$

where, p_i is the known position ith reference node, t is the ray parameter, which measures the distance along the ray, and $\overrightarrow{v_i}$ is the direction vector. Also, note that we are using the arrow (→) to denote the vectors that interact with tangent plane $T(v)$ of the known surface function for $v = (x, y, z)$. We can compute the ray parameter t for the intersection point of $\overrightarrow{rr_i}(t)$ to the tangent plane as:

$$T(\overrightarrow{rr_i}) = 0$$

$$t = \frac{(R_i - p_i)\cdot\overrightarrow{n}}{\overrightarrow{v_i}\cdot\overrightarrow{n}} \tag{13.16}$$

such that R_i is the intersection reflection point on the tangent plane that satisfies the equation and $\overrightarrow{n}$ is the normal vector at the intersection point. The normal vector to the known water surface function can be determined by picking out three points (P_1, P_2, P_3) that lie on the tangent plane on the water surface:

$$\overrightarrow{n} = (P_1 - P_2) \otimes (P_1 - P_3) \tag{13.17}$$

where $\otimes$ is the cross-product operator. We then calculate the transmitted reflected vector $\overrightarrow{tr_i} = [x_{ti} y_{ti} z_{ti}]^T$ at the intersection normal vector as follows:

$$\overrightarrow{tr_i} = \overrightarrow{rr_i} - 2(\overrightarrow{rr_i} * \overrightarrow{n})\overrightarrow{n} \tag{13.18}$$

We can now approximate our NLOS AOA ranging from the reflection point to the LD node as:

$$\theta_{Ri}\ (\text{azimuth}) = \tan^{-1}\left(\frac{y_{ti}}{x_{ti}}\right) \tag{13.19}$$

$$\phi_{Ri}\ (\text{elevation}) \approx \cos^{-1}\left(\frac{z_{ti}}{|\overrightarrow{tr_i}|}\right) \tag{13.20}$$

where $|\overrightarrow{tr_i}|$ is the magnitude of the transmitted reflected vector. Hence, if we have a collection of $\boldsymbol{\theta_R} = [\theta_{R1}\theta_{R2} \ldots \ \theta_{RN}]^T$ and $\boldsymbol{\phi_R} = [\phi_{R1}\phi_{R2} \ldots \ \phi_{RN}]^T$ angles, we can estimate the (x, y) position of the lost node $\boldsymbol{p} \approx \tilde{\boldsymbol{p}} = [x\ y\ ?]^T$ by utilising the closed-form least squares expression derived earlier.

$$\tilde{\boldsymbol{p}}_{\mathbf{NLOS}} = (\boldsymbol{H}(\boldsymbol{\phi_R}\boldsymbol{\theta_R})^T\boldsymbol{H}(\boldsymbol{\phi_R}, \boldsymbol{\theta_R}))^{-1}\boldsymbol{H}(\boldsymbol{\phi_R}, \boldsymbol{\theta_R})^T\boldsymbol{b}(\boldsymbol{\phi_R}, \boldsymbol{\theta_R}) \tag{13.21}$$

In a similar fashion to LOS positioning, the z-coordinate of the lost node is determined by utilising Equation 13.12 and substituting a single reference reflected elevation angle ϕ_{Ri} into (13.13) for β_i.

13.4 Prototype Evaluation

In this section, we validate the performance of the localization algorithm for both LOS and NLOS AOA ranging using the prototype shown in Figure 13.5, which consists of a

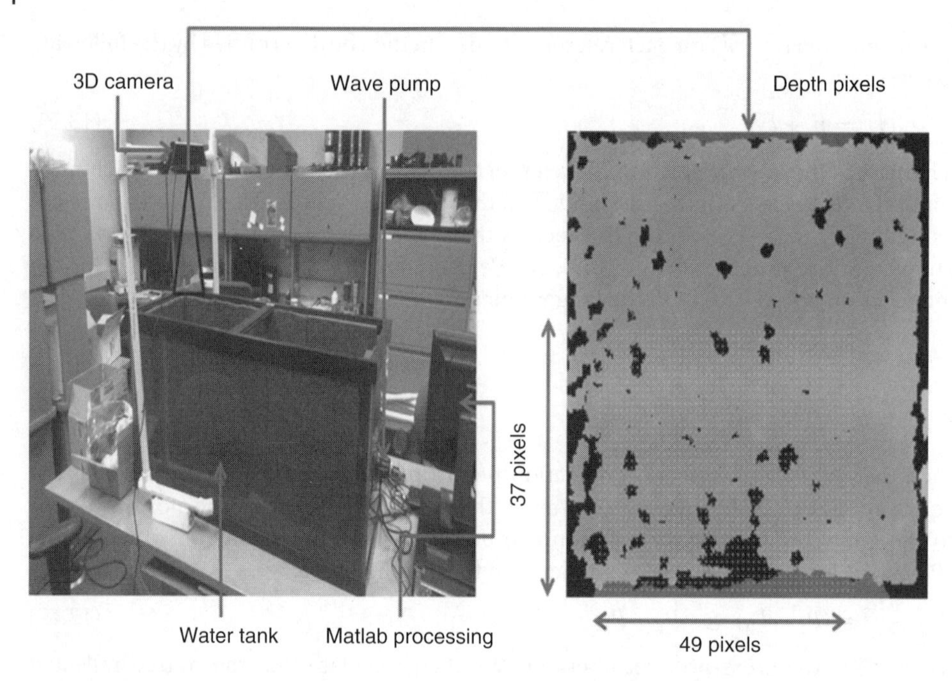

Figure 13.5 Experimental setup with a tank, wave pump and 3D camera. The projected view of the 3D Kinect sensor onto the tarp is used to create a scaled 3D underwater environment.

water tank, Microsoft 3D Kinect camera, water pump and a computer running Matlab. The water pump will be used to generate waves, which will be characterised by the tarp placed on top of the water. The tarp was also positioned directly below the 3D camera, which could then measure the distance to the tarp over time, giving an estimate of the water surface roughness. We define the root-mean-squared (RMS) roughness σ_{RMS} of the 2D sampled water surface as follows:

$$\sigma_{\mathrm{RMS}} = \sqrt{\frac{1}{XY}\sum_{i=1}^{X}\sum_{j=1}^{Y}\sigma_{ij}^{2}(t)} \tag{13.22}$$

such that $\sigma_{ij}(t) = (h_{ij}(t) - \overline{h})$; $h_{ij}(t)$ is the water surface height for the ith and jth sample relative to the mean value of the rough surface height $\overline{h}$. The 3D camera was then calibrated to only see the tarp, which uses 49×37 pixels. The 3D camera also gives the distance to each pixel in millimetres at a sampling rate of 30 Hz. The sampled water surface was then used in the Matlab simulation, taking the entire 49×37 pixels and forming a 3D underwater environment consisting of a $49 \times 37 \times 50$-m cube, a scaled view of our tank setup, where 50 m was the chosen depth of the underwater environment relative to the water surface.

We then simulated an ASN with one lost node and up to eight reference nodes, which includes both BS and GP nodes. All the nodes where randomly placed inside the defined 3D cube. Each node had the same LOS transmission range k_{LOS}, which was used to determine connectivity between nodes. The lost node then broadcast a message, which reached a subset of the reference nodes either through LOS or surface-reflected NLOS.

In both LOS and NLOS cases, the azimuth θ and elevation ϕ AOAs were measured by each reference node; these figures were corrupted by errors with a known variance. We then evaluated the LOS, NLOS and combined LOS and NLOS localization performance by defining the least-squared error function $E(P)$:

$$E(P) = \sum_{i=1}^{M} (\tilde{p}_i - p_i)^2 \tag{13.23}$$

where, $\boldsymbol{p} = [p_1 p_2 \dots p_M]^T$ is a vector of the true lost node position for M simulation runs and $\tilde{\boldsymbol{p}} = [\tilde{p}_1 \tilde{p}_2 \dots \tilde{p}_M]^T$ is another vector of the estimated lost node position after applying the closed-form regression analysis defined in (13.11) and (13.21).

Figure 13.6a shows the observed LOS localization error as we vary the AOA error variance for an average of eighty runs. For this test, we assume that both the azimuth and elevation AOAs will experience the same error variance. We see from Figure 13.6a that as we increase the variance, the localization error generally increases, which is expected, since errors induced in both the azimuth θ and elevation ϕ AOAs will affect the closed-form estimation in (13.11). On the other hand, we notice that as we obtain more reference nodes, the localization error will decrease, despite high AOA variance. This is because we now have more data points, which will be used to increase the position accuracy. Furthermore we note that the localization performances with 6, 9, and 10 reference nodes are relatively close to each other, mainly due to the chosen cubic dimensions of $49 \times 37 \times 50$ m, which limits the number of unique references that can be used for localization. Due to space constraints, we were not able to show results for larger underwater environments. Nevertheless, the results depict a linear improvement in the localization error.

The NLOS localization performance can be seen in Figure 13.6b for an AOA variance range of 0–20°. Similar to the LOS performance, we note that as we increase the number of reference nodes the localization error decreases. More interestingly, we note that the NLOS error for 6, 9 and 12 reference nodes is better than that of the LOS. The increased number of reference points possible for NLOS over LOS explains this phenomenon. Recall from Figure 13.5 that during NLOS positioning, the acoustic signal from the LD node will reflect onto the water's surface at multiple intersection points. This means that the set of intersection points $\{R_1, R_2, \dots, R_r\}$ will be larger than the set of true reference sensor nodes $\{S_1, S_2, \dots, S_s\}$, especially when the surface is not flat (in other words the water is rough). Thus NLOS localization (21) will outperform LOS when we have more reflection points than true reference sensor nodes.

Figure 13.7 shows the combined LOS and NLOS localization error as we increase the AOA variance, when averaged over three runs. The plot essentially shows that when combining LOS and NLOS, the localization performance will be bounded by the NLOS, especially if the water surface is rough. In other words, the multipath AOA variation of the acoustic channel will dominate any variation seen in the direct path. Another interesting study would be to see the effects of the water surface roughness on the localization error by utilising the sampled water surface function obtained from the 3D camera. Figure 13.8 gives an estimated localization performance of the NLOS error for 9 and 12 reference nodes. The error assumes that the AOA variance is proportional to the water-surface roughness. The left-hand image gives the water surface sample from the 3D camera for the tenth frame, the top-right plot gives the water surface roughness over

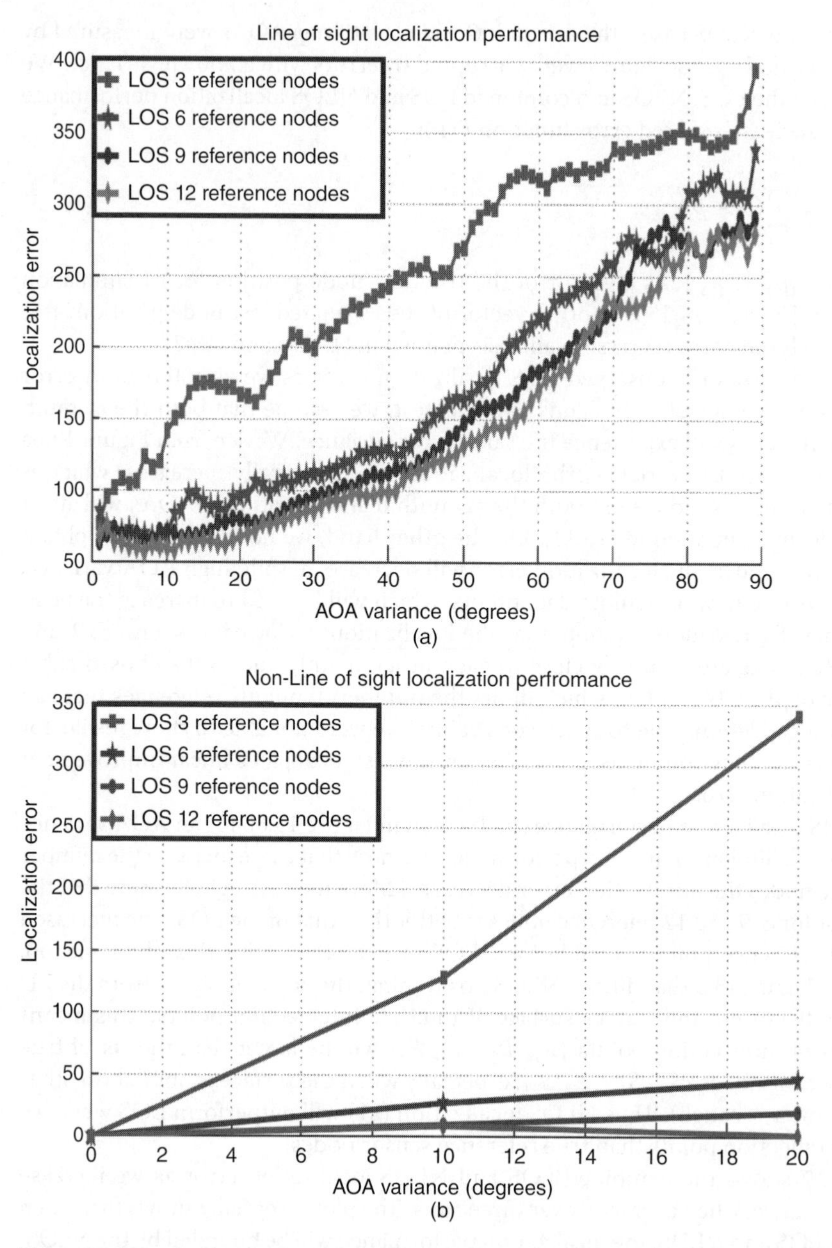

Figure 13.6 Localization performance as we vary the AOA variance: (a) for LOS operation (b) for NLOS operation. Results depict low localization error when we have enough reference nodes. The NLOS performance is better than LOS since the number of reflection points exceeds the true number of reference nodes.

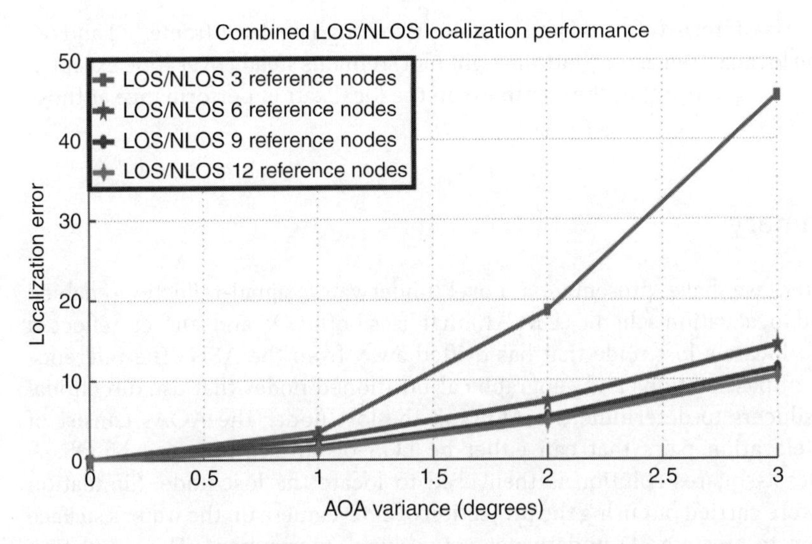

Figure 13.7 Combined localization error for varying number of reference nodes.

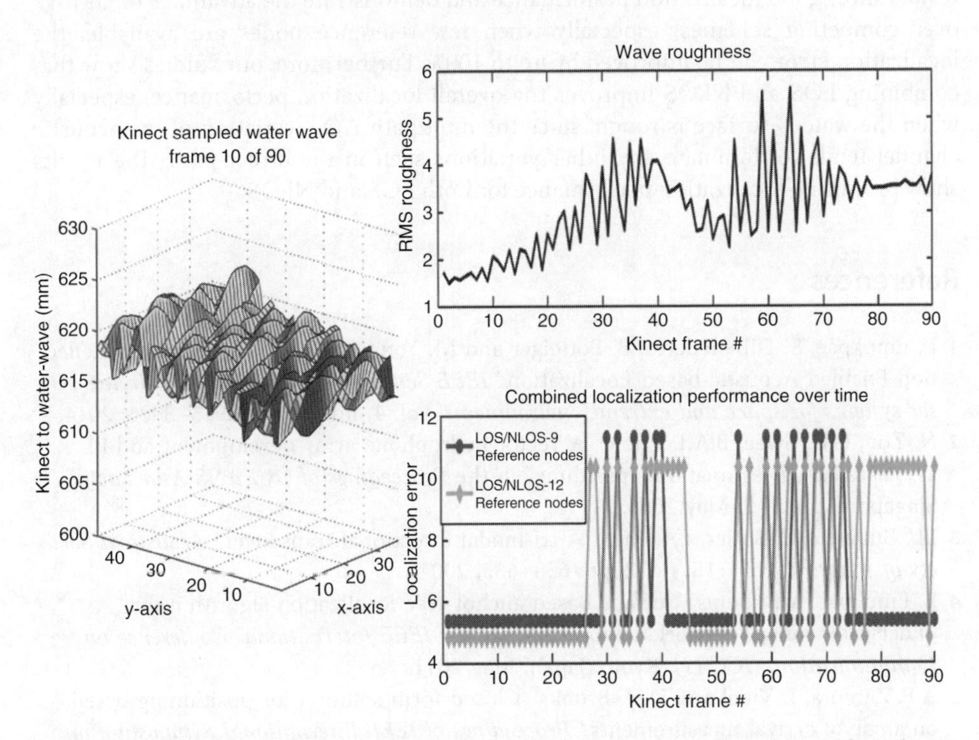

Figure 13.8 Localization error projection over time: left, sampled water surface obtained from the 3D camera; top right, projected effects of the water surface roughness on the localization error. bottom right, localization error per frame, which coincides with changing RMS roughness.

time, which starts off relatively mild and contains two rough peaks at frames 32 and 64. We see that the localization error (bottom-right plot) remains steady over time except at those peaks, which is a result of the variability in the localization performance at those frames.

13.5 Summary

In this chapter, we have presented a novel underwater signal-reflection-enabled acoustic-based localization scheme (UREAL) that uses both LOS and surface-reflected NLOS links to locate a lost node that has drifted away from the ASN. The reference nodes consist of base stations and geographical-positioned nodes that use directional acoustic transducers to determine the AOAs at the lost node. The AOAs consist of azimuth and elevation pairs that can either be LOS or surface-reflected NLOS. A closed-form least-squares solution is then used to locate the lost node. Simulation experiments were carried out using the projection of a 3D camera on the water's surface in an aquarium to create a 3D underwater networking environment. The simulation results show good localization performance and demonstrate the advantage of UREAL over competing schemes, especially when few reference nodes are available; the localization error can be improved by up to 100%. Furthermore, our studies show that combining LOS and NLOS improves the overall localization performance, especially when the water's surface is rough, since the multipath AOA variation of the acoustic channel tends to dominate any other variations seen in the direct path. The results show promising localization performance for both LOS and NLOS.

References

1 L. Emokpae, S. DiBenedetto, B. Potteiger and M. Younis, 'UREAL: Underwater Reflection Enabled Acoustic-based Localization,' *IEEE Sensors, Special Issue on Wireless sensor systems for space and extreme environments*, vol. **4**, no. 11, pp. 3915–3925, 2014.
2 N. Zou, C.C. Swee, B.A.L Chew, 'A vector hydrophone array development and its associated DOA estimation algorithms,' in the *Proceedings of OCEANS Asia Pacific*, Singapore, pp. 1–5, May 2006.
3 J.L. Butler, A.L. Butler, J.A. Rice, 'A tri-modal directional transducer,' *Acoustical Society of America*, vol. **115**, no. 2, pp. 658–665, 2004.
4 L. Emokpae, M. Younis, 'Surface Based anchor-free localization algorithm for underwater sensor networks', *Proceedings of the IEEE International Conference on Communication. (ICC'11)*, Kyoto, Japan, June 2011.
5 A.P. Zamora, J. Vidal and D.H. Brooks, 'Closed-form solution for positioning based on angle of arrival measurements,' *Proceedings of IEEE International Symposium on Personal, Indoor and Mobile Radio Communications (PIMRC '02)*, Lisboa, Portugal, 2002.

14

Coordinates Determination of Submerged Sensors with a Single Beacon Using the Cayley–Menger Determinant

Anisur Rahman[1] *and Vallipuram Muthukkumarasamy*[2]

[1] *East West University, Dhaka, Bangladesh*
[2] *Griffith University, Gold Coast, Australia*

14.1 Introduction

Underwater wireless sensor networks (UWSNs) are expected to enable oceanographic data collection and offshore exploration. Despite the many varieties of UWSN applications, the idea of submerged wireless communication may still seem far-fetched but it has nevertheless attracted researchers in recent decades [1]. In addition to underwater sensors, UWSNs may also incorporate surface stations and autonomous underwater vehicles.

The usual number of sensors deployed underwater to collect location-based data varies from several to thousands. The collection of data and the monitoring of the deployed sensors needs to be done in a dynamic fashion, if possible without a preinstalled infrastructure. Having a single boat or mobile station at the surface of the water to collect data, or to monitor and control the deployed underwater nodes, is a practical configuration. It is obvious that the locations of the submerged sensors will define the data validity, so determining the coordinates precisely is crucial. There are many distance measurement techniques for terrestrial applications based on signal strength, but these are not readily applied in underwater environments, for a number of reasons.

While various localization algorithms have been proposed for terrestrial wireless sensor networks (WSNs), there are relatively few efficient localization schemes for UWSNs, most of them being rather impractical. The characteristics of UWSNs are fundamentally different from those of terrestrial networks. Conventional underwater localization is mainly performed using acoustic signals since radio waves do not propagate well; this is despite the fact that the acoustic channel is characterised by major bandwidth limitations. Moreover, the variable speed of sound and the long propagation delays pose a unique set of challenges for localization in UWSNs. The localization schemes that use reference nodes can be broadly classified into two categories: range-based schemes (schemes that use range or bearing information), and range-free schemes (schemes that do not use range or bearing information). The former apply inter-node distances to multilateration or triangulation schemes, while the latter rely on profiling. In range-based schemes, range measurement using acoustic signals can be considerably more accurate with additional distance-measurement hardware [2, 3].

Wireless Sensor Systems for Extreme Environments: Space, Underwater, Underground, and Industrial.
First Edition. Edited by Habib F. Rashvand and Ali Abedi.

Despite the limitations of both radio and acoustic signals in underwater environments, we propose to exploit to our advantage the merits of each in order to minimize distance-measurement errors associated with timing, while at the same time compensating for coverage area. If the problem domains are at depth of less than 200 m, which covers most shallow water, the limited range of radio signals will not affect our localization scheme. In this method, radio signals will be used to measure the flight time of the acoustic signals for determining in-situ underwater acoustic speed for a vertical water column; similarly, the acoustic signals will indirectly be used for communication purposes. Even though the speed of radio signals underwater is a little less than in a vacuum ($\approx 3 \times 10^8$ m/s), in relation to the problem domain, the variation in speed will not have a significant effect on the proposed localization method. Moreover, the speed of acoustic signals, which varies as a function of:

- the temperature and salinity of the water between the surface node (beacon) and the deployed underwater sensors
- the depth of the water column.

Therefore, to determine the coordinates in a dynamic ad-hoc manner, a mechanism to calculate the in-situ average acoustic speed will also be devised.

In this chapter, a single beacon is used to determine the coordinates of the sensors. Having a mobile beacon – a boat – is more practical than having multiple beacons in the localization process. To determine the coordinates of these sensors in such a dynamic configuration, without a preinstalled reference point, the proposed mathematical model first partially incorporates the Cayley–Menger determinant, followed by linearisation to solve a system of non-linear equations produced by the determinant. To simplify procedures, it is assumed that the submerged sensors are static during the period of the computation – the time required to measure the distances of the sensors from different positions of the beacon. A solvable configuration of one beacon with three submerged sensors is described in a later section. The model computes the coordinates and the bearings with respect to the beacon node, which alleviates a number of problems in the domain of localization. Simulation results suggest that if the distances between the beacon and the sensors are true Euclidean, then the positional errors are negligible. For a problem domain of 150 m depth, theoretical positional errors found are in the 10^{-12}–10^{-14} m range. Considering the size of the deployed sensors, the generated error is almost negligible, which in turn validates the proposed mathematical model.

14.2 Underwater Wireless Sensor Networks

Sensed data can only be interpreted meaningfully when referenced to the location of the sensor, making localization an important problem. While global positioning system (GPS) receivers are commonly used to do this in terrestrial WSNs, it is infeasible in UWSNs as GPS signals do not propagate through water. Acoustic communication is the most promising modes of communication in underwater environments. However, underwater acoustic channels are characterised by harsh physical layer conditions with low bandwidth, high propagation delays and high bit-error rates. Here, the techniques and challenges of in UWSNs are categorised for:

- range-based and range-free techniques
- static reference nodes and mobile reference nodes
- single-stage and multi-stage schemes.

In addition to underwater sensor nodes, the network may also comprise surface stations and autonomous underwater vehicles (AUVs). Regardless of the type of deployment (outdoor, indoor, underground or underwater), the location of the sensors needs to be determined for meaningful interpretation of the sensed data. Since RF communications are significantly attenuated underwater [4], the use of GPS is restricted to surface nodes only. Therefore, localization underwater requires message exchanges between submerged UWSN nodes and surface nodes (or other reference nodes with known locations) by alternative means, usually acoustic communications. Unfortunately, underwater acoustic channels are characterised by long propagation delays, limited bandwidth, motion-induced Doppler shift, phase and amplitude fluctuations, multipath interference, and so on [4]. These characteristics pose severe challenges for designing localization schemes that fulfil the following desirable properties:

- *High accuracy:* The location of the sensor for which sensed data is derived should be accurate and unambiguous for meaningful interpretation of data. Localization protocols usually minimize the distance between the estimated and true locations.
- *Fast convergence:* Since nodes may drift in water currents, the localization procedure should be fast so that it reports the actual location when data is sensed.
- *Low communication costs:* Since the nodes are battery-powered and may be deployed for long durations, the communication overhead should be minimized.
- *Good scalability:* The long propagation delay and relatively high power attenuation in the underwater acoustic channel pose a scalability problem –performance is highly affected by the number of nodes in the network. Consequently, an underwater acoustic localization protocol should be distributed, rely on as few reference nodes as possible and the algorithm complexity at each node should be invariant with the network size.
- *Wide coverage:* The localization scheme should ensure that most of the nodes in the network can be localized.

In addition to these quantifiable properties, practical considerations such as ease and cost of deploying reference nodes and other required infrastructure should be taken into account too.

14.3 Dynamicity of Underwater Environment

While node deployment in terrestrial networks is relatively straightforward, the corresponding deployment underwater brings the following challenges:

14.3.1 Reference Deployment in the Deep Sea

To localize underwater nodes that have been deployed in the sea, terrestrial localization techniques would require a reference node to be deployed underwater, in addition to references attached to surface buoys. This is challenging, particularly in deep-sea applications, where reference nodes may need to be deployed on the seafloor at 3–4 km depth.

Moreover, as replacement of batteries for submerged modems is difficult, short-range, low-power communication to achieve reasonable data transmission rates is preferred, which may limit the localization coverage.

14.3.2 Node Mobility

While it is reasonable to assume that nodes in terrestrial networks remain static, underwater nodes will inevitably drift due to underwater currents, winds, shipping activity, and so on. In fact, nodes may drift in different directions, as oceanic currents are spatially dependent. While reference nodes attached to surface buoys can be precisely located through GPS updates, it is difficult to maintain submerged underwater nodes at precise locations. This may affect localization accuracy, as some distance measurements may have become obsolete by the time the node position is estimated. Moreover, motion of the sensor nodes may create the Doppler effect, which is due to the relative motion of the transmitter or the receiver. In underwater applications, mobile platforms such as AUVs can move at a speed of several knots, while untethered, free-floating equipment can drift with the ocean currents, which are generally slower than 1 knot [5]. The Doppler effect is related to the ratio of the relative transmitter–receiver velocity and the speed of the signal. Since the speed of sound in water is slower than speed of electromagnetic waves in the air, the Doppler effect can be more significant in UWSNs than in WSNs. Mobility also mandates that the localization process be repeated at certain intervals so that the node locations do not become obsolete. Therefore, mobility introduces another challenge from the viewpoint of communications overhead and energy-efficiency. Energy-efficiency is required since underwater equipment is expected to be left in the ocean for several weeks or months before it is collected and recharged for its next mission.

Underwater objects are moving continuously with water currents and dispersion. Research in hydrodynamics shows that the movement of underwater objects is closely related to many environmental factors, such as the water current and water temperature [6]. In different environments, the mobility characteristics of underwater objects are different. For example, the mobility patterns of objects near the seashore have a semiperiodic property because of tides, in contrast to objects in rivers. While it is almost impossible to devise a generic mobility model for underwater objects in all environments, some mobility models for underwater objects in specific environments based on hydrodynamics have been devised [7]. This indicates that the movement of underwater objects is not a totally random process. Temporal and spatial correlations are inherent in such movements, which makes their mobility patterns predictable. Zhou et al. [8] investigated the mobility characteristics of objects in shallow seashore areas. Tidal areas are characterised by their shallowness and their strong tidal currents. The non-linear interaction of tidal currents and bottom topography produces currents that give nonzero contributions to the tidally averaged currents. These so-called residual currents are important for the transport and mixing properties of the tidal flow.

If the mobility of the nodes can be determined, the localization process will interpolate the future positions of the nodes from their present positions. Mobility patterns in a specific area can be determined from a study of the area or it can be generalised from the difference of the present and immediate next positions of the nodes in a time frame. This is how mobility patterns can be determined, but the mechanism increases the communication overhead of the localization process.

14.3.3 Inter-node Time Synchronization

Since GPS signals are severely attenuated underwater, they cannot be used to synchronize time between nodes deployed underwater to compensate for clock drift due to both offset and skew [9]. Consequently, the accuracy of range measurements based on time of arrival (TOA) may be affected. Any scheme that relies on TOA or time difference of arrival (TDOA) requires tight time synchronization between the transmitter and the receiver clocks. One simple way to achieve this in terrestrial networks is to use a radio signal. Kwon et al. [10] use the difference in propagation times of acoustic and radio signals for calculating the distance. This works because the propagation speed of radio signals is many orders of magnitude higher than that of acoustic signals. The luxury of using radio signals for time synchronization has not been available in underwater scenarios until recently because radio waves do not propagate well underwater. However, Che et al. [11] have showed that it is now time to re-evaluate radio signals for underwater communications. Recently, with the advances in sensitivity of sensors, we have proposed a scheme to measure the distances between beacons and sensors.

14.3.4 Signal Reflection due to Obstacles and Surfaces

In near-shore or harbour environments, where obstacles may exist between nodes, non-line-of-sight (NLOS) signals reflecting from objects (such as the sea surface or a harbour wall) can be mistaken for line-of-sight (LOS) signals, and may significantly impact the accuracy of range measurements.

14.4 Proposed Configuration

14.4.1 Problem Domain

Underwater localization usually involves a boat at the surface and sensors deployed underwater, as depicted in Figure 14.1. As the boat is usually mobile and our proposed method also requires a mobile reference point, this normal configuration is readily applicable.

Moreover, a static reference point is cumbersome and time-consuming to achieve for dynamic localization. As most marine exploration takes place in shallow waters (100–200 m in depth), it would be possible to use radio signal for synchronization. In the proposed method we need at least three sensors and a floating beacon, and the only information used is the distance between them. In the marine environment, the deployed sensors can be at any level in the water; some sensors float freely, submerged or on the bottom. On the other hand, AUVs or unmanned underwater vehicles (UUVs) are sometimes deployed and the can cruise at a certain desired height. In any configuration, so long as the depth of the problem domain remains within radio-signal range (1.8–323 m) [11], our proposed distance-determination technique can be used. At greater depths, two-way message transfer could be used for synchronization instead. However, we are mainly focused on shallow depths, so that we can use radio signals for synchronization.

The beacon node (boat) at the surface of the water will be generating signals to measure the distances, using the procedures described in the following sections.

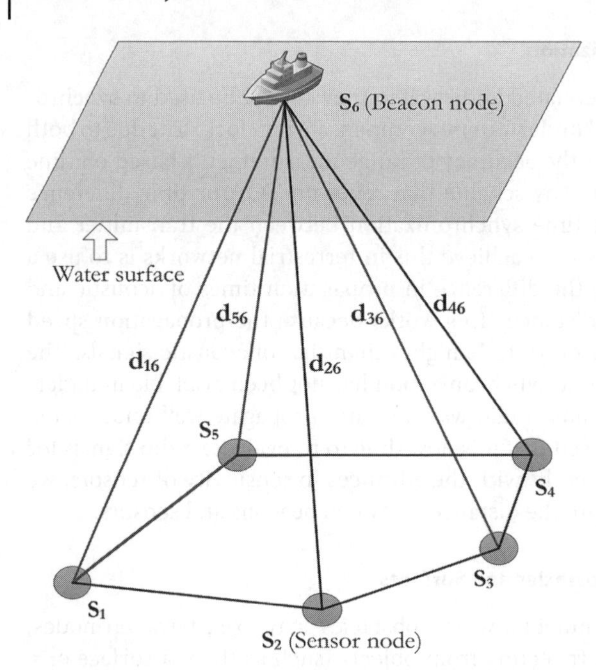

Figure 14.1 Configuration consisting of one beacon and submerged sensors.

Our proposed method requires one beacon at the surface and at least three deployed sensors; a solvable configuration of one beacon with three submerged sensors is shown in Figure 14.2. As UWSNs can have from a few to even thousands of sensors, three deployed data-collection sensors or nodes that need to be localized is a practical and representative number. If higher numbers of sensors need to be localized, they should be grouped in threes. For the sake of simplicity, at present we assume that the submerged sensors are static at the time of computation – the time that is required to measure the distances from different positions of the beacon.

As we are using both signals in the water, we are restricted to be shallow depths, although this is not problematic because most marine exploration takes place in shallow waters. The deployed nodes will be equipped with temperature sensors, which are very cost effective compared to equipment for sensing other parameters.

14.4.2 Environmental Constrains

Normally, underwater environments are more difficult to work in than terrestrial ones, but they have some features that can be exploited in determining coordinates. Water bodies are relatively homogeneous, because the usual obstacles present in water are smaller than those in terrestrial environments. The region of interest on the ground is more likely to be occupied by buildings and trees, which are the major factors causing multipath propagation.

Regarding signal propagation in water, acoustic signals propagate much further than radio signals, but the speed of an acoustic signal is much slower than that of a radio signal. Table 14.1 shows some limitations and typical measurements for radio and acoustic signals. The main environmental variable that we assume in our method to determine distances is the speed of acoustic signals in water. It depends on the temperature, salinity and permeability.

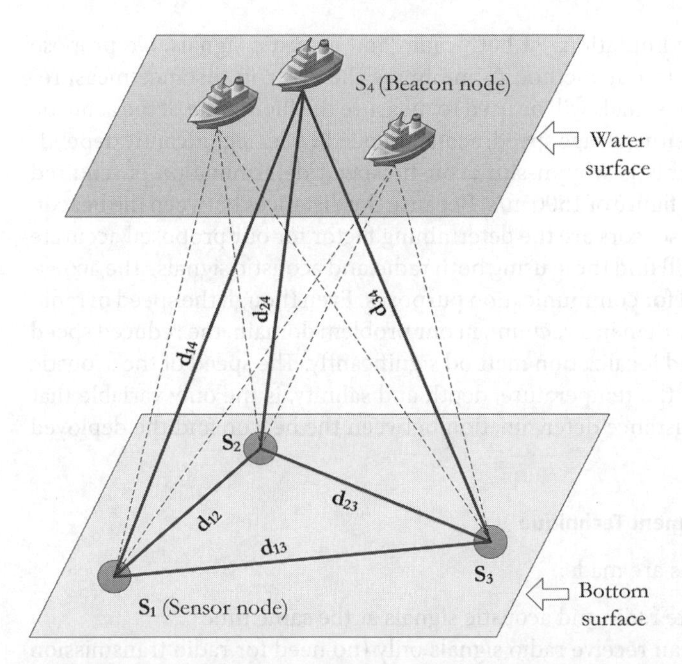

Figure 14.2 Solvable subset configuration with a mobile beacon and three submerged sensors

Table 14.1 Properties of radio and acoustic signals.

	Radio		Acoustic	
	Vacuum	**Water**	**Vacuum**	**Water**
Velocity	3×10^8 m/s	$\approx2.25\times10^6$ m/s	–	$\approx$1500 m/s
Range	–	1.8–323 m	–	1–20 km

14.5 Distance Determination

To localize underwater sensor nodes with a single mobile beacon in dynamic fashion, two aspects play important roles: how precisely the distances are measured and how mathematically accurate the coordinate-determination process is. As range-based methods are more precise than range-free methods, we intend to explore range-based techniques.

Signal propagation and the type of medium used underwater to measure distances is another factor that contributes to accurate localization. Acoustic signals have been the proven technology for the last few decades in underwater communications and distance measurements; on the other hand, light waves perform well where paths are clear and distances are short. With the incorporation of fourth-generation computing into the sensors as well as the improved efficiency that sensors have gained, it is now indispensable to reconsider the use of radio signals to enhance the ability to measure distance precisely. Radio signals travel much faster, over limited propagation distances, whereas acoustic signals travel slower but over longer distances.

Despite the underwater limitations of both radio and acoustic signals, we propose to use the merits of both in our method to minimize the error in distance measurement. In our method, radio signals will be used to measure the flight time of the acoustic signals to calculate the in-situ average speed; acoustic speed varies significantly depending on ambient factors. This is why in-situ acoustic speed determination is required rather than using a default figure of 1500 m/s. Because the distances between the beacon and deployed underwater sensors are the determining factor for our proposed accurate localization scheme, we will find these using both radio and acoustic signals. The acoustic signals will also be used for communication purposes. Even though the speed of radio signals is little less in water than in a vacuum, in our problem domain, the reduced speed will not affect the proposed localization method significantly. The speed of the acoustic signals, which varies with the temperature, depth and salinity, is the only variable that we need to consider for distance determination between the beacon and the deployed sensors.

14.5.1 Distance-measurement Technique

The following assumptions are made:

- The beacon can generate radio and acoustic signals at the same time.
- The deployed sensors can receive radio signals only (no need for radio transmission by the sensors, which would consume lot of power), but can generate and receive acoustic signals.
- Acoustic signals will be used for communications.
- The environmental factors that affect the acoustic signals will be considered while measuring inter-node distances.
- Sensor nodes are stationary during the short measurement period.
- The beacon (boat or buoy at water level) and sensor nodes (at the bottom) are in parallel or non-parallel states.
- Each sensor will have a unique ID.

The steps in the method are as follows:

1) There is simultaneous generation of radio and acoustic signals by beacon S_j, $j = 4,5,\ldots.$ at t_0 (here, S_j are different positions of the beacon at the same water level.
2) For any submerged sensors S_i, $i = 1,2,3$:
   a) the sensors receive the radio signals immediately at $t_{Ra(rec)} = t_0 + \varepsilon$, where ε is the travelling time of radio signals from the beacon to the sensors
   b) The sensor receives the acoustic signals after a while at $t_{Ac(rec)}$, where $t_{Ac(rec)} - t_0 \gg t_{Ra(rec)} - t_0$ due to speed of radio signals (2.25×10^6 m/s).
3) The time for the acoustic signals to travel from beacon to sensors is:

$$T_{ij(travel),\ i=1,2,3;j=4,5,6\ldots} = t_{Ac(rec)} - t_{Ac(tra)} = t_{Ac(rec)} - t_{Ra(tra)}$$

$$\because t_{Ac(tra)} = t_{Ra(tra)}$$

$$\therefore T_{ij(travel)} \approx t_{Ac(rec)} - t_{Ra(rec)} \because t_{Ra(rec)} = t_0 + \varepsilon \approx t_{Ra(tra)}$$

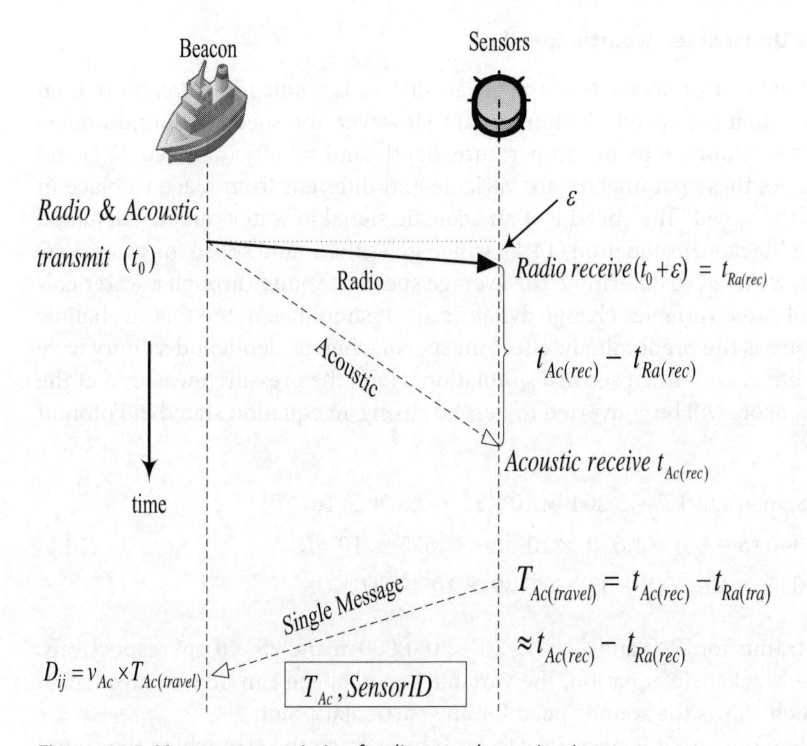

Figure 14.3 Message transmission for distance determination.

4) Sensor nodes send the time $T_{ij(travel)}$ and the individual sensor's ID back to the beacon using acoustic signals.
5) The beacon node computes the distance between the beacon and sensors:

$$d_{ij} = v_A \times T_{ij(travel)},$$

where v_A is the average speed of acoustic signals for the water column.

Figure 14.3 shows the sequence of the message transfer between the beacon and underwater sensor and the time frame. The beacon generates both radio and acoustic signals at the same time, but the radio signal will be received by the sensors immediately. In theory a period ε would be required, but this is negligible for radio signals to cover only ~200 m. The acoustic signal will be received after a substantial amount of time. Then the flight time of the acoustic signal can be calculated as the difference between the acoustic and the radio signals. This recorded time, along with the specific sensor's ID and values for the other parameters that affect underwater acoustic speed, will be sent to the beacon at the surface for further calculations. All the remaining calculations, from the average acoustic speed to the coordinate determination, will be carried out at the beacon (boat) as it has an unlimited power supply. The whole procedure of determining the flight time takes a very short time, so that any mobility of the sensor will have only a minimal effect.

14.5.2 Average Underwater Acoustic Speed

A typical speed of acoustic waves near the ocean surface is about 1500 m/s, more than four times faster than the speed of sound in air. However, the speed of sound underwater is affected predominantly by temperature, depth, and salinity (denoted T, D and S, respectively). As these parameters are variable and different from place to place in the water; so is the speed. The speed v of an acoustic signal in water can be calculated according to the Mackenzie equation (14.1), which gives the sound speed for given T, D and S. However, we need to determine the average speed of sound through a water column in which all three variables change dynamically. It should be noted that in shallow water temperature is the predominant effect on speed, whereas depth and salinity have only a minimal effect, as we can see in a simulation. Here, the pressure measured at the bottom by the sensors will be converted to depth D using an equation stated in Fofonoff and Millard [12].

$$\begin{aligned} v = {} & 1448.96 + 4.591T - 5.304 \times 10^{-2}T^2 + 2.374 \times 10^{-4}T^3 \\ & + 1.340\,(S - 35) + 1.630 \times 10^{-2}D + 1.675 \times 10^{-7}D^2 \\ & - 1.025 \times 10^{-2}T(S - 35) - 7.139 \times 10^{-13}TD^3 \end{aligned} \tag{14.1}$$

where the constraints for T, D and S are 2–30°C, 0–8000 m and 25–40 ppt respectively.

Following the Mackenzie equation, the variable sound speed can be portrayed as in Figure 14.4, which shows the sound speed for any particular point.

For our problem domain it is necessary to find the average speed at which the sound travels from beacon to the sensors for a particular water column. Our proposed average speed of sound for the problem domain can be calculated using Equation 14.2.

$$\begin{aligned} \overline{v} &= f_{avg}(T, D, S) \\ &= \frac{1}{A}\iiint_R f(T, D, S)dA \\ &= \frac{1}{A}\int_{S_i}^{S_f}\int_{0}^{D_f}\int_{T_i}^{T_f} f(T, D, S)dTdDdS \end{aligned} \tag{14.2}$$

where A is the area created by the limit of T, D and S and $f\,(T, D, S)$ is the Mackenzie equation.

The derivative of the multivariate Mackenzie equation [13] is used to compute the average speed of acoustic signals from beacon to sensors instead of using a figure of 1500 m/s, as most localization methods do. Average acoustic speed is simulated for various water columns with different parameter values. For example, Figure 14.5 shows the average acoustic speed for a 200-m water column with 20°C surface and 10°C bottom temperatures, and with salinity varying by 0.5 ppt from beacon to sensors. We have incorporated Gaussian noise ($\mu = 0$, $\sigma = 1$) to the 10°C bottom temperature and to the flight time because both of these values are more uncertain than the measured surface temperature. Over 100 iterations, the mean value of the average sound speed calculated over the column was found to be 1507.6 m/s, with a standard deviation of 1.97. The results suggest the model avoids multipath fading and is capable of producing in-situ results.

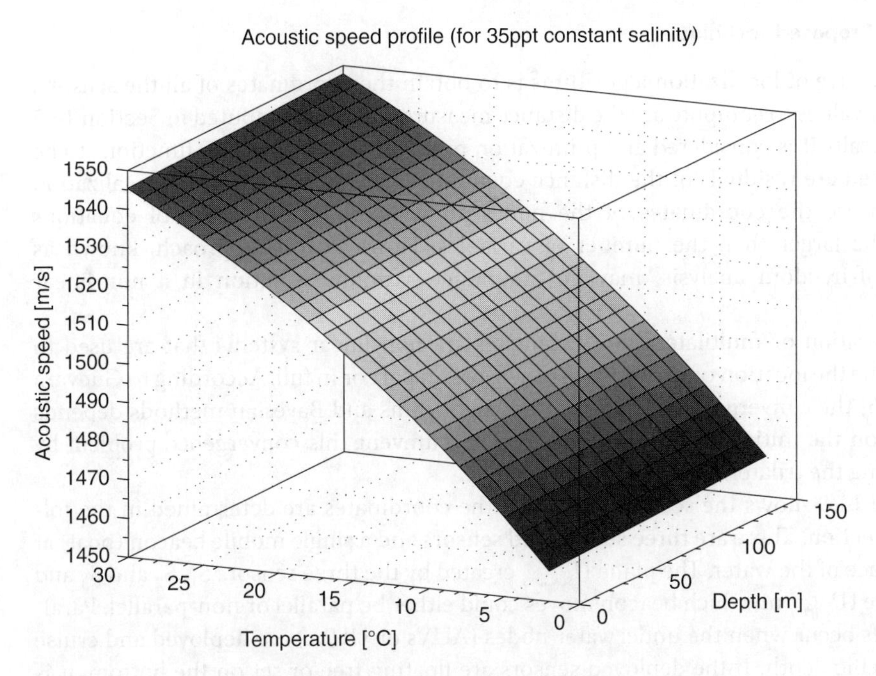

Figure 14.4 Acoustic speed profile according to the Mackenzie equation

14.6 Coordinate Determination

In recent years, researchers have shown great interest in underwater localization schemes. The precise coordinates of each sensor with respect to the other sensors and the beacon are needed to comprehend the validity of the gathered data; without the knowledge of its actual origin, data has limited value.

Coordinate determination is therefore a key part of underwater localization and various underwater applications have been researched. A plethora of techniques have been proposed to determine the coordinates of the sensors; some using mobile sensors and some using multiple beacon nodes. However, traditional monitoring systems are expensive and complicated as most of them use preinstalled reference points [14]. Moreover, multilateration techniques are mostly used to determine the location of the sensors with respect to three or more known beacon nodes, with non-linear distance equations then solved. However, the number of degrees of freedom means that a unique solution is not guaranteed.

In contrast, in our range-based solution, the Cayley–Menger determinant and linearised trilateration are used to determine the coordinates of the nodes in a schem in which none of nodes has a-priori knowledge of its location.

A practical dynamic approach to localizing submerged nodes with minimal logistics has been set out here. Our proposed technique is as simple as possible, with only a single beacon (boat/buoy) at the surface used to determine the coordinates of multiple underwater deployed sensors.

14.6.1 Proposed Technique

The objective of localization algorithms is to obtain the coordinates of all the sensors. The only values to compute are the distance measurements as computed in Section 14.5 and typically it is considered an optimization problem where objective functions to be minimized are residuals of the distance equations [15]. The variables in a localization problem are the coordinates of the nodes; in principle, the number of equations should be larger than the number of variables. However, this approach, known as degree-of-freedom analysis, may not guarantee a unique solution in a non-linear system.

Trilateration or multilateration techniques are non-linear systems that are used to determine the location or coordinates of sensors, in part or in full. According to Guevara et al. [16] the convergence of optimization algorithms and Bayesian methods depends heavily on the initial conditions used. They circumvent this convergence problem by linearising the trilateration equations.

Figure 14.6 shows the scenario in which the coordinates are determined in the following section. There are three submerged sensors and a single mobile beacon (boat) at the surface of the water. The plane $\Pi_{sensors}$ created by the three sensors S_1, S_2 and S_3 and the plane Π_{beacon} on which beacon moves could either be parallel or non-parallel. Parallel planes occur when the underwater nodes (AUVs or UUVs) are deployed and cruise at a specific depth. If the deployed sensors are floating free or set on the bottom, it is more likely that planes $\Pi_{sensors}$ and Π_{beacon} will not be parallel. Which of these situations applies in practice can be determined from the depth measurements of the sensors. It is convenient to incorporate a pressure sensor on all the sensors or nodes, so that the depth of the nodes can be calculated according to 14.3.

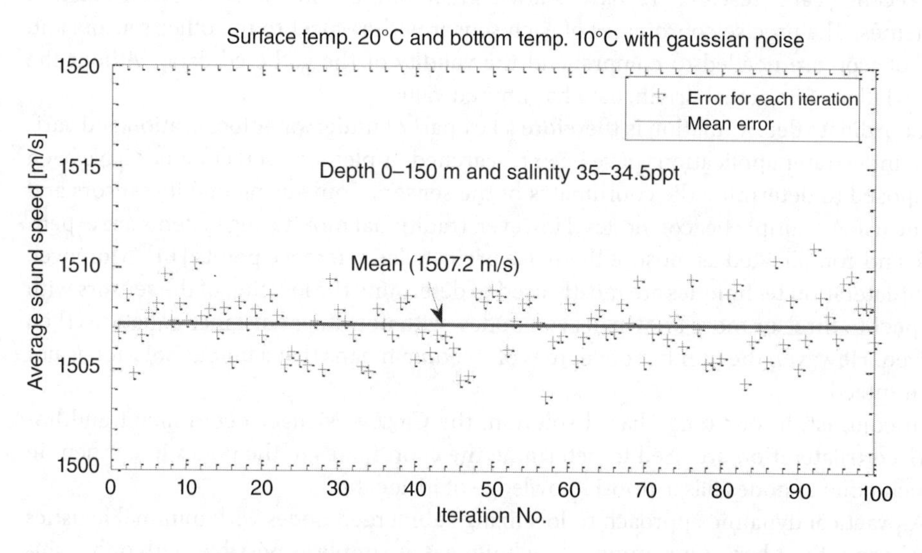

Figure 14.5 Average speed with 20 °C surface temperature for a200-m water column

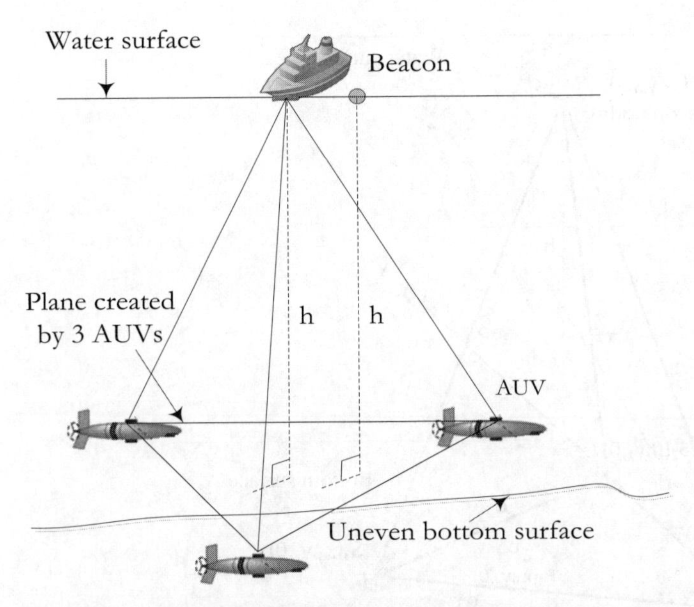

Figure 14.6 Parallel states scenario

$$D_s = \frac{9.7266 \times 10^2 P - 2.512 \times 10^{-1} P^2 + 2.279 \times 10^{-4} P^3 - 1.82 \times 10^{-7} P^4}{g(\emptyset) + 1.092 \times 10^{-4} P} \tag{14.3}$$

where P is pressure and Ø is latitude, where g(Ø) is given by the international formula for gravity.

14.6.2 Coordinates of the Sensors

Figure 14.7 shows the initial subset composed of the beacon node $S_j, j = 4, 5 \ldots .9$, and three sensor nodes $S_i, i = 1, 2, 3$. Without loss of generality, a coordinate system can be defined using one of the sensor nodes $S_i, i = 1, 2, 3$, as the origin (0, 0, 0)of the coordinate system. Now the trilateration equations can be written as a function of two groups of distance measurements: the distance between beacon and sensors $d_{14}, d_{24}, d_{34} \ldots$, which are measured data (known), and the inter-sensor distances d_{12}, d_{13}, d_{23} and the volume of tetrahedron V_t (here, t is the tetrahedron formed by the beacon and the three deployed sensors), which are unknown.

Based on the local positioning system configuration of Figure 14.7, we need to write an equation that includes all known and unknown distances. For that purpose, we express the volume of tetrahedron V_t using the Cayley–Menger determinant as follows:

$$288\, V_t^2 = \begin{vmatrix} 0 & 1 & 1 & 1 & 1 \\ 1 & 0 & d_{12}^2 & d_{13}^2 & d_{14}^2 \\ 1 & d_{12}^2 & 0 & d_{23}^2 & d_{24}^2 \\ 1 & d_{13}^2 & d_{23}^2 & 0 & d_{34}^2 \\ 1 & d_{14}^2 & d_{24}^2 & d_{34}^2 & 0 \end{vmatrix} \tag{14.4}$$

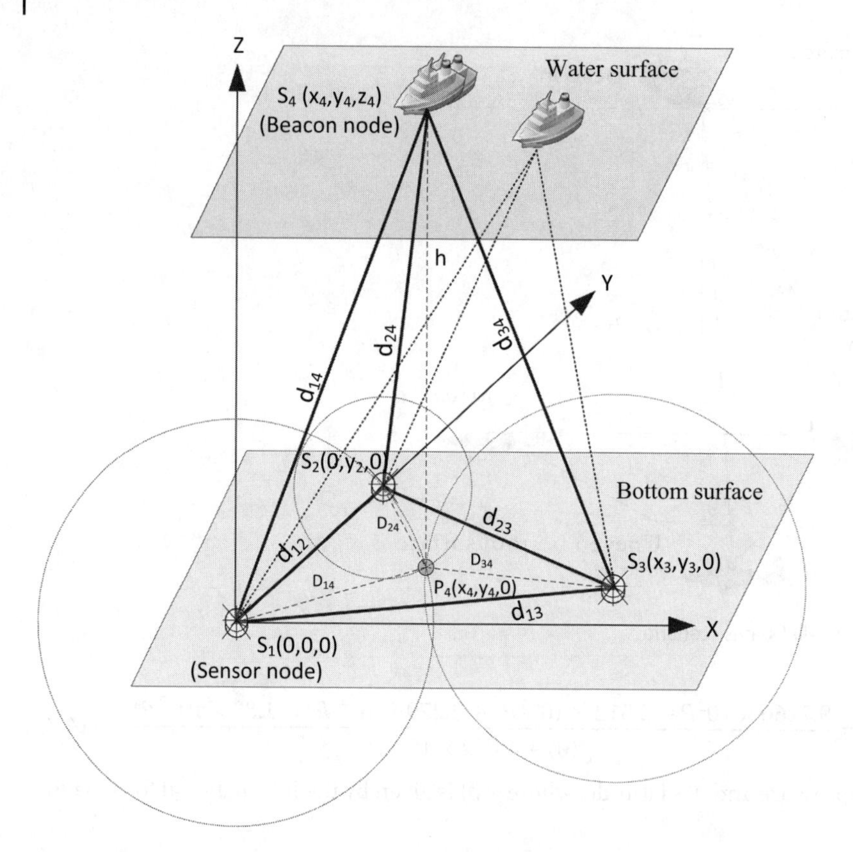

Figure 14.7 Subset of three sensors and a mobile beacon for coordinates determination.

By expanding 14.4, we get:

$$d_{34}^2 d_{23}^2 - d_{34}^2 d_{12}^2 + d_{34}^2 d_{13}^2 - \frac{d_{14}^2 d_{23}^4}{d_{12}^2} + d_{23}^2 d_{14}^2 + \frac{d_{13}^2 d_{14}^2 d_{23}^2}{d_{12}^2} - \frac{d_{24}^2 d_{13}^4}{d_{12}^2} + \frac{d_{13}^2 d_{23}^2 d_{24}^2}{d_{12}^2}$$
$$+ d_{13}^2 d_{24}^2 - d_{13}^2 d_{23}^2 - 144\frac{V_t^2}{d_{12}^2} + \frac{d_{14}^2 d_{23}^2 d_{24}^2}{d_{12}^2} + \frac{d_{14}^2 d_{23}^2 d_{34}^2}{d_{12}^2} - \frac{d_{23}^2 d_{24}^2 d_{34}^2}{d_{12}^2} - \frac{d_{14}^4 d_{23}^2}{d_{12}^2}$$
$$+ \frac{d_{13}^2 d_{24}^2 d_{34}^2}{d_{12}^2} - \frac{d_{13}^2 d_{14}^2 d_{34}^2}{d_{12}^2} + \frac{d_{13}^2 d_{14}^2 d_{24}^2}{d_{12}^2} - \frac{d_{13}^2 d_{24}^4}{d_{12}^2} - d_{34}^4 + d_{24}^2 d_{34}^2 + d_{14}^2 d_{34}^2 - d_{14}^2 d_{24}^2 = 0$$

Grouping known-unknown variables, we get:

$$d_{34}^2(d_{12}^2 - d_{23}^2 - d_{13}^2) + d_{14}^2\left(\frac{d_{23}^4}{d_{12}^2} - d_{23}^2 - \frac{d_{13}^2 d_{23}^2}{d_{12}^2}\right) + d_{24}^2\left(\frac{d_{13}^4}{d_{12}^2} - \frac{d_{13}^2 d_{23}^2}{d_{12}^2} - d_{13}^2\right)$$
$$- \left(d_{14}^2 d_{24}^2 + d_{14}^2 d_{34}^2 - d_{24}^2 d_{34}^2 - d_{14}^4\right)\frac{d_{23}^2}{d_{12}^2} - \left(d_{34}^2 d_{24}^2 - d_{14}^2 d_{34}^2 + d_{14}^2 d_{24}^2 - d_{24}^4\right)\frac{d_{13}^2}{d_{12}^2}$$
$$- \left(144\frac{V_t^2}{d_{12}^2} + d_{13}^2 d_{23}^2\right) = \left(d_{24}^2 d_{34}^2 - d_{34}^4 + d_{14}^2 d_{34}^2 - d_{14}^2 d_{24}^2\right)$$

Rearranging the terms

$$d_{14}^2\left(\frac{d_{23}^4}{d_{12}^2} - d_{23}^2 - \frac{d_{13}^2 d_{23}^2}{d_{12}^2}\right) + d_{24}^2\left(\frac{d_{13}^4}{d_{12}^2} - \frac{d_{13}^2 d_{23}^2}{d_{12}^2} - d_{13}^2\right) + d_{34}^2\left(d_{12}^2 - d_{23}^2 - d_{13}^2\right)$$
$$-\left(d_{14}^2 d_{24}^2 + d_{14}^2 d_{34}^2 - d_{24}^2 d_{34}^2 - d_{14}^4\right)\frac{d_{23}^2}{d_{12}^2} - \left(d_{34}^2 d_{24}^2 - d_{14}^2 d_{34}^2 + d_{14}^2 d_{24}^2 - d_{24}^4\right)\frac{d_{13}^2}{d_{12}^2}$$
$$+\left(144\frac{V_t^2}{d_{12}^2} + d_{13}^2 d_{23}^2\right) = \left(d_{24}^2 - d_{34}^4\right)\left(d_{34}^2 - d_{14}^2\right)$$

here,

$$\left(\frac{d_{23}^4}{d_{12}^2} - d_{23}^2 - \frac{d_{13}^2 d_{23}^2}{d_{12}^2}\right), \quad \left(\frac{d_{13}^4}{d_{12}^2} - \frac{d_{13}^2 d_{23}^2}{d_{12}^2} - d_{13}^2\right), \quad \left(d_{12}^2 - d_{23}^2 - d_{13}^2\right),$$
$$\frac{d_{23}^2}{d_{12}^2}, \quad \frac{d_{13}^2}{d_{12}^2}, \quad \text{and} \quad \left(144\frac{V_t^2}{d_{12}^2} + d_{13}^2 d_{23}^2\right)$$

are unknown terms.

The above expansion can be rewritten as:

$$\begin{aligned} &d_{14}^2 X_1 + d_{24}^2 X_2 + d_{34}^2 X_3 - \left(d_{14}^2 - d_{34}^2\right)\left(d_{24}^2 - d_{14}^2\right) X_4 \\ &\quad - \left(d_{24}^2 - d_{14}^2\right)\left(d_{34}^2 - d_{24}^2\right) X_5 + X_6 = \left(d_{24}^2 - d_{34}^2\right)\left(d_{34}^2 - d_{14}^2\right) \end{aligned} \tag{14.5}$$

where,

$$\left(\frac{d_{23}^4}{d_{12}^2} - d_{23}^2 - \frac{d_{13}^2 d_{23}^2}{d_{12}^2}\right) = X_1$$
$$\left(\frac{d_{13}^4}{d_{12}^2} - \frac{d_{13}^2 d_{23}^2}{d_{12}^2} - d_{13}^2\right) = X_2$$
$$\left(d_{12}^2 - d_{23}^2 - d_{13}^2\right) = X_3$$
$$\frac{d_{23}^2}{d_{12}^2} = X_4$$
$$\frac{d_{13}^2}{d_{12}^2} = X_5$$
$$\left(144\frac{V_t^2}{d_{12}^2} + d_{13}^2 d_{23}^2\right) = X_6$$

which resembles the following linear form:

$$a_1 x_1 + a_2 x_2 + \cdots + a_n x_n = b_1.$$

As we have six unknowns in (14.5), we need at least six measurements, which could be done following the procedure described earlier, steering the beacon node $S_j, j = 4, 5 \ldots .9$, to six different places and measuring the distances in the vicinity of S_4.

Eventually we get a system of linear equations of the form:

$$\begin{aligned}
a_{11}x_1 + a_{12}x_2 + \cdots + a_{1n}x_n &= b_1,\\
a_{21}x_1 + a_{22}x_2 + \cdots + a_{2n}x_n &= b_2,\\
&\vdots\\
a_{m1}x_1 + a_{m2}x_2 + \cdots + a_{mn}x_n &= b_m
\end{aligned} \tag{14.6}$$

If we omit reference to the variables, then system (14.6) can be represented by the array of all the coefficients, known as the augmented matrix of the system, where the first row of the array represents the first linear equation and so on. This could be expressed in an $AX = b$ linear form.

After doing so, for our system, we have

$$A = \begin{bmatrix} d_{14}^2 & d_{24}^2 & d_{34}^2 & -(d_{14}^2 - d_{34}^2)(d_{24}^2 - d_{14}^2) & -(d_{24}^2 - d_{14}^2)(d_{34}^2 - d_{24}^2) & 1 \\ d_{15}^2 & d_{25}^2 & d_{35}^2 & -(d_{15}^2 - d_{35}^2)(d_{25}^2 - d_{15}^2) & -(d_{25}^2 - d_{15}^2)(d_{35}^2 - d_{25}^2) & 1 \\ \vdots & \vdots & \vdots & \vdots & \vdots & \vdots \\ d_{19}^2 & d_{29}^2 & d_{39}^2 & -(d_{19}^2 - d_{39}^2)(d_{29}^2 - d_{19}^2) & -(d_{29}^2 - d_{19}^2)(d_{39}^2 - d_{29}^2) & 1 \end{bmatrix}.$$

$$X = \begin{bmatrix} \left(\dfrac{d_{23}^4}{d_{12}^2} - d_{23}^2 - \dfrac{d_{13}^2 d_{23}^2}{d_{12}^2}\right) \\ \left(\dfrac{d_{13}^4}{d_{12}^2} - \dfrac{d_{13}^2 d_{23}^2}{d_{12}^2} - d_{13}^2\right) \\ (d_{12}^2 - d_{23}^2 - d_{13}^2) \\ \dfrac{d_{23}^2}{d_{12}^2} \\ \dfrac{d_{13}^2}{d_{12}^2} \\ \left(144\dfrac{V_t^2}{d_{12}^2} + d_{13}^2 d_{23}^2\right) \end{bmatrix}, \qquad b = \begin{bmatrix} (d_{24}^2 - d_{34}^2)(d_{34}^2 - d_{14}^2) \\ (d_{25}^2 - d_{35}^2)(d_{35}^2 - d_{15}^2) \\ \vdots \\ (d_{29}^2 - d_{39}^2)(d_{29}^2 - d_{19}^2) \end{bmatrix}.$$

From the above representation, knowing X_1, X_2, X_3, X_4, X_5 and X_6 we calculate d_{12}, d_{13} and d_{23} as following:

$$d_{12}^2 = \frac{X_3}{(1 - X_4 - X_5)},\ d_{13}^2 = \frac{X_3 X_5}{(1 - X_4 - X_5)},\ \text{and } d_{23}^2 = \frac{X_3 X_4}{(1 - X_4 - X_5)}.$$

If the coordinates of the submerged sensors S_1, S_2 and S_3 are $(0, 0, 0)$, $(0, y_2, 0)$ and $(x_3, y_3, 0)$, respectively, then according to Figure 14.7, the inter-sensor distances can be written with respect to coordinates of the sensors as:

$$d_{12}^2 = y_2^2,\ d_{13}^2 = x_3^2 + y_3^2,\ \text{and } d_{23}^2 = x_3^2 + (y_2 - y_3)^2.$$

From the above values we get the unknown variables y_2, x_3 and $y_3$computed as:

$$y_2 = d_{12},\ y_3 = \frac{d_{12}^2 + d_{13}^2 - d_{23}^2}{2d_{12}},\ x_3 = \sqrt{\left(d_{13}^2 - \left(\frac{d_{12}^2 + d_{13}^2 - d_{23}^2}{2d_{12}}\right)^2\right)}$$

Table 14.2 Coordinates of the sensors with known measurements.

Sensor	Coordinates
S_1	$(0,0,0)$
S_2	$(0, d_{12}, 0)$
S_3	$\left(\sqrt{\left(d_{13}^2 - \left(\frac{d_{12}^2 + d_{13}^2 - d_{23}^2}{2d_{12}}\right)^2\right)}, \frac{d_{12}^2 + d_{13}^2 - d_{23}^2}{2d_{12}}, 0\right)$

where d_{12}, d_{13} and d_{23} are the computed distances. Table 14.2 summarises the coordinates of the sensors for this system.

14.6.3 Coordinates of the Sensors with Respect to the Beacon

Until now we have been able to find the coordinates of the sensor nodes; to find the coordinates of the beacon we follow the steps described below.

After measuring the vertical distance h between the beacon node $S_4(x_4, y_4, z_4)$ and the XY plane (the plane of the sensor nodes) we can assume the projected coordinate of the beacon node $S_4(x_4, y_4, z_4)$ on the plane XY is $P_4(x_4, y_4, 0)$. To find x_4 and y_4, we can apply trilateration in the following manner, assuming the distances between S_1, S_2, S_3 and P_4 are D_{14}, D_{24} and D_{34} respectively.

$$D_{14}^2 = x_4^2 + y_4^2, \tag{14.7}$$

$$D_{24}^2 = x_4^2 + (y_4 - y_2)^2, \tag{14.8}$$

$$D_{34}^2 = (x_4 - x_3)^2 + (y_4 - y_3)^2, \tag{14.9}$$

$$D_{14}^2 = d_{14}^2 - h^2, \tag{14.10}$$

$$D_{24}^2 = d_{24}^2 - h^2, \tag{14.11}$$

$$D_{34}^2 = d_{34}^2 - h^2. \tag{14.12}$$

From (14.7)–(14.9) we obtain the projected beacon coordinates $P_4(x_4, y_4, 0)$, where

$$x_4 = \frac{\sqrt{4d_{12}^2 D_{14}^2 - \left(D_{14}^2 - D_{24}^2 + d_{12}^2\right)^2}}{2d_{12}}, \quad y_4 = \frac{1}{2d_{12}}\left(D_{14}^2 - D_{24}^2 + d_{12}^2\right)$$

As d_{14}, d_{24} and d_{34} are the hypotenuses of the $\Delta S_1P_4S_4$, $\Delta S_2P_4S_4$ and $\Delta S_3P_4S_4$ respectively, so it is possible to obtain D_{14}, D_{24}, and D_{34} using Pythagoras' theorem; that is, D_{14}, D_{24} and D_{34} are calculated from (14.10), (14.11) and (14.12), respectively. Therefore, the coordinate of the beacon node $S_4(x_4, y_4, z_4)$ is $S_4(x_4, y_4, h)$, where x_4, y_4 and h are known elements.

$$\therefore S_4(x_4, y_4, z_4)$$

$$= S_4\left(\left(\frac{\sqrt{4d_{12}^2 D_{14}^2 - \left(D_{14}^2 - D_{24}^2 + d_{12}^2\right)^2}}{2d_{12}}\right), \left(\frac{1}{2d_{12}}\left(D_{14}^2 - D_{24}^2 + d_{12}^2\right)\right), h\right)$$

Table 14.3 Coordinates of the sensors with respect to beacon

Sensor	Coordinates
S_4	$(0, 0, 0)$
S_1	$(-x_4, -y_4, -z_4)$
S_2	$(-x_4, y_2 - y_4, -z_4)$
S_3	$(x_3 - x_4, y_3 - y_4, -z_4)$

Table 14.4 Coordinates of the sensors with known and computed values

Sensor	Coordinates
S_4	$(0, 0, 0)$
S_1	$\left(\frac{\sqrt{4d_{12}^2D_{14}^2 - (D_{14}^2 - D_{24}^2 + d_{12}^2)^2}}{2d_{12}}, \quad -\frac{1}{2d_{12}}(D_{14}^2 - D_{24}^2 + d_{12}^2), \quad -h\right)$
S_2	$\left(\frac{\sqrt{4d_{12}^2D_{14}^2 - (D_{14}^2 - D_{24}^2 + d_{12}^2)^2}}{2d_{12}}, \quad \frac{1}{2d_{12}}(d_{12}^2 - D_{14}^2 + D_{24}^2), \quad -h\right)$
S_3	$\left(\left(\sqrt{\left(d_{13}^2 - \left(\frac{d_{12}^2 + d_{13}^2 - d_{23}^2}{2d_{12}}\right)^2\right)} - \frac{\sqrt{4d_{12}^2D_{14}^2 - (D_{14}^2 - D_{24}^2 + d_{12}^2)^2}}{2d_{12}}\right), \frac{1}{2d_{12}}(d_{13}^2 - d_{23}^2 - D_{14}^2 + D_{24}^2), -h\right)$

So if the origin of the Cartesian system is transferred on to the coordinate of the beacon node, it is possible to find the coordinates of the other sensors with respect to S_4, the beacon node. A linear transformation would give the results as in Tables 14.3 and 14.4 in known values:

14.7 Simulation Results

In order to validate the mathematical model, the proposed method was simulated using Matlab for the problem domain depicted earlier at 150 m depth with a single beacon node capable of determining the coordinates and bearings of the submerged sensors. A group of three sensors was placed at (0,0,0), (0,75,0), and (80,40,0). The coordinates of the sensors were randomly chosen, while for computational simplicity one of them was placed at the origin and the other one on the axis of the problem domain as discussed earlier. The mobile beacon was moved randomly in a plane parallel to the bottom plane defined by the positions of the sensors. While computing the coordinates of the sensors S_2 and S_3 with respect to S_1, Gaussian noise was added to the true Euclidean distances between the beacon and sensors.

14.7.1 Coordinates with Euclidean Distances

In our proposed approach, the number of beacons required was just one, floating on the surface of the water, and there was a minimum of three sensors – a standard number in environment monitoring with sensors. If there are more sensors, three at a time will be localized. Our method is capable of determining 3D coordinates with respect to the beacon node along with bearing information, which gives a better understanding of the location of the sensors, since the coordinates of the beacon node are knowable with the help of GPS.

In order to validate the mathematical model, a group of three sensors is placed randomly in the XY plane and the mobile beacon is steered above, assumed to be in a plane parallel to the bottom plane where the sensors are deployed. While the coordinates of the sensors are chosen randomly, for computational simplicity one of the sensors is marked as the origin and the other one on the y-axis of the problem domain. The third sensor could be positioned at any point of x–y plane. To get a distance measurement from six different positions of the beacon, it was randomly moved to six different coordinates, in different orientations. Mobility of the submerged sensors is not considered in the proposed mathematical model.

To prove the mathematical model, true Euclidean distances between the sensors and beacon are considered while computing the coordinates of the sensors S_2 and S_3 with respect to S_1.

The simulation results suggest that if the distances between beacon and sensors are true Euclidean then the positional errors are negligible. For a problem domain of 150 m depth, the positional errors are in the 10^{-12}–10^{-14} m range. For a sensor that is 0.5 m to several meters in length, the generated error is negligible, thus validating the proposed mathematical model. Figure 14.8 shows the accuracy and precision of the position

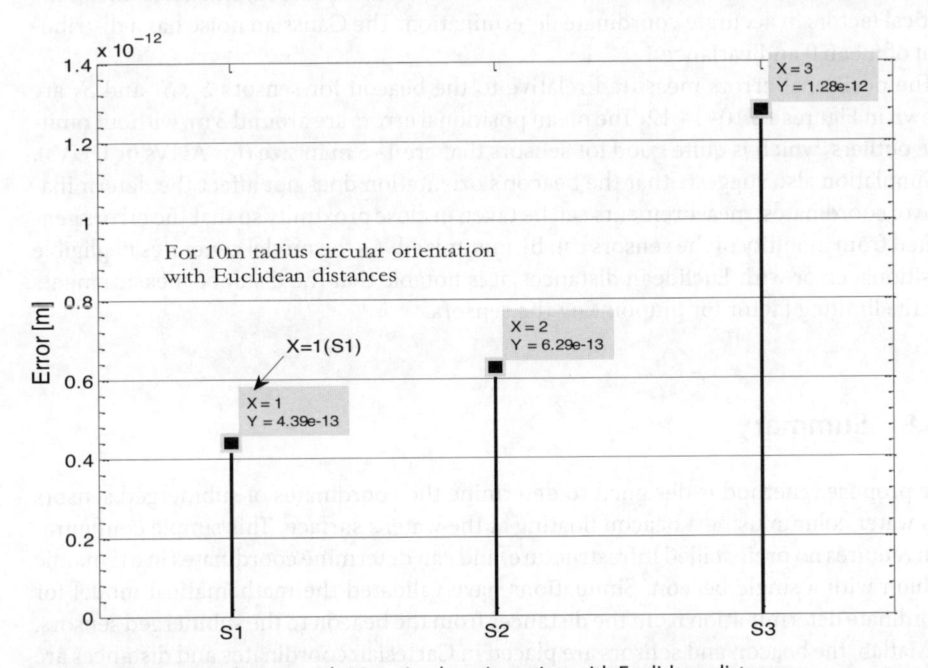

Figure 14.8 Positional errors with 10 m circular orientation with Euclidean distances.

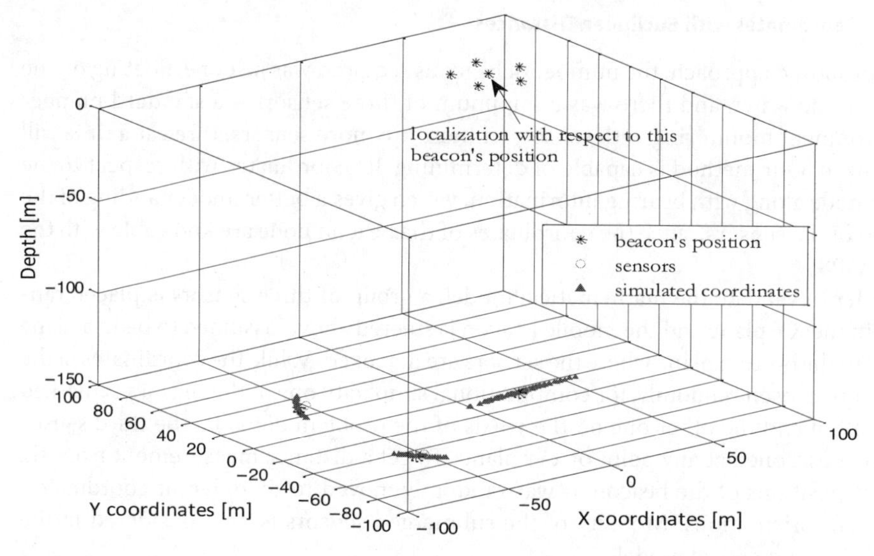

Figure 14.9 Calculated sensor positions with respect to actual coordinates.

detection. Positional errors are negligible for different orientations of the beacon; these negligible errors are generated by linearisation of a system of non-linear equations.

14.7.2 Coordinates with Gaussian Noise

Figure 14.9 shows the computed coordinates of the sensors with Gaussian noise added to the distance measurements; the precision of the distance measurements is one of the critical factors in accurate coordinate determination. The Gaussian noise has a distribution of mean 0 and variance 1.

The positional errors measured relative to the beacon for sensors S_1, S_2 and S_3 are shown in Figures 14.10–14.12. The mean positional errors are around 3 m, without omitting outliers, which is quite good for sensors that are 1–5 m in size (for AUVs or UUVs).

Simulation also suggests that the beacon's orientation does not affect the determination of coordinates; measurements can be taken in close proximity so that the errors generated from mobility of the sensors can be minimized. As the model generates negligible positional error with Euclidean distances, it is notable that the distance measurements are the limiting factor for pinpointing the sensors.

14.8 Summary

The proposed method is designed to determine the coordinates of submerged sensors in a water column using a beacon floating at the water's surface. This simple configuration requires no preinstalled infrastructure, and can determine coordinates in a dynamic fashion with a single beacon. Simulations have validated the mathematical model for coordinate determination from the distances from the beacon to the submerged sensors. In Matlab, the beacon and sensors are placed in Cartesian coordinates and distances are

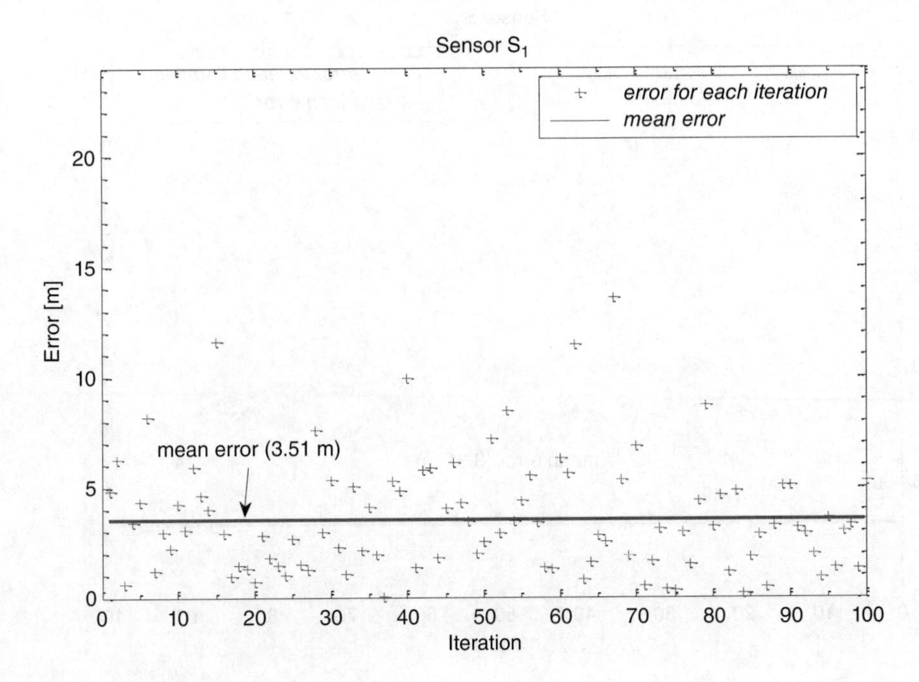

Figure 14.10 Distance error for sensor S_1.

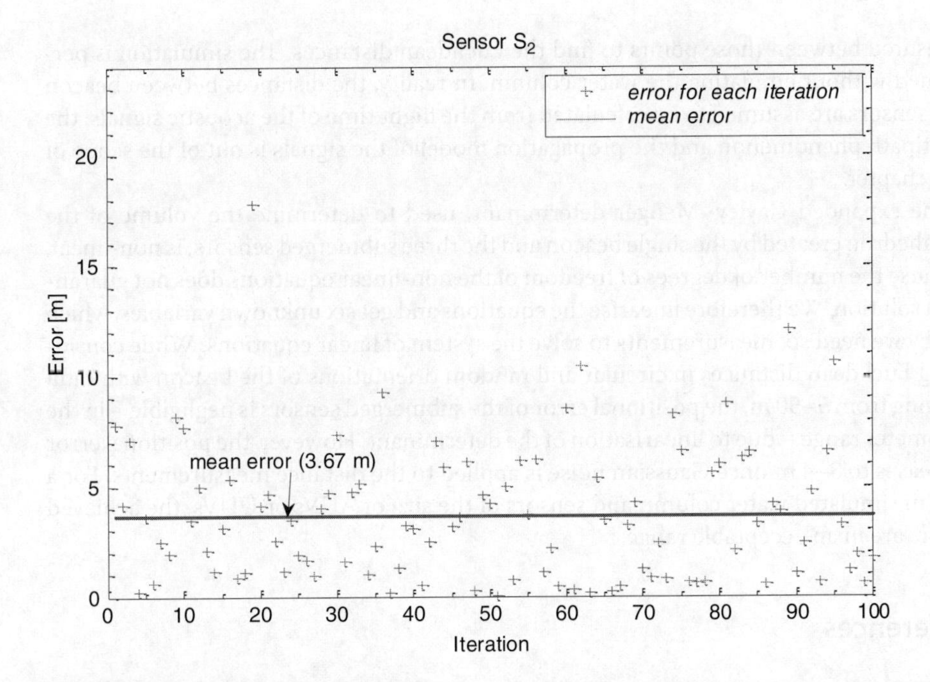

Figure 14.11 Distance error for sensor S_2.

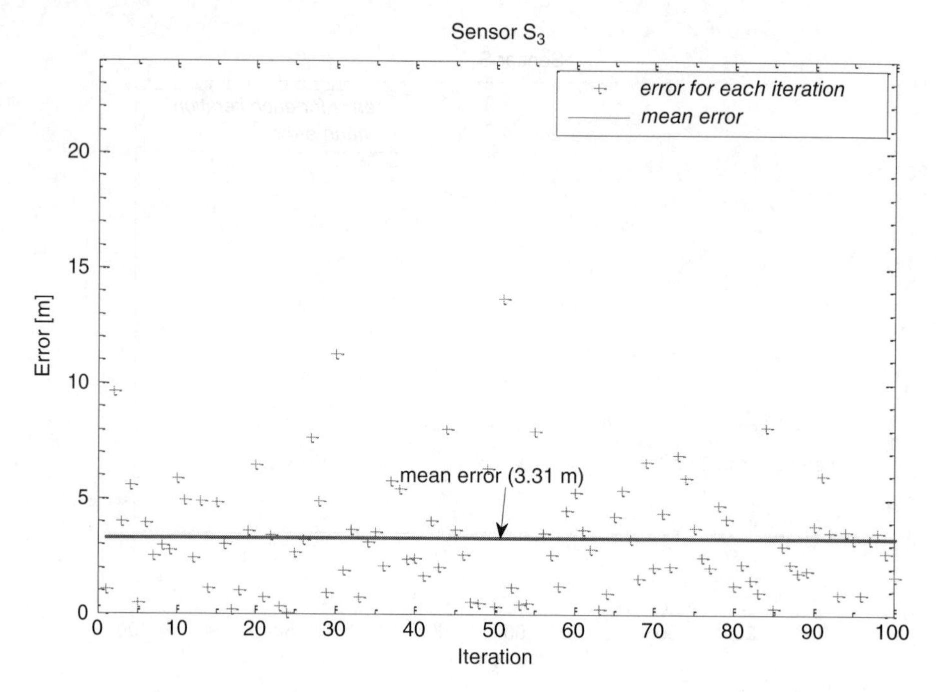

Figure 14.12 Distance error for sensor S_3.

measured between those points to find the Euclidean distances. The simulation is performed without emulating the water column. In reality, the distances between beacon and sensors are assumed to be calculated from the flight time of the acoustic signals; the multipath phenomenon and the propagation model of the signals is out of the scope of this chapter.

The expanded Cayley–Menger determinant, used to determine the volume of the tetrahedron created by the single beacon and the three submerged sensors, is non-linear, because the number of degrees of freedom of the non-linear equations does not guarantee a solution. We therefore linearise the equations and get six unknown variables, which is why we need six measurements to solve the system of linear equations. While considering Euclidean distances in circular and random orientations of the beacon with radii ranging from 5–50 m, the positional error of the submerged sensors is negligible – in the picometer range – due to linearisation of the determinant. However, the positional error increases to 3–4 m once Gaussian noise is applied to the distance measurements. For a 150-m simulated water column and sensors of the size of AUVs or UUVs, the achieved errors are in an acceptable range.

References

1 I.F. Akyildiz, D. Pompili, and T. Melodia, 'Underwater acoustic sensor networks: research challenges,' *Ad Hoc Networks*, vol. **3**, pp. 257–279, 2005.

2 J.H. Cui, J. Kong, M. Gerla, and S. Zhou, 'The challenges of building mobile underwater wireless networks for aquatic applications,' *Network, IEEE*, vol. **20**, pp. 12–18, 2006.

3 P. Xie, J.H. Cui, and L. Lao, 'VBF: vector-based forwarding protocol for underwater sensor networks,' *Networking 2006. Networking Technologies, Services, and Protocols; Performance of Computer and Communication Networks; Mobile and Wireless Communications Systems*, pp. 1216–1221, 2006.

4 W.S. Burdic, *Underwater Acoustic System Analysis* vol. **113**. Peninsula Publishing, 2002.

5 M. Erol-Kantarci, H.T. Mouftah, and S. Oktug, 'A survey of architectures and localization techniques for underwater acoustic sensor networks,' *Communications Surveys & Tutorials, IEEE*, vol. **13**, pp. 487–502, 2011.

6 A. Novikov and A.C. Bagtzoglou, 'Hydrodynamic model of the lower Hudson river estuarine system and its application for water quality management,' *Water Resources Management*, vol. **20**, pp. 257–276, 2006.

7 A.C. Bagtzoglou and A. Novikov, 'Chaotic behavior and pollution dispersion characteristics in engineered tidal embayments: a numerical investigation,' *Journal of the American Water Resources Association*, vol. **43**, pp. 207–219, 2007.

8 Z. Zhou, Z. Peng, J.-H. Cui, Z. Shi, and A.C. Bagtzoglou, 'Scalable localization with mobility prediction for underwater sensor networks,' *Mobile Computing, IEEE Transactions on*, vol. **10**, pp. 335–348, 2011.

9 H.P. Tan, R. Diamant, W.K.G. Seah, and M. Waldmeyer, 'A survey of techniques and challenges in underwater localization,' *Ocean Engineering*, vol. **38**, pp. 1663–1676, 2011.

10 Y.M. Kwon, K. Mechitov, S. Sundresh, W. Kim, and G. Agha, 'Resilient localization for sensor networks in outdoor environments,' in *Distributed Computing Systems, 2005. ICDCS 2005. Proceedings. 25th IEEE International Conference on*, 2005, pp. 643–652.

11 X. Che, I. Wells, G. Dickers, P. Kear, and X. Gong, 'Re-evaluation of RF electromagnetic communication in underwater sensor networks,' *Communications Magazine, IEEE*, vol. **48**, pp. 143–151, 2010.

12 N.P. Fofonoff and R.C. Millard, 'Algorithms for computation of fundamental properties of seawater,' *Unesco Technical Papers in Marine Science*, vol. **44**, p. 53, 1983.

13 K.V. Mackenzie, 'Nine-term equation for sound speed in the oceans,' *The Journal of the Acoustical Society of America*, vol. **70**, p. 807, 1981.

14 G. Han, J. Jiang, L. Shu, Y. Xu, and F. Wang, 'Localization algorithms of underwater wireless sensor networks: A survey,' *Sensors*, vol. **12**, pp. 2026–2061, 2012.

15 G. Borriello and J. Hightower, 'A survey and taxonomy of location systems for ubiquitous computing,' *IEEE Computer*, vol. **34**, pp. 57–66, 2001.

16 J. Guevara, A. Jiménez, J. Prieto, and F. Seco, 'Auto-localization algorithm for local positioning systems,' *Ad Hoc Networks*, vol. **10**, pp. 1090–1100, 2012.

15

Underwater and Submerged Wireless Sensor Systems: Security Issues and Solutions

Kübra Kalkan[1], *Albert Levi*[2] *and Sherali Zeadally*[3]

[1] *Boğaziçi University, Istanbul, Turkey*
[2] *Sabancı Üniversitesi, Istanbul, Turkey*
[3] *University of Kentucky, Lexington, US*

15.1 Introduction

One of the areas that has attracted considerable attention from the network research community recently is the use of wireless sensor systems for underwater aquatic applications. Some researchers have worked on the specific challenges of aquatic environments [1–3], whereas others have proposed new network protocols [4], network designs [5] and simulators of aquatic environments [6]. The application areas of these studies include military surveillance, oceanographic data collection, pollution control, climate recording and industrial products [7].

An underwater wireless sensor system (UWSS) can be defined as a wireless communication system that consists of several battery-powered underwater sensor nodes [23]. The most important features of the underwater environment are the existence of currents, high pressure and the presence of marine organisms, and these result in specific challenges. One of these relates to communication frequency. Radio frequency, which is used for airborne wireless communication, is not suitable for the underwater environment. For this reason, acoustic waves must be used in all underwater communications, which brings particular challenges [2, 8, 9] such as large latency, low bandwidth and high error rates. These all must be considered in underwater modeling [1]. In addition, since there are currents and marine organisms underwater, mobility also becomes a significant issue. As UWSSs represent a fairly recent research area, the most common issues that have been addressed by researchers to date are synchronization [11], data gathering [10], localization [12], routing protocols [13, 14], energy minimization and Media Access Control (MAC) [15, 16].

If sensitive data is collected, then it must be strictly protected from unauthorised access by third parties. For instance, an oil company may be looking for a place to drill. This company will not want to share the data with its competitors. In addition, the military uses UWSSs in order to detect enemy forces within its marine territories. Ecological and oceanographic data can sometimes be crucial for government policies. For these reasons, the security of UWSSs is vital. However, UWSSs are prone to node-capture and other kinds of malicious attack due to the underwater channel characteristics

Wireless Sensor Systems for Extreme Environments: Space, Underwater, Underground, and Industrial.
First Edition. Edited by Habib F. Rashvand and Ali Abedi.

(such as high bit-error rates, propagation delays and low bandwidth). In this chapter, we discuss security breaches in UWSSs and the security solutions that can prevent them.

The rest of the chapter is organized as follows. In Section 15.2, we present some background information about UWSSs and discuss some of its design and security-related challenges. In Section 15.3, we discuss security issues of UWSSs. In Section 15.4, we give some future research challenges and opportunities for security in UWSSs. Finally, we make some concluding remarks in Section 15.5.

15.2 Underwater Wireless Sensor Systems

UWSSs are similar to terrestrial wireless sensor systems in terms of their functions, structure and energy limitations [17]. For instance, they can measure different types of physical properties, such as temperature, sound or pressure [18]. They can track an object or monitor the surrounding environment to collect data [19, 22]. They have limited battery lives, limited memory and very limited data-processing capacity, just as terrestrial nodes do. However, in addition to these features they have other characteristics because of their deployment underwater:

More expensive hardware: Since UWSSs can be exposed to high pressure, humidity, water currents and marine organisms, the hardware of the sensors should be more robust. Consequently, they can become very expensive.

Scarcity in deployment: Since sensors for UWSS are more expensive, they are not used as readily as terrestrial ones, resulting in smaller production [17].

Longer-distance communication: Since the usage of UWSS sensors is scarce, to provide coverage and connectivity between nodes, they need to communicate over longer distances.

Higher-power requirement: Since the sensors for UWSSs need to communicate over longer distances, they need to have more power. In addition, they cannot be charged with solar power, as is the case for terrestrial sensors.

Acoustic frequency usage: Radio frequency does not propagate well underwater, because water absorbs energy [24]. UWSS sensor nodes should therefore communicate using acoustic waves.

High propagation delay: One of the most unfortunate features of acoustic waves is their speed. Since the speed of the wave decreases in water, the propagation delay increases. The propagation speed of acoustic signals in water is five-fold smaller than the propagation speed of radio waves in free space [17].

Higher multipath fading: Since the network topology changes due to the existence of currents, multipath fading effects increase rapidly.

Higher attenuation: Related to their speed in water, acoustic waves are attenuated more than they are in free space.

Low bandwidth: The available bandwidth is reduced when distances increase [25]. For this reason, as UWSSs need to communicate over large distances, they have low bandwidth.

High bit-error rate: Since the quality of the channel decreases with high propagation delays, multipath fading, attenuation and bit-error rates increase.

Mobility: In underwater networks, there are external factors, such as currents, winds and underwater organisms, which can cause the nodes to drift. For this reason, modeling of underwater networks should consider mobility. In addition, the nodes' positions cannot be detected via GPS, the signal of which is not suitable for acoustic environments. Caruso et al. proposed a mobility model for underwater sensor networks called the meandering current mobility model [7], in which nodes are moved by the effects of meandering sub-surface currents and vortices. This model is for large ocean environments that span several kilometres. The authors consider the paths of nodes to be deterministic and that nearby sensors move in a similar way to each other. To simulate the mobility of nodes, it is important to model the movement of the ocean in which they are immersed. The movement of the nodes over three days is depicted in Figure 15.1. This model is more realistic than other group mobility models for wireless sensor networks because the nodes drift according to the movement of the ocean.

The need for suitable network architecture: Technologies such as GPS, which are used to control the location of nodes, cannot be used underwater. In most localization schemes for UWSSs, some reference nodes are used. These reference nodes' locations are known and they are used to determine other nodes' positions by calculating the distances from the reference nodes [15]. This allows the structure to be calculated in a hierarchical way. In this structure, some nodes are special. One such hierarchical underwater sensor network scheme was proposed by Zhou et al. [27]. This scheme consists of three types of node: surface buoys, anchor nodes and ordinary sensor nodes. Each surface buoy is equipped with a GPS device. All the anchor nodes can estimate their positions by directly contacting the surface buoys. Localization of the ordinary nodes is also determined through the anchor nodes. In another scheme, 'Dive and Rise' (DNR) positioning was proposed [28]. In this scheme, DNR beacons are equipped with GPS equipment. The beacons move along the y-axis. When they come to the surface, they learn their locations from the GPS system. When they dive into the water, they broadcast their position information in order to help the other nodes determined their own positions. Another scheme consists of four types of node: surface buoys, detachable elevator transceivers (DETs), anchor nodes and ordinary nodes [29]. In this scheme, surface buoys are equipped with GPS. The DETs are attached to a surface buoy, which can rise up an 'elevator' to broadcast its position, before falling again. This scheme increases the localization accuracy and decreases the cost of the system. Figure 15.2 shows the hierarchical design used. There are three types of nodes organized in a heirarchical way. Surface buoys communicate amongst themselves through the air. Each surface buoy is attached to an elevator and can moves up and down it. Each group of nodes can only communicate with its own elevator. Also, each group of nodes can communicate within its group. Thus, it is obvious that there is a hierarchical design.

Dynamic network topology: Due to the mobility of the nodes, the network topology changes. Underwater objects may move at speeds of 3–6 km/h in typical underwater conditions [7]. While routing the communication packets, sensors should be able handle the dynamic conditions under the water. Geographical routing is promising for UWSSs because of its scalability and limited signalling properties [17].

All the challenges described above should be considered when deploying an underwater sensor network. In addition, many of the characteristics of UWSSs require efficient and robust security solutions. For instance, poor network conditions can result in significant loss of security packets. Additionally, physical security is important issue because the nodes are deployed in a hostile environment, where eavesdropping of communications is possible and information exchanged can be modified. Therefore, securing UWSSs is vital. In Section 15.3, we discuss security issues of UWSSs and present solutions to address them.

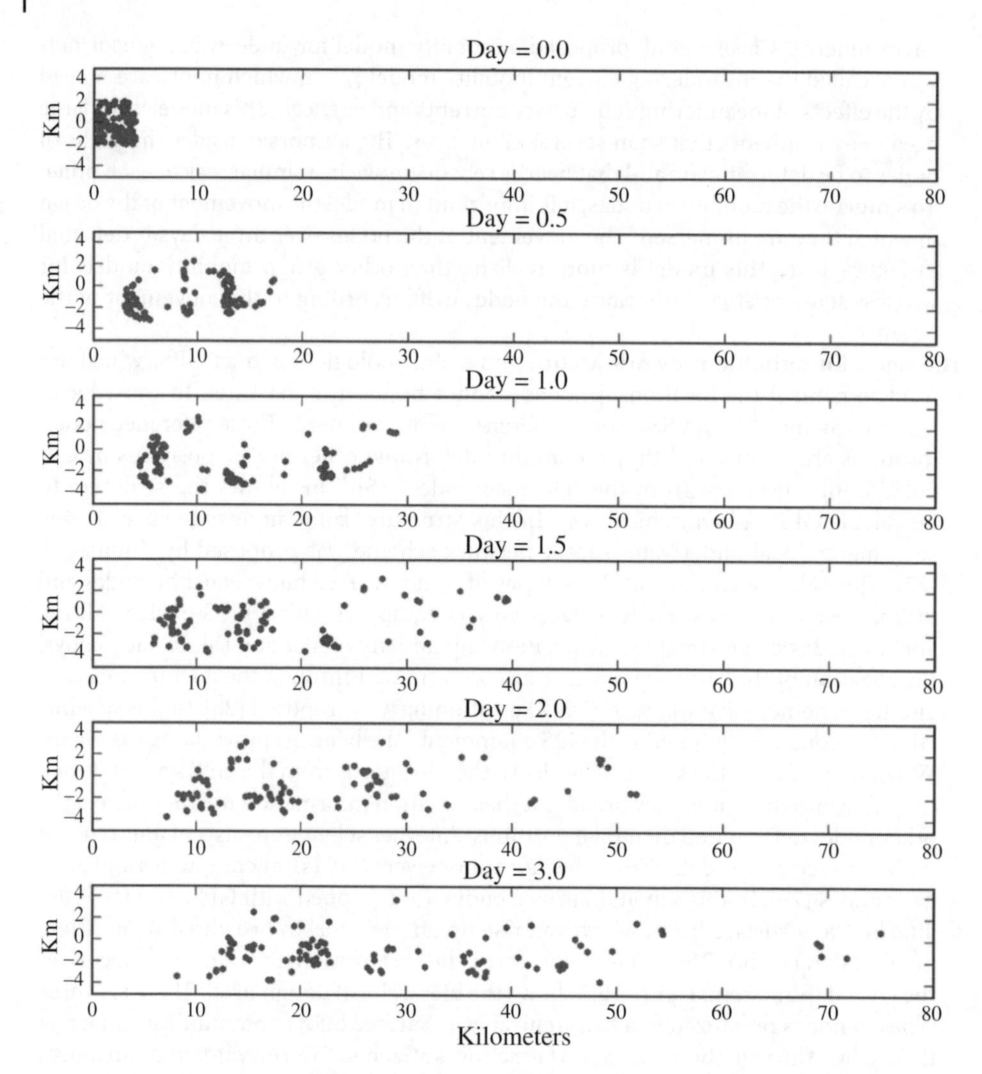

Figure 15.1 Time evolution of the position of one hundred sensors randomly released in a square 4 km on each side [7].

15.3 Security Requirements, Issues and Solutions

In this section, we describe the security requirements that need to be satisfied to secure UWSSs. This is followed by a discussion of the related security issues and the solutions to address them.

15.3.1 Security Requirements

For wireless applications, an adversary node can not only eavesdrop on the traffic but can also can interrupt and modify messages [18, 20]. Security requirements that apply to terrestrial wireless applications are also valid for underwater sensor networks [30].

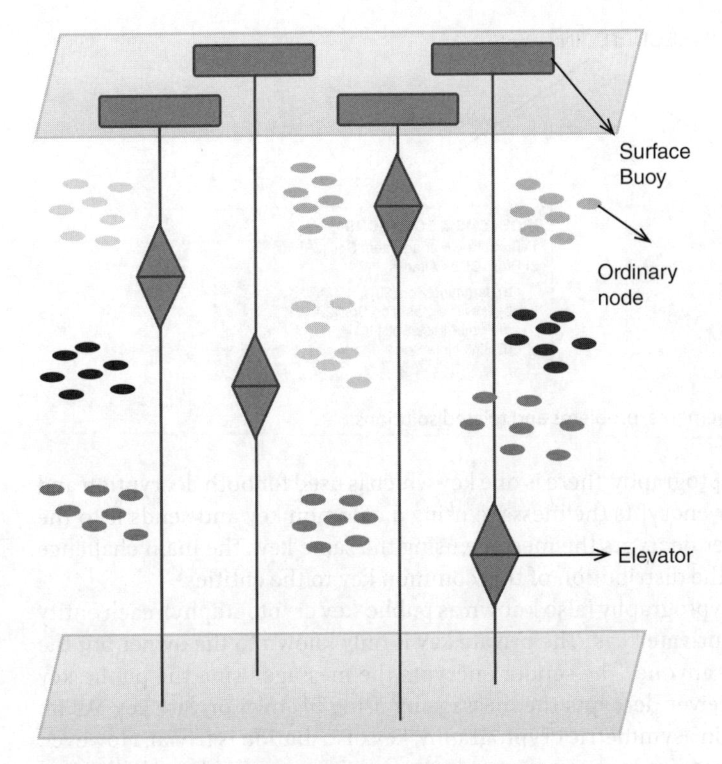

Figure 15.2 Network design of a hierarchical UWSS [39].

These are:

- *Data confidentiality* is the protection of data from unauthorised parties against eavesdropping. It is ensured by encryption of messages with a secret key.
- *Integrity* is the assurance that the message received is exactly same as the message that is sent by the authorised party. In other words, if integrity is provided, then there is no insertion into or deletion or modification of the message.
- *Freshness* means that the data in the message is recent and that it is not simply an old message resent.
- *Availability* means that a wireless sensor network can provide service whenever it is needed.
- *Authentication* is the assurance that the communicating entity is the one it claims to be.

Security issues related to these requirements and proposed solutions are presented in Figure 15.3 and explained in Section 3.2.

15.3.2 Security Issues and Solutions

15.3.2.1 Key Management

Cryptographic mechanisms are used to enforce authentication, data confidentiality and integrity requirements. There are two types of cryptographic mechanism for encryption: asymmetric key cryptography and symmetric key cryptography.

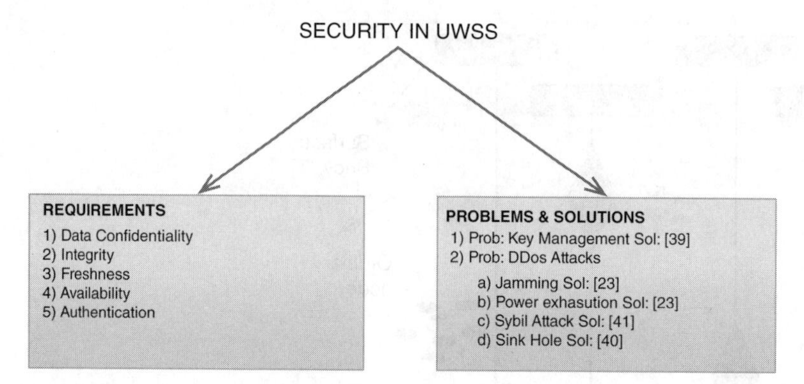

Figure 15.3 Security requirements, problems and related solutions.

In symmetric key cryptography, there is one key which is used for both decryption and encryption. The sender encrypts the message using a common key and sends it to the other party. The receiver decrypts the message using the same key. The main challenge with symmetric key is the distribution of this common key to the entities.

In asymmetric key cryptography (also known as public-key cryptography), each entity has its own public and private keys. The private key is only known to the owner but the public key is known to anyone. The sender encrypts the message using the public key of the sender. Then receiver decrypts the message by using his own private key. As no common keys are used in asymmetric cryptography, key distribution is trivial. However, public-key operations require more energy and computational power. Due the limited battery life of sensor nodes, public key cryptography is not preferred for wireless sensor networks, so symmetric keys are used for wireless sensor networks and similarly for UWSSs [31, 32].

Many researchers have studied distribution of symmetric keys and have proposed schemes to address it [18, 33–38]. It is not a trivial question, because there is a trade-off between memory and resilience. If only one pairwise key is used in the entire network, then the adversary can compromise all nodes by capturing only one node. This means that the network is not resilient to mitigate attacks. However, if different pairwise keys are generated for each pair, the sensor network becomes more resilient to capture attacks because if a node is captured it cannot learn anything about other links. However, in this model each node should store n-1 keys, where n is the number of nodes in network. As nodes have limited memory, it is not possible for a node to store this amount of key information. Hence, it is not easy to handle resilience and memory issues in key distribution. In addition to these challenges, UWSSs face additional security challenges, as explained in Section 15.2.

A solution to the key management problem is proposed by Kalkan and Levi [39]. A key distribution model for UWSSs was applied for two group mobility models:

- the nomadic mobility model
- the meandering current mobility model.

The nomadic mobility key-distribution scheme works in three dimensions. It is suitable only for small coastal areas. On the other hand, the meandering mobility key-distribution model is two dimensional and spans several kilometres in the open

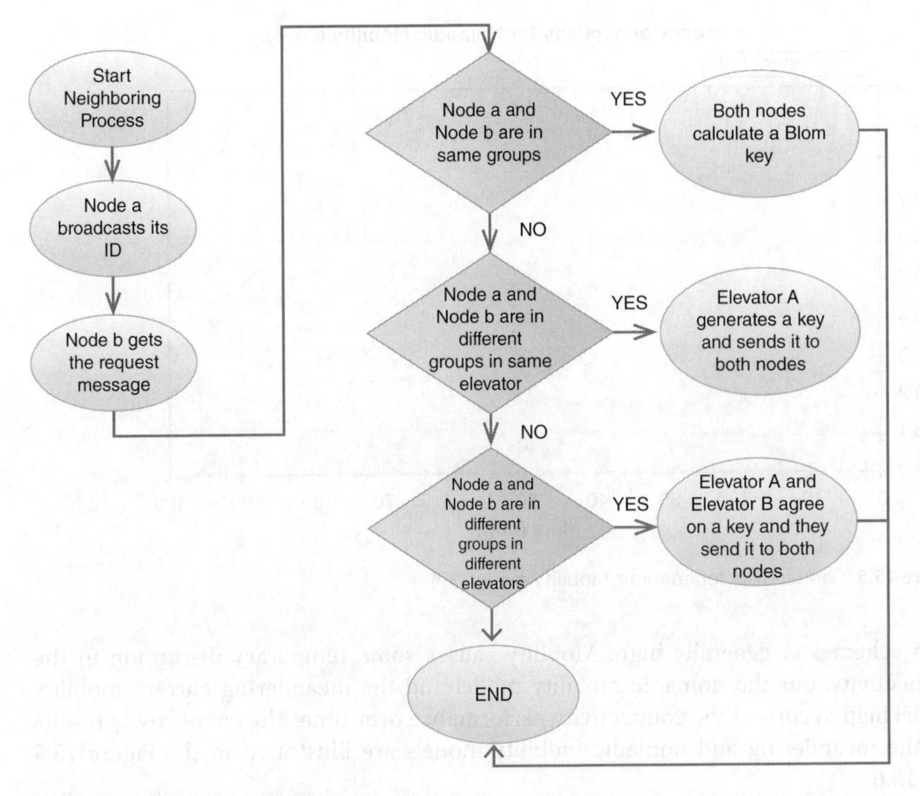

Figure 15.4 Node communication process.

sea. In both schemes, a hierarchical structure is used, with surface buoys, elevators and ordinary nodes. Communicating nodes can be in the same group, in different groups in same elevator, or in different groups on different elevators. The communication scheme for two nodes, a and b, is illustrated in Figure 15.4. If the nodes are in same group, they calculate their own key. If they are in different groups on the same elevator, their elevator generates a key and sends it to both. If the nodes are on different elevators, then the elevators agree on a key and send it to both nodes. Secure and resilient group communication is handled via the well-known Blom key-distribution scheme [37]. In random pool-based schemes, there is no guarantee that two neighbors will share a common key and this means that they need to communicate via other secure links. This in turn increases the communication overhead. In contrast, Blom's scheme guarantees that any two nodes in a group can generate a common key. This is a matrix-based solution. Also, Blom's scheme has the λ-secure property, where λ is a threshold for a captured number of nodes. This property suggests that the network remains perfectly secure until λnodes are captured. However, if more than λ nodes are captured, all the keys in the group are revealed. By using the Blom scheme, the negative effects of the dynamic network topology are reduced.

Secure connectivity and resilience are normal measures of the performance of a key-management scheme. Secure connectivity is the probability of sharing a common key between any two neighbor nodes. Resilience is given in terms of a compromised-links ratio. Kalkan and Levi [39] showed that secure connectivity for

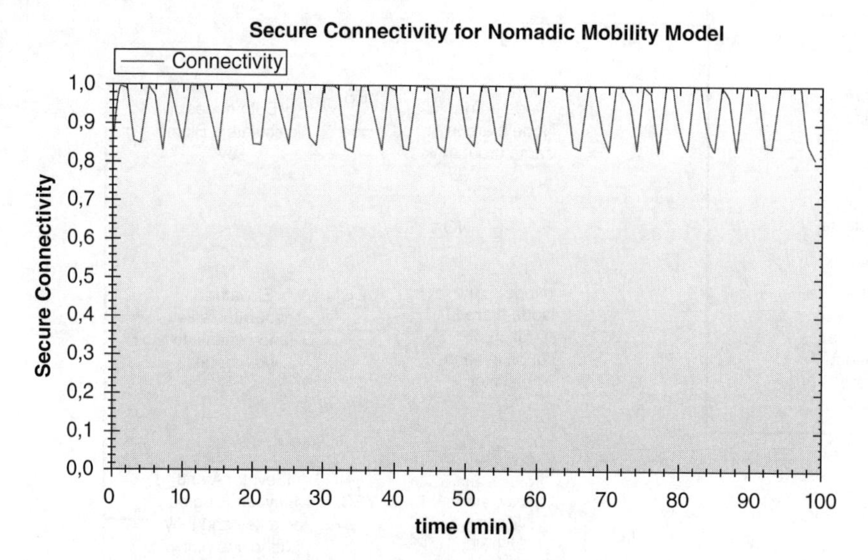

Figure 15.5 Connectivity for nomadic mobility model [39]

both schemes is generally high. Mobility causes some temporary disruption in the connectivity, but the nomadic mobility model and the meandering current mobility model help to correct the connectivity performance over time. The connectivity results for the meandering and nomadic mobility models are illustrated in the Figure 15.5 and 15.6.

Moreover, these models give good resilience because even if some nodes are captured by an adversary, only a small number of links between uncaptured nodes is compromised.

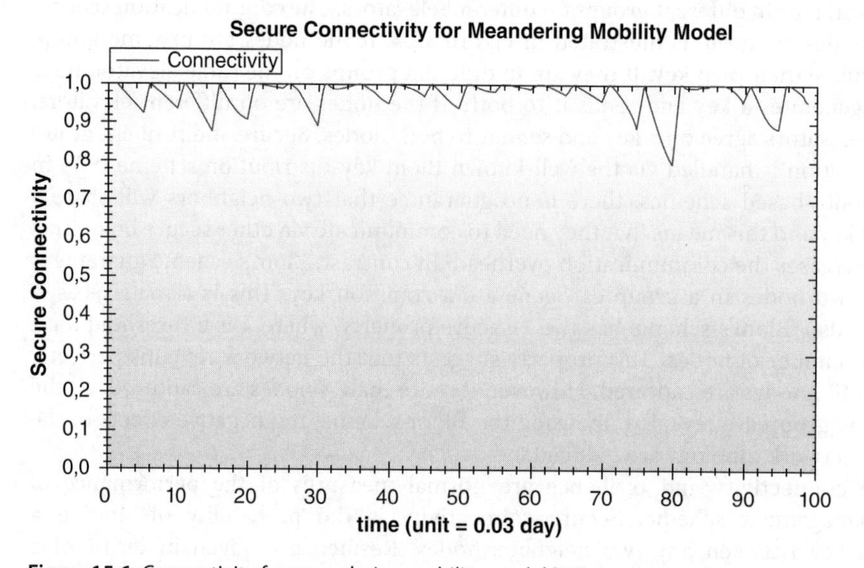

Figure 15.6 Connectivity for meandering mobility model [39]

15.3.2.2 Denial of Service Attacks

Distributed Denial of Service (DDoS) attacks involve flooding a system from several other machines. The aim of a DDoS attack is to use all available services and prevents innocent users' accessing the system. Since attackers are also appear to be acting as innocent users and attack packets do not contain any malicious parts, mitigating DDoS attacks remains a significant challenge. There are several types of DDoS attacks, affecting different layers of the system. Since UWSSs have security vulnerabilities, as discussed in Section 15.2, they are prone to DDoS attacks. We discuss some of these attacks and possible solutions to mitigate them below.

Jamming: This attack targets the physical layer. Attackers emit noise signals onto the communication channel. Other users in the area cannot send or receive packets because their communication bands are being used by the attackers. Since UWSSs operate at low frequencies and have high latencies, they can be seriously affected by jamming attacks. The solution is to have sensor nodes in UWSSs use lower duty cycles [23]. Since the available power is limited for sensor nodes, they can sleep during jamming attacks. They awaken only when they need to send or receive data; otherwise, they sleep. When nodes need to send/receive data, they wake up and check if the jamming is still happening. When nodes are not sending data, they continue to sleep and periodically wake up to check if data is available for transmission or receipt.

Power exhaustion: Attackers make continuous Request To Send (RTS) requests and repeated retransmissions in order to reduce the lifetimes of the nodes. This attack is aimed at the data link layer. In this case, the MAC protocol is kept busy with additional retransmissions. Since the power of UWSS nodes is limited, this attack is quite harmful to the nodes. The solution is to limit the number of retransmissions [18, 23].

Sybil attack: In this type of attack, the attacker has multiple identities, each pretending to be in a different place [17]. In this way, a DDoS attack can be launchead fairly easily. This kind of attack can generally be mitigated using location verification techniques [42, 43]. However, the UWSS environment is not suitable for this type of prevention because mobility cannot be easily handled in location verification techniques. Identity authentication is another prevention technique. Mukhopadhyay and Indranil [41] analysed end-to-end authentication techniques for UWSSs. They focused on digital signature techniques. Digital signatures are an authentication mechanism that shows a message has been created by a known user. It also provides non-repudiation, such that the sender cannot deny that he sent the message. The authors' results showed that digital signature schemes that perform well with terrestrial schemes do not necessarily perform well for UWSSs. Their results can be used to determine the properties that are needed for a technique that will work on UWSSs. However, to the best of our knowledge, there is no authentication technique suitable for UWSSs. This remains an open area of future research.

Sinkhole attack: Here, the attacker attracts most of the traffic by pretending to have a high-quality route [17]. Geographic routing has been proposed to mitigate this kind of attack. Here, the source sends a message using the geographic position of the destination instead of using the network address of the destination. Since mobility is an important issue that needs to be considered for UWSSs, secure routing protocols are needed. Several secure routing protocols have been proposed in the literature and they have been described by Souiki et al, [40]. Secure routing protocols that are specifically designed for UWSSs have been classified according to the packet-forwarding strategies used: greedy, restricted directional flooding and hierarchical [40]. These protocols are appropriate for UWSSs since they consider reliability, mobility and the three-dimensional environment of UWSSs.

15.4 Future Challenges and Research Directions

In this section, we discuss some of the security challenges that are still to be addressed by researchers and designers of UWSSs. These challenges are related to security of localization, cross-layer design and time synchronization. All these issues are explained in following subsections.

15.4.1 Secure Localization

Localization is used in packet-routing decisions. Sensor nodes decide how to relay packets according to the positions of neighbor nodes. Terrestrial techniques cannot be used for UWSSs because of the long propagation delays, multi-path propagation and fading [17]. In addition, the mobility of nodes in the underwater environment makes position estimation difficult. Several models have been proposed for UWSSs [44–47]. However, none has focused on the security of these localization techniques. As stated in Section 15.3, DDoS attacks such as the Sybil and sinkhole attack abuse the location information of nodes. To prevent such attacks, cryptographic mechanisms need to be used in the positioning techniques. In addition, resilient algorithms can be used that give true position estimation even when the system is under Sybil or sinkhole attack [17]. Secure localization is therefore one of the most important open issues in UWSSs.

15.4.2 Secure Cross-layer Design

Current security research in UWSSs has focused on only individual layers of the sensor nodes. However, Zeadally and Siddiqui [18] argue that layered security results in several inadequacies.

First, the layered-security approach results in redundant security services. Each layer tries to provide security services on its own. For instance, the network layer deals with secure localization, routing and time synchronization, whereas the TCP layer and the physical layer protect themselves from DDoS attacks. As each layer focuses on its own problems, all layers are left to deal with key management and various cryptographic issues individually. However, if all these layers are considered simultaneously, this redundancy in security services can be reduced.

Second, layered security is non-adaptive, which makes it unsuitable for complex and unstable environments such as UWSSs. Cross-layer design can address some of the dynamic demands of the underwater environment. In addition, a layered architecture cannot provide protection against cross-layer attacks. In contrast, cross-layer designs can better handle these types of attack. Finally, a cross-layer design means lower energy consumption, because there is an overall view of the entire system. Consequently, the appropriate security services can be chosen, thus minimizing energy consumption. In the future, more research efforts are needed to develop secure cross-layer design architectures and protocols for UWSSs.

15.4.3 Secure Time Synchronization

Time synchronization is essential in scheduling protocols such as time division multiple access, which requires precise timing information to determine sleep and wake-up intervals [17]. Due to the conditions of the underwater environment, terrestrial models cannot be applied to UWSSs. Several models have been proposed for time synchronization in UWSSs [48–50]. However, none has considered security. Time-related attacks, such as replay attacks, can be mitigated using cryptographic techniques [26]. However, delay attacks, in which transmission of time synchronization messages are deliberately delayed, cannot. To prevent such attacks, other types of countermeasure should be applied. Song et al.[26] looked at various terrestrial countermeasures, but these cannot be applied to UWSSs due to the long propagation delays and the mobility of the nodes. Thus, in the future, efficient, novel secure time-synchronization schemes will be required to counter delay attacks.

15.5 Summary

One very recent application area of sensors is the underwater environment. Underwater communications are very different from airborne communication, mainly due to the fact that radio frequencies cannot be used. Instead, acoustic waves are used but they come with new and additional security challenges that need to be addressed.

Highly variable water currents, high pressure, and marine organisms are also challenges. In addition, security in UWSSs is of great importance because the sensitive underwater environment means that UWSS are liable to malicious attacks. We have discussed various characteristics of UWSSs and identified the security requirements and associated vulnerabilities that arise in the underwater environment. We have also discussed some possible solutions that can mitigate against attacks on UWSSs. Finally, we have highlighted some future challenges that still need to be addressed by security researchers and designers focusing on the security aspects of UWSSs.

References

1 J.H. Cui, J. Kong, M. Gerla, and S. Zhou, 'The challenges of building scalable mobile underwater wireless sensor networks for aquatic applications', *IEEE Network*, **20** (3), 12–17, 2006.

2 I.F. Akyildiz, D. Pompili, and T. Melodia, 'Challenges for efficient communication in underwater acoustic sensor networks', *ACM SIGBED Review*, **1** (1), 3–8, 2004.

3 J. Heidemann, W. Ye, J. Wills, A. Syed, and Y. Li, '*Research challenges and applications for underwater sensor networking*', In IEEE Wireless Communications and Networking Conference, Las Vegas, Nevada, USA, April 2006.

4 G.G. Xie and J. Gibson, A networking protocol for underwater acoustic networks. Technical report TR-CS-00-02, Department of Computer Science, Naval Postgraduate School, 2000.

5 J. Proakis, E.M. Sozer, J.A. Rice, and M. Stojanovic, 'Shallow water acoustic networks', *IEEE Communications Magazine*, **39** (11), 114–119, 2001.

6 P. Xie, Z. Zhou, Z. Peng, H. Yan, T. Hu, J.H. Cui, Z. Shi, Y. Fei and S. Zhou, 'Aqua-Sim: an NS-2 based simulator for underwater sensor networks'. In *OCEANS 2009, MTS/IEEE Biloxi-Marine Technology for Our Future: Global and Local Challenges*, 2009.
7 A. Caruso, F. Paparella, L.F.M. Vieira, M. Erol, and M. Gerla, 'The meandering current mobility model and its impact on underwater mobile sensor networks', In *Proceedings of International Conference on Computer Communications 2008*, pp. 221–225.
8 J. Partan, J. Kurose, and B.N. Levine, 'A survey of practical issues in underwater networks', in *International Conference on UnderWater Networks and Systems (WUWNet'06)*, Los Angeles, CA, USA, 2006, pp. 17–24.
9 J. Kong, J. Cui, D. Wu, and M. Gerla, 'Building underwater ad hoc networks and sensor networks for large scale real-time aquatic applications', in *IEEE Premier Military Communications Event (MILCOM'05)*, Atlantic City, NJ, USA, 2005.
10 I. Vasilescu, K. Kotay, D. Rus, M. Dunbabin, and P. Corke, 'Data collection, storage, and retrieval with an underwater sensor network', in *Embedded Networked Sensor Systems (SenSys'15)*, San Diego, California, USA, 2005, pp. 154–165.
11 A. Syed and J. Heidemann, 'Time synchronization for high latency acoustic networks', in *Proceedings of International Conference on Computer Communications (Infocom)*, Barcelona, Spain, April 2006, pp. 1–12.
12 V. Chandrasekhar, W.K. Seah, Y.S. Choo, and H.V. Ee, 'Localization in underwater sensor networks: survey and challenges', in *International Conference on Underwater Networks and Systems (UWNet06)*, Los Angeles, CA, USA, 2006, pp. 33–40.
13 D. Pompili and T. Melodia, 'Three-dimensional routing in underwater acoustic sensor networks', in *PE-WASUN '05: Proceedings of the 2nd ACM International Workshop on Performance Evaluation of Wireless Ad hoc, Sensor, and Ubiquitous Networks'*, Montreal, Quebec, Canada, 2005, pp. 214–221.
14 P. Xie, J. Cui, and L. Lao, 'VBF: Vector-based forwarding protocol for underwater sensor networks', in *Proceedings of IFIP Networking'06*, Portugal, May 2006, pp. 1216–1221.
15 N. Chirdchoo, W.-S. Soh, and K.C. Chua, 'Aloha-based MAC protocols with collision avoidance for underwater acoustic networks', in *International Conference on Computer Communications, INFOCOM 2007*, Anchorage, Alaska, USA, May **2007**, pp. 2271–2275.
16 D. Makhija, P. Kumaraswamy, and R. Roy, 'Challenges and design of MAC protocol for underwater acoustic sensor networks', in *4th International Symposium on Modeling and Optimization in Mobile, Ad Hoc and Wireless Networks*, Boston, Massachusetts, USA, **3–6** April 2006, pp. 1–6.
17 M. Domingo, 'Securing underwater wireless communication networks', *IEEE Wireless Communications*, **18** (1), 22–28, 2011.
18 S. Zeadally and F. Siddiqui, 'Security protocols for wireless local area networks (WLANs) and cellular networks', *Journal of Internet Technology*, **8** (1), 11–25, 2007.
19 I.F. Akyildiz, W. Su, Y. Sankarasubramaniam, and E. Cayirci, 'Wireless sensor networks: a survey', *Computer Networks*, **38** (4): 393–422, 2002.
20 D.P. Agrawal and Q-A. Zeng, *Introduction to Wireless and Mobile Systems*, Brooks/Cole Publishing, 2003.

21 N. Jain and D.P. Agrawal, 'Current trends in wireless sensor network design', *International Journal of Distributed Sensor Networks*, **1** (1), 101–122, 2005.

22 D.W. Carman, P.S. Kruus, and B.J. Matt, 'Constraints and approaches for distributed sensor network security', Technical Report 00-010, NAI Labs, 2000.

23 Y. Cong, G. Yang, Z. Wei, and W. Zhou, 'Security in underwater sensor network', *IEEE 2010 International Conference on Communications and Mobile Computing (CMC), April* 2010, vol. **1**, pp. 162–168.

24 J. Llor and M.P. Malumbres, 'Modeling underwater wireless sensor networks', Wireless sensor networks: application-centric design, *InTech Open*, pp. 185–203.

25 M. Stojanovic, 'On the relationship between capacity and distance in an underwater acoustic communication channel', *ACM Mobile Computing and Communications Review* **11**, 34–43, 2007.

26 H. Song, S. Zhu, and G. Cao, 'Attack-resilient time synchronization for wireless sensor networks', *Ad Hoc Networks*, **5** (1), 112–125, 2007.

27 Z. Zhou, J.-H. Cui, and S. Zhou, 'Localization for large-scale underwater sensor networks', Technical report UbiNet-TR06-04, University of Connecticut Computer Science and Engineering, 2004.

28 M. Erol, L.F. Vieira, and M. Gerla, 'Localization with Dive 'n' Rise (DNR) beacons for underwater acoustic sensor networks', *Proceedings of the Second Workshop on Underwater Networks*, 2007, pp. 97–100.

29 K. Chen, Y. Zhou, and J. He, 'A localization scheme for underwater wireless sensor networks', in *International Journal of Adanced Science and Technology* **4**, 9–16, 2009.

30 X. Chen, K. Makki, and N. Pissinou, 'Sensor network security: a survey', *IEEE Communications Surveys and Tutorials*, **11** (2), 52–73,2009.

31 Y. Cheng and D.P. Agrawal, 'Improved pairwise key establishment for wireless sensor networks', *2006 IEEE International Conference on Wireless and Mobile Computing, Networking and Communications*, **2006**, pp. 442–448.

32 H. Cam, S. Ozdemir, D. Muthuavinashiappan, and P. Nair, 'Energy efficient security protocol for wireless sensor networks', *Vehicular Technology Conference*, 2003.

33 L. Eschenauer and V.D. Gligor, 'A key-management scheme for distributed sensor networks', in: *Proceedings of the 9th ACM Conference on Computer and Communications Security*, 2002.

34 H. Chan, A. Perrig, and D. Song, 'Random key pre-distribution schemes for sensor networks', *Proceedings of IEEE Symposium on Security and Privacy*, Berkeley, California, May 11–14 2003, pp. 197–213.

35 D. Liu and P. Ning, 'Improving key pre-distribution with deployment knowledge in static sensor networks', *ACM Transactions on Sensor Networks (TOSN)*, **1** (2), 204–239, 2005.

36 D. Liu, P. Ning, and W. Du, 'Group-based key pre-distribution in wireless sensor networks', *Proceedings of 2005 ACM Workshop on Wireless Security (WiSe 2005)*, September 2005.

37 R. Blom, 'An optimal class of symmetric key generation systems', in T. Beth, N. Cot, I. Ingemarsson (eds), *Advances in Cryptology: Proceedings of EUROCRYPT 84*, Lecture Notes in Computer Science, vol. 209, Springer-Verlag, 1985.

38 W. Du, J. Deng, Y.S. Han, and P.K. Varshney, 'A pairwise key pre-distribution scheme for wireless sensor networks', *Proceedings of the 10th ACM Conference*

on Computer and Communications Security (CCS), Washington, DC, USA, 27–31 October 2003, pp. 42–51.

39 K. Kalka and A. Levi, 'Key distribution scheme for peer-to-peer communication in mobile underwater wireless sensor networks', *Peer-to-Peer Networking and Applications,* 7 (4), 698–709, 2014.

40 S. Souiki, M. Feham, and N. Labraoui, 'Geographic routing protocols for underwater wireless sensor networks: a survey'. arXiv preprint arXiv:1403.3779, 2014.

41 E. Souza, H.C. Wong, I. Cunha, A. Loureiro, L.F. Vieira, and L.B. Oliveira, 'End-to-end authentication in under-water sensor networks', *2013 IEEE Symposium on Computers and Communications (ISCC),* July 2013, pp. 299–304.

42 D. Mukhopadhyay and S. Indranil, 'Location verification based defense against Sybil attack in sensor networks', *International Conference on Distributed Computing and Networkin,* pp. 509-521, 2006.

43 J. Newsome, E. Shi, D. Song, and A. Perrig, 'The Sybil attack in sensor networks: analysis and defenses', In *Proceedings of the 3rd International Symposium on Information Processing in Sensor Networks,* April 2004, pp. 259–268.

44 M. Erol and S. Oktug, 'A localization and routing framework for mobile underwater sensor networks', *Proceedings IEEE International Conference on Computer Communications INFOCOM,* 2008.

45 W. Cheng, A.Y. Teymorian, L. Ma, X. Cheng, X. Lu, and Z. Lu. 'Underwater localization in sparse 3D acoustic sensor networks', *Proceedings of IEEE International Conference on Computer Communications INFOCOM,* 2008.

46 Z. Zhou, J.-H. Cui, and A. Bagtzoglou, 'Scalable localization with mobility prediction for underwater sensor networks', *Proceedings of Proceedings of IEEE International Conference on Computer Communications INFOCOM,* 2008.

47 Y. Zhou, B.J. Gu, K. Chen, J.B. Chen, and H.B. Guan, 'A range-free localization scheme for large scale underwater wireless sensor networks', *Journal of Shanghai Jiaotong University (Science)* **14**, 562–568, 2009

48 C. Tian, H. Jiang, X. Liu, X. Wang, W. Liu, and Y. Wang, 'Trimessage: a lightweight time synchronization protocol for high latency and resource-constrained networks', in *Proceedings of IEEE International Conference on Communications (ICC 2009),* Dresden, Germany, 2009, pp. 5010–5014.

49 C. Tian, W. Liu, J. Jin, Y. Wang, and Y. Mo, 'Localization and synchronization for 3D underwater acoustic sensor networks', In *International Conference on Ubiquitous Intelligence and Computing,* 2007, pp. 622-631.

50 W. Chirdchoo, S. Soh, and K. Chua, 'MU-Sync: A time synchronization protocol for underwater mobile networks', *Proceedings of WUWNet,* 2008.

Part IV

Underground and Confined Environments WSS Solutions

16

Achievable Throughput of Magnetic Induction Based Sensor Networks for Underground Communications

Steven Kisseleff[1], Ian F. Akyildiz[2] and Wolfgang H. Gerstacker[1]

[1] *Institute for Digital Communications, Friedrich-Alexander-Universität Erlangen-Nürnberg, Germany*
[2] *Broadband Wireless Networking Lab, Georgia Institute of Technology, Atlanta, USA*

16.1 Introduction

Communication through the underground medium is a challenging research area, and has been an open issue for more than a century. This type of communication enables efficient sensor networking for a wide variety of applications, such as

- soil condition monitoring
- earthquake prediction
- communication in mines, tunnels, and oil reservoirs
- construction health monitoring
- localization of objects that interact with the ground, such as humans and machines.

All of these applications require the deployment of a sensor network infrastructure below the ground. The sensors become part of the sensed environment and might deliver more precise sensing information than an above-ground deployment. In recent years, wireless underground sensor networks (WUSNs) have been proposed. These have many beneficial properties [1, 2]:

- avoidance of possible collisions with landscaping equipment such as tractors
- real-time information retrieval
- self-healing in the event of device failures
- ease of deployment and extensibility.

One of the factors that has slowed down the evolution of WUSNs is the challenging underground communication channel, which is heavily affected by the heterogeneity of soil. It may consist of rock, sand and even water. In particular the water content is crucial for the performance of WUSNs, since traditional wireless signal propagation techniques using electromagnetic (EM) waves can only be applied over very small transmission ranges underground due to the large path loss and vulnerability to changes in soil moisture [3].

Magnetic induction (MI) has been considered as an alternative to EM waves. Magnetic induction based communication has been already investigated in various studies,

Wireless Sensor Systems for Extreme Environments: Space, Underwater, Underground, and Industrial.
First Edition. Edited by Habib F. Rashvand and Ali Abedi.

mostly in the context of near-field communication [4] and wireless power transfer [5]. These studies have provided insights into the design of point-to-point MI-based signal transmission systems. Furthermore, several attempts have been made to extend the MI-based point-to-point transmission to systems with multiple receivers [6], multiple transmitters [7], or even multiple relays [8]. Recently, the concept of MI waveguides has been introduced. This exploits the effect of passive relaying (sometimes referred to as MI waves), if multiple magnetic relays implemented as resonant coils are combined in waveguide structures and deployed between two transceiver nodes [9–11]. This solution is supposed to benefit from a lower equivalent path loss.

MI-WUSNs were proposed by Sun and Akyildiz [9]. Channel models for MI-WUSNs with frequency-selective path loss have also been described [9, 12]. These models incorporate the losses due to the propagation medium and power reflections between the coils. In parallel to these theoretical studies, some research groups have conducted experiments in order to verify the correctness of the assumptions in the most common system models [13].

In general, there are two deployment strategies for the transmission links of MI-WUSNs: MI waveguides with a high relay density ($\approx \frac{1 \text{ Relay}}{3\, m}$) and direct MI transmission with no passive relays [12]. The first deployment strategy exploits the effect of strong coupling between any two neighboring magnetic devices, such that the overall path loss is heavily reduced. The second deployment strategy corresponds to traditional MI-based signal propagation, where the coils are separated by long distances and a conductive medium, such that a very weak coupling results. Despite a larger path loss using this deployment scheme, the frequency-selectivity of the resulting transmission channel is much lower than with MI waveguides [14]. Therefore, direct MI transmission based WUSNs may outperform MI waveguide based WUSNs in some cases [15].

Previous studies have mostly considered the channel capacity or the path loss as a performance measure. In contrast, Kisseleff et al. [14] looked at the optimal methods of digital transmission for point-to-point MI-WUSNs, where the achievable data rate using realistic transmit/equalization/receive filters was maximized. However, their proposed approaches were only valid for point-to-point transmissions and would need to be modified accordingly, in order to be used in more general network structures.

The network throughput, also called network capacity, reflects the behaviour of the busiest communication link of the network, also called the bottleneck link [16]. This performance metric has been intensively studied for different network paradigms: cognitive radio networks [17], ad-hoc networks with directional antennas [18], and even MI-WUSNs in [19], where a scaling law was provided by adopting a channel model from Sun and Akyildiz's paper [9]. One of the assumptions in that paper is a weak coupling between the coils in an MI waveguide, independently of the system parameters. However, as shown in Kisseleff et al. [12], for an MI waveguide with high relay density, the coupling between the coils is very strong and the system parameters can be adjusted to maximize the channel capacity for a given waveguide.

Other differences to traditional wireless networks are interference propagation and variations of the channel and noise characteristics depending on the topology of the network, which were not considered by Sun and Akyildiz [9]. This leads to a significant difference in channel models and in network design. As a result, a novel throughput maximization procedure has been proposed [15], in which these effects are taken into account. However, only information theoretic performance bounds were obtained [15]. Thus, the achievable network throughput using practical system components may significantly deviate from the sophisticated theoretical bounds.

The focus of this chapter is on maximizing the achievable network throughput for MI-WUSNs using realistic signal-processing components. For this, the optimization techniques described in earlier papers [14, 15] are combined. This advanced optimization is described in Section 16.2. Section 16.3 is a performance comparison of the two deployment schemes (MI waveguides and direct MI transmission) using the optimized set of system parameters. With the discussion on system performance in Section 16.4, this chapter is concluded.

16.2 Throughput Maximization for MI-WUSNs

In this section, the optimal selection of system parameters for maximizing the achievable throughput of the bottleneck link is explained. For this, the fundamentals of MI-based signal transmissions using the two most common deployment schemes (direct MI transmission and MI waveguides) for MI-WUSNs are discussed in Section 16.2.1. The practical aspects of signal processing and the network structure are described in Sections 16.2.2 and 16.2.3, respectively. Based on the resulting system model, the general optimization problem is formulated in Section 16.2.4. This problem is then solved for direct MI transmission based WUSNs in Section 16.2.5 and for MI waveguide based WUSNs in Section 16.2.6, respectively.

16.2.1 Signal Transmission in MI-WUSNs

In order to reduce the design and implementation costs of practical MI-WUSNs, we assume that all devices used in the WUSNs are mass produced. Hence, each resonance circuit of an MI-WUSN includes a magnetic antenna (which can be realized by an air core coil), a capacitor C and a resistor R (which models the copper resistance of the coil). In addition, each transceiver contains a load resistor R_L, which is used for data reception. Depending on the deployment strategy, each link may contain $k-1$ passive relays (MI waveguide), see Figure 16.1. The inductivity of a coil is given by [9]:

$$L = \frac{1}{2}\mu\pi N^2 a, \tag{16.1}$$

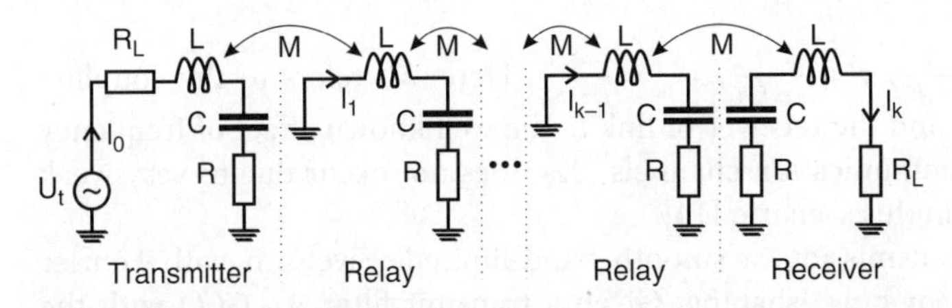

Figure 16.1 Block diagram of MI waveguide with transmitter, receiver, and $(k-1)$ relays.

where N denotes the number of windings, a is the radius of the coil, and μ denotes the permeability of the soil. The copper resistance of the coil is given by

$$R = \rho \cdot \frac{l_w}{A_w} = \rho \cdot \frac{2aN}{r_w^2}, \tag{16.2}$$

where $\rho \approx 1.678 \cdot 10^{-2}\ \Omega \cdot \mathrm{mm}^2/\mathrm{m}$ is the copper resistivity, l_w denotes the total wire length, A_w is the cross-sectional area of the wire and r_w is the radius of the wire. The capacitance of the capacitor is chosen to make each circuit resonant at a frequency f_0 [9]; in other words, $C = \frac{1}{(2\pi f_0)^2 L}$. With increasing resonance frequency, a very low capacitance of the capacitor may be required, which imposes an additional difficulty for the implementation. Hence, the capacitance is bounded below C_0 [12].

In general, the induced voltage is related to the coupling between the coils, which is determined by the mutual inductance [19]:

$$M = \mu \pi N^2 \frac{a^4}{4d^3} \cdot J \cdot G, \tag{16.3}$$

where d denotes the transmission distance. J represents the polarisation factor, which depends on the alignment of the coils. The orientation of the coil axes can be optimized in order to avoid interference from the adjacent links and to reduce the path loss for the useful signals [15]. However, the adjustments of the coil orientations may significantly increase the deployment costs. Hence, we assume that all coil axes are parallel and point in the direction of the ground surface, which yields $J = 1$. In addition, a loss factor $G = \exp\left(-\frac{d}{\delta}\right)$ due to eddy currents is introduced [12, 20], where δ represents the skin depth, which depends on the signal frequency, conductivity and the permittivity of the soil [12]. For the load resistor, we select R_L, which minimizes the losses due to power reflections [15]. Since all involved signal mappings are linear for MI-based transmission, a linear channel model results.

16.2.1.1 Direct MI Transmission Based WUSNs

First, we consider the channel and noise models for direct MI transmission based WUSNs. Here, the transmitter and receiver of each transmission link communicate directly, with no passive relays between them, such that $k = 1$ holds. The channel transfer function $H(f)$ is defined as the received voltage at the load impedance R_L of the receiver divided by the transmit voltage U_t, see Figure 16.1. For link i, the channel transfer function $H_i(f)$ can be obtained from [21]:

$$H_i(f) = \frac{x_{L,i}}{(x_i + x_{L,i})^2 - 1}, \tag{16.4}$$

where $Z = R + j2\pi fL + \frac{1}{j2\pi fC}$, $x_i = \frac{Z}{j2\pi fM_i}$, $x_{L,i} = \frac{R_L}{j2\pi fM_i}$. Here, M_i refers to the coupling between the transmitter and the receiver of link i. The well known effect of frequency splitting in MI-based communication channels [22] does not occur due to very weak couplings between coils in this scenario [14].

For practical signal transmission, a smooth band-limited waveform with Fourier transform $G(f)$ is used for pulse shaping. Given a transmit filter $A_i \cdot G(f)$ with the amplification coefficient A_i for link i, the consumed transmit power is given by [23]:

$$P_{S,i} = \frac{1}{2}\int_B \frac{|A_i \cdot G(f)|^2}{|j2\pi f M_i|} \frac{|x_i + x_{L,i}|}{|(x_i + x_{L,i})^2 - 1|} \mathrm{d}f, \tag{16.5}$$

where B is the bandwidth of the transmitted waveform. This consumed transmit power corresponds to the apparent transmit power [24], which incorporates not only the active power, but also the reactive power released by the electric circuit. This reactive power needs to be taken into account, since it typically limits the performance of the communication system and may therefore impose additional constraints on the system design [25].

The noise signal received at the load resistor of the receiver can be approximated by its strongest component, which originates from the resistors of the receiver circuit. This approximation is valid for direct MI transmission based WUSNs due to weak couplings between the receiver circuit and any other device. Hence, the receive noise power density spectrum can be approximated by [23]:

$$P_{N,i}(f) \approx \frac{4K_B T_K R_L (R + R_L)}{|Z + R_L|^2}, \tag{16.6}$$

where $K_B \approx 1.38 \cdot 10^{-23}$ J/K is the Boltzmann constant and T_K is the temperature in Kelvin (290 K $\triangleq$ 17°C). In addition to the noise signals, the transmission of the useful signal is disturbed by interfering signals from other nodes, which are scheduled for transmission in the same time slot. For direct MI transmission based WUSNs, we model these interference signals as transmissions from the interfering nodes to the receiver of the considered link using (16.4) with the respective mutual inductance. Moreover, the interference power depends on the amplification coefficients $A_k, \forall k \neq i$ assigned to the particular interfering nodes.

The optimal bandwidth of direct MI transmission based WUSNs is typically very large, up to one third of the optimal carrier frequency [14, 15]. However, the low-path-loss region for the direct MI channel is very narrow, so that the channel impulse response is very long (up to several tens of thousand taps). Such a channel cannot be equalized in a practical system. As a result, a three-band approach has been proposed [14], in which the total band is split into three orthogonal sub-bands with independent processing, similar to traditional frequency-division multiplexing. The mid sub-band is selected, such that the length of the equivalent discrete-time channel impulse response is less or equal to 100 taps. The width of the left and right sub-bands can be chosen via optimization. Moreover, power can be allocated among the three sub-bands, in order to maximize the achievable data rate.

16.2.1.2 MI Waveguide Based WUSNs

For MI waveguide based WUSNs, the channel transfer function for link i is heavily affected by the signal reflections from the neighboring coils, both transceivers and passive relays. In principle, there are infinitely many reflections, such that a closed-form expression for the resulting signal propagation is very hard to obtain. However, an approximation method has been provided [15] and the value can be obtained via summation of the strongest reflections (so-called inter-waveguide reflections [15]), which result from the shortest signal paths; see Figure 16.2. Hence, we consider the

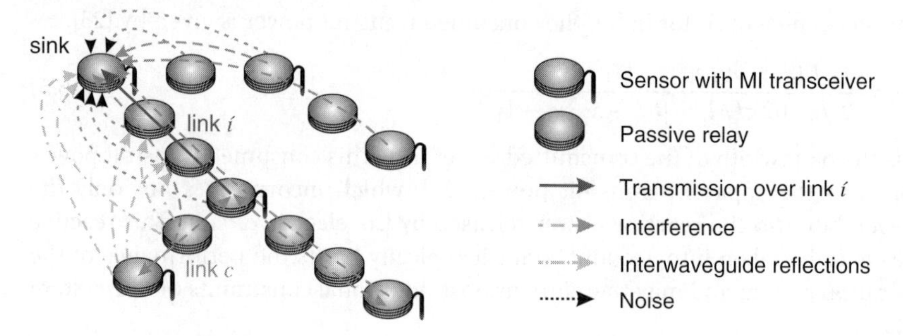

Figure 16.2 Signal propagation in MI waveguide based WUSNs.

neighboring waveguides connected to the transmitter of link i. We denote $M'_{n_i,n_c,c}$ as the mutual inductance between relay n_i in the considered waveguide i and relay n_c in the neighboring waveguide c of length k_c. Correspondingly, $x'_{n_i,n_c,c} = \frac{Z}{j2\pi f M'_{n_i,n_c,c}}$ is defined. In addition, we adopt the function $S(x_i, x_{L,i}, n)$, as defined in [12]:

$$\begin{aligned} S(x_i, x_{L,i}, n) &= F(x_i, n) + x_{L,i} \cdot F(x_i, n-1) \\ F(x_i, n) &= x_i \cdot F(x_i, n-1) - F(x_i, n-2), n \geq 2 \\ F(x_i, 0) &= 1, F(x_i, 1) = x_i. \end{aligned} \tag{16.7}$$

Hence, the approximation of the channel transfer function for link i can be derived using the equations provided by Kisseleff et al. [15]:

$$H_i(f) \approx \frac{R_L}{j2\pi f M_i Q_i} \left(1 + \sum_{c=1}^{N_{c,i}-1} \sum_{n_i=1}^{k_i} \sum_{n_c=1}^{\min(n_i,k_c)} \frac{x_i^{n_i-n_c}}{x'_{n_i,n_c,c}} \right), \tag{16.8}$$

where $N_{c,i}$ stands for the number of waveguides connected to the transmitter of link i (including waveguide i). The factor Q_i is defined as [15]:

$$Q_i = (x_i + x_{L,i}) \cdot S(x_i, x_{L,i}, k_i) - N_{c,i} \cdot S(x_i, x_{L,i}, k_i - 1). \tag{16.9}$$

For simplicity, we assume that the transmit power consumed in the transmitter is significantly affected only by the closest relays of the connected waveguides. In fact, the influence of the relays further away from the transmitter is negligible due to much larger path loss. In this case, the current in the transmit circuit of link i can be approximately given by

$$I_{t,i} = \frac{U_{t,i}}{j2\pi f M_i} \frac{x_i + x_{L,i}}{(x_i + x_{L,i})^2 - N_{c,i}}. \tag{16.10}$$

Then, similarly to (16.5), the consumed transmit power using a realistic transmit filter for pulse shaping is obtained:

$$P_{S,i} \approx \frac{1}{2} \int_B \frac{|A_i \cdot G(f)|^2}{|j2\pi f M_i|} \frac{|x_i + x_{L,i}|}{|(x_i + x_{L,i})^2 - N_{c,i}|} df. \tag{16.11}$$

For MI waveguide based WUSNs, the received noise is calculated [15] as the sum of noise signals coming from all resistors in all waveguides connected to the receiver of link i; see Figure 16.2. Then, the total noise power density is approximately given by

$$P_{N,i}(f) \approx P_{N,i,\mathrm{Rx}}(f) + \sum_{c_r=1}^{N_{r,i}} P_{N,i,c_r,R}(f), \tag{16.12}$$

where $N_{r,i}$ stands for the number of waveguides connected to the receiver of link i. Here, the noise signal from the wire resistance R of the resonance circuit c_r is denoted as $P_{N,i,c_r,R}(f)$ and the noise signal from the load resistor R_L at the receiver yields a noise power density $P_{N,i,\mathrm{Rx}}(f)$. These noise contributions have been calculated elsewhere [15] and are not reproduced here due to space limitations.

The transmissions in MI waveguide based WUSNs are also disturbed by the interfering signals from other sensor nodes. Due to the strong coupling between the coils, these signals do not propagate omnidirectionally as they do in direct MI transmission based WUSNs, but are guided through the network via passive relaying of the MI waveguides [15]. Therefore, the interference propagation can be viewed as a signal transmission via an extended MI waveguide, which corresponds to the path between the interfering node and the receiver of the considered link. This extended MI waveguide consists of several MI waveguides (sub-links). Hence, we model the channel transfer function for the interference signals similarly to (16.8), where the sum of lengths of the sub-links is used instead of k_i.

Each relay of an MI waveguide attenuates the slopes of the resulting channel frequency response [14]. Hence, a very narrow transmission band results. However, within this narrow band, a very high signal-to-noise ratio (SNR) can be observed due to low path loss close to the resonance frequency. Interestingly, in many cases, the SNR is high enough to enable a modulation with 14–18 bit/symbol. Due to the vast complexity of such modulation schemes and correspondingly an increased energy consumption in the low-power sensor nodes, a lower modulation order is preferable. However, a straightforward choice of a smaller signal constellation leads to heavy losses of the achievable data rate; for example, up to ≈45%, if 10 bit/symbol is selected instead of 18 bit/symbol. Hence, a bandwidth expansion approach has been proposed [14], where the bandwidth is optimized under the constraint of a given maximum constellation size $M_{\max}$.

16.2.2 Practical Aspects of System Design

For a practical system design, the signal generation, reception and processing components need to comply with the design of realistic signal filters. Hence, the information-theoretical bounds given by the Shannon channel capacity [19] are not the proper measures for the achievable data rate. Specifically, the transmit pulse is usually assumed to comply with the water-filling rule [26]. Such pulses can hardly be generated and are not applicable in practice. Instead, a smooth band-limited waveform (typically a square-root Nyquist filter, such as a root-raised cosine filter) is used for pulse shaping. For receive filtering we employ the whitened matched filter [27]. Since the total transmission channel is frequency-selective, an equalization scheme is needed for signal detection. For this, we use a decision-feedback equalization scheme, which minimizes the mean-squared error of the output signal. The resulting SNR for transmissions over link i is denoted as $\mathrm{SNR}_{\mathrm{eq},i}$. For a given modulation scheme with constellation size $M_{\text{mod},i}$, the symbol error rate (SER_i) is obtained using the equations from [28]

$$\text{SER}_i = \begin{cases} Q\left(\sqrt{2\text{SNR}_{\text{eq},i}K}\right), & \text{BPSK}, M_{\text{mod},i} = 2, \\ 2Q\left(\sqrt{\text{SNR}_{\text{eq},i}K}\right), & \text{QPSK}, M_{\text{mod},i} = 4, \\ \frac{4(\sqrt{M_{\text{mod},i}}-1)}{\sqrt{M_{\text{mod},i}}} Q\left(\sqrt{\frac{3\text{SNR}_{\text{eq},i}K}{M_{\text{mod},i}-1}}\right), & \text{rectangular } M_{\text{mod},i} - \text{QAM}, \\ 4Q\left(\sqrt{\frac{3\text{SNR}_{\text{eq},i}K}{M_{\text{mod},i}-1}}\right), & \text{non} - \text{rectangular } M_{\text{mod},i} - \text{QAM}, \end{cases} \tag{16.13}$$

where $Q(\cdot)$ is the complementary Gaussian error integral, and K stands for the coding gain of the employed channel code [29]. Hence, assuming a channel code with a code rate R_c, the achievable data rate of link i is given by

$$R_{a,i} = \frac{R_c \log_2 M_{\text{mod},i}}{T}, \tag{16.14}$$

where T denotes the symbol interval.[1] In the following, we consider uncoded transmission, such that $K = 1$ and $R_c = 1$ holds. Hence, not a single error can be corrected at the receiver. Instead, the whole data block may need to be retransmitted. As known from the literature, the error propagation in decision-feedback equalizers can dramatically increase SER_i and reduce the system performance. Hence, a training sequence needs to be attached to each data block, in order to cleanse the content of the shift registers of equalizer filters by replacing the decided register states with the correct ones. This is of particular importance after each erroneous data block. Assuming a training sequence of length N_{TS} (equal for all network links), the optimally chosen total length of the transmission block (data + training) can be given by [14]:

$$N_{BL,i} = \left\lfloor \frac{N_{TS}}{2}\left(1 + \sqrt{1 - \frac{4}{N_{TS}\log\left(1 - \text{SER}_i\right)}}\right) \right\rfloor, \tag{16.15}$$

where the $\lfloor \cdot \rfloor$ operator is applied in order to restrict the block length to integer numbers. Of course, the use of a training sequence reduces the data rate. Therefore, we introduce an effective achievable data rate (similar to Kisseleff et al. [14]), which is given by

$$R_{\text{eff},i} = R_{a,i}\frac{(N_{BL,i} - N_{TS})(1 - \text{SER}_i)^{(N_{BL,i}-N_{TS})}}{N_{BL,i}}. \tag{16.16}$$

Obviously, for large values of SER_i, we obtain $N_{BL,i} \approx N_{TS}$ and $R_{\text{eff},i} \approx 0$, due to the aforementioned block retransmissions.

16.2.3 Network Specification

In this section, we focus on tree-based networks with N_{nodes} sensors, including a single sink node. The sink node collects sensor information from all nodes and uploads it wirelessly or via a wired connection to an above-ground device. This network structure is appropriate for most target applications of WUSNs. In order to be able to receive the

1 Due to assumed mass production of the transceivers, the bandwidth and the symbol interval also need to be identical for all transmission links.

data from nodes far away from the sink, each node not only transmits its own information, but also relays all received data from other nodes. The decode-and-forward relaying concept is selected in this work.

In order to avoid the loss of data packets due to strong interference coming from nodes transmitting simultaneously, a multi-node scheduling scheme needs to be established. The transceivers are operated in half-duplex mode. The signal reception and transmission is carried out in different time slots by means of time-division multiple access (TDMA). Other multiple access schemes, such as frequency-division or code-division multiple access are not suitable for MI-WUSNs [15], due to issues related to the high frequency-selectivity of the transmission channels. For simplicity, we assume that the synchronization of transmissions for TDMA can be perfectly done using well known approaches, such as that of Sivrikaya and Yener [30].

The simplest multi-node scheduling schem would assign all sensor nodes to disjoint time slots, such that no interference between adjacent data streams was imposed. As much as this scheduling scenario may simplify the design and implementation of the network, it also reduces the achievable data rate due to the degradation of the bandwidth efficiency. Obviously, if some of the nodes are far away from the receiver of a particular transmission link, the respective interference signals hardly influence the transmission. In such cases, an orthogonal separation of the data streams in terms of TDMA is not necessary. Hence, a multi-node scheduling scheme can be chosen, which maximizes the network throughput. The optimal set of nodes that need to be scheduled for transmissions in MI-WUSNs can be determined using the method described in [15]. This strategy is based on the separation of the nodes' indices into two sets, $D_{i,1}$ and $D_{i,2}$. The indices in $D_{i,2}$ represent strong interferers for transmissions of link i that need to be taken into account in the multi-node scheduling. $D_{i,1}$ contains the indices of the remaining nodes. These remaining nodes, of course, disturb the transmission and reduce $\mathrm{SNR}_{\mathrm{eq},i}$ and the achievable data rate. However, their influence is limited compared to the influence of the nodes assigned to $D_{i,2}$. Due to the similar frequency-selectivity of all links of the same network (narrow low-path-loss region around the center frequency [14]),

$$\max_{k \in D_{i,1}} \int_{-\infty}^{\infty} P_{I,i,k}(f)\mathrm{d}f \leq \min_{k \in D_{i,2}} \int_{-\infty}^{\infty} P_{I,i,k}(f)\mathrm{d}f \tag{16.17}$$

holds, where $P_{I,i,k}(f)$ represents the interference power density that comes from node k and arrives at the receiver of link i. Based on this observation, the optimal sets $D_{i,1}$ and $D_{i,2}$ can be found iteratively, by storing all interferers in $D_{i,1}$ and moving the strongest interferers one by one to $D_{i,2}$. Then, in each iteration, the throughput of link i is calculated and the partitioning with the maximum throughput is selected.

In order to avoid the loss of data packets, traffic load for each link has to be lower than the achievable data rate of the respective link [16]. In addition, the traffic load of a link equals the throughput of an information stream multiplied by the number of data streams to be served by the node (compare with [16, 19]). This is due to a frequent assumption that the amount of sensed information is equal for all sensor nodes, which is reasonable for WUSNs, too. Hence, some nodes may compete and disturb each others' transmissions. Hence, we obtain for the throughput of link i

$$\mathrm{THR}_i \leq \frac{R_{\mathrm{eff},i}}{N_{\mathrm{streams},i}(1 + N_{\mathrm{interf.},i})}, \tag{16.18}$$

where $N_{\text{streams},i}$ denotes the number of data streams served by link i and $N_{\text{interf.},i} = \sum_{k \in D_{i,2}} 1$. In the case of equality in (16.18), we obtain the upper bound for the network throughput. This upper bound is referred to as the throughput of link i throughout the remainder of this chapter.

16.2.4 Throughput Maximization

For e throughput maximization, we select the available system parameters that do not influence the implementation/deployment costs. These parameters are:

- resonance frequency f_0 and number of coil windings N
- amplification coefficient A_i for each link i
- modulation order $M_{\text{mod},i}$ and the optimal block length $N_{BL,i}$ for each link i
- optimal bandwidth B and, correspondingly, the symbol duration T.

Among different network topologies, the minimum spanning tree seems promising due to the reduced average and maximum transmission distance between the transceivers [31, 32], which is especially beneficial for the maximization of the bottleneck throughput. Obviously, the topology of MI waveguide based WUSNs can be optimized, since the amount of interference experienced by MI waveguide based links depends on the number of waveguides connected to each other. However, the topology optimization requires a very high computational effort [15], while the achievable throughput gain is limited. Therefore, we do not utilise the deployment optimization algorithms proposed in [15]. Instead, we focus on the system parameters that are more related to the signal processing [14]. Using the throughput metric introduced in (16.18), the throughput maximization problem can be formulated as follows:

$$\max_{f_0, N, A_i, M_{\text{mod},i}, N_{BL,i}, B} \min_i \{\text{THR}_i\}, \tag{16.19}$$

$$\text{s.t.:} \quad 1)\ P_{S,i} = P\ \forall i, \qquad 2)\ M_{\text{mod},i} \leq M_{\max}\ \forall i,$$

$$3)\ \frac{1}{(2\pi f_0)^2 L} \geq C_0, \quad 4)\ N \leq N_{\max},$$

where all nodes transmit with the same output power P (constraint 1)). Moreover, the constellation size is restricted to $M_{\max}$ (constraint 2)) and the capacitance of the capacitor should be larger than the minimum allowed capacitance C_0 (constraint 3)) [12]. Furthermore, very large coils (especially single-layer coils) with thousands of turns are not applicable in WUSNs. In particular, for MI waveguide based WUSNs with relay density $\approx \frac{1\ \text{relay}}{3\ \text{m}}$, the deployment of a coil of 1 m length can be more expensive than connecting all the sensors via cable. Hence, we restrict the number of coil windings to $N_{\max}$ (constraint 4)).

Finding the optimal system parameters for maximizing the channel capacity of an MI-link is a non-convex problem, [12], which cannot be solved using convex optimization tools [33]. But since the problem in [12] can be approximately viewed as a subproblem of (16.19), (16.19) is also non-convex (similar to the problem in [15]). Hence, we adopt the suboptimal optimization approaches proposed in [14], because these methods typically provide a significant performance gain compared to the respective baseline schemes.

Due to the different nature of signal propagation for the two deployment strategies (direct MI transmission and MI waveguides), the optimization techniques proposed for these schemes significantly differ from each other. In the following, we describe the optimization strategy for direct MI transmission based WUSNs in Section 16.2.5 and for MI waveguide based WUSNs in Section 16.2.6, respectively.

16.2.5 Throughput of Direct MI Transmission Based WUSNs

For direct MI transmission based WUSNs, there are three amplification coefficients ($A_{i,1}$, $A_{i,2}$, and $A_{i,3}$ stacked in a vector $\mathbf{A}_i$) due to the mentioned three-band approach and the corresponding power allocation. Typically, all three sub-bands provide different signal quality in terms of $\text{SNR}_{\text{eq},i}$. Hence, the optimal modulation scheme can vary from sub-band to sub-band. We denote the optimally selected modulation schemes $M_{\text{mod},i,1}$, $M_{\text{mod},i,2}$ and $M_{\text{mod},i,3}$ (stacked in a vector $\mathbf{M}_{\text{mod},i}$) for sub-bands 1, 2 and 3, respectively. Also, these three sub-bands may have different bandwidths (B_1, B_2 and B_3 stacked in a vector $\mathbf{B}$). The respective bandwidths are identical for all transmission links due to the mass production of the transceiver hardware, as mentioned in Section 16.2.2. The corresponding symbol durations are denoted as T_1, T_2 and T_3. We define three different block lengths, $N_{BL,i,1}$, $N_{BL,i,2}$ and $N_{BL,i,3}$ for sub-bands 1, 2 and 3, respectively. Furthermore, we assume $N_{BL,i,1}T_1 = N_{BL,i,2}T_2 = N_{BL,i,3}T_3$ in order to ensure that the transmissions start and end simultaneously in all sub-bands. Otherwise, one of the sub-bands would finish transmission earlier, which may reduce the bandwidth efficiency and the achievable data rate. Also, in order to avoid some of the nodes scheduled for transmission in the same time slot finishing the transmission earlier than others, the block length for the same sub-band should be equal for all network links: $N_{BL,i,1} = N_{BL,1}$, $N_{BL,i,2} = N_{BL,2}$ and $N_{BL,i,3} = N_{BL,3}$ $\forall i$, where the optimal block lengths are denoted as $N_{BL,1}$, $N_{BL,2}$ and $N_{BL,3}$ for the sub-bands 1, 2 and 3, respectively.

Due to the three-band approach in direct MI transmission based WUSNs, the complexity of optimization problem (16.19) is very high. In particular, the choice of the amplification coefficients $A_{i,1}$, $A_{i,2}$ and $A_{i,3}$ per node is subject to resource allocation. However, the cost function in (16.19) depends non-linearly on these factors. Further, the non-convex relationship between the cost function and the resonance frequency, bandwidth and other parameters, prevents use of convex optimization tools [33], as mentioned in Section 16.2.4.

In order to reduce the complexity of the problem, we exploit the dependencies between some of the system parameters. Due to the symmetry of the transfer functions and noise power densities of the two side-bands [14], the optimal solution for B_1, T_1, $A_{i,1}$ and $N_{BL,1}$ is equal to B_3, T_3, $A_{i,3}$ and $N_{BL,3}$, respectively. In addition, the amplification coefficient $A_{i,2}$ can be calculated directly using $A_{i,1}$ and $A_{i,3}$ for a given total transmit power $P_{S,i}$ of each link i. The number of coil windings is set to its maximum value $N_{\max}$, as suggested by [15]. Also, the bandwidth B_2 of the mid sub-band can be determined, such that the corresponding equivalent discrete-time channel impulse response (CIR) is shorter than 100 taps [14].

Among the available optimization parameters, we select the following independent parameters: f_0, B_1, $N_{BL,1}$ and $A_{i,1}$ $\forall i$. Further system parameters can be easily specified for the given values of f_0, B_1, B_2, $N_{BL,1}$, $A_{i,1}$ $\forall i$ and $A_{i,2}$ $\forall i$. For example, using B_1, B_2 and $N_{BL,1}$, the block length for the mid sub-band $N_{BL,2}$ can be calculated via $N_{BL,2} = N_{BL,1}\frac{B_2}{B_1}$. Then, using $N_{BL,2}$ and $A_{i,2}$, the optimal modulation order for the mid sub-band is determined, such that the effective achievable data rate $R_{\text{eff},i}$ is maximized. Unfortunately, no technique for the joint optimization of f_0, B_1, $N_{BL,1}$ and $A_{i,1}$ $\forall i$ exists in the literature. Hence, a grid search in f_0, B_1 and $N_{BL,1}$ is inevitable. The optimization of $A_{i,1}$ $\forall i$ (power allocation) is obviously an NP-hard problem, since the number of potential solutions increases exponentially with the number of sensor nodes. Hence, an exhaustive search is not applicable and we propose the following suboptimal solution based on an iterative algorithm; see Algorithm 1.

Algorithm 1 Power allocation for direct MI transmission based WUSNs

---

1: Input: A tree-based WUSN, f_0, N, B_1, B_2, $N_{BL,1}$, $N_{BL,2}$;
2: Output: $A_{i,1}$, $A_{i,2}$ $\forall i$;
3: Initialize: $\text{THR}_{opt} \leftarrow 0$, $A_{i,1,old} \leftarrow 0$, $A_{i,2,old} \leftarrow 0$, $\forall i$;
4: **for all** i **do**
5:   assume $A_{k,1} \leftarrow 0$, $A_{k,2} \leftarrow 0$, $\forall k \neq i$ (no interference);
6:   determine $A_{i,1}$ and $A_{i,2}$ that maximize THR_i (via full search);
7:   calculate THR_i;
8: **end for**
9: $ind1 \leftarrow \arg\min_i\{\text{THR}_i\}$;
10: $A_{i,1,old} \leftarrow A_{i,1}$, $A_{i,2,old} \leftarrow A_{i,2}$, $\forall i$ (store current state);
11: $\epsilon \leftarrow 0.5$;
12: **while** $\epsilon \geq \epsilon_{\min}$ **do**
13:   determine $A_{ind1,1}$ and $A_{ind1,2}$ that maximize $\min_i\{\text{THR}_i\}$;
14:   calculate THR_i, $\forall i$ (with interference);
15:   $ind2 \leftarrow \arg\min_i\{\text{THR}_i\}$;
16:   **if** $(ind1 == ind2) \bigwedge (\min_i\{\text{THR}_i\} > \text{THR}_{opt})$ **then**
17:     $\text{THR}_{opt} \leftarrow \min_i\{\text{THR}_i\}$;
18:     $A_{i,1,old} \leftarrow A_{i,1}$ and $A_{i,2,old} \leftarrow A_{i,2}$, $\forall i$;
19:     **for all** $k \neq ind1$ **do**
20:       $A_{k,1} \leftarrow A_{k,1}(1+\epsilon)$;
21:       calculate $A_{k,2}$;
22:     **end for**
23:   **else if** $(ind1 == ind2) \bigwedge (\min_i\{\text{THR}_i\} \leq \text{THR}_{opt})$ **then**
24:     $\epsilon \leftarrow -\epsilon$;
25:     **for all** $k \neq ind1$ **do**
26:       $A_{k,1} \leftarrow A_{k,1}(1+\epsilon)$;
27:       calculate $A_{k,2}$;
28:     **end for**
29:   **else if** $(\min_i\{\text{THR}_i\} > \text{THR}_{opt})$ **then**
30:     $\text{THR}_{opt} \leftarrow \min_i\{\text{THR}_i\}$;
31:     $A_{i,1,old} \leftarrow A_{i,1}$ and $A_{i,2,old} \leftarrow A_{i,2}$, $\forall i$;
32:   **else**
33:     $A_{i,1} \leftarrow A_{i,1,old}$ and $A_{i,2} \leftarrow A_{i,2,old}$, $\forall i$ (restore the previous state);
34:     $\epsilon \leftarrow 0.5\epsilon$;
35:   **end if**
36:   $ind1 \leftarrow ind2$;
37: **end while**

The basic idea of this approach is to select the bottleneck link and to optimize its amplification coefficients for a larger bottleneck throughput. If the amplification coefficients of this link are already optimal, we try to reduce the impact of interference by scaling the factors of the remaining nodes. At first, the maximum bottleneck throughput THR_{opt} is initialized to zero and the step size ϵ is set to 0.5. Furthermore, the optimal amplification coefficients $A_{i,1}$ and $A_{i,2}$ $\forall i$ are found; these maximize the respective THR_i under the assumption that the adjacent links do not interfere. This seems to be a good starting point, since the exact amount of interference and the corresponding multi-node scheduling are unknown at this stage. Among all links, the preliminary bottleneck link $ind1$ is selected, for which THR_i is minimum. In each iteration of the algorithm, at first the amplification coefficients $A_{ind1,1}$ and $A_{ind1,2}$ are determined via grid search, such that the bottleneck throughput $\min_i\{\mathrm{THR}_i\}$ is maximized. Here, the interference powers and the optimal scheduling are obtained for each point of the grid. Then, the bottleneck link $ind2$ and its throughput are determined. If this throughput is larger than THR_{opt}, all factors $A_{i,1}$ and $A_{i,2}$ $\forall i$ are temporarily stored in $A_{i,1,old}$ and $A_{i,2,old}$, respectively, as a potential solution for the optimization problem. Also, THR_{opt} is updated with the new value of the bottleneck throughput. If $ind1$ and $ind2$ correspond to the same link, the optimization of $A_{ind1,1}$ and $A_{ind1,2}$ does not increase the throughput of this link. Hence, we scale the amplification coefficients $A_{k,1}$ $\forall k \neq ind1$ with $(1+\epsilon)$ in order to reduce the interference in the first (and third) sub-band ($\epsilon < 0$) or in the second sub-band ($\epsilon > 0$). Unfortunately, by reducing the interference in the first (and third) sub-band, the interference in the second sub-band is automatically enhanced (and vice versa). This means that a positive step size ϵ does not necessarily improve the performance. This situation occurs if $ind1$ is equal to $ind2$, while the achievable bottleneck throughput is reduced compared to the previous step. Therefore, we change the sign of the step size by setting $\epsilon \leftarrow -\epsilon$. If $ind1$ and $ind2$ belong to different nodes, we expect that the bottleneck throughput has been increased in the previous iteration. Otherwise, probably the selected step size was too large, such that the performance of link $ind2$ degraded. Hence, we go back to the previous successful iteration by restoring all amplification coefficients that have been considered a potential solution. Also, we reduce the step size via $\epsilon \leftarrow 0.5\epsilon$. After each iteration, the index of the current link with the worst throughput $ind2$ is stored in $ind1$, such that this link is used as a reference for the next iteration. The algorithm is terminated if the step size becomes too small; that is $\epsilon < \epsilon_{\min}$, where $\epsilon_{\min} = 0.1$ is used in this study.

The overall throughput maximization (including the grid search) for direct MI transmission based WUSNs is described in Algorithm 2. In the initialization phase, the maximum achievable bottleneck throughput $\mathrm{THR}_{\max}$ is set to zero. As mentioned earlier in this section, the optimal value for the number of windings is $N = N_{\max}$. Also, the bandwidth of the mid sub-band (B_2) is determined, such that the length of CIR is lower than 100 taps. Then, an exhaustive search in variables f_0, B_1 and $N_{BL,1}$ is performed. For each point of this search, at first, $N_{BL,3}$ and $N_{BL,2}$ are calculated. Using these values, Algorithm 1 is executed in order to obtain the optimal values for the amplification coefficients. Hence, all parameters except for the modulation schemes $M_{\mathrm{mod},i}$ are known. These modulation schemes are determined in order to maximize the respective throughput metrics THR_i. Then, the bottleneck throughput is obtained via $\min_i\{\mathrm{THR}_i\}$. If this throughput is larger than $\mathrm{THR}_{\max}$, all current parameters are stored as a potential solution for the global optimization problem and $\mathrm{THR}_{\max}$ is updated with the new value of the bottleneck throughput.

Algorithm 2 Throughput maximization for direct MI transmission based WUSNs

1: Input: A tree-based WUSN;
2: Output: f_0, N, $\mathbf{A}_i$, $\mathbf{M}_{\mathrm{mod},i}$, $N_{BL,1}$, $N_{BL,2}$, $N_{BL,3}$, $\mathbf{B}$;
3: initialize: $\mathrm{THR}_{\max} \leftarrow 0$, $N \leftarrow N_{\max}$;
4: **for all** f_0 **do**
5: determine maximum B_2, for which the length of equivalent CIR is below 100 taps;
6: **for all** B_1 **do**
7: $B_3 \leftarrow B_1$;
8: **for all** $N_{BL,1}$ **do**
9: $N_{BL,3} \leftarrow N_{BL,1}$;
10: $N_{BL,2} \leftarrow N_{BL,1}\frac{B_2}{B_1}$;
11: execute Algorithm 1;
12: **for all** i **do**
13: determine $M_{\mathrm{mod},i,1}$, $M_{\mathrm{mod},i,2}$, $M_{\mathrm{mod},i,3}$ that maximize THR_i;
14: calculate THR_i;
15: **end for**
16: **if** $\min_i\{\mathrm{THR}_i\} > \mathrm{THR}_{\max}$ **then**
17: store B_1, B_2, and B_3 in $\mathbf{B}$, $\forall i$;
18: store $A_{i,1,old}$, $A_{i,2,old}$, and $A_{i,3,old}$ in $\mathbf{A}_i$, $\forall i$;
19: store $M_{\mathrm{mod},i,1}$, $M_{\mathrm{mod},i,2}$, and $M_{\mathrm{mod},i,3}$ in $\mathbf{M}_{\mathrm{mod},i}$, $\forall i$;
20: $\mathrm{THR}_{\max} \leftarrow \min_i\{\mathrm{THR}_i\}$;
21: store f_0, N, $\mathbf{A}_i$, $\mathbf{M}_{\mathrm{mod},i}$, $N_{BL,1}$, $N_{BL,2}$, $N_{BL,3}$, $\mathbf{B}$;
22: **end if**
23: **end for**
24: **end for**
25: **end for**

Due to the vast computational complexity of Algorithms 1 and 2, the optimization should be carried out offline. Of course, the channel conditions may change over time, such that the system performance may degrade. In order to prevent this, the system parameters need to be adapted in real time. However, these adjustments are beyond the scope of this work.

16.2.6 Throughput of MI Waveguide Based WUSNs

The throughput maximization for MI waveguide based WUSNs is less complex than for direct MI transmission based WUSNs due to there being only one transmission band instead of three sub-bands. Hence, the amplification coefficients A_i can be calculated directly for the given values of f_0, N and B. In addition, constraint 3 in (16.19) (the so-called capacitor constraint [12]) becomes an equality for MI waveguides due to the strong coupling principle [15]. Hence, the resonance frequency can be expressed by inserting (16.1) into the capacitor constraint. This yields

$$f_0 = \frac{1}{N\sqrt{2\pi^3 \mu a C_0}}, \tag{16.20}$$

which is very beneficial, since N is an integer number. Therefore, a full search in N is possible and may lead to a more accurate solution than a grid search in continuous variable f_0 (as is done for direct MI transmission based WUSNs). Unfortunately, the optimal bandwidth that maximizes the network throughput cannot be computed in a direct manner and needs to be optimized via a grid search. Similarly to the approach used for direct MI transmission based WUSNs, we assume that all block lengths are equal for all network links: $N_{BL,i} = N_{BL}\ \forall i$. Hence, the choice of the optimal block length N_{BL} is not trivial. Therefore, we propose the following optimization strategy, see Algorithms 3 and 4.

Algorithm 3 Block length optimization for MI waveguide based WUSNs

1: Input: A tree-based WUSN, f_0, N, B, A_i, $\mathrm{SNR}_{\mathrm{eq},i}$, $\forall i$;
2: Output: $N_{BL,\min}$, $N_{BL,\max}$;
3: **for all** i **do**
4: calculate SER_i, $N_{BL,i}$ and THR_i for different $M_{\mathrm{mod},i}$;
5: select $M_{\mathrm{mod},i}$ that maximizes THR_i;
6: store the maximum THR_i;
7: **end for**
8: store $N_{BL,i}\ \forall i$ (optimal N_{BL} from the perspective of link i);
9: $ind1 \leftarrow \arg\min_i\{\mathrm{THR}_i\}$;
10: $N_{BL,\min} \leftarrow N_{BL,ind1}$;
11: **for all** i **do**
12: calculate THR_i using $N_{BL,\min}$ instead of $N_{BL,i}$ for different $M_{\mathrm{mod},i}$;
13: select $M_{\mathrm{mod},i}$ that maximizes THR_i;
14: store the maximum THR_i;
15: **end for**
16: $ind2 \leftarrow \arg\min_i\{\mathrm{THR}_i\}$;
17: $N_{BL,\max} \leftarrow N_{BL,ind2}$;

We perform a full search over different block lengths and select the N_{BL} that maximizes the bottleneck throughput of the network. In order to reduce the computational complexity, the ranges of the block length need to be determined. In Algorithm 3, for the given values of f_0, N, B, A_i and $\mathrm{SNR}_{\mathrm{eq},i}\ \forall i$, we select the values $N_{BL,\min}$ and $N_{BL,\max}$ as the ranges for the full search. The idea is to obtain one of the bounds (say, $N_{BL,\min}$) from the bottleneck link, if the throughput is maximized for each link independently. Unfortunately, using $N_{BL} = N_{BL,\min}$ for the whole network, may lead to a different bottleneck throughput and correspondingly to a different bottleneck link. The throughput of this new bottleneck link might be maximized by a different block length, which gives us the second bound (that is, $N_{BL,\max}$). Intuitively, we assume that the optimal block length N_{BL} lies between between these two bounds. At first, $M_{\mathrm{mod},i}$, SER_i and $N_{BL,i}$ are optimized for each link independently using (16.13)–(16.15), such that the respective throughput metric THR_i is maximized. We store the values $N_{BL,i}\ \forall i$, in order to reuse them later. Then, the link *ind*1 with the lowest THR_{ind1} is preliminarily selected as the bottleneck link. The corresponding block length is $N_{BL,ind1}$. Using $N_{BL,\min} = N_{BL,ind1}$ instead of $N_{BL,i}$, the optimal values $M_{\text{mod},i}$ and SER_i are determined for each link i to maximize the throughput metric THR_i. The throughput metric is updated for all network links. Then, the link *ind*2 with the lowest throughput is selected. The corresponding block length $N_{BL,\max} = N_{BL,ind2}$ is obtained from the previously stored values $N_{BL,i}, \forall i$. Within the range $[N_{BL,\min}, N_{BL,\max}]$, a full search among different block lengths is executed in Algorithm 4.

Algorithm 4 Throughput maximization for MI waveguide based WUSNs

1: Input: A tree-based WUSN;
2: Output: Optimal values for $M_{\text{mod},i}, \forall i, f_0, N, B, N_{BL}$;
3: $\text{THR}_{\max} \leftarrow 0$;
4: **for** $N \leftarrow 1$ **to** $N_{\max}$ **do**
5: $\quad f_0 \leftarrow \frac{1}{N\sqrt{2\pi^3 \mu a C_0}}$;
6: $\quad$ **for all** B **do**
7: $\quad\quad$ calculate A_i using (16.11), calculate $\text{SNR}_{\text{eq},i}$;
8: $\quad\quad$ execute Algorithm 3;
9: $\quad\quad$ **for** $N_{BL} \leftarrow N_{BL,\min}$ **to** $N_{BL,\max}$ **do**
10: $\quad\quad\quad$ **for all** i **do**
11: $\quad\quad\quad\quad$ determine $M_{\text{mod},i}$ that maximizes THR_i using N_{BL};
12: $\quad\quad\quad\quad$ calculate THR_i;
13: $\quad\quad\quad$ **end for**
14: $\quad\quad\quad$ **if** $\min_i\{\text{THR}_i\} > \text{THR}_{\max}$ **then**
15: $\quad\quad\quad\quad$ $\text{THR}_{\max} \leftarrow \min_i\{\text{THR}_i\}$;
16: $\quad\quad\quad\quad$ store $M_{\text{mod},i}, \forall i, f_0, N, B, N_{BL}$;
17: $\quad\quad\quad$ **end if**
18: $\quad\quad$ **end for**
19: $\quad$ **end for**
20: **end for**

The overall throughput maximization for MI waveguide based WUSNs is as follows. First, we initialize the maximum achievable bottleneck throughput $\text{THR}_{\max} = 0$. For each value of N in the range between 1 and $N_{\max}$, the corresponding resonance frequency f_0 is calculated using (16.20). Then, a grid search in B is performed. This grid spans the range $[B_{\min}, B_{\max}]$. For each point of the grid, the amplification coefficients A_i and the signal-to-noise ratios $\text{SNR}_{\text{eq},i}$ are calculated for all the links. Thereafter, Algorithm 3 is executed in order to determine the ranges for the block length search. If $N_{BL,ind1}$ is equal to $N_{BL,ind2}$, no full search is actually needed, since $N_{BL,ind1}$ is already the optimal value for the block length. If $N_{BL,ind1}$ is not equal to $N_{BL,ind2}$, a full search among the different block lengths N_{BL} is performed in the range $[N_{BL,ind1}, N_{BL,ind2}]$. For each value of N_{BL}, the modulation scheme $M_{\text{mod},i}$ and the symbol error rate SER_i that maximize the throughput metric THR_i for link i are obtained. The bottleneck throughput is selected, which is the minimum among all metrics $\text{THR}_i\ \forall i$. If this bottleneck throughput is larger than $\text{THR}_{\max}$, the respective parameters $M_{\text{mod},i}, \forall i, f_0, N, B$, and N_{BL} are stored as a potential solution for the global optimization problem. In addition, the bottleneck throughput is stored as the new value for THR_{max}; see Algorithm 4. Using this strategy, a solution for the optimization problem (16.19) that satisfies all constraints is obtained. Of course, this proposed solution is a-priori suboptimal due to the aforementioned non-convexity of the original problem. However, due to the full search using most of the available parameters, a close-to-optimum performance can be expected.

16.3 Results

In this section, the numerical results in terms of achievable throughput for randomly generated networks are discussed.

In our simulations, we assume a total transmit power of $P = 10\,\text{mW}$ per node. We use coils with wire radius 0.5 mm and coil radius 0.15 m. The conductivity of dry soil $\sigma = 0.01\,\text{S/m}$ is selected [34]. Since the permeability of soil is close to that of air, we use $\mu = \mu_0$ with the magnetic constant $\mu_0 = 4\pi \cdot 10^{-7}\,\text{H/m}$. We set the lowest capacitance of the capacitor to $C_0 = 1\,\text{pF}$. Furthermore, we restrict the total number of coil windings to $N_{\text{max}} = 100$.

In most applications of WUSNs, the density of sensor nodes needs to be high to ensure that enough sensed information is collected. However, with a small distance between two nodes, MI-based transmission is outperformed by EM wave based transmission [12, 35]. Therefore, we consider randomly deployed nodes with the minimum distance between any two nodes not less than 21 m in order to provide a scenario in which MI-based transmission yields a better performance than traditional EM wave based transmission. We assume a square field of size $F_x \times F_x$, yielding a total area F_x^2. Within this field, a random uniformly distributed set of N_{nodes} sensor nodes is generated for each network optimization. In this set, a sink node is selected; this is the closest node to the lower left-hand field corner. We show the bottleneck throughput distributions for 100 randomly deployed sensor networks using either MI waveguide or direct MI transmissions. Furthermore, these 100 networks are obtained for four different scenarios:

- $N_{\text{nodes}} = 10$, $F_x^2 = 0.01\,\text{km}^2$
- $N_{\text{nodes}} = 10$, $F_x^2 = 0.04\,\text{km}^2$
- $N_{\text{nodes}} = 20$, $F_x^2 = 0.04\,\text{km}^2$
- $N_{\text{nodes}} = 40$, $F_x^2 = 0.04\,\text{km}^2$.

This allows for conclusions with respect to the impact of an increase in the number of sensors while keeping a constant field size, and an increase in the field size while keeping a constant number of sensors.

In addition, we investigate the behaviour of the system performance for the different values of the maximum constellation size M_{max}. Here, M_{max} corresponds to QPSK, 16-QAM or 64-QAM.

Although the goal of this work is maximizing the bottleneck throughput, an important measure for the system performance is the amount of information that can be collected by the sink node per second. For this, the throughput is multiplied by the number of packets that can be successfully received by the sink. Since the length of packets/data blocks is equal for all network links, and since block errors are rare due to the optimization, the number of packets corresponds to the number of streams served by the sink

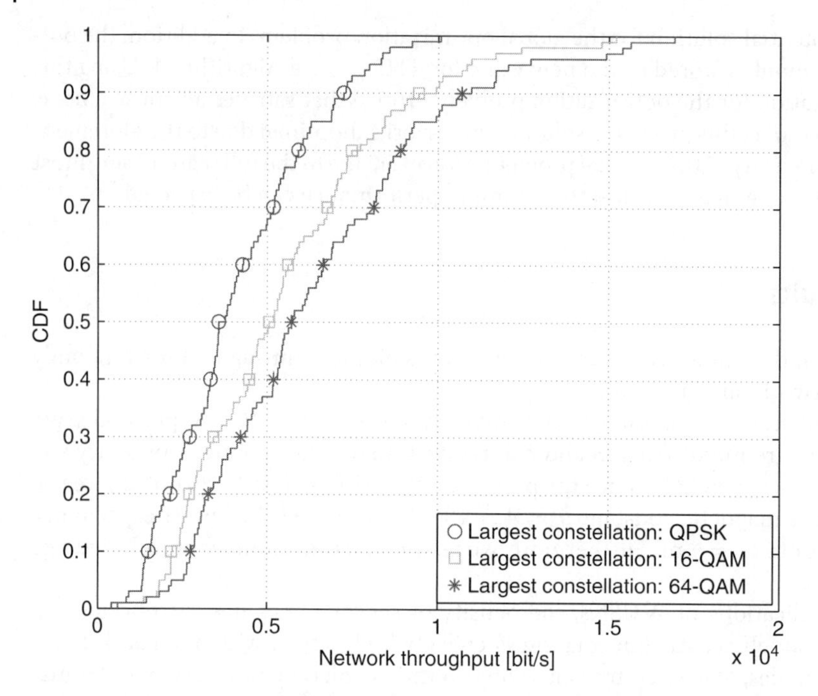

Figure 16.3 Cumulative distribution of network throughput for direct MI transmission based WUSNs with 10 nodes deployed in a 0.01 km^2 area.

node, which equals the number of nodes N_{nodes} in the network. Therefore, in the considered scenarios, we multiply the bottleneck throughput by 10, 20 or 40, in order to obtain the available data rate of the sink node.

First, we show the results for the direct MI transmission based WUSNs and MI waveguide based WUSNs separately. Then, a comparison of the two schemes is provided.

16.3.1 Direct MI Transmission Based WUSNs

In Figure 16.3, a cumulative distribution function of the throughput for a small network of 10 nodes deployed in field of 0.01 km^2 size is shown for direct MI transmission based WUSNs. The optimal resonance frequencies range between 65 kHz and 270 kHz and total bandwidths[2] between 530 Hz and 190 kHz. With increasing size of the largest enabled signal constellation, we observe a considerable increase of the achievable throughput. With increasing constellation size $M_{\mathrm{max}} = \{4, 16, 64\}$, the average throughput increases from 4.32 kbit/s to 5.86 kbit/s to 6.65 kbit/s, respectively.

16.3.2 MI Waveguide Based WUSNs

For MI waveguide based WUSNs, a cumulative distribution function of the throughput for a network with 10 nodes deployed in field of area 0.01 km^2 is shown in Figure 16.4. Here, the throughput increase f using $M_{\mathrm{max}} = 16$ compared to $M_{\mathrm{max}} = 4$ and using $M_{\mathrm{max}} = 64$ compared to $M_{\mathrm{max}} = 16$ is more significant than for direct MI transmission

2 Here, the sum $B_1 + B_2 + B_3$ of all three bandwidths is calculated for each network.

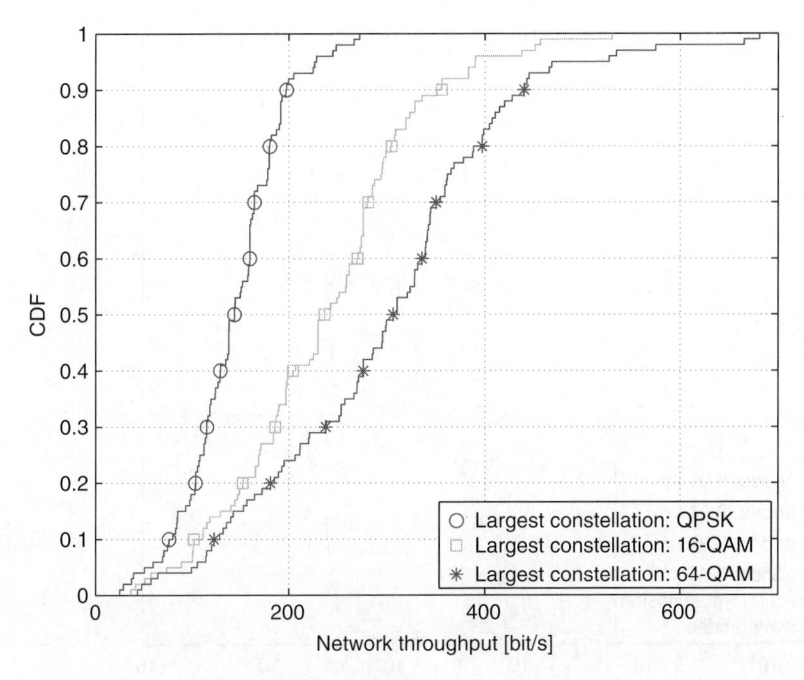

Figure 16.4 Cumulative distribution function of network throughput for MI waveguide based WUSNs with 10 nodes deployed in a 0.01 km^2 area.

based WUSNs using the same values of M_{max}. Interestingly, the average achievable throughputs for MI waveguide based WUSNs are 141 bit/s, 236 bit/s and 287 bit/s for $M_{max} = \{4, 16, 64\}$, respectively, which is much less than for direct MI transmission based WUSNs.

16.3.3 Comparison

Figure 16.5 shows the cumulative distribution functions for the achievable data rates of the sink node for different networks. We observe that with increasing field size (from 0.01 km^2 to 0.04 km^2), the achievable throughput decreases dramatically for both deployment schemes (direct MI transmission and MI waveguides), which is due to much larger average transmission distances between any two sensor nodes. This leads to the degradation of the achievable data rate per link. In particular, for MI waveguide based WUSNs with 10 nodes in an area of 0.04 km^2, ≈50% of the networks can only reach a total achievable data rate at the sink node of less than 1 bit/s. With increasing numbers of nodes per network (from 10 to 20 to 40), the achievable data rate at the sink node increases, although the number of data streams that need to be served by the bottleneck link is dramatically higher and the number of interfering signals increases as well. The reason for this behaviour is the shorter average transmission distances between adjacent nodes, such that the path loss for the useful signal reduces. Obviously, this effect is much stronger than the throughput decrease due to the aforementioned increased number of interferers and data streams.

Furthermore, we observe that direct MI transmission based WUSNs markedly outperform MI waveguide based WUSNs in terms of achievable data rate. Similar

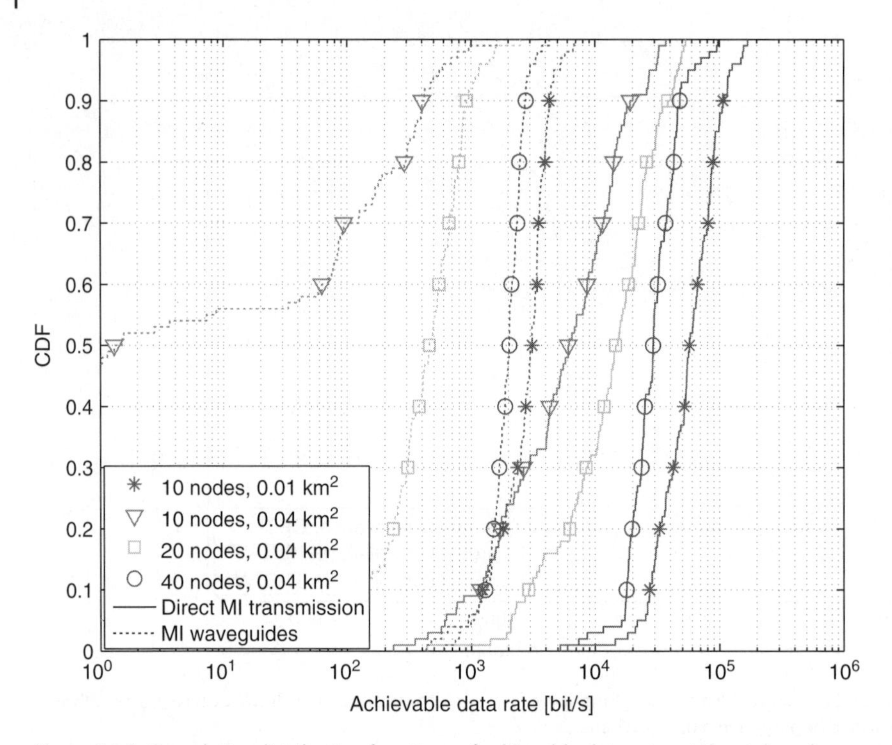

Figure 16.5 Cumulative distribution functions of achievable data rates at the sink node.

results have been obtained in [15] and implicitly in [14]. One of the reasons for this unusual behaviour is the strong coupling between sensor nodes connected via MI waveguides, which means that the interfering signals propagate through the whole network and yield a much larger interference power, which reduces the achievable data rate. Also, the capacitor constraint prevents using very high resonance frequencies (above 100–200 MHz), which might be the optimal frequency range for operating MI waveguides [12]. As a result, the use of MI waveguides is only justified for average transmission distances above 45 m [12], which is, however, not desirable due to the very large path loss. For 10 nodes in an area of 0.01 km^2, the average throughput gain of direct MI transmission compared to MI waveguide transmission is more than 2460% with a peak value of 15 880%. A similar result is obtained for 40 nodes in an area of 0.04 km^2, where the average throughput gain of direct MI transmission is 1500%, with a peak value of 7369%. This result shows that deployment of MI waveguide based WUSNs is not practical. Hence, the direct MI transmission based WUSNs are preferable and yield achievable data rates at the sink node between 6 kbit/s and 169 kbit/s for 10 nodes in an area of 0.01 km^2, between 440 bit/s and 53 kbit/s for 20 nodes in 0.04 km^2, and between 52 kbit/s and 97 kbit/s for 40 nodes in 0.04 km^2.

16.4 Discussion

Using the proposed optimization techniques, the network throughput has been determined for direct MI transmission based WUSNs and for MI waveguide based WUSNs.

From our observations we can conclude that MI waveguides are markedly outperformed by the direct MI transmissions in all considered scenarios. Furthermore, due to much higher number of devices that must be deployed for MI waveguide based WUSNs, two more disadvantages of this scheme arise. First of all, the system becomes less flexible, in that the topology and routing cannot be changed easily. This may yield some adaptation problems, as will be discussed later. The second disadvantage is the increased impact of misalignment, which becomes crucial with increasing numbers of deployed magnetic devices. In particular, due to possible deviation of any passive relay within a waveguide, the actual transmission channel can dramatically change and heavily affect the overall achievable data rate [14]. In direct MI transmission based WUSNs, the system is highly flexible, since the topology can be easily modified by changing the signal processing and scheduling. In addition, any deviation of the actual transmission channels from the assumed ones are insignificant due to large transmission distances between any two MI transceivers, such that the overall performance mostly remains unchanged.

Using the optimized system parameters, the final system configuration is fully determined. Of course, the achievable data rate heavily depends on the time-varying properties of the medium, because the whole system has been optimized for a given conductivity of the soil. In practice, the conductivity may slowly change over time, especially due to rainfall or earthquakes [3]. This deviation may result in packet loss and create connectivity problems in the sensor network. Due to a high flexibility of direct MI transmission based WUSNs, this problem can be dealt with by adapting the topology and channel amplification to the varying mutual inductances. This can be done using channel estimation (conventional or transmitter-side [21]) approaches and possibly by issuing a rescheduling procedure [30]. However, system adaptation to environmental changes is beyond the scope of this work and remains an open issue for future investigations.

Although the focus of this chapter is on information transmission, it should be noted that throughput maximization may not be the only criterion for the system design. In particular, the problem of charging the batteries of the sensor nodes may impose additional constraints on the parameter selection. Obviously, due to their deployment in a dense underground medium, sensor batteries are mostly inaccessible. Hence, wireless charging of the nodes is needed. For the weak couplings between transceivers in direct MI transmission based WUSNs, the resulting wireless power-transfer efficiency might be very low, such that the overall energy efficiency of the system (defined as the achievable network throughput divided by the power used for charging the network) is limited [36]. For the strong coupling between any two coils in MI waveguide based WUSNs, the passive relaying of the power signals can be exploited in order to guide the energy to the distant sensors, such that a much higher energy efficiency may result. In future, a comparison of these two deployment schemes based on the joint maximization of the network throughput and energy efficiency may need to be carried out.

16.5 Summary

We have presented optimization techniques for the two most important cases of magnetic induction based WUSNs: MI waveguides and direct MI transmission based WUSNs. The main goal of this work is to provide the achievable throughput of such networks. For this purpose, we utilised the channel, noise and interference models, which incorporate all relevant signal propagation effects discussed in earlier studies of MI-based communication systems. As a big step towards the practical implementation of MI-WUSNs, signal processing using realistic transmit, receive and equalization filters has been investigated. By taking into account the properties of MI-WUSNs (both in networking and in signal processing) a throughput maximization problem was formalised. It should be noted that for direct MI transmission based WUSNs, a recently proposed three-band approach has been used. In addition, some practical constraints have been imposed; for example the maximum size of the signal constellation and an equal block length for all network links. Unfortunately, the resulting optimization problem is non-convex, such that a closed-form solution cannot be found. Moreover, most of the system parameters can only be determined via grid search due to the non-linear dependency of the achievable data rate on these parameters. Nevertheless, an iterative power allocation algorithm has been proposed for direct MI transmission based WUSNs in order to reduce the computation complexity. For MI waveguide based WUSNs, the full search bounds for the block length optimization have been determined. Using the proposed optimization approaches, the achievable throughput has been obtained for both schemes. MI waveguide based WUSNs turn out to be outperformed by direct MI transmission based WUSNs. Furthermore, future work on system adaptation to the time-varying channel conditions and on charging of sensor nodes in the dense underground medium have been considered.

References

1 I.F. Akyildiz, W. Su, Y. Sankarasubramaniam and E. Cayirci (2002) Wireless sensor networks: A survey. *Computer Networks,* **38** (4), 393–422.

2 I.F. Akyildiz and E.P. Stuntebeck (2006) Wireless underground sensor networks: Research challenges. *Ad Hoc Networks,* **4** (6), 669–686.

3 M.C. Vuran and A.R. Silva (2009) Communication through soil in wireless underground sensor network: theory and practice, in *Sensor Networks* (ed. G. Ferrari), Springer.

4 R. Bansal (2004) Near-field magnetic communication. *IEEE Antennas and Propagation Magazine,* **46** (2), 114–115.

5 A. Karalis, J.D. Joannopoulos and M. Soljacic (2008) Efficient wireless non-radiative mid-range energy transfer. *Annals of Physics,* **323** (1), 34–48.

6 J.J. Casanova, Z.N. Low and J. Lin (2009) A loosely coupled planar wireless power system for multiple receivers. *IEEE Transactions on Industrial Electronics,* **56** (8), 3060–3068.

7 I.J. Yoon and H. Ling (2011) Investigation of near-field wireless power transfer under multiple transmitters. *IEEE Antennas and Wireless Propagation Letters,* **10**, 662–665.

8 M. Masihpour and J.I. Agbinya (2010) Cooperative relay in near field magnetic induction: a new technology for embedded medical communication systems, in *Proceedings of IB2Com*, pp. 1–6.

9 Z. Sun and I.F. Akyildiz (2010) Magnetic induction communications for wireless underground sensor networks. *IEEE Transactions on Antennas and Propagation,* **58** (7), 2426–2435.

10 E. Shamonina, V.A. Kalinin, K.H. Ringhofer and L. Solymar (2002) Magneto-inductive waveguide. *Electronics Letters*, **38** (8), 371–373.

11 R.R.A. Syms, I.R. Young and L. Solymar (2006) Low-loss magneto-inductive waveguides. *Journal of Physics D: Applied Physics*, **39** (18), 3945–3951.

12 S. Kisseleff, W.H. Gerstacker, R. Schober, Z. Sun and I.F. Akyildiz (2013) Channel capacity of magnetic induction based wireless underground sensor networks under practical constraints, in *Proceedings of IEEE WCNC 2013*.

13 X. Tan, Z. Sun and I.F. Akyildiz (2015) A testbed of magnetic induction-based communication system for underground applications. *arXiv:1503.02519*.

14 S. Kisseleff, I.F. Akyildiz and W.H. Gerstacker (2015) Digital signal transmission in magnetic induction based wireless underground sensor networks. *IEEE Transactions on Communications*, **63** (6), 2300–2311.

15 S. Kisseleff, I.F. Akyildiz and W.H. Gerstacker (2014) Throughput of the magnetic induction based wireless underground sensor networks: Key optimization techniques. *IEEE Transactions on Communications*, **62** (12), 4426–4439.

16 P. Gupta and P.R. Kumar (2000) The capacity of wireless networks. *IEEE Transactions on Information Theory*, **46** (2), 388–404.

17 C.X. Wang, X. Hong, H.H. Chen and J. Thompson (2009) On capacity of cognitive radio networks with average interference power constraints. *IEEE Transactions on Wireless Communications*, **8** (4), 1620–1625.

18 A. Spyropoulos and C.S. Raghavendra (2003) Capacity bounds for ad-hoc networks using directional antennas, in *Proceedings of IEEE ICC 2003*.

19 Z. Sun and I.F. Akyildiz (2012) On capacity of magnetic induction-based wireless underground sensor networks, in *Proceedings of IEEE INFOCOM 2012*, pp. 370–378.

20 J.R. Wait (1952) Current-carrying wire loops in a simple inhomogeneous region. *Journal of Applied Physics*, **23** (4), 497–498.

21 S. Kisseleff, I.F. Akyildiz and W. Gerstacker (2014) Transmitter-side channel estimation in magnetic induction based communication systems, in *Proceedings of IEEE BlackSeaCom 2014*.

22 Y. Zhang, Z. Zhao and K. Chen (2013) Frequency splitting analysis of magnetically-coupled resonant wireless power transfer, in *Proceedings of IEEE ECCE 2013*.

23 S. Kisseleff, I.F. Akyildiz and W.H. Gerstacker (2014) Disaster detection in magnetic induction based wireless sensor networks with limited feedback, in *IFIP Wireless Days*.

24 H. Akagi, E.H. Watanabe and M. Aredes (2007) *Instantaneous Power Theory and Applications to Power Conditioning*, Wiley-IEEE Press.

25 N. Tal, Y. Morag and Y. Levron (2015) Design of magnetic transmitters with efficient reactive power utilization for inductive communication and wireless power transfer, in *Proceedings of IEEE International Conference on Microwaves, Communications, Antennas and Electronic Systems (COMCAS)*.

26 D. Tse and P. Viswanath (2005) *Fundamentals of Wireless Communication*, Cambridge University Press.

27 J.G. Proakis (2001) *Digital Communications*, McGraw-Hill Higher Education.

28 A. Goldsmith (2005) *Wireless Communications*, Cambridge University Press.

29 T.S. Rappaport (2002) *Wireless Communications: Principles and Practice*, Prentice Hall, 2nd edn.

30 F. Sivrikaya and B. Yener (2004) Time synchronization in sensor networks: A survey. *IEEE Networks,* **18** (4), 45–50.

31 R.C. Prim (1957) Shortest connection networks and some generalizations. *Bell System Technical Journal,* **36** (6), 1389–1401.

32 W. Qin and Q. Cheng (2009) The constrained min-max spanning tree problem, in *Proceedings of ICIECS 2009.*

33 S. Boyd and L. Vandenberghe (2004) *Convex Optimization,* Cambridge University Press.

34 A. Markham and N. Trigoni (2012) Magneto-inductive networked rescue system (MINERS) taking sensor networks underground, in *Proceedings of IPSN 2012.*

35 I.F. Akyildiz, Z. Sun and M.C. Vuran (2009) Signal propagation techniques for wireless underground communication networks. *Physical Communication,* **2** (3), 167–183.

36 S. Kisseleff, X. Chen, I.F. Akyildiz and W.H. Gerstacker (2016) Efficient charging of access limited wireless underground sensor networks. *IEEE Transactions on Communications,* **64** (5), 2130–2142.

17

Agricultural Applications of Underground Wireless Sensor Systems: A Technical Review

Saeideh Sheikhpour[1], Ali Mahani[1] and Habib F. Rashvand[2]

[1] Department of Electrical Engineering, Shahid Bahonar University of Kerman, Kerman Province, Iran
[2] Advanced Communication Systems, University of Warwick, UK

17.1 Introduction

Unique features of wireless sensor networks (WSNs) such as their low-power, low-cost and large-scale production make them a very attractive technology for wide range of applications in many areas such as industrial sites (production control, structural health monitoring, leak monitoring, surveillance, automation, robotics, natural resource monitoring), agriculture and environment (Earth and environmental monitoring, biological studies, animal tracing, forest fire monitoring), health (tracking and monitoring patients and doctors, drug administration in hospitals), homes (home automation, smart environment, safety, old age assistance) and the public sector (smart transport, traffic tracking, unmanned vehicles, vehicle tracking, surveillance) [1, 2].

On the agricultural front a whole new trend in research and development has emerged as WSNs have been applied in the agricultural and farming industries. WSNs have been extensively used in agriculture to boost yields and reduce harmful impacts on the environment in recent years. These networks are composed of a set of small-scale nodes deployed above ground, underground, or attached to plant, to gather real-time information about fields and crops that is essential for decision-making [3, 4].

The use of WSNs has been extended to other agricultural activities including irrigation management [5–9], pest management [10–14], fertiliser management [15, 16], field monitoring [17–21], climate monitoring [22], greenhouse control [23–28], and animal monitoring [29, 30]. Demand for these networks is rapidly growing in all these agricultural applications, due to their unique advantages.

Combining WSNs with agricultural principles can help to achieve higher yields, reduced pesticides use, improved quality and yield, reduced water usage, faster crop cycles, less labour and fewer harmful effects on the environment compared to traditional agriculture [31]. However, WSNs come with their own features and constraints, including low processing capability, low battery power, short-range communication, low transmission rates and the limited memory of the sensor nodes. Depending on their precise specifications, WSNs for agricultural applications bring different challenges. Thus, many research studies have been developed in order to improve agricultural WSNs, creating new algorithms and protocols, introducing new research topics and applications and developing new design concepts [1, 2, 32].

Wireless Sensor Systems for Extreme Environments: Space, Underwater, Underground, and Industrial.
First Edition. Edited by Habib F. Rashvand and Ali Abedi.

The main aim of this chapter is to present a survey of research proposals for WSNs in various agricultural activities and some of the solutions for solving their limitations and constraints. In summary, the chapter covers:

- a review of the recent literature on WSNs in agriculture
- the fundamentals and challenges of WSN-based agricultural applications
- current sensor technologies for agriculture
- the opportunities for future research.

The remainder of this chapter is organized as follows: Section 17.2 gives a brief introduction to WSNs and their technologies in the agricultural domain, and the related challenges in deployed environmental networks are presented in Section 17.3, Section 17.4 discusses the design issues of WSNs for agricultural activities, and then an overview of WSN applications in agriculture is provided in Section 17.5. Finally, Section 17.6 concludes this chapter and discusses a few future research directions.

17.2 WSN Technology in Agriculture

WSNs are composed of a set of sensor nodes that have capabilities to measure some physical attributes and convert them into signals for the observer. They are becoming ubiquitous in almost every field of day-to-day life because of advances in technology and cost reductions. In this section, the architecture of a sensor node, wireless communication technologies for agricultural WSN-based systems and available sensor nodes for agricultural applications are presented.

17.2.1 Sensor Node Architecture

Wireless sensor nodes are devices with capabilities of sensing data, processing it and networking. Sensor nodes are the basis of a WSN. Each sensor node in the network measures some physical quantity, such as temperature, humidity, sunlight or pressure, and converts it into digital data. Then, the sensor node processes the sensed data and transmits it. A sensor node could be comprised of the following components (Figure 17.1):

- array of MEMS sensors (analog or digital)
- analogue to digital converter (ADC)
- microcontroller (to process sensor signals)
- transceiver
- power supply
- external memory.

The selection of these units largely depends on the particular application. External memory is optional and may not be required in all cases.

17.2.2 Wireless Communication Technologies and Standards

Wireless communication technologies that are used in the agricultural domain – including ZigBee, Bluetooth, WiFi, WiMAX, and GPRS – are briefly discussed in this subsection.

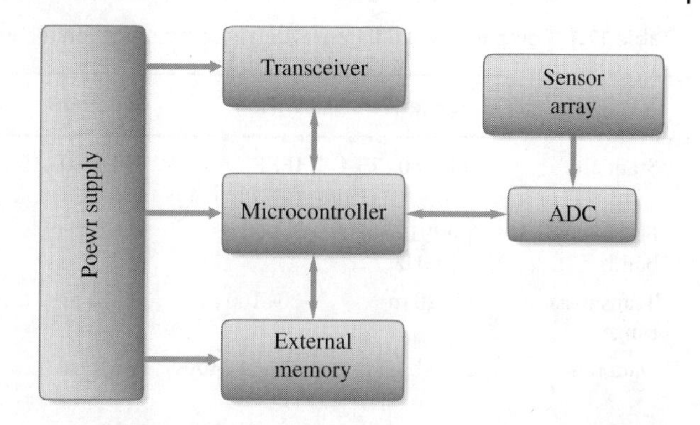

Figure 17.1 Sensor node architecture.

ZigBee: ZigBee [33] is a low-cost and low-powered wireless networking technology based on the IEEE 802.15.4 standard. It was introduced in May 2003 by the ZigBee Alliance. It works in the 2.3 GHz Industrial, Scientific and Medical (ISM) band globally and consists of four layers: physical layer, medium access control (MAC) layer, network layer and application layer. The MAC- and physical-layer specifications are provided in IEEE 802.15.4, while ZigBee provides specifications for the application and network layers. ZigBee is considered most promising for some WSN-based agricultural applications due to its energy-efficiency, simple networking configuration and reliability. Kalaivani et al. [34] have surveyed recent research on ZigBee-based WSNs in agriculture.

WiFi: WiFi [35] is a wireless local area network (WLAN) based on the IEEE 802.11 family. It allows devices such as computers and smartphones (within a range about 20 m from the access point) to connect to the network. It consists of a physical layer and a data-link layer and operates on two frequency bands: 2.4 GHz and 5 GHz ISM. WiFi-based WSNs can be applied in agriculture for intelligent farming [36, 37].

Bluetooth: Bluetooth [38] is a wireless personal area network (WPAN) based on the IEEE 802.15.1 standard. It was developed by Ericsson in 1994 for exchanging data over short distances (~10 m). It works in the 2.4 GHz and the 915 and 868 MHz radio bands. Bluetooth provides connectivity to other mobile users (up to eight at a time) for sharing data without wires. Bluetooth can be used for short-range communication in agricultural communication.

WiMAX: WiMAX [39] is a relatively new communication technology based on the IEEE 802.16 standards family. It was developed for creating metropolitan area networks (MANs). It defines physical and MAC layers for wireless interfaces. WiMAX technology provides long-range communication (up to a few kilometres) in the 2–66 GHz frequency band with high data rates. While WiMAX is an expensive technology, it can be used in several agricultural activities. For example, Rani et al. [40] used a WiMAX-based WSN for agricultural field monitoring.

GPRS: The General Packet Radio Service [41] is a wireless technology used on GSM networks. It was developed by the European Telecommunications Standards Institute for the periodic transmission of small data messages. It has a high transmission rate and the ability to transfer real-time data and operates in the 850, 900, 1800, and 1900 MHz frequency bands. GPRS is used in agricultural applications for sending information from target areas to monitoring centers [7].

Table 17.1 Comparison of different wireless communication technologies.

	ZigBee	WiFi	Bluetooth	WiMAX	GPRS
Standard	IEEE 802.15.4	IEEE 802.11a,b,g,n	IEEE 802.15.1	IEEE 802.16a,e	–
Frequency band	868/915 MHz, 2.4 GHz	2.4 GHz	2.4 GHz	2–66 GHz	1800/1900, 850/900 MHz
Transmission range	10–20 m	20–100 m	8–10 m	0.3–49 km	2–35 km
Data rate	250 kbps	11–54 Mbps	1 Mbps	0.4–1 Gbps (stationary), 50–100 Mbps (mobile)	50–100 kbps
Modulation/ protocol	DSSS, CSMA/CA	DSSS/CCK, OFDM	FHSS	OFDM, OFDMA	TDMA, DSSS
Energy consumption	Low	High	Medium	Medium	Medium
Cost	Low	High	Low	High	Medium
Success metrics	Reliability, power, cost	Speed, flexibility	Cost, convenience	Throughput, speed, range	Range, cost, convenience,

Each of the wireless technologies has different features and capabilities (Table 17.1), which will define its applicability in any particular field or application. The cost of each communication protocol depends on its complexity and hardware and software requirements and its energy consumption depends on the bit rate and the current required in reception and transmission modes [42].

17.2.3 Available Sensor Node for Agricultural Activities

Different wireless sensor platforms are available for use in agricultural applications [3, 4]; there is a wide variety of sensor platforms because of the many WSN applications in different fields of science and technology. Table 17.2 compares some available sensor platforms suitable for agricultural applications. The use of each sensor platform for any particular soil, plant or environmental monitoring application depends on the parameters that it can sense and its own features.

Table 17.3 provides a list of existing sensor nodes and their measurement parameters.

Soil sensors measure parameters such as moisture levels, temperature and salinity. So they are appropriate for agricultural applications such as irrigation and farmland monitoring.

Environment sensors are suitable for decision-making in agricultural activities. Leaf/plant category sensors are usually attached to a plant, and are suitable for agricultural applications such as greenhouse control, crop monitoring and pesticide management.

Table 17.2 Comparison of some commercial sensor platforms.

	MICA2	MICAz	IRIS	Imote2	TinyNode584	TinyNode184	TelosB	Tmote Sky	LOTUS
Microcontroller	ATmega128L	ATmega128L	ATmega128L	Marvell/ XScalePXA271	MSP430	MSP430F2417	TIMSP430	TI MSP430	Cortex M3 LPC 17xx
Clock speed (MHZ)	7.373	7.373	7.373	13–416	8	16	6.717	6.717	10–100
Power									
Battery voltage	3.3	3.3	3.3	3.3	3.6	3.6	3.6	3.6	3.3
Current: sleep	15 μA	15 μA	15 μA	390 μA	4 μA	22 μA	3.3–1.8 μA	5.1 μA	10 μA
Current: active	8 mA	8 mA	8 mA	66 mA	77 mA	2.2 A	1.7–5.1 mA	23 mA	50 mA
Memory									
RAM			4 kB	256 kB SRAM 32 MB SDRAM	10 kB	10 kB	10 kB	10 kB	64kB SRAM
ROM	128 kB	128 kB	128 kB						
Flash	512 kB	512 kB	512 kB	32 MB	512 kB	512 kB	548 kB	48 kB	512 kB 64 MB serial
EEPROM	4 kB	4 kB	4 kB				16 kB		
Operational frequency band(MHZ)	868/915	2400	2400	2400	868/915	868/915	2400–2483.5	2400	2400
Transceiver chip	CC1000	CC2420	RF230	CC2420	XE1205	SX1211	CC2420	CC2420	Atmel RF231
I/O interface	51 pin interface to other extension, GPIO, I2C, SPI, ADC, UART	51 pin interface to other extension, GPIO, I2C, SPI, ADC, UART	51 pin interface to other extension, GPIO, I2C, SPI, ADC, UART	UART 3x, GPIO, SPI, DIO, JTAG I^2S, USB, AC'97, Camera, IMB400 multimedia	Factory- made extension board for custom interface electronics, GPIO, I2C, SPI, ADC, UART	Factory-made extension board for custom interface electronics, GPIO, I2C, SPI, ADC, UART	6-pin and 10-pin to other extension, UART, I^2S, SPI, DIO	on-board sensors, no manufacture built extension board, GPIO, I2C, SPI, ADC, UART	51 pin interface to other extension, 3xUART, SPI, I2C, I2S, GPIO, ADC
Data rate(Kbps)	38.4 (Baud)	250	250	250	153	200	250	250	250
Max range (m)									
Outdoor	300	75–100	>300	30	Up to 2000	150	75–100	125	>500
Indoor		20–30	>50			50	20–30	50	>50
Manufacture	Crossbow	Crossbow	Crossbow	Crossbow	Shockfish SA	Shockfish SA	Crossbow	Moteiv	MEMSIC

Table 17.3 Available sensor nodes for agricultural applications

Sensing device	Available sensors	Category	References
Pogo portable	Soil moisture, rain/water flow, soil temperature, conductivity, salinity	Soil	http://www.stevenswater.com
Tipping bucket rain gauge	Rain/water flow	Soil	http://www.stevenswater.com
Hydra probe II	Soil moisture, rain/water flow, soil temperature, water level, conductivity, salinity	Soil	http://www.stevenswater.com
VH-400	Soil moisture, water level	Soil	http://www.vegetronix.com
MP406	Soil moisture, soil temperature	Soil	www.ictinternational.com.au
EC250	Soil moisture, rain/water flow, soil temperature, conductivity, salinity	Soil	www.stevenswater.com
THERM200	Soil temperature	Soil	http://www.vegetronix.com
WET-2	Soil temperature, conductivity, salinity	Soil	http://www.dynamax.com
ECRN-100 high-REC rain	Rain/water flow	Soil	http://www.decagon.com
Cl-340 hand-held photosynthesis	Photosynthesis, moisture, temperature, hydrogen, wetness, CO_2	Leaf/plant	http://www.solfranc.com
LW100	Moisture, wetness, temperature	Leaf/plant	http://www.globalw.com
Leaf wetness sensor	Moisture	Leaf/plant	http://www.decagon.com
237-L	Moisture, wetness, temperature	Leaf/plant	http://www.campbellsci.com
LT-2 M	Temperature	Leaf/plant	http://www.solfranc.com
PTM-48A photosynthesis monitor	Photosynthesis, moisture, temperature, wetness, CO_2	Leaf/plant	http://phyto-sensor.com
TT4 multi-sensor thermocouple	Moisture, Temperature	Leaf/plant	www.ictinternational.com
TPS-2 portable photosynthesis	Photosynthesis, moisture, temperature, wetness, CO_2	Leaf/plant	www.ppsystems.com
Field scout CM1000TM	Photosynthesis	Leaf/plant	http://www.specmeters.com
Met One Series 380 rain gauge	Rain fall	Environment	http://www.stevenswater.com
CM-100 compact weather station	Temperature, humidity, atmospheric pressure, wind speed, wind direction	Environment	http://www.stevenswater.com

Table 17.3 (Continued)

Sensing device	Available sensors	Category	References
WXT520 compact weather station	Temperature, humidity, atmospheric pressure, solar radiation, wind speed, wind direction, rain fall,	Environment	http://www.stevenswater.com
XFAM-115KPASR	Temperature, humidity, atmospheric pressure	Environment	http://www.pewatron.com
Met Station One (MSO)	Temperature, humidity, atmospheric pressure, wind speed, wind direction	environment	http://www.stevenswater.com
SHT71	Temperature, humidity, atmospheric pressure	Environment	http://www.sensirion.com
HMP45C	Temperature, humidity, atmospheric pressure	Environment	http://www.campbellsci.com
RG13/RG13H	Rain fall	Environment	http://www.vaisala.com
LI-200 Pyranometer	Solar radiation	Environment	http://www.stevenswater.com

17.3 WSNs for Agriculture

Agricultural applications use three types of WSNs: terrestrial wireless sensor networks (TWSNs), wireless underground sensor networks (WUSNs) and hybrid wireless sensor network (HWSNs). In this section, these variants of WSNs are discussed.

17.3.1 Terrestrial Wireless Sensor Networks

A terrestrial wireless sensor network (TWSN) is formed by a large number of sensor nodes, which are spatially deployed above the ground and cooperate with each other. This makes it possible to collect needed information and to control processes that take place above the ground. Figure 17.2 shows an example of a TWSN deployed on the

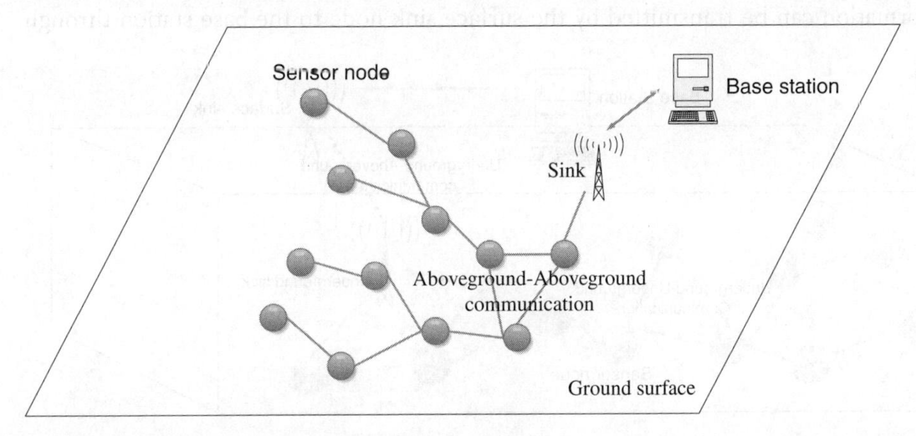

Figure 17.2 A typical terrestrial wireless sensor network

ground (such as farmland). This network consists of twelve sensor nodes and a network sink and a base station that uses radio frequency (RF) communication over distances of a few hundred metres. In TWSNs, the nearest nodes should be connected in a reliable manner, so the distance between the nearest nodes should be less than a certain value in the deployment strategy.

TWSNs are extensively used and researched for applications in the agricultural domain. For example, they can be installed on farmland for sensing real-time soil and environment conditions: light levels, temperature and moisture levels. This information informs tasks such as pest control, irrigation management and fertiliser control. Autonomous sensor nodes with suitable sensing elements are required. After deployment each sensor node starts to interact and exchange data. Upon receiving data from the sensor nodes an appropriate decision can be made at the control unit by a remote user.

17.3.2 Wireless Underground Sensor Networks

In TWSNs, it is very difficult to obtain the real-time information required for precision agriculture such as the water content deeper in the soil and data the status of underground crops. TWSNs therefore cannot fully meet all requirements of agricultural applications. Wireless underground sensor networks (WUSNs) are another type of WSN that is employed in agriculture. The use of WUSNs in the agriculture is a relatively new field, but has become an attractive topic for research. In WUSNs, sensor nodes are buried in the soil at the desired depth and wirelessly transmit data through underground communication. Wireless communication in underground environments is a challenging issue in WUSN-based system design because of signal losses, high bit-error rates, multipath fading, noise and high levels of attenuation. Akyildiz et al. [32] surveyed the design problems and challenges in this area, for example underground wireless channels.

A WUSN comprising fourteen underground sensors, an underground sink node, a surface sink and a base station is depicted in Figure 17.3. The sensor nodes collect the information of interest and send it to the underground sink using underground–underground communication. Then, the underground sink node transmits the information to the surface sink node via underground–aboveground communication. Information can be transmitted by the surface sink node to the base station through

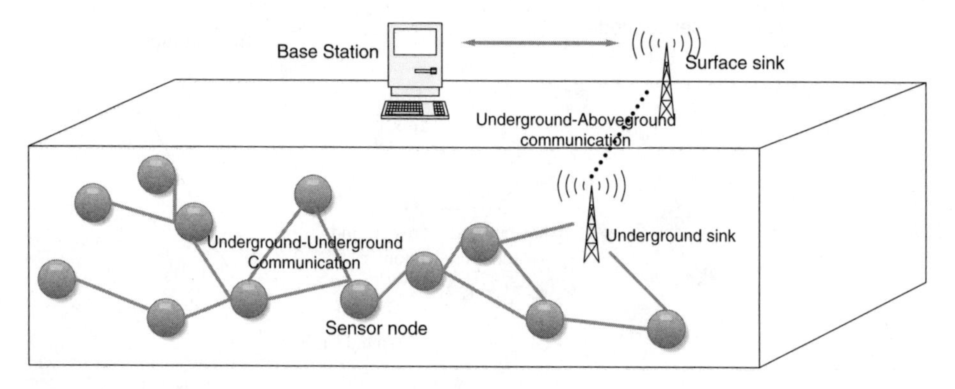

Figure 17.3 A typical wireless underground sensor network

aboveground–above ground communication. In general, WUSNs are more expensive than TWSNs in terms of the required equipment, deployment and maintenance costs.

17.3.3 Hybrid Wireless Sensor Networks

Some agricultural applications like time determination of irrigation are unsuitable for TWSNs or WUSNs. Hybrid wireless sensor networks (HWSNs), which combine TWSNs and WUSNs, have therefore been introduced to better and more precisely accomplish such applications in the agricultural domain. HWSNs combine the advantages of TWSNs and WUSNs. An HWSN contains both underground and aboveground sensor nodes that collaboratively perform a specific task. They have been suggested for field monitoring [43]. Their use reduces the need for human involvement in information collection and provides more accurate information than single TWSNs or WUSNs.

17.4 Design Challenges of WSNs in Agriculture

As mentioned in Section 17.3, the design of WSN-based systems for agricultural applications is an exciting research topic because of the unique possibilities for automation. In this section, seven considerations for agricultural WSN design are discussed: energy consumption, power resources, fault tolerance, scalability, network architectures, coverage and connectivity, and wireless underground communication.

17.4.1 Energy Consumption

WSNs are expected to stay operational for long periods in applications such as field monitoring . The energy provided by batteries is very limited, so battery-life is a major stumbling block in the development of WSNs. Many approaches have been suggested to manage the energy consumption of battery-powered sensor nodes. Some of these have focused on harvesting energy from external environment. However, these energy sources are intermittent, so that a battery is required as well. Another suggested solution is used of rechargeable batteries, but in many of the harsh environments in which WSNs are deployed, batteries are often not accessible.

In general, the energy consumption of sensor nodes in a WSN is caused by the sensing, communication and computation tasks. Communication (transmission and reception of data) is a major source of energy consumption. The energy consumption of the sensing task depends on the specific sensor and application. In many cases, the energy consump tion of computation tasks is negligible compared to that of other tasks. Therefore, energy saving techniques should be incorporated in the design of WSNs in order to decrease the energy consumption. This will prolong the lifetime of the network. For example, the lifetime of a sensor node such as the Telos or Mica2, powered by two AA batteries and using current radio technologies, is just three days in continuous receive mode, and is perhaps a week or two if the radio is turned off but the CPU remains active. Therefore, it is vital to minimize energy use in both communication and computation in order to implement long-lived applications [44].

Currently, techniques to give low power consumption have attracted much attention in WSN-based system design. Examples include low-power routing protocols, data aggregation approaches, low-power MAC protocols and sleep/wakeup protocols. Rault

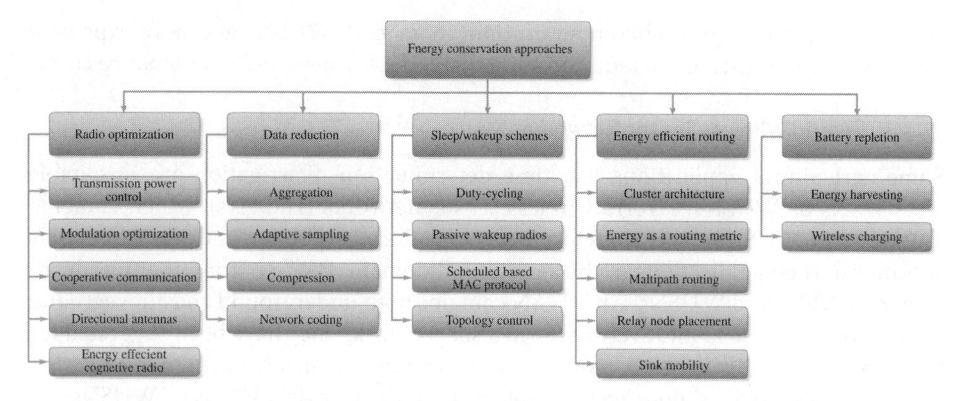

Figure 17.4 Proposed classification of energy conservation approaches [43].

et al. [45] presented a new classification of energy conservation approaches, as found in the recent literature. They classified energy conservation approaches into five categories (as shown in Figure 17.4): radio optimization, data reduction, sleep/wakeup schemes, energy efficient routing and battery repletion.

Radio optimization: The radio consumes much more energy than other parts of the sensor. Many researchers have proposed radio-parameter optimization techniques to reduce the energy consumption of the radio.

Data reduction: Some of the approaches suggested by researchers can reduced the amount of data that is actually sent to the base station, and this results in considerable energy saving.

Sleep/wakeup schemes: Some energy-conservation approaches use periodic sleep/wakeup schemes. Sensor nodes periodically go to sleep by switching off their radios.

Energy efficient routing: Routing is a vital feature in multi-hop sensor networks and occurs frequently. Energy-efficient routing protocols can conserve a significant amount of energy.

Battery repletion: Several researchers have suggested energy harvesting and wireless charging as promising solutions for special cases of WSNs (for more detail refer to the paper by Rault et al. [45].)

17.4.2 Power Sources

Sensor-node power sources can be categorised into two general types: batteries and energy-harvesting. This classification is shown in Figure 17.5. Battery-powered sensor nodes have a very limited capacity and are soon exhausted, so they are not practical for applications that take a very long time, such as field monitoring in agriculture. In the second type of power sources, the energy for sensor node is harvested from the external environment. Energy harvesting is a process of capturing, storing and using energy from an external source.

As mentioned in Section 17.3, both TWSNs and WUSNs are used in agriculture. It is possible for sensor nodes to scavenge energy from the sun, wind or from vibrations

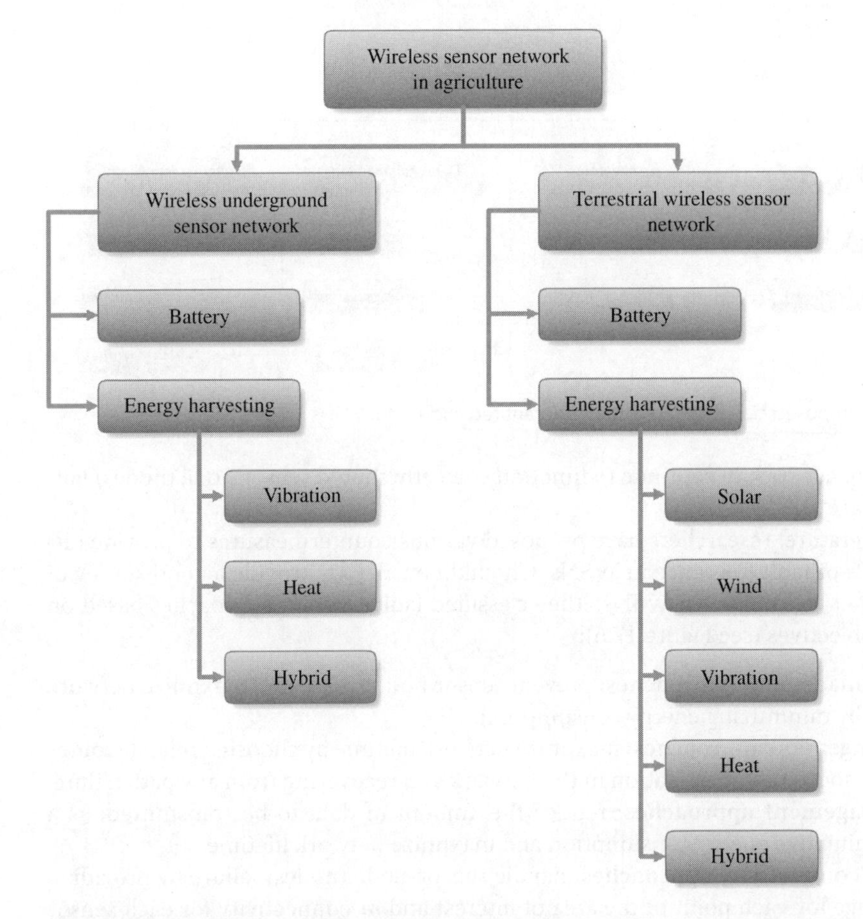

Figure 17.5 Classification of WSN power sources.

in TWSNs. Hwang et al. [46] used solar-powered sensor nodes in a server system that could be used in agricultural applications. Morais et al. [47] integrated solar, wind and hydroelectric energy as power sources for sensor nodes in agriculture. Wind energy is used by Nayak et al. [48] to power sensor nodes used in agricultural environmental monitoring.

Vibrations are a promising source for energy scavenging in WUSNs. Kahrobaee and Vuran [49] integrated energy-harvesting components and underground sensor nodes in order to harvest energy from the underground vibrational energy in applications such as field and mobile irrigation systems.

17.4.3 Fault Tolerance

Sensor node failure is inevitable in a typical WSN due to the resource-constrained nature of sensor nodes and the harsh environment in which WSNs are normally deployed. Failures can have different causes, such as energy depletion, physical damage and hardware and software problems. Generally, WSNs are vulnerable and prone to failure, so it is necessary to incorporate appropriate fault-tolerant mechanisms in their design,

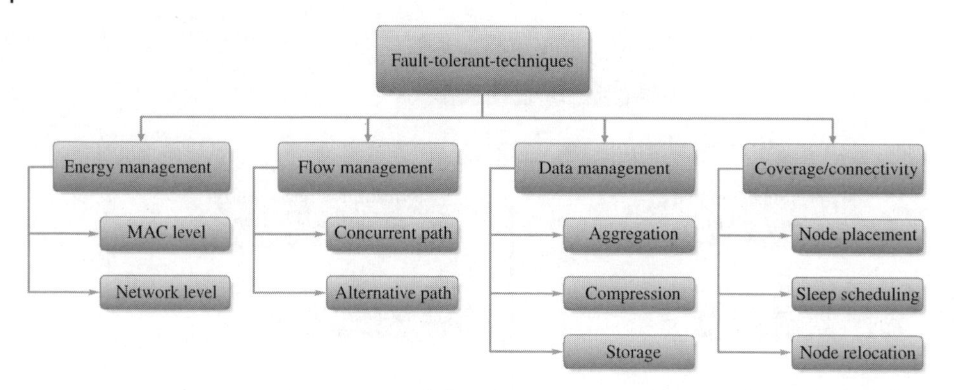

Figure 17.6 Proposed classification of fault tolerant techniques for WSNs [48].

allowing the network to continue to function correctly should sensor node (nodes) failure take place.

In the literature, researchers have proposed various countermeasures to provide different levels of fault tolerance in WSNs. Chouikhi et. al. [50] provide a good survey of fault-tolerant techniques in WSNs. They classified fault-tolerant approaches based on the main objectives (see Figure 17.6):

Energy management approaches: prevent sensor node failure and maximize network lifetime by minimizing energy consumption.

Flow management approaches: maximize network lifetime by choosing reliable routes between source and destination in the network and recovering from any path failure.

Data management approaches: reduce the amount of data to be transmitted; as a result, minimize energy consumption and maximize network lifetime.

Coverage/connectivity approaches: handle sensor-node (nodes) failure by providing k-coverage for each point of the area of interest and m-connectivity for each sensor node.

Most of these techniques can be considered for agricultural applications in order to provide the desired level of fault tolerance.

17.4.4 Scalability

Depending on the requirements of the application, the number of sensor nodes in the network and the size of network can be varied. Therefore, there is a need to design protocols for WSNs that are independent of their size and sensor-node count and which are therefore applicable in any network.

Specifically, the fields in which WSNs are deployed can be very large (several tens of hectares). In these applications, the number of sensor nodes can vary from hundreds to dozens. Therefore, scalability is an important and critical factor that should be considered when developing WSN protocols for agricultural applications [51].

17.4.5 Network Architecture

In terms of the interactions between sensor nodes, WSN architectures can be classified into three groups: star, mesh and tree (see Figure 17.7) [52].

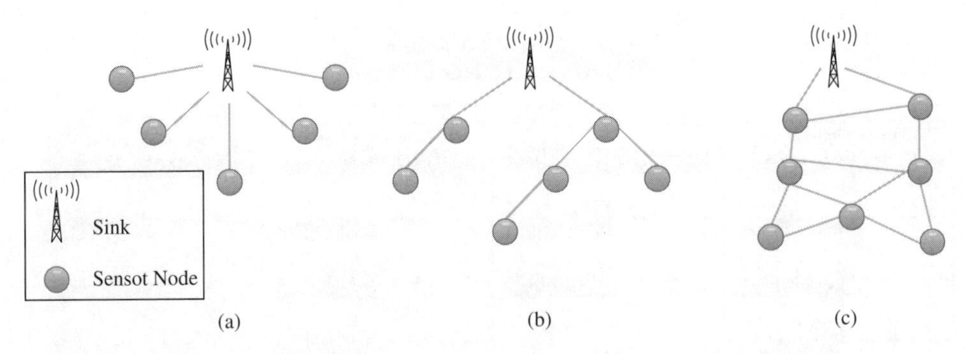

Figure 17.7 Network topologies: (a) star; (b) tree; (c) mesh.

Star topology: In star topologies, all sensor nodes are directly connected to the sink and sensor nodes have no interactions with each other (Figure 17.7a). This topology is simple and does not need any routing protocol. The disadvantages of this topology are its high consumption of energy due to the need for one-hop communication, and the limit on the maximum number of sensor nodes in the network and maximum number of connections supported by the sink.

Tree topology: A tree topology (Figure 17.7b) is virtually hierarchical. The sensor nodes that are located near the sink directly exchange data with it and the sensors that are located farther away from the sink interact with the nearer ones. In a tree topology, all sensor nodes can directly interact with each other. Sensor nodes in lower levels (the 'leaves') can be transmit data through the sensor nodes in higher levels (the 'root' or 'branches'). The lifetime of the network with such a topology is very short since the nodes nearest the sensor consume much more energy than the others due to the high traffic load they have to carry.

Mesh topology: In this topology, sensor nodes can be connected to each other, and interactions take place between each sensor node and every nearest sensor node (shown in Figure 17.7c). Many communication protocols have been developed for this type of topology. A mesh topology is energy efficient because of the short data transmission paths and the small number of retransmissions required. Mesh topologies are used in many WSN-based applications.

WSNs can also be classified into two classes based on the system hierarchy: flat and hierarchical architectures.

Flat architecture: All the sensor nodes in a network with a flat architecture play the same role. Sensor nodes in this architecture can connect to each other, as in the mesh architecture, or only to the sink, as in a star architecture. The routing protocols for flat architectures are usually simple. They are suitable for small networks.

Hierarchical architecture: In flat architectures, sensor nodes are organized in different levels and play different roles. Several star networks (clusters) contain sensor nodes in the lower levels of the hierarchy (cluster members) with the nearest sensor nodes in a higher level of the hierarchy (cluster-head). In each cluster, cluster members connect to the cluster-head through inter-cluster communications and the cluster head connects to other cluster heads or the sink through intra-cluster communications. Hierarchical architectures are suitable for large networks.

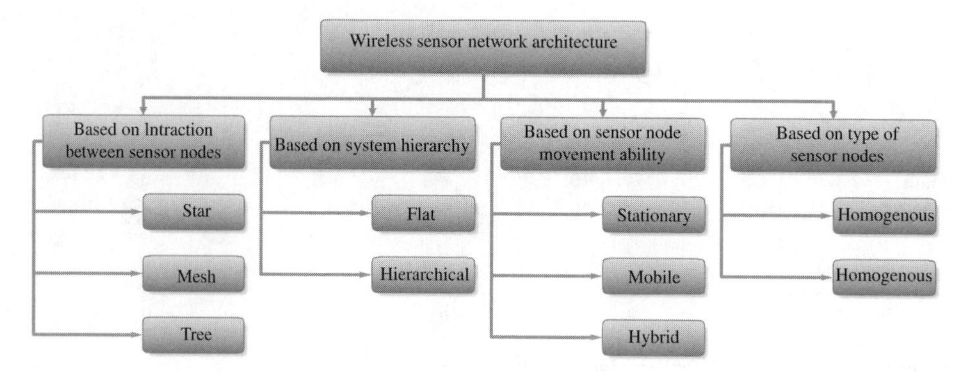

Figure 17.8 Different classifications of WSN architectures.

WSN architectures can also be classified depending on whether they have sensor nodes that are:

- stationary
- mobile
- hybrid.

Hybrid architectures have both stationary and mobile sensor nodes.

WSNs can also be classified according to the types of sensing device used in the network:

- homogenous architectures: in which all the sensor nodes are identical
- heterogeneous architectures, in which there are different types of sensor node.

Figure 17.8 depicts the classification of WSN architectures with respect to parameters discussed in this subsection.

17.4.6 Coverage and Connectivity

Coverage and connectivity have been always two vital concerns in the design of WSNs, and particularly in the agricultural domain because of their direct effect on efficiency and the accuracy of the information gathered.

The coverage is the area that is monitored by at least one sensor node in the network. It is necessary that at least one sensor node be located within transmission range of another node in the network to establish connectivity, because if the sensor nodes in the network are disconnected from each other or from the sink, the network is useless.

There are several solutions to provide coverage and connectivity in sensor networks: optimized deployment strategies, mobile sensor nodes to fill in gaps, and use of sensor nodes with variable sensing and transmission radii.

Two survey articles describing solutions to WSN coverage and connectivity problems were published in 2012 and 2008 by Zhu et al. [50] and Ghosh et al. [51], respectively. In the earlier survey, the authors discussed the fundamental concepts of coverage and connectivity, which were then considered from three energy-efficiency points of view: deployment strategy, sleep-scheduling mechanisms and adjustable coverage radius. They also reviewed previous papers in the area. In the later survey, the authors provided a comprehensive survey of recent research, from application-specific studies, such as those for event/intruder detection and for robotic applications, to generic ones for monitoring or mobile network applications.

17.4.7 Wireless Underground Communication

In agricultural WSNs, sensor nodes are deployed either on the ground surface (TWSNs), or at a specific depth underground (WUSNs), as discussed in Section 17.3. Different environmental factors present different challenges for each type of WSN. Hence, it is necessary to select appropriate solutions that will deal with these conditions, so that the network can perform its assigned task correctly. Wireless communication is a critical component of the WSNs. While aboveground wireless communication in TWSNs has been available for many years, wireless communication underground is one a new and challenging issue for system designers. Radio electromagnetic wave propagation (EM) [32], magnetic induction (MI) [52] and hybrids of the two can be used in agricultural WUSNs as the communication technology.

Underground communication with EM waves suffers from high levels of path loss and signal attenuation, reflection/refraction, multipath fading, high bit-error rates, reduced propagation velocity, and noise. These factors depend on:

- the frequency of the EM wave (higher frequency leading to higher attenuation)
- the depth of the sensor
- the soil properties:
  - moisture (more moisture leading to higher signal attenuation)
  - temperature (higher temperature leading to higher attenuation)
  - composition (larger particles leading to higher attenuation)
  - density (higher density leading to higher attenuation) [32].

Vuran et al. [53] studied the dependence of path loss, bit-error rate and maximum transmission distance of electromagnetic waves on factors such as soil type, moisture content, deployment depth of the sensor nodes, internode distance and the range of frequencies.

In addition, various properties of the soil, such as its makeup, the dynamics of its moisture content and its density, can cause dynamic communication channel conditions. Efficient propagation of EM waves at lower frequencies (and therefore with lower attenuation) requires larger antennas, which are incompatible with the miniaturisation that underground sensor nodes require. Therefore, dynamic channel conditions and large antennas are another two serious problems for EM-based underground communication [54].

It is obvious that EM waves are inappropriate for transfer of information underground, especially at depth. Magnetic induction (MI) waveguide structures [55–57] are another wireless communication technology that is a possible alternative to EM waves for underground deployments [32]. In MI communications, multiple resonant relay coils are deployed between two sensor nodes. The MI waveguide uses the induction coupling between these small independent coils instead of the large antennas used for EM wave propagation [58]. The potential advantages of MI-based communication for underground communications are:

- uniform attenuation regardless of medium
- resilience to multipath fading

- longer transmission range
- higher energy efficiency
- relatively simpler transmission and reception schemes.

Despite these advantages, MI-based communication cannot completely replace EM in WUSNs because of:

- bandwidth limitations (a few kilohertz)
- difficulty in establishing communication between underground and aboveground environments
- an absence of commercial transceivers for this recent technology.

Parameswaran et al. [59] explored the possibility of using underground MI-based communication for irrigation control in horticulture.

The implementation of a hybrid of EM and MI technology is a promising solution for underground communications. EM and MI technologies can be used for communicating between sensor nodes located on the ground and those located in the soil near the surface, and between sensor nodes located underground [60].

17.5 WSN-based Applications in Agriculture

Use of WSNs in modern agriculture has become very popular in recent years. As shown in Figure 17.9, agricultural applications aided by WSNs can be categorised into three groups:

- environmental monitoring
- resource management
- facility control.

In this section this taxonomy is overviewed.

17.5.1 Environmental Monitoring

A common WSN application is to monitor environmental conditions of a given area. WSNs are widely used for field [17–21], storage [61], crop [62, 63], climate [22] and animal [29, 30] monitoring (see Figure 17.9).

Roy et al. [19] developed a low-power and low-data-rate environmental monitoring system. Their proposed monitoring system comprises different battery-powered sensor nodes, statically deployed across farmland for environmental monitoring (temperature, humdity) and soil-parameter monitoring (pH, moisture and humus content). Their aim was to achieve higher yields and lower waste. The monitoring system included several heterogeneous sensor nodes and a monitoring station and used a wireless multi-hop mesh-based network topology based on the IEEE 802.15.4/ZigBee standard. Each sensor node wakes up periodically, collects data for an interval of time, aggregates it, and transmits it to the next hop, before going to sleep. Data aggregation and sleep-wake up scheduling mechanisms were used to minimize energy consumption. It was found that changes in the humidity and environmental conditions caused significant changes in

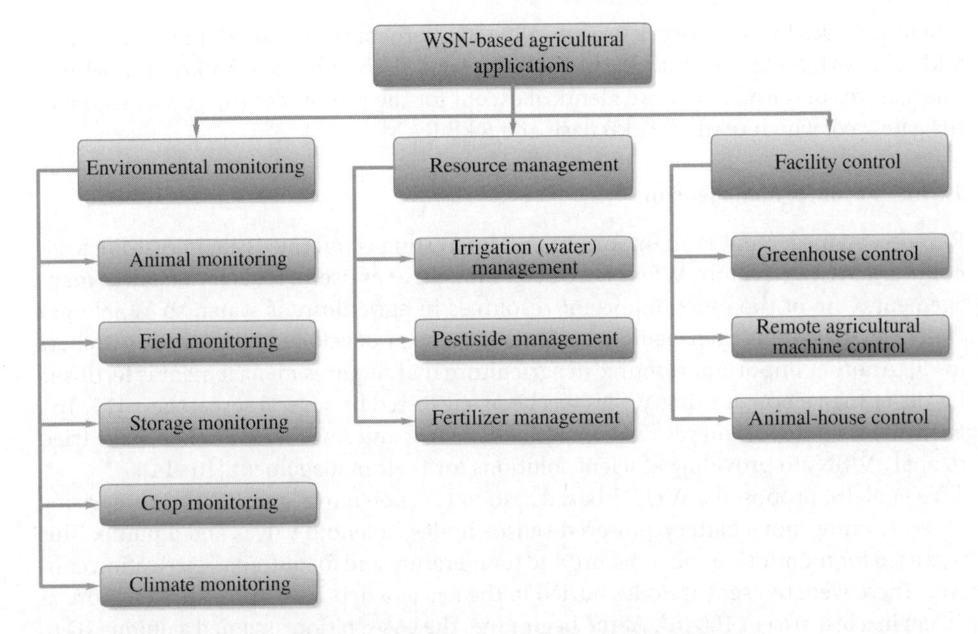

Figure 17.9 Classification of WSN-based applications in the agricultural domain.

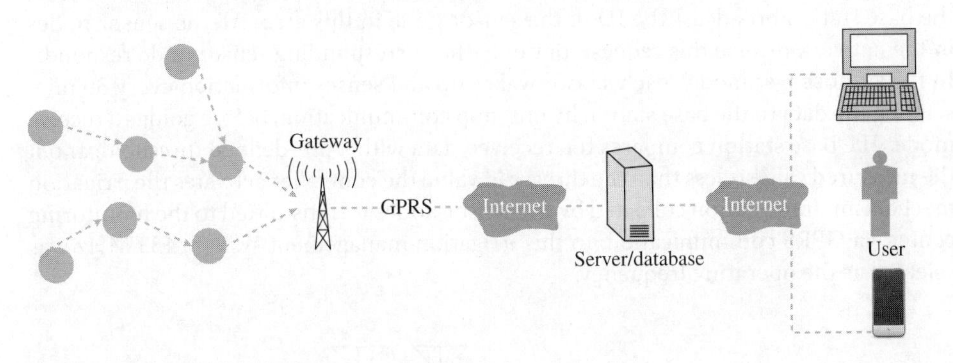

Figure 17.10 WSN-based system overview [61].

the transmission range. The authors proposed using a set of routers to mitigate sudden weather damage.

Juul et al. [61] proposed an environmental monitoring system comprising a WSN and a user interface. This was used for crop-storage monitoring in order to avoid losses and to achieve higher quality in the crops. Their suggested WSN involves a gateway to collect data from the sensor nodes in the storage trough, with a single-hop (star topology) or multi-hop communication mode (tree topology). This information – temperature and humidity – was passed to the central server. Information collected by the network is transmitted from the gateway to the central server and control messages and commands come back to the gateway using GPRS. Figure 17.10 illustrates the system. An alarm can be issued by the user interface if the storage becomes too hot or too cold or if the measurement of a small number of nodes deviates too much from the mean of all the nodes.

Sensor nodes in the network use an MSP430 microcontroller, an SX1231 radio chip, 5 kB of RAM, 55 kB onboard flash, 1 MB external flash and a 8.5 Ah lithium battery. The features of the all nodes are identical except for the root nodes (those connected to the gateway), which used of 92 kB flash and 8 kB RAM.

17.5.2 Resource Management

Resource management is an important and interesting economic issue in real-life agricultural activity. Recently, WSNs have been thought of as useful tools for resource management. One of the most important resources in agriculture is water, so WSN-base water- or irrigation- management systems have been developed by many researchers [5–9]. Another important resource in agriculture that requires management is fertiliser. Fertiliser management using WSNs has been suggested by several researchers [15, 16]. Pesticide is an important resource in agriculture too, and several researchers have tried to apply WSNs to providing efficient solutions for their management [10–14].

Yu et al. [6] proposed a WUSN-based system (as shown in Figure 17.11), comprising several homogenous battery-powered sensor nodes, solenoid valves and a pump. This was used for monitoring soil moisture and temperature and for automatic irrigation control. There were ten sensor nodes buried in the soil at a depth of 20 cm and ten more at 40 cm in a test area of 100 m^2. At the beginning, the base station assigned a unique ID to each underground sensor. When it needed to collect information from a particular area, the base station broadcast the ID of the sensor node in this area. All the sensor nodes in the network receive this request, but only the corresponding sensor node responds. In the author's test model, each sensor wakes up and senses information every 30 min, sending the data to the base station by one-hop communication, before going to receive mode. The base station compares the received data with a pre-defined threshold, and if the measured value is less than the threshold value the controller activates the irrigation mechanism. Information collected by the sensor nodes is transmitted to the monitoring center via GPRS communication. In this irrigation management system, 433 MHz was selected as the operating frequency.

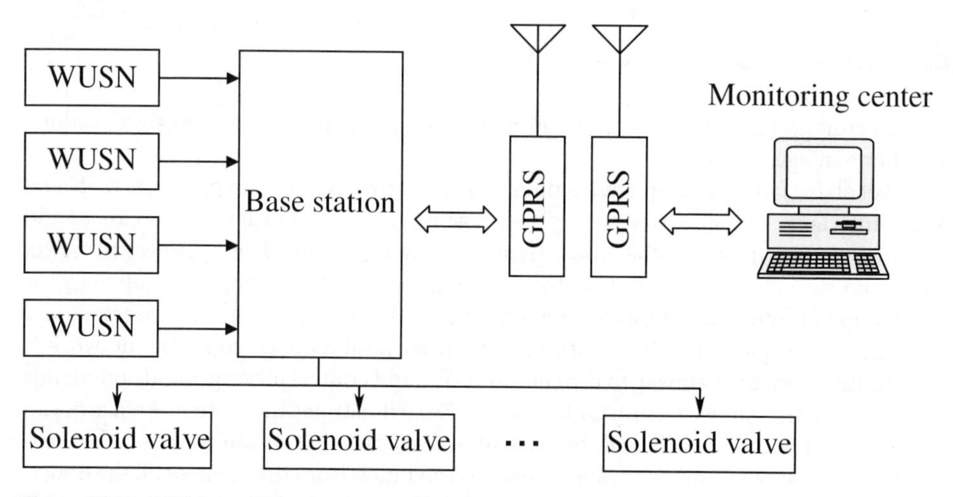

Figure 17.11 WSN-based irrigation system [61].

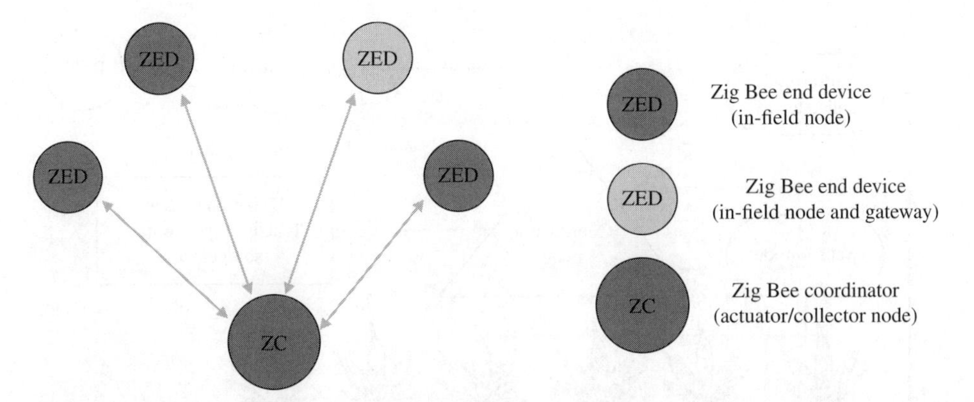

Figure 17.12 Matufa's network topology [7].

Mafuta et al. [7] described the implementation of an automatic irrigation management system based on a WSN. This comprised a ZigBee coordinator (ZC) and several other ZigBee end devices (ZED) in a single-hop star topology (Figure 17.12).

Each ZED was formed of three in-field nodes and the gateway, and the ZC was one controller node. All of the in-field sensor nodes was equipped with soil moisture and temperature sensors, a solar panel and a battery as power sources, and a communication module (ZigBee and GPRS modules for gateway nodes and ZigBee modules for the other sensor nodes in the network). In their experiment, four in-field nodes were placed in the test-bed – two in each 8×7 m plot – as ZEDs, and one of these four nodes was assigned the role of gateway node. The gateway is responsible for relaying information to the remote station for diagnostic purposes. Each sensor node measure soil moisture and temperature and its own battery level; every 30 min if the system is in the idle state or every 2 min if system is in irrigation mode. The data are send to the ZC, then the sensor switches to sleep mode for energy-conservation purposes. When the ZC receives data from the in-field sensor nodes, it aggregate it and sends the results back to the gateway through ZigBee communication. Depending on the level of the received data, the irrigation system decides whether to irrigate or not. Afterwards, the gateway nodes transmit the data from the controller node – indicating status and fault alarms – to the remote monitoring station (RMS) through a cellular network via GPRS. The architecture of this irrigation management system is illustrated in Figure 17.13.

Sakthipriya [15] proposed a system for crop monitoring. The goal is to monitor soil moisture and pH for water and fertiliser management. In their proposed system, several homogenous sensor nodes, to measure leaf wetness, soil moisture, soil pH, and atmospheric pressure, and a data acquisition board are deployed in a multi-hop ad-hoc mesh network. An MPR2400CA radio platform and ATMega128L microcontroller are used. All the sensor nodes collect information from their surroundings. At each sampling, depending on the measured value of soil moisture, each sensor node triggers the water sprinkler in that area (or not). The value of the soil pH is forwarded to a central coordinator/sink. Then, the coordinator aggregates this information and reports it to the farmer using the SMS system via GSM, to inform a decision about whether to fertilise the particular region. The XMesh routing protocol, which minimizes the total number of transmissions in delivering a data packet, is used to transmit data from source to destination in multi-hops for energy efficiency purposes.

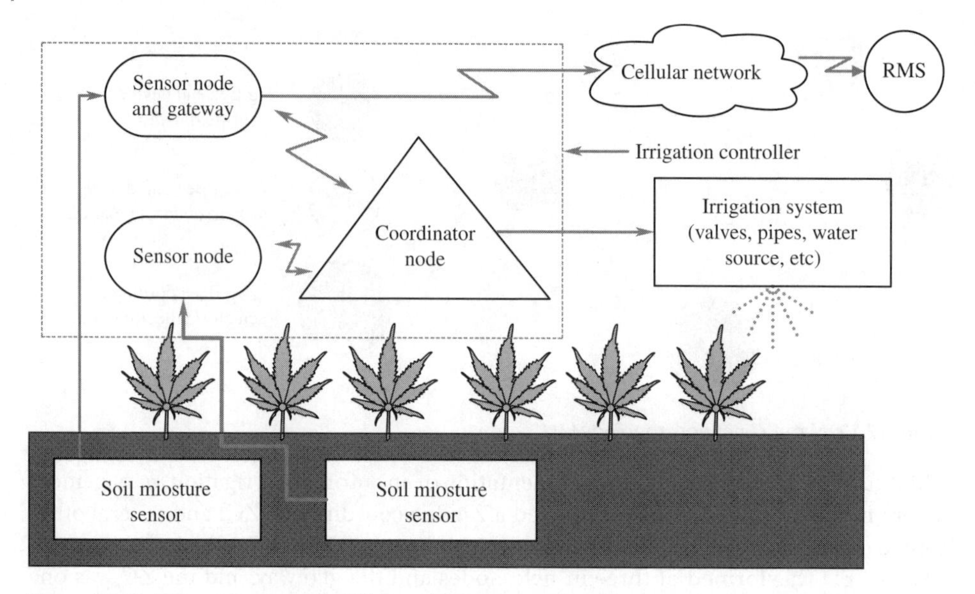

Figure 17.13 The architecture of WSN-based irrigation system [7].

Santos et al. [13] developed a WSN-based agricultural system for three different tasks relating to cost effective pesticide management:

- evaluation of environmental conditions
- correction of the crop spraying path during operation
- evaluation of spraying efficiency.

A set of homogenous and battery-powered sensor nodes is grouped into clusters and cluster-head selection is mostly based on the sensor node's proximity to the aerial crop-spraying vehicle. Sensor nodes can transmit collected data directly, or using multi-hop communication. The WSN firstly supports decision on whether the field needs spraying or not, with the data about the environmental conditions collected by the sensor nodes and transmitted to the farmer. Secondly, it monitors the direction and speed of the wind and changes the vehicle path if necessary. This task makes for uniform spraying in the area and avoids spraying outside it. Thirdly, the WSN evaluates the quality of spraying by determining the pesticide amount on the crop leaf after spraying using a leaf witness sensor. Any gaps in the spray coverage due to drift or the spray path correction are found, and then further spraying can be performed in that area. The architecture of the system is shown in Figure 17.14. This system was implemented by OMNET++ using the MiXiM framework. In the simulation, sensor modules were developed using an RF transceiver model number CC2420 from Texas Instruments, at 2.4 GHz and using the IEEE 802.15.4 standard.

17.5.3 Facility Control

The use of sensor networks for control of agricultural facilities such as greenhouses [23–28], remote devices (tractors, sprayers and agricultural robots) [64] and animal houses [65–67]. decreases labour, improves efficiency and reduces costs.

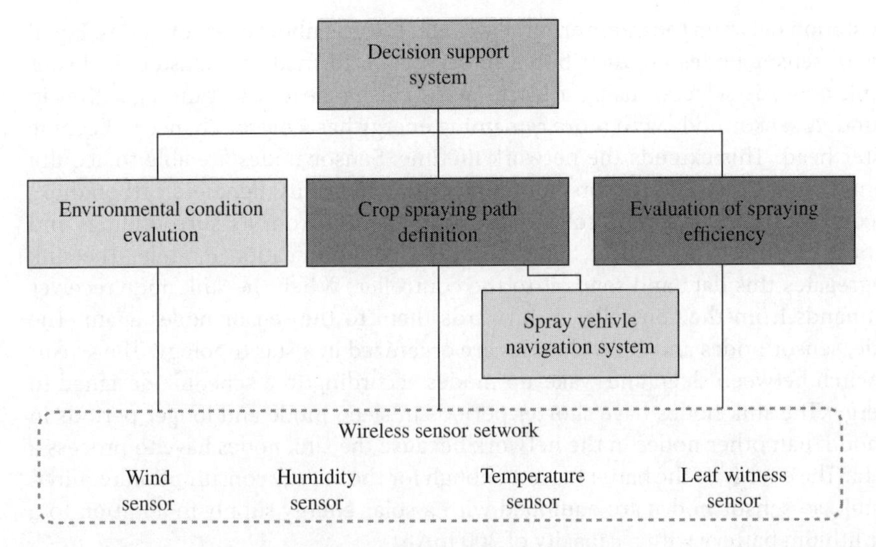

Figure 17.14 Crop-spraying support system architecture [13].

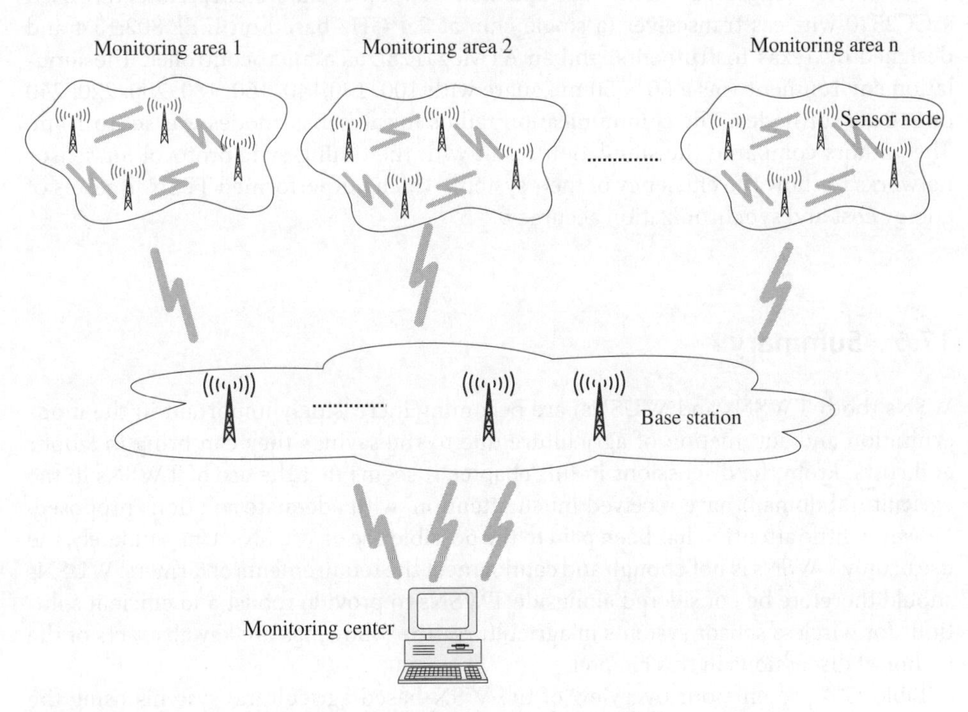

Figure 17.15 Structure of WSN-based greenhouse control system [26].

Song et al. [26] developed a WSN-based system comprising a monitoring center, sensor nodes and control equipment for automation irrigation control. The ensor nodes in this system are deployed randomly in the greenhouse. The system is illustrated in Figure 17.15. The greenhouse is divided into non-overlapping identical areas, each called a monitoring area. Each monitoring area is managed by a base station. The base station is a relay station between the monitoring center and the greenhouse sensor nodes. Equal numbers of sensor nodes are distributed in each area, formed as a cluster. A cluster head (sink node) is selected using a leach-based cluster-head-selection algorithm in each round. A sensor node with more remaining energy has a bigger chance to become the cluster head. This extends the network lifetime. Sensor nodes are able to acquire data from different installed sensors; for temperature, humidity, light and carbon dioxide concentrations. Each sensor collects appropriate data from its surroundings and transmits it directly to its cluster head through a communication module. The sink node aggregates this data and sends it to the controller. When the sink node receives the commands from the controller, it forwards them to the sensor nodes again. The sink node, sensor nodes and control nodes are organized in a star topology. The sensor nodes switch between sleep and wake-up modes according to a schedule designed to save energy. The sink nodes have shorter periods in sleep mode and longer periods in active mode than other nodes in the network because the sink nodes have to process a lot of data. The energy of the battery is not enough for the energy consumption required by the sink, so sensor nodes are equipped with a solar energy supply in addition to a polymer lithium battery with a capacity of 300 mAh.

The authors' suggested system was implemented by NS2. The system architecture used a CC2530 wireless transceiver (a single chip at 2.4 GHz based on IEEE.802.15.4 and designed by Texas Instruments) and an ATMega128L as a microcontroller. The simulation environment was a 50 × 50 m square with 100, 120,140, 160, 180, 200, 220, 240 or 260 sensor nodes. The communication radius for all sensor nodes was set to 10 m. The authors compared the simulation result with the timing-sync protocol for sensor networks to show the efficiency of their system, which outperformed TPSN in terms of energy cost and synchronization accuracy.

17.6 Summary

WSNs (both TWSNs and WUSNs) are becoming increasingly important in the modernisation and automation of agriculture due to the savings they can bring in labour and costs. From the discussions in this chapter, it seems that the use of TWSNs in the agricultural domain have received much attention, with adequate solutions proposed. However, little attention has been paid to the possible use of WUSNs. Unfortunately, the use of only TWSNs is not enough and cannot meet the requirements of farmers. WUSNs should therefore be considered alongside TWSNs to provide robust and efficient solutions for wireless sensor systems in agriculture. The followings are key abstracts of the technical discussions in this chapter.

Table 17.4 presents our overview of the WSN-based agricultural systems using the classification in Section 17.4; each row corresponds to one WSN design challenge.

Based on our survey of the literature, the following potential trends merit further research:

- *Cost reductions and miniaturisation:* Sensor nodes should be able to operate in large networks with numerous sensors and be easy deployable and replaceable for easy-to-setup applications.

Table 17.4 Main features of WSN-based system for different agricultural applications

		Roy et al. [19]	Juul et al. [53]	Yu et al. [6]	Mafuta et al. [7]	Sakthipriya [15]	Santos et al. [13]	Song et al. [26]
Application		Field monitoring	Storage monitoring	Irrigation management	Irrigation management	Water and fertiliser management	Pesticide management	Greenhouse control
Communication technology		ZigBee	GPRS	GPRS	ZigBee-GPRS	GSM	ZigBee	ZigBee
Operating frequency		2.4 GHz	433 kHz	433 kHz	2.4 GHz	2.4 GHz	2.4 GHz	2.4 GHz
Power source		Battery	Battery	Battery	Solar-battery	Battery	Battery	Battery-solar
Energy conservation approach		Sleep-wake up schedule Data aggregation	Energy efficient routing	Sleep-wake up schedule	Sleep-wake up schedule Data aggregation	Data aggregation Energy efficient routing	Energy efficient routing	Sleep-wake up schedule Data aggregation
Architecture	Type of sensor nodes	Homogenous	Heterogeneous	Homogenous	Homogenous	Homogenous	Homogenous	Homogenous
	System hierarchy	Flat	Flat	Flat	Flat	Flat	Hierarchical	Hierarchical
	Interaction between sensor nodes	Mesh	Star	Star	Star	Mesh	Tree	Star
	Sensor node movement	Stationary	Stationary	Stationary	Stationary	Stationary	Stationary	Stationary
Fault tolerant scheme		Data management Energy management	Energy management	Energy management	Data management Energy management	Data management Energy management	Energy management	Data management Energy management
Scalability		Good	Good	Good	Good	Good	Good	Good
Type		Terrestrial	Terrestrial	Underground	Terrestrial	Terrestrial	Terrestrial	Terrestrial

- *Generalised cost-effective schemes for different agricultural services:* Agricultural products are sold in a very competitive market, so costs and economy play important roles. Cost-effective solutions are needed to make WSNs affordable, and any new agricultural project should stay within a reasonable budget.
- *Energy conservation:* Some agricultural applications need continuous monitoring of physical phenomena for long periods. It is therefore necessary to extend the network lifetime sufficiently. One way to achieve this objective is utilise energy-harvesting techniques. Sensor nodes in agricultural TWSNs can harvest energy from natural sources like the sun and wind. Also, some specific energysources are available to sensor nodes in agricultural applications, such as vibration and heat. In general, the use of rechargeable batteries in sensor nodes can improve network energy efficiency.
- *Maintenance:* Low maintenance is a primary concern in the design of WSNs for agriculture, so as to minimize overall costs and enable large-scale networks to be deployed.
- *Fault tolerance:* In networks consisting of numerous independent sensor nodes, the occurrence of fault(s) is unavoidable. There are many reasons for faults to occur. Examples are energy depletion, physical damage, and hardware and software problems. To get the desired level of functionality in the presence of these faults is vital. So, besides energy-conservation techniques, other techniques to promote fault tolerancce should also be considered in the design process of WSNs.
- *Cross-layer design:* Cross-layer approaches are introduced in WSNs to overcome their constraints. These approaches minimize the overhead by sharing data among layers, providing cheaper and more efficient solutions than traditional layer approaches. WSNs for agriculture can benefit from such approaches.
- *Mobile nodes:* Mobile nodes can be introduced into networks to improve the sensing and control capabilities. Typical examples of their use are: automatic sensor node deployment, fast recovery from disconnections, dynamic event coverage, mobile object tracking, flexible topology adjustment, and efficient data collection and processing. Mobile sensor nodes are resource-rich devices in terms of energy, communication power, sensing and computing capabilities. They can play key roles in prolonging network lifetimes and enhancing network capabilities.
- *Agricultural automation:* Use of WSNs for automation of agricultural services is much discussed, particularly with respect to preparing the ground for plantation, ploughing, managing the agricultural calendar (time of rest, time of planting, time of harvesting), and harvesting and post-harvesting activities.

References

1 J. Yick, B. Mukherjee, and D. Ghosal, 'Wireless sensor network survey', *Computer Networks*, vol. 52, no. 12, pp. 2292–2330, August 2008.
2 I.F. Akyildiz, W. Su, Y. Sankarasubramaniam, and E. Cayirci, 'Wireless sensor networks: a survey', *Computer Networks*, vol. 38, no. 4, pp. 393–422, 2002.
3 A. Rehman, A.Z. Abbasi, N. Islam, Z.A. Shaikh, 'A review of wireless sensors and networks' applications in agriculture', *Computer Standards & Interfaces*, vol. 3, no. 2, pp.102–135, 2011.

4 T. Ojhaa, S. Misraa, N.S. Raghuwanshib, 'Wireless sensor networks for agriculture: The state-of-the-artin practice and future challenges', *Computers and Electronics in Agriculture*, vol. 118, pp. 66–84, 2015.
5 H. Navarro-Hellín, R. Torres-Sánchez, F. Soto-Valles, C. Albaladejo-Pérez, J.A López-Riquelme, R. DomingoMiguel, 'A wireless sensors architecture for efficient irrigation water management', *Agricultural Water Management*, vol. 9, no. 151, pp. 64–74, 2015.
6 X. Yu, W. Han, Z. Zhang, 'Study on wireless underground sensor networks for remote irrigation monitoring system', *International Journal of Hybrid Information Technology*, vol. 8, no.1, pp. 409–416, 2015.
7 M. Mafuta, M. Zennaro, A. Bagula, G. Ault, H. Gombachika and T. Chadza, 'Successful deployment of a wireless sensor network for *precision agriculture* in Malawi', *International Journal of Distributed Sensor Networks*, vol. 9, no. 5, pp. 1–13, 2013.
8 P. Patil, B.L. Desai, 'Intelligent irrigation control system by employing wireless sensor networks', *International Journal of Computer Applications*, vol. 79, no. 11, 2013.
9 S.A. Nikolidakis, D. Kandris, D.D. Vergados, C. Douligeris, 'Energy efficient automated control of irrigation in agriculture by using wireless sensor networks', *Computers and Electronics in Agriculture*, vol. 113, 154–163, 2015.
10 R.Z. Ahmed and R.C. Biradar, 'Redundancy aware data aggregation for pest control in coffee plantation using wireless sensor networks', Signal Processing and Integrated Networks (SPIN), 2nd International Conference on, Noida, pp. 984–989, 2015.
11 E. Ferro, V.M. Brea, D. Cabello, P. López, J. Iglesias and J. Castillejo, 'Wireless sensor mote for snail pest detection', Sensors, 2014 IEEE, Valencia, pp. 114–117, 2014.
12 S. Datir and S. Wagh, 'Monitoring and detection of agricultural disease using wireless sensor network', *International Journal of Computer Applications*, vol. 87, no.4, 2014.
13 I.M. Santos, F.G. da Costa, C.E. Cugnasca and J. Ueyama, 'Computational simulation of wireless sensor networks for pesticide drift control', *Precision Agriculture*, vol. 15, no. 3, pp 290–303, 2014.
14 S. Azfar, A. Nadeem and A. Basit, 'Pest detection and control techniques using wireless sensor network: A review', *Journal of Entomology and Zoology Studies*, vol. 3, no. 2, pp. 92–99, 2015.
15 N. Saktipriya, 'An effective method for crop monitoring using wireless sensor network', *Middle-East Journal of Scientific Research*, vol. 20, no. 9, pp. 1127–1132, 2014.
16 Y. Song, J. Ma and X. Zhang, 'Design of water and fertilizer measurement and control system for seedlings soil based on wireless sensor networks', *Applied Mechanics and Materials*, Vols. 241–244, pp. 86–91, 2013.
17 Y. Zhu, J. Song, and F. Dong, 'Applications of wireless sensor network in the agriculture environment monitoring', *Procedia Engineering*, vol. 16, pp. 608–614, 2011.
18 A. Camaa, F.G. Montoyaa, J. Gomeza, J.L.D.L. Cruzb and F.M. Agugliaro, 'Integration of communication technologies in sensor networks to monitor the Amazon environment', *Journal of Cleaner Production*, vol. 59, pp. 32–42, 2013.
19 S. Roy and So. Bandyopadhyay, 'A test-bed on real-time monitoring of agricultural parameters using wireless sensor networks for precision agriculture', In Proceedings of International Conference on Intelligent Infrastructure, 2012.

20 M. Zhang, M. Li, W. Wang, C. Liu and H. Gao, 'Temporal and spatial variability of soil moisture based on WSN', *Mathematical and Computer Modelling*, vol. 58, pp. 826–833, 2012.
21 J. Xu, J. Zhang, X. Zheng, X. Wei and J. Han, 'Wireless sensors in farmland environmental monitoring', Cyber-Enabled Distributed Computing and Knowledge Discovery (CyberC), International Conference on, Xi'an, pp. 372–379, 2015.
22 R.H. Ma, Y.H. Wang and C.Y. Lee, 'Wireless remote weather monitoring system based on MEMS technologies', *Sensors*, vol. 11, pp. 2715–2727, 2011.
23 Y.Q. Jiang, T. Li, M. Zhang, S. Sha and Y.H. Ji, 'WSN-based control system of CO_2 concentration in greenhouse', *Intelligent Automation and Soft Computing*, vol. 21, no. 3, pp. 285–294, 2015.
24 V.S. Jahnavi and S.F. Ahamed, 'Smart wireless sensor network for automated greenhouse', *IETE Journal of Research*, vol. 61, no. 2, pp. 180–185, 2015.
25 T. Gomes, J. Brito, H. Abreu, H. Gomes and J. Cabral, 'GreenMon: An efficient wireless sensor network monitoring solution for greenhouses', Industrial Technology (ICIT), 2015 IEEE International Conference on, Seville, pp. 2192–2197, 2015.
26 Y. Song, J. Ma, X. Zhang and Y. Feng, 'Design of wireless sensor network-based greenhouse environment monitoring and automatic control system', *Journal of Networks*, vol. 7, no. 5, pp. 838–844, 2012.
27 J. Baviskar, A. Mulla, A. Baviskar, S. Ashtekar and A. Chintawar, 'Real time monitoring and control system for green house based on 802.15.4 wireless sensor network', Communication Systems and Network Technologies (CSNT), 2014 Fourth International Conference on, Bhopal, pp. 98–103, 2014.
28 J.J. Roldán, G. Joossen, D. Sanz, J. del Cerro and A. Barrientos, 'Mini-UAV based sensory system for measuring environmental variables in greenhouses', *Sensors*, vol. 15, no. 2, pp. 3334–3350, 2015.
29 P.K. Mashoko Nkwari, S. Rimer and B.S. Paul, 'Cattle monitoring system using wireless sensor network in order to prevent cattle rustling', *IST-Africa Conference Proceedings, Le Meridien Ile Maurice*, 2014, pp. 1–10, 2014.
30 A. Kumar and G.P. Hancke, 'A ZigBee-based animal health monitoring system', in *IEEE Sensors Journal*, vol. 15, no. 1, pp. 610–617, Jan. 2015.
31 S. Thessler, L. Kooistra, F. Teye, H. Huitu and A.K. Bregt, 'Geosensors to support crop production: Current applications and user requirements', *Sensors*, vol. 11, pp. 6656–6684, 2011.
32 I.F. Akyildiz and E.P. Stuntebeck, 'Wireless underground sensor networks: Research challenge', *Ad Hoc Networks Journal*, vol. 4, pp. 669–686, July 2006.
33 Institute of Electrical and Electronics Engineers, IEEE Standard for Information Technology – Telecommunications and information exchange between systems – Local and metropolitan area networks – Specific requirements – Part 15.4: Wireless LAN Medium Access Control (MAC) and Physical Layer (PHY) Specifications for Low-Rate Wireless Personal Area Networks (LR-WPANS)' IEEE 802.15.4–2003, 2003.
34 T. Kalaivani, A. Allirani and P. Priya, 'A survey on ZigBee based wireless sensor networks in agriculture', Trends in Information Sciences and Computing (TISC), 2011 3rd International Conference on, Chennai, pp. 85–89, 2011.
35 Institute of Electrical and Electronics Engineers, 'IEEE Standard for Information Technology – Telecommunications and Information Exchange between

Systems – Local and Metropolitan Area Networks – Specific Requirements – Part 11: Wireless LAN Medium Access Control (MAC) and Physical Layer (PHY) Specifications', IEEE 802.11–1997, 1997.

36 G.R. Mendez, M.A. Md Yunus and S.C. Mukhopadhyay, 'A WiFi based smart wireless sensor network for monitoring an agricultural environment', Instrumentation and Measurement Technology Conference (I2MTC), 2012 IEEE International, Graz, pp. 2640–2645, 2012.

37 G.R. Mendez, M.A. Yunus and Dr. S.C. Mukhopadhyay, 'A Wi-Fi based smart wireless sensor network for an agricultural environment', Fifth International Conference on Sensing Technology, pp. 405–410, January 2011.

38 Institute of Electrical and Electronics Engineers, 'IEEE Standard for Information Technology – Telecommunications and Information Exchange between Systems – LAN/MAN – Specific Requirements – Part 15: Wireless Medium Access Control (MAC) and Physical Layer (PHY) Specifications for Wireless Personal Area Networks (WPANs)', IEEE 802.15.1–2002, 2002.

39 Institute of Electrical and Electronics Engineers, 'IEEE Standard for Local and Metropolitan Area Networks – Part 16: Air Interface for Fixed Broadband Wireless Access Systems', IEEE 802.16–2001, 2001.

40 M. Usha Rani, C. Suganya, S. Kamalesh and A. Sumithra, 'An integration of wireless sensor network through Wi-max for agriculture monitoring', Computer Communication and Informatics (ICCCI), 2014 International Conference on, Coimbatore, pp. 1–5, 2014.

41 General Packet Radio Service. http://www.3gpp.org/.

42 C. Saad, and B. Mostafa, 'Comparative performance analysis of wireless communication protocols for intelligent sensors and their applications,' *International Journal of Advanced Computer Science and Applications*, vol. 5, no. 4, pp. 76–85, 2014.

43 X. Yu, P. Wu, W. Han, Z. Zhang, 'A survey on wireless sensor network infrastructure for agriculture', *Computer Standards & Interfaces*, vol. 35, no. 1, pp. 59–64, 2013.

44 P. Desnoyers. *Distributed Data Collection: Archiving, Indexing, and Analysis*. ProQuest, 2008.

45 T. Rault, A. Bouabdallah,and Y. Challal, 'Energy efficiency in wireless sensor networks: A top-down survey', *Computer Networks*, vol. 67, pp. 104–122, 2014.

46 J. Hwang, C. Shin and H. Yoe, 'Study on an agricultural environment monitoring server system using wireless sensor networks', *Sensors*, vol. 10, pp. 11189–11211, 2010.

47 R. Morais, S. Matos, M. Fernandes, A. Valente, S. Soares, P. Ferreira, and M. Reis, 'Sun, wind and water flow as energy supply for small stationary data acquisition platforms', *Computers and Electronics in Agriculture*, vol. 64, no. 2, pp. 120–132, 2008.

48 A. Nayak, G. Prakash and A. Rao, 'Harnessing wind energy to power sensor networks for agriculture', In: International Conference on Advances in Energy Conversion Technologies, pp. 221–6, 2014.

49 S. Kahrobaee and M.C. Vuran, 'Vibration energy harvesting for wireless underground sensor networks', in *Proceedings of IEEE ICC '13*, Budapest, Hungary, Jun. 2013.

50 S. Chouikhi, I. El Korbi, Y. Ghamri-Doudane, and L. Azouz Saidane, 'A survey on fault tolerance in small and large scale wireless sensor networks', *Computer Communications*, vol. 69, pp. 1–16, 2015.

51 P. Corke, T. Wark, R. Jurdak, W. Hu, P. Valencia and D. Moore, 'Environmental wireless sensor networks', *Proceedings of the IEEE*, vol. 98, no. 11, pp. 1903–1917, 2010.

52 S.-H. Yang, *Wireless Sensor Networks: Principles, Design and Applications.* Springer, 2014.

53 C. Zhu, C. Zheng, L. Shu, and G. Han, 'A survey on coverage and connectivity issues in wireless sensor networks', *Journal of Network and Computer Applications*, vol. 35, pp. 619–632, 2012.

54 A. Ghosh and S. Das, 'Coverage and connectivity issues in wireless sensor networks: A survey', *Pervasive and Mobile Computing*, vol. 4, pp. 303–334, 2008.

55 Z. Sun and I.F. Akyildiz, 'Underground wireless communication using magnetic induction', Communications, IEEE International Conference on, Dresden, pp. 1–5, 2009.

56 M.C. Vuran, I.F. Akyildiz, 'Channel model and analysis for wireless underground sensor networks in soil medium', *Physical Communication*, vol. 3, no. 4, pp. 245–254, 2010.

57 L. Li, M.C. Vuran, and I.F. Akyildiz, 'Characteristics of underground channel for wireless underground sensor networks', Med-Hoc-Net'07, Corfu, Greece, 2007.

58 E. Shamonina, V.A. Kalinin, K.H. Ringhofer, and L. Solymar, 'Magneto-inductive waveguide', *Electronics Letters*, vol. 38, no. 8, pp. 371– 373, 2002.

59 E. Shamonina, V.A. Kalinin, K.H. Ringhofer, and L. Solymar, 'Magneto-inductive waves in one, two, and three dimensions', *Journal of Applied Physics*, vol. 92, no. 10, pp. 6252–6261, 2002.

60 R.R. A Syms, I.R. Young, and L. Solymar, 'Low-loss magneto inductive waveguides', *Journal of Physics D: Applied Physics*, vol. 39, pp. 3945–3951, 2006.

61 X. Tan, Z. Sun and I.F. Akyildiz, 'Wireless underground sensor networks: MI-based communication systems for underground applications', in *IEEE Antennas and Propagation Magazine*, vol. 57, no. 4, pp. 74–87, 2015.

62 V. Parameswaran, H. Zhou and Z. Zhang, 'Irrigation control using wireless underground sensor networks,' Sensing Technology (ICST), 2012 Sixth International Conference on, Kolkata, pp. 653–659, 2012.

63 A.R. Silva, 'Channel characterization for wireless underground sensor networks', Master's thesis, University of Nebraska at Lincoln, 2010.

64 J.P. Juul, O. Green and R.H. Jacobsen, 'Deployment of wireless sensor networks in crop storages', *Wireless Personal Communications*, vol. 81, no. 4, pp. 1437–1454, 2015.

65 C-R. Rad, O. Hancu, I-A. Takacs and G. Olteanu, 'Smart monitoring of potato crop: a cyber-physical system architecture model in the field of precision agriculture', *Agriculture and Agricultural Science Procedia*, vol. 6, pp. 73–79, 2015.

66 M.A. Fernandes, S.G. Matos, E. Peres, C.R. Cunha, J.A. López, P. Ferreira, M.J.C.S. Reis and R. Morais, 'A framework for wireless sensor networks management for precision viticulture and agriculture based on IEEE 1451 standard', *Computers and Electronics in Agriculture*, vol. 95, pp. 19–30, 2013.

67 J.A. Gazquez, N. Novas, F. Manzano-Agugliaro, 'Intelligent low cost tele-control system for agricultural vehicles in harmful environments', *Journal of Cleaner Production*, vol. 113, pp. 204–215, 2016.

68 C.E. Cugnasca, A.M. Saraiva, I.D.A. Naas, D.J. de Moura and G.W. Ceschini, 'Ad Hoc wireless sensor networks applied to animal welfare research', Proceedings

of Livestock Environment VIII, ASABE Eighth International Symposium, Iguassu Falls, Brazil, 4 September 2008.
69 M.J. Darr, L. Zhao, 'A wireless data acquisition system for monitoring temperature variations in swine barns', Proceedings of Livestock Environment VIII. ASABE Eighth International Symposium, Iguassu Falls, Brazil, Septemebr 2008.
70 F. Llario, S. Sendra, L. Parra and J. Lloret, 'Detection and protection of the attacks to the sheep and goats using an intelligent wireless sensor network', Communications Workshops (ICC), IEEE International Conference on, Budapest, pp. 1015–1019, 2013.

Part V

Industrial and Other WSS Solutions

18

Structural Health Monitoring with WSNs

Chaoqing Tang[1], Habib F. Rashvand[2], Gui Yun Tian[1], Pan Hu[3], Ali Imam Sunny[1] and Haitao Wang[3]

[1] *School of Electrical and Electronic Engineering, Newcastle University, Newcastle upon Tyne, UK*
[2] *Advanced Communication Systems, University of Warwick, UK*
[3] *College of Automation Engineering, Nanjing University of Aeronautics and Astronautics, China*

18.1 Introduction

One truly successful and expanding application of wireless sensor networks (WSNs) is structural health monitoring (SHM) systems. Its continuous growth trends are driven by three socio-economic dimensions. The first is safety requirements and concerns over:

- deteriorating old buildings and structures mostly due to the age and corrosion
- ever increasing level of traffic, often far beyond the levels most bridges and other buildings were designed for
- fear of seismic waves, earthquakes and hurricanes
- continuously changing climates, and climate extremes, with significant negative impacts on structures predicted to be safe not long ago.

The second dimension is the recent advances in smart and low-energy wireless sensors, which are now easily produced at highly competitive costs and with improved performance. The third dimension is globalization and the advent of the Internet. The Internet, and its extended version, the Internet of Things (IoT), means easy automation and remote control bringing new possibilities of providing very high quality monitoring and surveillance services at fraction of the cost previously required.

Along with the development of science and technology, more and more applications have been developed through the interaction and penetration from different disciplines. In the last decade, a new vision has been developed: objects can be equipped with identifying, sensing, networking and processing capabilities that will allow them to communicate with one another and with other devices and services over the Internet to accomplish certain objectives. This vision has produced a paradigm being referred to as the IoT [1]. The ability to encode and track objects through IoT technologies has allowed individuals and organizations to become more efficient, speed up processes, reduce errors, prevent theft, and incorporate complex and flexible organizational systems [2].

Wireless Sensor Systems for Extreme Environments: Space, Underwater, Underground, and Industrial.
First Edition. Edited by Habib F. Rashvand and Ali Abedi.

The IoT was a term coined by Kevin Ashton in a presentation at Procter & Gamble in 1999 [3]. After a decade of development, the concept has now matured to the point at which it can be a new paradigm for wireless telecommunication industries [4]. Immense progress has been made in equipment monitoring, healthcare, manufacturing and many other fields. For example, cost reductions have allowed for frequent monitoring of structures and flexible systems without extensive cabling make it possible to bridge the gap between non-destructive testing and SHM [5] for conditional maintenance and intelligent lifecycle assessment.

WSNs, or more precisely wireless smart sensor systems, as an enabling technology for SHM applications, have been described in many other chapters of this book as well in the Introduction. We will therefore only mention them very briefly here.

As shown in Figure 18.1, a WSN normally consists of a large number of sensor nodes distributed over a large area, equipped with powerful sink nodes to gather readings from the sensor nodes [6, 7].

With the inclusion of some autonomy or limited intelligence at certain nodes and complementary components, WSNs are able to do some in-network computing for simple processing tasks. The management center can help with both precision and speed for SHM applications. Intelligent monitoring reduces repetitive human inspection requirements through automated routines. It can often reduce the maintenance load through early detection of faults and disparities, leading to improved safety and increased reliability. In general, a WSN is designed and optimized for a specific application [8]. As illustrated in Figure 18.2, there are four typical monitoring application areas of WSNs, processing data upon 'predict and prevent' events using intelligent algorithms.

In order to gauge the trends in three of the main areas we queried scholarly databases for the terms 'wireless sensor networks', 'wireless sensor networks: intelligent monitoring', and 'IoT' the 20-year period from 1996 to 2015. The databases included were Google Scholar, IEEE Xplore and Web of Science. The results are shown in Figure 18.3, which shows that whilst intelligent-monitoring-based WSNs is growing fast, driven by ever increasing new applications, interest in both the IoT and WSNs passed their peaks some years ago.

The remainder of this chapter is organized as follows. In Section 18.2, we introduce the state of the art of some key issues in WSN research. Section 18.3 reviews intelligent monitoring and SHM, and we present recent trends in IoT-SHM. In Section 18.4, the network overlay is briefly discussed before our chapter summary is presented in Section 18.5.

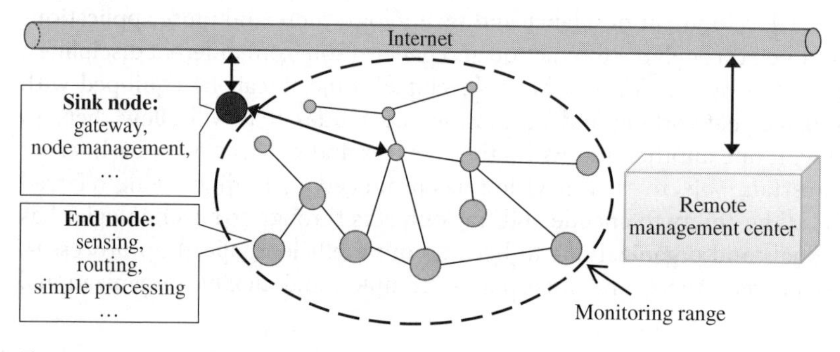

Figure 18.1 A typical WSN system.

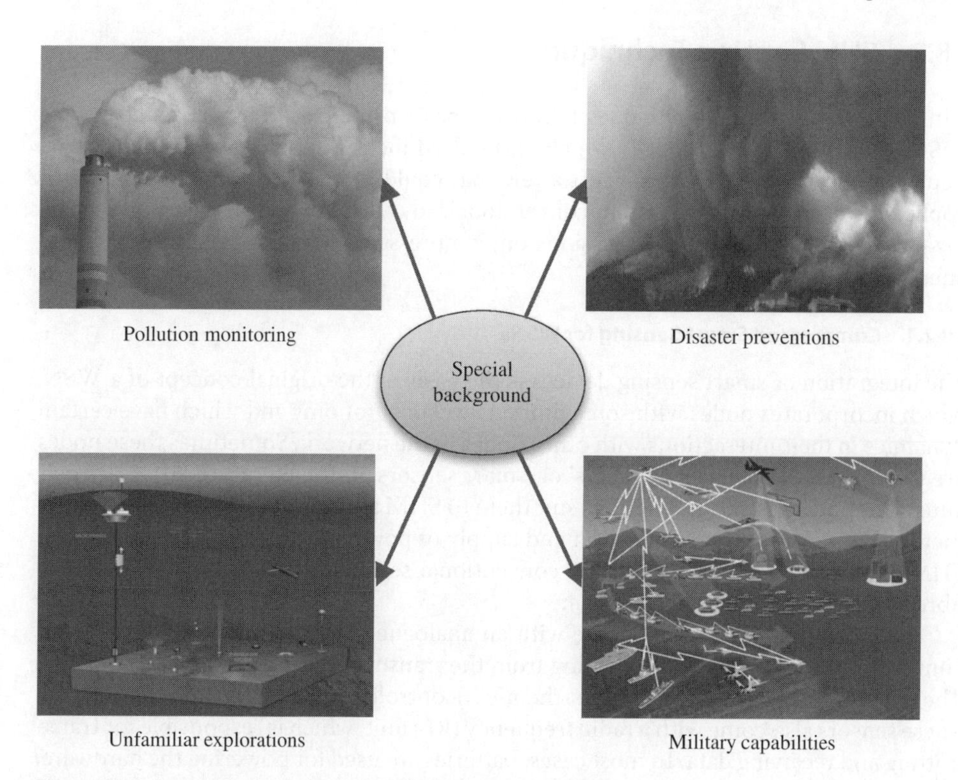

Figure 18.2 Four typical monitoring application areas of WSNs.

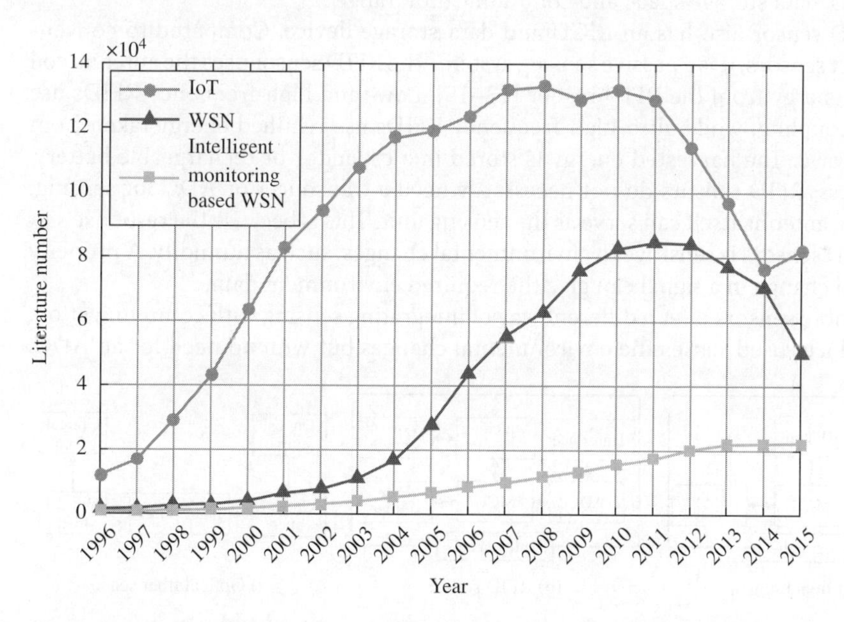

Figure 18.3 Statistical results of scholarly literature retrieval queries for IoT, WSNs and SHMs.

18.2 SHM Sensing Techniques

Although SHM sensing techniques follow the basic principles of smart functions for WSNs, it is quite hard to specify which particular function should be used in any particular application. Here, we suggest a few that could be used in various WSN-SHM application scenarios more than in conventional, so called 'ubiquitous sensing' scenarios [9]. Here are some brief discussions on compressed smart sensing for WSNs and energy consumption.

18.2.1 Compressed Smart Sensing for WSNs

The integration of smart sensing devices is the basis of the original concept of a WSN, which incorporates nodes with some limited level of autonomy and which have certain dynamics in their interactions with other nodes in the network. Sometimes these nodes are referred to as 'intelligent sensors' or 'smart sensors' in the literature [10, 11]. Their autonomy and dynamic interactions suit them to SHM applications [12]. Based on their methods of sensing, communication and supply of power, we divide smart sensors for SHM applications into three groups: conventional sensors, RFID sensors, and optical fibre sensors, as shown in Figure 18.4.

Conventional smart sensors come with an analogue-to-digital converter (ADC) for converting the analogue signal coming from the transducer into a binary digital signal. The digital signal can be processed at the microcontroller unit (MCU) and then stored. These sensors also come with a radio frequency (RF) unit, which is responsible for transmitting and receiving data. In most cases, batteries are used for powering the hardware, which means the power resource is usually limited. This limits the computational ability of the MCU, data storage space and communication range.

The RFID sensor also has an MCU and data-storage device. Compared to conventional smart sensors, it is a passive sensing system. The RFID sensor uses the antenna coil to harvest energy from the RFID reader [13–15]. Low- and high-frequency RFIDs use inductive coupling, while ultra-high-frequency RFIDs use rectified energy taken from the radio wave. The harvested energy is stored in a capacitor or rechargeable battery. Nevertheless, RFID sensors do not necessarily need a transducer or ADC for sensing. Instead, the antenna itself can serve as the sensing unit. This is because the response signal for RFID sensors is sensitive to environmental changes, such as humidity. A properly interpreted change in a signal can give the required environment data.

Optical fibre sensors offer a different way of integrating sensing with communication. The transducer used senses the environmental changes but with no need for an ADC.

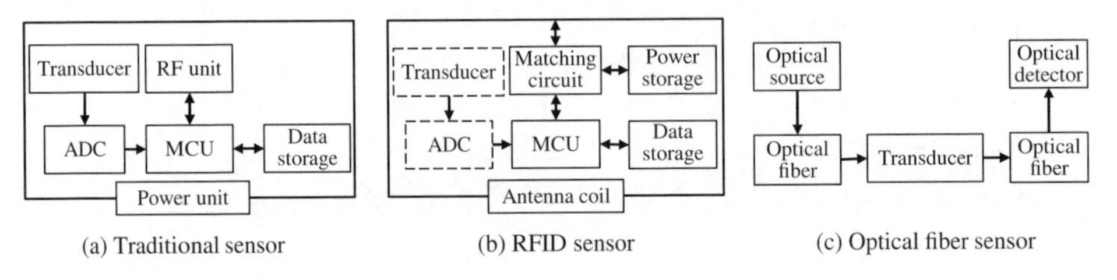

Figure 18.4 Three groups of smart sensors.

An optical source generates a light beam, which propagates along the optical fibre. The distortion in the transducer can affect five optical properties of light, namely intensity, phase, polarisation, wavelength and spectral distribution. Any of these changes of the light properties carries information about the measured parameter.

18.2.1.1 Compressed Sensing

All of the aforementioned sensors are normally fixed at a sampling rate estimated from the Nyquist–Shannon's sampling theorem, which states that the sampling rate should greater than twice the greatest signal frequency. However, the actual flow of sensing information varies extensively, thus creating huge excessive flows of redundant sampled data. This causes an unnecessary overload of most system resources: ADC capacity, storage, and communication. The technique of compressed sensing (CS) – also known as compressive sensing, compressive sampling, or sparse sampling – requires much less energy and other resources. CS is a signal-processing technique for efficiently acquiring and reconstructing a signal, finding solutions such as feature extraction and quantitative evaluation of structure health.

The amount of saving achieved through CS depends on the statistical nature of the sensing signal. For a regular-rate signal there is no significant saving because of the extra complexity engendered by using CS, but for high-rate signals whose power is dominated by limited samples (also called 'sparse' signals) the gains are extremely significant. The sparsity in a signal refers to its proportion of non-zero values. The signal may not be sparse itself but only in other domains (such as the frequency domain or the wavelet domain) that still apply to CS. Donoho [16] was first to propose the CS concept, after which Candès and Tao [17] developed the mathematical modeling and demonstrated the rationale of the theory.

Due to the versatile nature of SHM applications in practice, accurate estimation based on massive statistical data collection has become the norm for this industry. Therefore, developments of WSNs are directly usable in continuous monitoring systems such as SHMs.

As mentioned earlier, compressed sensing also helps to overcome high data-rate overheads and uneven energy consumption, sending only small weighted sums of sensor readings to the sink nodes for data processing. Haupt et al. [18] were the first to extend CS to a new approach of decentralised compression for WSNs. Then Luo et al. [19] considered a densely deployed massive WSN, and used a CS-based scheme to detect abnormal readings from sensors. This relied on the fact that abnormalities are sparse in the time-domain. As a result, the heavy demand for intensive computation and complicated transmission control in the WSN was eliminated.

The various phenomena monitored by large-scale WSNs usually occur at scattered localized positions, and so can be represented by sparse signals in the spatial domain. Exploiting sparsity for applying CS leads to far more accurate detection of monitoring parameters and removes most redundant data [20–22], and thus saving energy. For example, Liang and Tian adopted compressive data collection, achieved by turning off a fraction of the sensors at a time using a simple random node-sleeping strategy [20]. Sparse signal recovery via decentralised in-network processing was developed to conserve sensing energy and prolong the network lifetime.

To address the problem of traffic metrics estimation in network engineering, Zhang et al. [21] proposed a spatio-temporal CS framework with two key components:

- an algorithm called SRMF that leverages the spatio-temporal sparsity of real-world traffic matrices
- a mechanism for combining low-rank approximations with local interpolation procedures.

The proposed framework was evaluated for many monitoring applications such as network tomography, traffic prediction and anomaly detection to confirm its flexibility and effectiveness. The results showed that they up to 70% of the values could be reconstructed even if some 98% of data is missing. Sartipi [22] presented a distributed compression framework that exploits spatial as well as temporal correlation within a WSN. CS can be also used for spatial compression among sensor nodes. Temporal compression is obtained by adjusting the number of measurements according to the temporal correlation among sensors.

Further improvements of WSNs for high-performance monitoring can be achieved through adaptive sampling, particularly using low sampling rates for high sensing quality. It has been shown that CS makes adaptive sampling even better [23, 24]. Kho et al. [23] develop a theory of adaptive CS for collecting information from WSNs in an energy-efficient manner. The key idea of the algorithm is to perform projections iteratively to maximize the amount of information gained per quantum of energy expended. They show that adaptive sampling performs approximately twice as well as fixed sampling. Similarly, an adaptive CS (ACS) scheme for sample-scheduling in WSNs has been proposed [24]. ACS estimates the minimum required sample rate for a given sensing quality on a per-sampling-window basis and adjusts the sampling rate of the sensors accordingly, hence achieving high sensing quality at an overall low sample rate.

18.2.2 Energy Consumption

The nodes in WSNs are generally located at inaccessible places that are suitable for collecting accurate data but at which they cannot be connected to the electricity mains or other source of power. This makes them naturally energy-limited. To make the overall network lifetime last for an acceptable period, the sensors must reduce their own energy consumption. Some nodes are designed to make use of surrounding environment for energy harvesting, while others minimize the consumption to an absolute minimum, relying on eventual recharging. SHM is not an exception, unless a particular application has been especially designed with energy in mind or is lucky enough to have an integrated power source. In general, applications such as old bridges, old monuments and historical buildings cannot be wired for energy. Passive sensing, energy harvesting and low power consumption techniques are therefore attempted.

18.2.2.1 Energy Conservation

From a single sensor node point of view, most of the energy is used for data acquisition, data processing and data transmission. When considering the energy saving of the node, we first need to determine the energy consumption of each part of the sensor nodes. Regarding data acquisition, the working current of the sensor cannot be ignored. In terms of data processing, the measured sensor signal needs to be filtered, amplified and processed to enter the ADC of the WSN node. The energy/power consumption of the analogue circuits cannot be ignored. For the data transmission part of the process, the communication module normally sends the signals to a routing node. In large networks, a coordinator node represents a large sector of nodes and therefore has the highest energy consumption. From the whole sensor network point of view, how the nodes interact with each other, normally called the communication protocol, also influences the network lifetime.

There are also various wireless communication standards with associated protocols designed for different data rates and different power consumptions. Figure 18.5 shows some common wireless communication standards, some with one or more protocols.

Among all these, RFID, BLE (Bluetooth low energy) and IEEE 802.15.4 are in the low-energy-consumption region, normally designed for close contact communications such as machine-to-machine, machine-to-human, and embedded monitoring devices. These technologies are thus suitable to be used in intelligent sensor nodes.

The RFID tag is a simple chip or label attached to provide an object's identity. The RFID reader transmits a query signal to the tag and receives a reflected signal back, which in turn is passed to the database. The database connects to a processing center to identify objects based on these reflected signals within a range between 10 cm and 200 m [25]. RFID tags can be active, passive or semi-passive/active. Active tags are powered by battery while passive ones do not need a battery. Semi-passive/active tags use board power when needed. The Bluetooth Special Interest Group has produced the Bluetooth 4.2 standard, which defines BLE as well as the high-speed IP connectivity needed to support the IoT and therefore remote SHM [26]. IP is not necessarily required in all IoT devices (many sensor nodes do not have IP), but when they exchange data with the Internet, the gateway device uses IP. IEEE 802.15.4 [27] only defines the Physical and MAC layers, and provides a low power and low speed solution for WSNs. It has advantages, such as simple networking, scalability to up to thousands of nodes, and low costs. Many

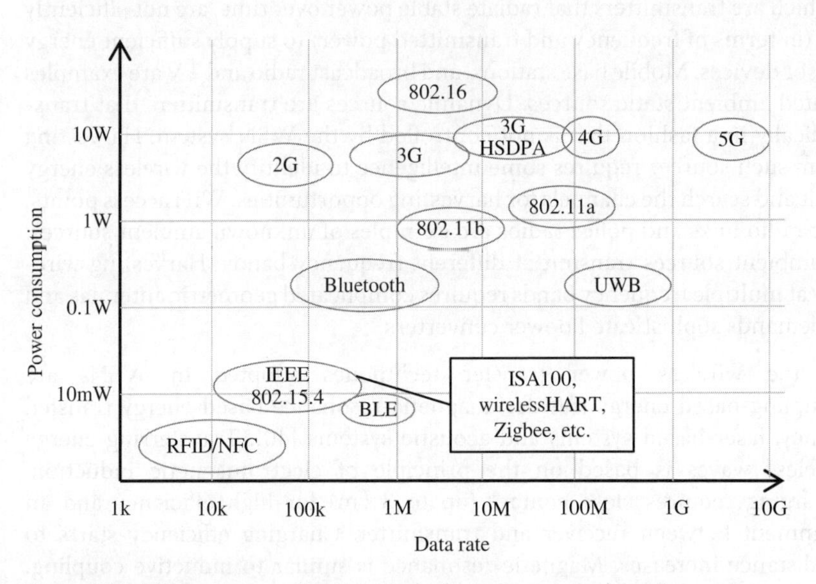

Figure 18.5 Wireless communication standards, energy versus data rate.

protocols based on it are widely accepted for WSN and IoT applications. Examples of such protocols include ISA100 [28], wirelessHART [29] and ZigBee [30]. In order to improve communication reliability while maintaining low energy consumption, many protocols adopt retransmission and other redundancy-based schemes [31]. For all these algorithms, if a sensor node in a network is responsible for most of the routing and communication tasks, it is likely to run out of power first [32]. Scheduling sleep modes is also helpful for energy conservation, and is easy to realize [33]. Therefore, the communication protocol needs to balance transmission reliability and low energy consumption.

18.2.2.2 Power Harvesting

On top of minimizing the energy consumption, harvesting energy, when available, can effectively prolong sensor node lifetimes. It is possible to harvest energy in various ways from the environment. Common examples are solar energy [34], wind energy, tidal energy, and, typical for SHM applications, vibrational energy [35, 36]. Often various combinations of these energy sources are used.

In some SHM cases other alternatives are through wireless power transfer of electromagnetic [37] or acoustic waves [38]. Some sensor nodes receiving continuous radio wave radiation through their antennas can convert the received RF or acoustic energy into stable DC energy to supply the sensor device. This can normally done using another transducer, which is usually made of some piezoelectric material [39]. Due to the limited wireless power source and low conversion ratio, the DC energy is stored in an energy storage device [40] such as a super-capacitor or a rechargeable battery.

In general, wireless energy sources can be classified into two categories [37]:

Dedicated sources are deployed to provide a predictable energy supply to the device and can be optimized in terms of frequency and maximum power to meet the requirements of the sensor devices. A sink node is an example of a dedicated source.

Ambient sources can be further divided into static and dynamic subcategories. Static sources, which are transmitters that radiate stable power over time, are not efficiently optimized (in terms of frequency and transmitted power) to supply sufficient energy for the sensor devices. Mobile base stations, and broadcast radio and TV are examples of anticipated ambient static sources. Dynamic sources are transmitters that transmit periodically in a fashion that is not controlled by the WSN system. Harvesting energy from such sources requires some intelligence to identify the wireless energy harvest unit and search the channels for harvesting opportunities. WiFi access points, microwave radio links and police radios are examples of unknown ambient sources. Different ambient sources transmit at different frequency bands. Harvesting wireless energy at multiple frequency bands requires complicated geometric antennas and normally demands sophisticated power converters.

Currently, the wireless power transfer techniques adopted in WSNs are inductive-coupling-based energy transfer, magnetic-resonance based-energy transfer, radio frequency, laser-based systems and acoustic systems [40]. Transferring energy through wireless waves is based on the principle of electromagnetic induction. Inductive charging requires close contact (up to 3 cm) for high efficiency and an accurate alignment between receiver and transmitter. Charging efficiency starts to drop as the distance increases. Magnetic resonance is similar to inductive coupling. So long as both coils resonate at the same frequency, they can exchange energy at a distance. This concept is also known as resonant coupling. Radio waves can be used to transfer energy from a transmitting antenna to a receiving antenna using far-field radiative waves. This method does not require a line of sight between transmitter and receiver because the energy can be radiated in any direction from the transmitting antenna. Because the radiation poses potential health risks to humans, the output power is restricted by government regulations so the technique can only be used for low-power applications [41]. Laser-based approaches convert the power source into an intense laser beam, focusing it on a panel of photovoltaics cells located at the receiver. Line of sight is required. A transfer efficiency of 98% over 50 m has been reported using such laser-based systems [42]. Acoustic energy (vibrations, sound, ultrasound) can be scavenged from the ambient environment or transferred wirelessly to a receiver. It is believed to have higher transfer efficiency over long distances than the inductive-coupling, magnetic-resonance and radio-frequency techniques because of the longer wavelengths involved [38].

18.3 WSN-enabled SHM Applications

As shown in Figure 18.3, research into the intelligent monitoring of structures and buildings is continuing to rise, in contrast to the falling away interest in the IoT and WSNs. Different wireless communication network and intelligent/smart sensor systems are associated with SHM applications, as detailed below.

- *Intelligent WSNs for SHM applications:* both WSNs and Internet technologies are needed for this new application paradigm. There is no room in this paradigm for non-intelligent sensors.
- WSNs are maturing to become powerful development platforms for uses such SHM. In other words, monitoring applications of WSNs count as the most significant applications of smart sensor technologies.
- There are no moves to apply IoT technologies to WSNs and SHM. This means that SHM applications cannot be counted as IoT applications. However, Internet standards and Internet resources can be used as facilitators to make SHM applications automatic and/or remotely controllable.

In this section, we briefly discuss integration of the IoT and SHM, so that the reader can see the possible future outcomes. We then explain some common trends and properties of SHM services before expanding on some key SHM sensing technologies that were mentioned in Section 18.2.

18.3.1 IoT–SHM Integration

The IoT (also called the Internet of Objects) refers to a network connection between objects; usually a self-configuring network. In the vision of the IoT, almost all things surrounding us are 'smart' and 'interconnected'. For structural monitoring applications, such as SHM, smart structures should play an important role. If the IoT is incorporated into smart structures it will change their use, maintenance, and support by responding intelligently to changes in the environment [4, 43].

Figure 18.6 is a basic block diagram for integrating Internet services into SHM. Here, sensing functions (RFID, classic smart sensors, optical fibres), data processing, and

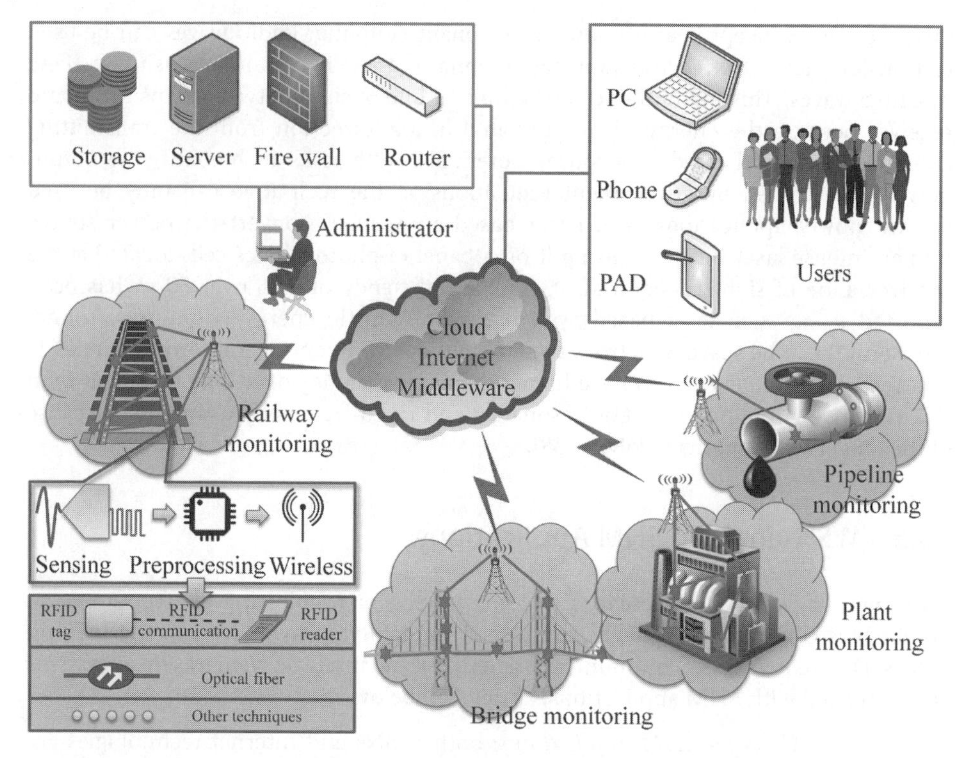

Figure 18.6 Typical use of the Internet for remote monitoring and other SHM services.

wireless communication techniques are deployed in structures to collect, process, transmit, and manage information. Information is forwarded to the data center after some preliminary processing (packaging, classification) at the gateways and other stages of the middleware.

Besides various benefits, the integration of SHM with the Internet (or the IoT in some respects) could bring with it many practical problems. One important one is the high sampling-rate requirement for some applications [44].

In line with big data management, more and more IoT–SHM systems will be developed for intelligent monitoring of structures in changing environments. This will require not only capitalising on the IoT, but further development of SHM as follows.

Low cost: Cost is always a consideration for deployment of new technologies. It becomes even more important for large-scale SHM applications. The development of semiconductor technology in this decade has reduced the cost of microchips greatly and there are all kinds of thriving open-source projects, which also contribute to cheaper deployments of SHMs. However, pursuit of lower costs of semiconductor microchips, analogue components, system devices, patent licensing, installation, and maintenance never comes to an end. These days, RFID antennas are being printed, a trend that promises to reduce the costs through greater availability and speed [45, 46].

Environmental adaption: In some extreme circumstances, often applicable to SHM applications, conventional systems and devices do not perform as expected. That is, for SHM applications in extreme or harsh environments, there will be special requirements. This may lead to make use of less environmentally sensitive systems and devices in general and specially designed products for sustainability and lower maintenance. Grudén et al. [47] mounted sensors on a train's bogies to monitor their temperature but found that the working environment was too harsh. The high acceleration and unpredictable shocks were too much for the electronics and other parts of the system to bear.

Holistic integrated systems: Holistic integrated systems, incorporating multiphysics sensing data, data transmission, data interpretation, management platform, IoT interface, sustainable power supply, or even monitoring systems integrated into structures, are the ultimate way to achieve real-time information and real-time decision-making. Conventionally, monitoring systems are post-installed on structures. This usually introduces additional installation costs and such systems are sometimes not able to withstand their environment (for example, vibration, strong wind or rain). Embedding monitoring systems into structures for lifetime monitoring is preferable. Prototypes already exist in the shape of smart structural materials [48–50] and smart pipelines [51, 52].

18.3.2 IoT–SHM Applications

Non-destructive testing and evaluation systems may thrive as IoT–SHM applications, benefiting from huge range of resources, protocols, and standards that the IoT–SHM concept brings. There are already many funded SHM applications trying to use the Internet and the IoT. For example, the HEMOW project [53] was funded under the EU's Seventh Framework Programme for offshore wind farm health monitoring and maintenance industries. This subsection reviews some typical IoT–SHM applications.

18.3.2.1 Traditional Sensor-based Applications

Lead (Pb) zirconate (Zr) titanate (Ti) materials (or simply PZT materials) are ceramics that change their shapes in electric fields, a phenomenon known as the piezoelectric effect. They have been used as sensing and actuating devices and are used heavily in many industrial applications. In general PZTs can be a transducer, an actuator or the original material. Their very interesting features are key to micro-electro-mechanical systems (MEMS) technologies.

PZT materials and acoustic technologies are simple, robust, and flexible, which guarantees good adaptation in harsh environments. These features, together with their low cost make them very common in SHM applications. Albert et al. [54] presented a novel piezoelectric harvester-based self-powered adaptive solution with wireless data transmission capability for SHM. They investigated the maximum power transfer condition under different loading conditions and different oscillation amplitudes/frequencies of the piezoelectric transducer. Gao et al. [44] presented a wireless piezoelectric sensor platform for distributed large-scale SHM. A block-diagram presentation of their approach is shown in Figure 18.7. In the proposed wireless PZT network, a set of PZT transducers are deployed on the surface of structures. The propagation characteristics of the excited ultrasonic signals within the structure are inspected to identify damage.

A distributed microelectronic system for the acoustic monitoring of areas of environmental interest was recently proposed [55]. It is based on a wireless network of

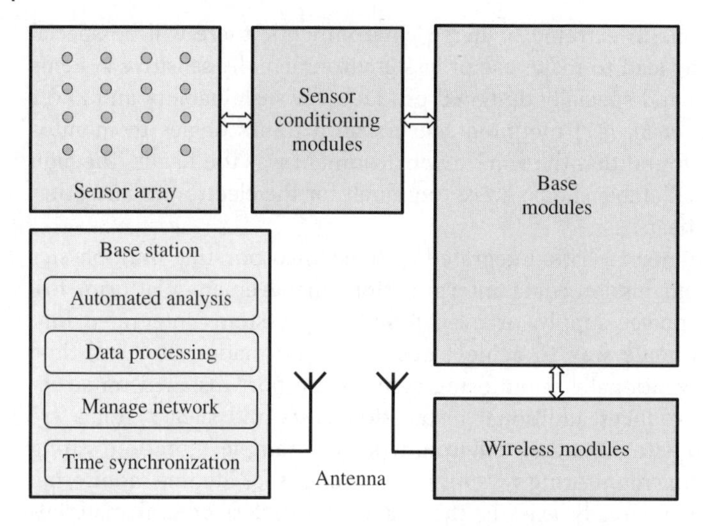

Figure 18.7 A block diagram of the wireless PZT SHM system.

acoustic sensors (microphones) and the automated generation of multi-level sound maps for environmental assessment. Similarly, a hybrid sensor network system has been described [56]. This is able to use any kind of commercially available acoustic emission (AE) sensor and consists of a wireless AE sensor and a wireless strain sensor as an excitation and sensing device. The collected data is forwarded to a remote monitoring center over the Internet using an embedded router. The design was field tested in a smaller structure of the University of Stuttgart, as shown in Figure 18.8.

18.3.2.2 Optical-fibre Integrated Applications

Optical-fibre integrated SHM systems enable passive sensing, which is preferable in some harsh environments because such systems can be embedded easily into structures. One example is submarine cable monitoring. A comprehensive review of SHM of civil infrastructure using optical fibre sensing can be found in the literature [57]. Optical-fibre-based sensing systems are usually integrated with other wireless systems for better flexibility. Qing et al. [58] and Raghupathi et al. [59] have explored creating a

Figure 18.8 'Concerto Bridge' in Brunswick equipped with wireless AE sensors (left) and with wireless strain sensors (right).

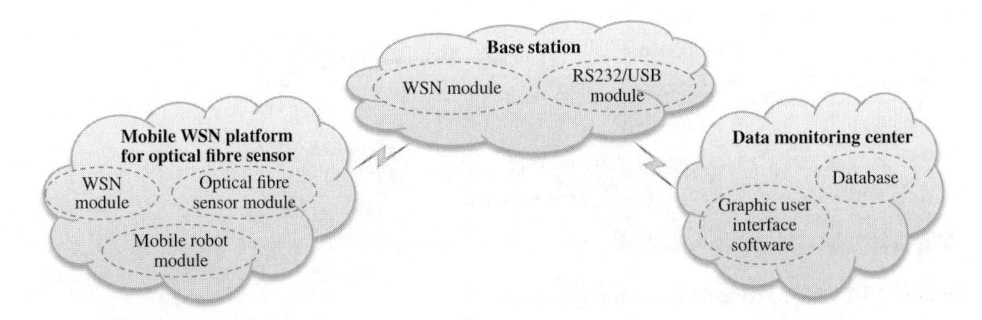

Figure 18.9 System architecture of an optical fibre sensor WSN platform.

hybrid WSN through the establishment of a generic WSN platform consisting of spatially distributed sensor nodes. Zhou et al. [60] presented a wireless mobile platform, as shown in Figure 18.9, to locate and gather data from different types of optical-fibre sensor.

18.3.3 RFID Technology for SHM

As mentioned previously, RFID is a cost-effective approach to SHM, and it is also an effective technology for track and trace applications. RFID tags can go beyond the ID in RFID and be used as low-cost sensors. This is done by mapping a change in some physical parameter of interest to controlled changes in RFID tag antenna properties. This section gives some SHM applications of systems based on passive RFID sensors. Such applications include low-frequency (LF) and ultra-high-frequency (UHF) corrosion monitoring in metallic structures and strain/crack monitoring in concrete structures.

Technically speaking, RFID systems consist of three main components. The RFID reader typically consists of a microcontroller and the RF circuitry – envelope detectors and filters – required to transmit and receive RF energy. The reader transfers enough power to the tag to activate it and receive the data stored in the tag's memory and then finally write the data to the tag's memory. Tags that have been used in this area are passive off-the shelf RFID tags. Passive tags get all their power from a near-field carrier signal generated by the reader. This signal is typically from strong magnetic resonant coupling or a radio wave. Finally, the middleware forwards data to other systems, typically autonomous control systems or a cloud platform.

A review of passive wireless sensors for SHM is to be found in the literature [61]. However, in this section we only consider RFID sensors. The UHF RFID called EPC1Gen 2 is a passive RFID with low costs, being made of commercial off-the-shelf components and is simply for measurement read-out. Applications are found in

- monitoring of physical parameters (temperature, pressure, humidity, motion, sound)
- non-invasive monitoring (extremely important in medical applications)
- controlling the integrity of an object.

A few passive RFID designs have been aimed at use as corrosion/deformation/crack sensors. These all exploit detuning of the tag due to the deformation of the structure.

Sunny et al. [13] used LF RFID sensors and selective transient feature extraction to characterise corrosion progress in mild steel samples. The set of samples consisted of

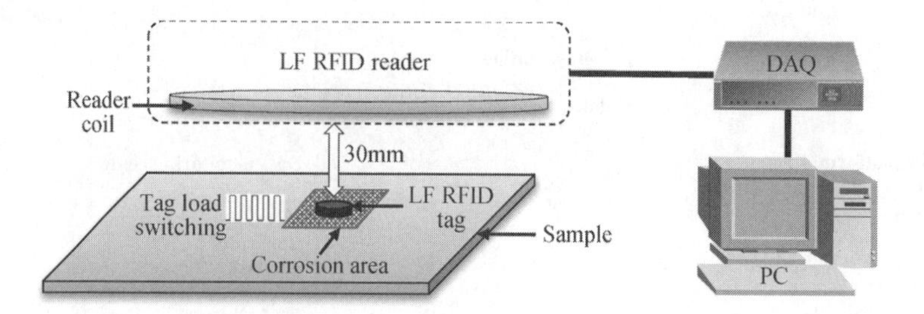

Figure 18.10 LF RFID corrosion monitoring system.

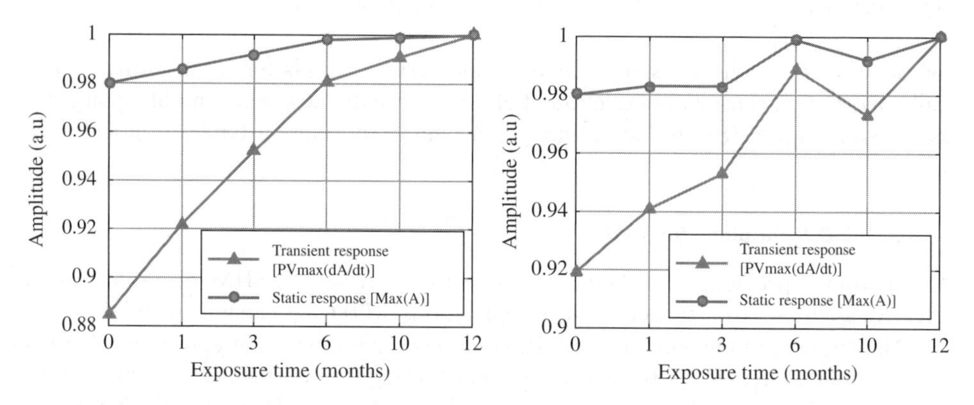

Figure 18.11 Static and transient analysis of corrosion progress for uncoated and coated samples.

coated and uncoated mild steel plates (grade S275), subject to different durations of atmospheric exposure (1, 3, 6, 10 and 12 months). The system diagram is shown in Figure 18.10.

As the corrosion expands with time, the rust layer loosens up and flakes fall off. This means that later the corrosion spreads rather than increasing in thickness. Some of the static and transient analysis results can be seen in the Figure 18.11. Two features – static and transient responses – have been selected as determining permeability and conductivity changes, respectively, as the corrosion progresses.

UHF RFID also finds an application in RFID sensing. One recent proposal was a UHF RFID patch antenna with a meander line impedance matching line for material corrosion sensing. The novelty of such an antenna is in minimizing the dependency on stress transfer and applying 3D shape meandering and folding methods to improve the resolution of corrosion sensing. The tag-antenna based sensing (TABS) is validated through experiment with an RFID development kit. The experimental results of corrosion sensing with the meander line antenna show that the minimum threshold power frequency (also known as the resonance frequency) shifts with different samples, as illustrated in Figure 18.12.

A rectangular microstrip patch antenna in combination with IC chips was proposed by Yi et al. [14]. An RFID strain sensor was deformed, its resonant frequency changing along with change in its dimension due to the strain. A meander dipole antenna was used by Occhiuzzi et al. [15] as a strain sensor. The RFID sensing units were embedded

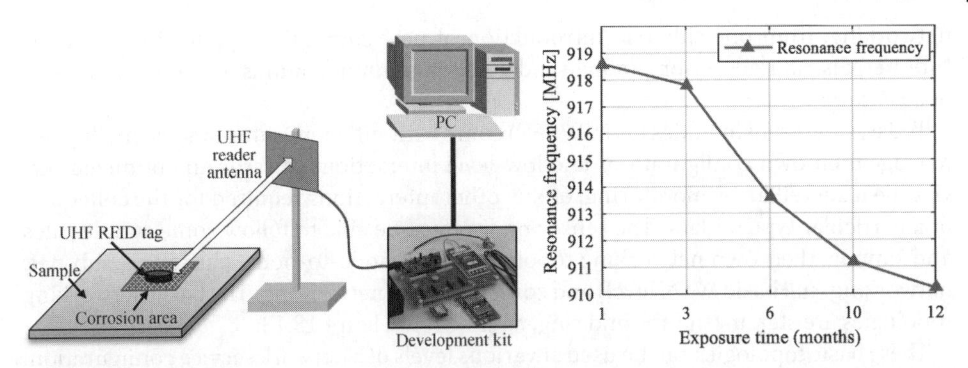

Figure 18.12 UHF RFID corrosion monitoring system (left); resonance frequency shift with corrosion thickness/stage change (right).

in the structure for lifetime monitoring. The embedded nature of the RFID tags eliminates any exposed cables on the surface of concrete structures. Electromagnetic waves are provided to the RFID tags by RFID reader equipment from outside the structure, driving the sensors to measure strain. Measurement can be performed easily by anyone with a computer and a special reader/writer. More recent works [62, 63] have explored the adaptation and optimization of RFID passive sensors for crack monitoring in environments considered to be RFID-harsh, either because of relatively high dielectric losses (concrete) or because of the presence of metallic materials.

The above discussion shows that RFID sensors incorporating antennas can be designed and developed for applications such corrosion, stress, crack and environmental monitoring, and measurement of other physical parameters. Further research should be undertaken in the areas of smart antennas, film and printing technologies for miniaturised sensor networks, and integration of RFID sensor networks and the IoT.

18.4 Network Topology and Overlays

In WSNs, nodes, clusters and subnet components of a network need to follow certain basic infrastructural routines to link with each other, and to manage their inner and interactive operations within the covered range. The methods used for applying these operations are defined by the network topology.

Beyond the network topologies helping the sensor nodes, network management also includes system maintenance such as periodic testing, dealing with faulty nodes and malfunctions of essential parts. And many advanced networked services can benefit from an overall coverage, a single umbrella of common care, which is referred to as the 'network overlay', contributes to a holistic integrated systems vision of WSNs in SHM.

This section briefly discusses network topologies, before explaining the network overlay.

18.4.1 Networking Topology

Traditionally, all communication networking topologies were rooted in standard telecommunications industry protocols. Newer data applications, have challenged this norm, requiring new protocols such as unstructured networking and ad hoc networking, from one side and introduction of new computing applications such as, Internet, distributed sensors, mobile and wireless communications over the last quarter of a century.

Today, smart sensor nodes (as in WSNs) come with sufficient autonomous intelligence manage their own configurations and low-level interactions towards an optimized service, be it surveillance, monitoring, or any other interactions required for the collection of a particular type of data. The sensor nodes must be able to follow some simple rules and support their own networking topologies, be it node-to-node, clustering, sub-net, networking (full basic WSN level) and complex multi-networking. The basic networking topologies are star, mesh, tree and ring, as shown in Figure 18.13.

These basic topologies can be used at various levels of a network service configuration. For WSNs built from smart sensor nodes, network viability creates very tough requirements with regard to energy, quality of data and performance. The topology needed for optimized operation depends on the nature of the service and therefore can differ from one application to another.

It is worth mentioning that networking services (for example, rules, protocols, interconnections, routings) in WSNs are internal, provided by components of the network. This is different to WSN *data* services, which deliver raw or partially processed data to databases, or storage media, through sinks.

To discuss the structure of WSNs at their various levels (node, cluster, and higher levels) we have to divide WSN networking services into three *network service groups* (NSGs):

1) *NSG1 – Network Service Group 1* represents physical, link and cluster-level static interconnection services. The recent name for these services is media access control (MAC) services, and they are similar to traditional telecoms services, where bandwidth, low error rate, reliability and coverage are the prime objectives.
2) *NSG2 – Network Service Group 2* represents dynamic ad hoc peer-to-peer (e.g. Internet) services.
3) *NSG3 – Network Service Group 3* is for network management, overlay and supervisory services.

In general, NSG1 services can accommodate static clustering [64–66] if networking services are static, but dynamic clustering is also possible for more critical services. Dynamic clustering requires autonomous and more intelligent sensing nodes, which demands more advanced and energy-efficient devices [67, 68]. NSG2 services [69] are where the networking services are achieved through traditional infrastructure and standards. Peer-to-peer services can be simplified if the network management services are

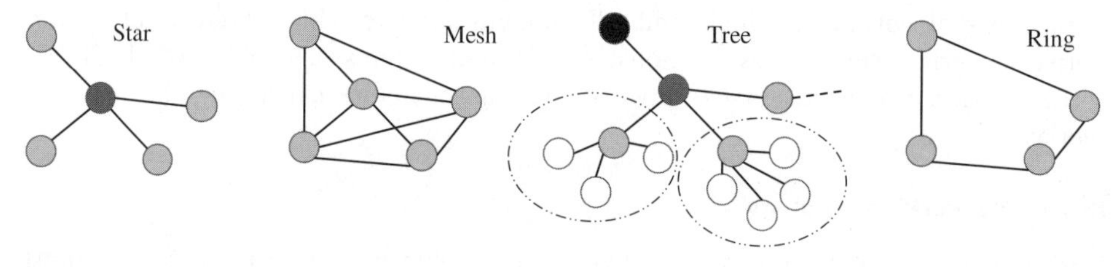

Figure 18.13 Basic networking topologies.

moved away from the lower layers. That is, if systems and devices operate under NSG1 services, and other network services are left to NSG3. Therefore, a new flexible NSG3 can now take care of all the remaining services.

Due to development of advanced and fault-free devices, NSG3s for WSNs are becoming very simple for most applications. However, for very sophisticated WSNs, especially for those used in extreme environments, more advanced applications are required. A new overlay-style NSG3 layer can now look after overall parametric efficiencies, reliability, security, and sustainability of the network. For example, if our WSN is designed to collect occasional samples of temperature or pressure changes over a large area, then the network lifetime becomes the main networking objective. Bandwidth and intelligence of the sensor nodes are secondary, and power consumption of the nodes can be significantly reduced. However, a heavy-duty video surveillance WSN application would have bandwidth and precision of the video images as prime objectives. The same considerations apply to WSNs used in extreme environments; a few unique parameters of the application and the working environment will take higher priorities than others.

In most SHM applications of WSNs, monitoring is the main objective. These applications can be classified as static or dynamic. Static applications require fewer topological changes, but monitoring of moving objects such as vehicles, trains or airplanes is quite different: monitoring the dynamics of a process increases the networking complexity. Under such circumstances the sensors must periodically report their current locations. As an example, Abdelhakim et al. [66] proposed a mobile-access coordinated WSN architecture. Such new architectures, however, impose long delays and make such WSNs unsuitable for time-sensitive applications.

18.4.2 Network Overlay as a Service

An overlay, as traditionally applied to a network, is a collection of network management services to take care of all its internal aspects, including faults, security and sustainability using advanced measurement techniques. It is an important value-added service layer over the operating network, enhancing reliability and ultimately delivering the required performance.

To investigate recent developments and common themes, we now briefly scan some recent studies to help us better understand overlay technologies. We also discuss some examples (see Table 18.1).

18.4.2.1 Cases Supporting the Need for NSG3-style Overlays

Some designers have tried to integrate traditional network management and new overlay technologies to enable management of dynamic-topology sensor networks. This happens in harsh environments where an automatic reconfiguration can help the system to regain optimum performance, which would be lost due to constant changes in the status of the devices. For example, Gelal et al. [70] propose a dynamic-topology control algorithm that optimizes the trade-off between node density and overall performance of the network. Equipped with embedded directional antennas and powerful neighbor-discovery functions, the overlay helps with dynamic clustering. It also enables the process to benefit from shared common functions such as resilience, maintenance and security. A related work is from Sethi et al. [71], who built a self-organizing overlay network using agent technologies, to perform the network management functions. In this case the agents analyse the environment and collaborate to optimize allocation of their own available resources.

The need for an overlay that can enable a resilient network in a generic format, as an essential requirement in distributed sensor systems, has been emphasised extensively in recent years. An example was mentioned in Rashvand and Calero [72], where the term 'resilience' includes agility, security, and sustainability of the service.

18.4.2.2 Cases Supporting Integration of Overlay with IP Technologies

Most practical WSNs make use of unstructured peer-to-peer overlay networks. For these cases, a well-defined overlay network can be used to match the peers' requirements. Unstructured peer-to-peer overlay networks can be used for many wireless SHM and IoT applications, but due to their complexity we do not recommend them for small applications.

In the early days of the Internet, Harvey et al. proposed a scalable networking approach [73] for transmission of sensitive data over peer-to-peer networks. Using a distributed hash table, this secure networking service was an improvement over earlier developments. This Internet-compatible overlay method was regarded as very efficient and reliable. However, as most SHM applications need to deal with poor-quality channels, it is less attractive for use in extreme environments. However, peer-to-peer overlaying can be used in some distributed SHM applications because of their simplicity and unique compatibility with the Internet. The early developments in this area have been described by Lua et al. [69]. They consider peer-to-peer overlay network models as a communication framework and as a distributed cooperative network in which the peers build their own self-organizing systems.

Classic functions deployed at this level include interface, bus and memory management. There are many examples. Wentzlaff et al. [74] outlined their factored operating system to meet the new challenges of scalability, faultiness and variability of demand for single-chip thousand-core many-core systems. Villari et al. [75] proposed a new software platform architecture for the management of smart environments associated with the IoT. To do this they considered the problems of using peer-to-peer processing for ubiquitous sensor network framework standards.

18.4.2.3 Cases of Supporting the Applications

Integration of traditional network management overlay technologies and new wireless sensing applications would greatly simplify deployment and delivery of new services. Costa et al. [76] claim that due to the especially harsh and hectic conditions during a disaster such as an earthquake or an industrial fire, used of a middleware-based overlay has significant advantages. In their work, they reason that due to the key challenges such as the heterogeneity, resource scarcity and dynamism of WSNs, using a middleware technology at the core of the process would enable the system to deal with the multiple challenges encountered during disasters. Obviously, this system requires sufficient intelligence to detect and to react in emergency situations, such as the fast spread of fire or a dangerous chemical spillage.

Table 18.1 Comparison of above cases for new overlay technologies.

Case analysed	IoT applicability	SHM applicability	Overlay remarks
Lua 2005 [69]	Fully compatible	Case compatible	Require adjustment
Gelal 2009 [70]	Fully compatible	Case compatible	Require adjustment
Sethi 2011 [71]	Fully compatible	Fully compatible	Overlay ready
HFR 2012 [72]	Fully compatible	Fully compatible	require expansion
Harvey 2003 [73]	Compatible	Not optimum	Require adjustment
Wentzlaff 2010 [74]	Limited compatibility	Limited use	Not useful
Villari 2014 [75]	Compatible	Case compatible	Require adjustment
Costa 2007 [76]	Limited compatibility	Limited use	Require adjustment
Wang 2008 [77]	Fully compatible	Case compatible	Require adjustment
Ahlborn 2010 [78]	Limited use	Fully compatible	Not useful
Lissenden 2009 [79]	Limited compatibility	Limited compatibility	Further adjustment
Borkowski 2011 [80]	Compatible	Case compatible	Further adjustment

An interesting study on use of an overlay in automatic software design is from Wang et al. [77]. Their combined software and firmware tools ('middleware') hide the complexity of the underlying hardware and network platforms from the system designers. Relating wireless sensing to pervasive computing, they use embedded operating systems to manage system resources and increase dynamic executions in programming.

Bridge health monitoring is another area where overlays are being deployed all over the world. This rapid expansion is due to the increase in vehicle numbers, overloading bridges with slow moving traffic. Older bridges were not designed for such heavy loading. The work of Ahlborn et al. is particularly interesting for detailed analysis of these cases [78]. They include an interesting analysis of traditional inspection methods, considering all of the associated uncertainties and labour costs. This area can benefit from the new wireless-Internet-based remote controlled overlay SHM applications.

Another interesting case that can benefit from a network overlay comes from Lissenden et al. [79]. Their work addresses critical sensors that impact safety through diagnostic processes that predict faults well ahead of any actual disastrous conditions. For example, when inspecting the condition of an airplane body structure, two complementary parameters of vibration and strain can be examined. Embedded fibre optic sensors measure the strain, whilst microelectromechanical accelerometers detect vibration levels. The authors use new techniques based on active ultrasonic, passive acoustic, and electromechanical impedance measurements with particular attention to guided ultrasonic waves for laminated composites.

One of the key technologies associated SHM is the accuracy of data produced by the sensing devices. Inaccuracies can easily compromise the effectiveness of the application. Borkowski et al. have studied the fundamentals of existing systems and sensors, including piezoelectric systems, to develop an improved electromechanical elastodynamic model [80]. This modeling can become extremely valuable because research and development of SHM applications is difficult, and experiments are costly, so modeling and simulations are highly desirable.

18.5 Summary

This chapter reviews the state-of-the art developments of WSNs for SHM. Internet technologies and integration with the IoT – for example, RFID-sensor-based SHM – are useful as SHM becomes a new service industry. The key progressions and future developments are:

Compressed sensing: Compressed sensing techniques offer a mathematical framework for the detection and allocation of sparse signals with a reduced number of samples. Although it is the powerful sink nodes or monitoring centers that reconstruct down-sampling signals, further research into fast reconstruction algorithms is necessary. These will compensate for the signal processing delay incurred by large numbers of iterations. Compressed sensing is still in its infancy in the SHM industry, and applications inspired by its used are just only just emerging. More advanced modalities are needed. This calls for more efforts in modeling, exploring sparsity, designing measurement matrices and developing hardware for specific applications.

Guarantee acceptable network lifetime: Power limitations remain a key challenge for WSNs and therefore their SHM applications. Lightweight wireless clustering and communication protocols are in demand and should be designed and optimized for certain classes of WSN-based SHM applications. IEEE 802.15.4 based protocols have achieved relatively low power consumption, but they need to be extended to support fault tolerant performance and flexible routing topologies, areas in which they are still inadequate for harsh environments such as nuclear power stations. They must also bring lightweight transmission burdens and data compression methods. From the hardware point of view, developing components with low power consumption contributes to power conservation. New WSN-specific processors as well as high-performance transducers will be needed.

Need for global IoT architecture: The IoT is an integration of distributed infrastructures and heterogeneous services. Currently the IoT lacks a theory to integrate the virtual and real worlds in a unified framework [81]. A global IoT architecture which provides an SHM interface is important if IoT–SHM applications are to be developed. The architecture should integrate different sensing systems and services as well as big data technologies [82]. This will allow a low-cost, reliable and hybrid monitoring network to be created. This will promote a data-driven economy and even community healthcare for a sustainable society.

Challenges of traditional, optical fibre and RFID-based systems: Traditional sensors are more mature and have good environmental adaption capabilities. More power-efficient transducers, chips and routing techniques are attractive for in-network computing. Optical fibre sensors are passive and robust to electromagnetic interference. However, the optical fibre also brings complexity in relation to installation and replacement. More cost-effective optical fibre demodulation instruments should be developed. Antennas and electromagnetic resonators are passive devices that can sense physical parameters, and receive and transmit data. Some pioneer literature on monitoring strain, cracks and corrosion are just appearing and more work needs to be done on better inductance and more powerful functions, such as detection of strain direction. Sensing data from backscattering signals is likely to be contaminated if there are other metallic elements in the host structure, thus leading to false results. Improvements in cost, stability and reliability are required.

Need for new style network management: Deployment of network-based services can be greatly simplified using overlay network services. These technologies are new to both researchers and industry, so new innovative solutions are currently not likely. Development of new standards may trigger more integrated solutions for better global-scale systems and services to help the industries and guide researchers to come up with new ideas.

Acknowledgment

The authors would like to thanks EPSRC, NSFC and the China Scholarship Council for their partial funding of the work. We also acknowledge our researcher Aobo Zhao for his contribution to the results in this chapter and Dr. Xuewu Dai and all the reviewers for their constructive comments.

References

1 A. Whitmore, A. Agarwal and D.L. Xu, 'The Internet of Things—A survey of topics and trends,' *Information Systems Frontiers*, vol. 17, pp. 261–274, 2015.

2 S. Madakam, R. Ramaswamy and S. Tripathi, 'Internet of Things (IoT): A literature review,' *Journal of Computer and Communications*, vol. 3, pp. 164–173, 2015.

3 K. Ashton, 'That 'internet of things' thing,' *RFID Journal*, vol. 22, pp. 97–114, 2009.

4 A. Al-Fuqaha, M. Guizani, M. Mohammadi, M. Aledhari and M. Ayyash, 'Internet of things: A survey on enabling technologies, protocols, and applications,' *IEEE Communications Surveys & Tutorials*, vol. 17, no. 4, pp. 2347–2376, 2015.

5 C. Perera, C.H. Liu and S. Jayawardena, 'The emerging internet of things marketplace from an industrial perspective: A survey,' *IEEE Transactions on Emerging Topics in Computing*, vol. 3, no. 4, pp. 585–598, 2015.

6 N. Shahid, I.H. Naqvi and S.B. Qaisar, 'Characteristics and classification of outlier detection techniques for wireless sensor networks in harsh environments: A survey,' *Artificial Intelligence Review*, vol. 43, pp. 193–228, 2015.

7 S.A. Basit and M. Kumar, 'A review of routing protocols for underwater wireless sensor networks,' *International Journal of Advanced Research in Computer and Communication Engineering*, vol. 4, no. 12, pp. 373–378, 2015.

8 V.J. Hodge, S.O. Keefe, M. Weeks and A. Moulds, 'Wireless sensor networks for condition monitoring in the railway industry: A survey,' *IEEE Transactions on Intelligent Transportation Systems*, vol. 16, no. 3, pp. 1088–1106, 2015.

9 D. Puccinelli and M. Haenggi, 'Wireless sensor networks: applications and challenges of ubiquitous sensing,' *IEEE Circuits and Systems Magazine*, vol. 5, pp. 19–31, 2005.

10 G.Y. Tian, Z.X. Zhao and R.W. Baines, 'A Fieldbus-based intelligent sensor,' *Mechatronics*, vol. 10, pp. 835–849, 2000.

11 G.Y. Tian, 'Design and implementation of distributed measurement systems using fieldbus-based intelligent sensors,' *IEEE Transactions on Instrumentation and Measurement*, vol. 50, no. 5, pp. 1197–1202, 2001.

12 G.Y. Tian, G. Yin and D. Taylor, 'Internet-based manufacturing: A review and a new infrastructure for distributed intelligent manufacturing,' *Journal of Intelligent Manufacturing*, vol. 13, pp. 323–338, 2002.

13 A.I. Sunny, G.Y. Tian, J. Zhang and M. Pal, 'Low frequency (LF) RFID sensors and selective transient feature extraction for corrosion characterisation,' *Sensors and Actuators A: Physical*, vol. 241, pp. 34–43, 2016.

14 X. Yi, C. Cho, C.H. Fang, J. Cooper, V. Lakafosis and R. Vyas, 'Wireless strain and crack sensing using a folded patch antenna,' in *6th European Conference on Antennas and Propagation*, Prague, 2012.

15 C. Occhiuzzi, C. Paggi and G. Marrocco, 'Passive RFID strain-sensor based on meander-line antennas,' *IEEE Transactions on Antennas and Propagation*, vol. 59, no. 12, pp. 4836–4840, 2011.

16 D.L. Donoho, 'Compressed sensing,' *IEEE Transactions on Information Theory*, vol. 52, no. 4, pp. 1289–1306, 2006.

17 E.J. Candès, J. Romberg and T. Tao, 'Robust uncertainty principles: Exact signal reconstruction from highly incomplete frequency information,' *IEEE Transactions on Information Theory*, vol. 52, no. 2, pp. 489–509, 2006.

18 J. Haupt, W.U. Bajwa, M. Rabbat and R. Nowak, 'Compressed sensing for networked data,' *IEEE Signal Processing Magazine*, vol. 25, pp. 92–101, 2008.

19 C. Luo, F. Wu, J. Sun and C.W. Chen, 'Compressive data gathering for large-scale wireless sensor networks,' in *Proceedings of the 15th annual international conference on Mobile computing and networking*, Beijing, 2009.

20 Q. Liang and Z. Tian, 'Decentralized sparse signal recovery for compressive sleeping wireless sensor networks,' *IEEE Transactions on Signal Processing*, vol. 58, no. 7, pp. 3816–3827, 2010.

21 Y. Zhang, M. Roughan, W. Willinger and L. Qiu, 'Spatio-temporal compressive sensing and internet traffic matrices,' in *ACM SIGCOMM Computer Communication Review*, Barcelona, 2009.

22 M. Sartipi, 'Low-complexity distributed compression in wireless sensor networks,' in *2012 Data Compression Conference*, Snowbird, 2012.

23 J. Kho, A. Rogers and N.R. Jennings, 'Decentralized control of adaptive sampling in wireless sensor networks,' *ACM Transactions on Sensor Networks*, vol. 5, no. 3, pp. 1–35, 2009.

24 J. Hao, B. Zhang, Z. Jiao and S. Mao, 'Adaptive compressive sensing based sample scheduling mechanism for wireless sensor networks,' *Pervasive and Mobile Computing*, vol. 22, pp. 113–125, 2015.

25 R. Want, 'An introduction to RFID technology,' *IEEE Pervasive Computing*, vol. 5, no. 1, pp. 25–33, 2006.

26 Bluetooth Special Interest Group, 'Bluetooth Core Specification 4.2,' 2 12 2014. Online.. Available: https://www.bluetooth.org/en-us/specification/adopted-specifications.

27 IEEE Computer Society, 'IEEE Standard for Local and metropolitan area networks--Part 15.4: Low-Rate Wireless Personal Area Networks (LR-WPANs),' 5 5 2014. Online.. Available: http://ieeexplore.ieee.org/xpl/articleDetails.jsp?arnumber=6809836.

28 ISA100 Committee, 'ISA100.11a Technology Standard: Wireless systems for industrial automation: Process control and related applications,' 4 5 2011. *Online*.. Available: http://www.nivis.com/technology/ISA100.11a.php.
29 Wireless Specialist, 'System Engineering Guidelines: IEC 62591 WirelessHART,' 2 2016. Online.. Available: http://www2.emersonprocess.com/siteadmincenter/PM%20Central%20Web%20Documents/EMR_WirelessHART_SysEngGuide.pdf.
30 ZigBee Alliance, 'ZigBee 3.0 specification,' 2015. Online. Available: http://www.zigbee.org/zigbee-for-developers/zigbee3–0/.
31 M.A. Mahmood, W.K. Seah and I. Welch, 'Reliability in wireless sensor networks: A survey and challenges ahead,' *Computer Networks*, vol. 79, pp. 166–187, 2015.
32 A. Ghaffari, 'Congestion control mechanisms in Wireless Sensor networks: A survey,' *Journal of Network and Computer Applications*, vol. 52, pp. 101–115, 2015.
33 A. Pughat and V. Sharma, 'A review on stochastic approach for dynamic power management in wireless sensor networks,' *Human-centric Computing and Information Sciences*, vol. 5, no. 4, pp. 1–14, 2015.
34 Y. Hu, J. Zhang, W. Cao, J. Wu, G.Y. Tian and S.J. Finney, 'Online two-section PV array fault diagnosis with optimized voltage sensor locations,' *IEEE Transactions on Industrial Electronics*, vol. 62, no. 11, pp. 7237–7246, 2015.
35 P. Havinga, 'WiBRATE - Wireless, self-powered vibration monitoring and control for complex industrial systems,' 2011–2015. Online.. Available: www.wibrate.eu.
36 B.V. Technology, 'SIRIUS - Wireless Sensing System for High-Performance Industrial Monitoring and Control,' 2013–2015. Online. Available: www.sirius-system.eu.
37 P. Kamalinejad, C. Mahapatra, Z. Sheng, S. Mirabbasi, V.C. Leung and Y.L. Guan, 'Wireless energy harvesting for the Internet of Things,' *IEEE Communications Magazine*, vol. 53, no. 6, pp. 102–108, 2015.
38 M.G. Roes, J.L. Duarte, M.A. Hendrix and E.A. Lomonova, 'Acoustic energy transfer: A review,' *IEEE Transactions on Industrial Electronics*, vol. 60, no. 1, pp. 242–248, 2013.
39 S.Y. Hui, W. Zhong and C.K. Lee, 'A critical review of recent progress in mid-range wireless power transfer,' *IEEE Transactions on Power Electronics*, vol. 29, no. 9, pp. 4500–4511, 2014.
40 F. Akhtar and M.H. Rehman, 'Energy replenishment using renewable and traditional energy resources for sustainable wireless sensor networks: A review,' *Renewable and Sustainable Energy Reviews*, vol. 45, pp. 769–784, 2015.
41 L. Xie, Y. Shi, Y.T. Hou and A. Lou, 'Wireless power transfer and applications to sensor networks,' *IEEE Wireless Communications*, vol. 20, no. 4, pp. 140–145, 2013.
42 M. Erol-Kantarci and H.T. Mouftah, 'Suresense: sustainable wireless rechargeable sensor networks for the smart grid,' *IEEE Wireless Communications*, vol. 19, no. 3, pp. 30–36, 2012.
43 K. Worden, E.J. Cross, N. Dervilis, E. Papatheou and I. Antoniadou, 'Structural health monitoring: From structures to systems-of-systems,' *IFAC-PapersOnLine*, vol. 48, pp. 1–17, 2015.
44 S. Gao, X. Dai, Z. Liu and G.Y. Tian, 'High-performance wireless piezoelectric sensor network for distributed structural health monitoring,' *International Journal of Distributed Sensor Networks*, vol. 2016, pp. 1–16, 2016.

45 B.S. Cook, J.R. Cooper and M.M. Tentzeris, 'An inkjet-printed microfluidic RFID-enabled platform for wireless lab-on-chip applications,' *IEEE Transactions on Microwave Theory and Techniques*, vol. 61, no. 12, pp. 4714–4723, 2013.

46 V. Lakafosis, A. Rida, R. Vyas, L. Yang, S. Nikolaou and M.M. Tentzeris, 'Progress towards the first wireless sensor networks consisting of inkjet-printed, paper-based RFID-enabled sensor tags,' *Proceedings of the IEEE*, vol. 98, no. 9, pp. 1601–1609, 2010.

47 M. Grudén, A. Westman, J. Platbardis, A. Rydberg and P. Hallbjomer, 'Reliability experiments for wireless sensor networks in train environment,' in *Wireless Technology Conference*, Rome, 2009.

48 B. Han, S. Ding and X. Yu, 'Intrinsic self-sensing concrete and structures: A review,' *Measurement*, vol. 59, pp. 110–128, 2015.

49 A. Ramos, D. Girbau, A. Lazaro and R. Villarino, 'Wireless concrete mixture composition sensor based on time-coded UWB RFID,' *IEEE Microwave and Wireless Components Letters*, vol. 25, no. 10, pp. 681–683, 2015.

50 T.J. Lesthaeghe, S. Frishman, S.D. Holland and T.J. Wipf, 'RFID tags for detecting concrete degradation in bridge decks,' Institute for Transportation at Digital Repository @ Iowa State University, Iowa, 2013.

51 Y. Huang, X. Liang, S.A. Galedar and F. Azarmi, 'Integrated fiber optic sensing system for pipeline corrosion monitoring,' *Pipelines*, vol. 2015, pp. 1667–1676, 2015.

52 J.H. Kim, G. Sharma, N. Boudriga, S. Iyengar and N. Prabakar, 'Autonomous pipeline monitoring and maintenance system: A RFID-based approach,' *EURASIP Journal on Wireless Communications and Networking*, vol. 2015, pp. 1–21, 2015.

53 G.Y. Tian, 'Health Monitoring of Offshore Wind Farms (HEMOW),' 2011–2015. Online.. Available: www.hemow.eu.

54 A.A. Carulla, J.C. Farrarons, J.L. Sanchez and P.M. Catala, 'Piezoelectric harvester-based self-powered adaptive circuit with wireless data transmission capability for structural health monitoring,' in *Design of Circuits and Integrated Systems*, Estoril, 2015.

55 S.M. Potirakis, B. Nefzi, N.A. Tatlas, G. Tuna and M. Rangoussi, 'A wireless network of acoustic sensors for environmental monitoring,' *Key Engineering Materials*, vol. 605, pp. 43–46, 2014.

56 C. Grosse, G. McLaskey, S. Bachmaier, S.D. Glaser and M. Krüger, 'A hybrid wireless sensor network for acoustic emission testing in SHM,' in *The 15th International Symposium on Smart Structures and Materials & Nondestructive Evaluation and Health Monitoring*, San Diego, 2008.

57 X. Ye, Y. Su and J. Han, 'Structural health monitoring of civil infrastructure using optical fiber sensing technology: A comprehensive review,' *The Scientific World Journal*, vol. 2014, pp. 1–11, 2014.

58 X. Qing, A. Kumar, C. Zhang, I.F. Gonzalez, G. Guo and K.K. Chang, 'A hybrid piezoelectric/fiber optic diagnostic system for structural health monitoring,' *Smart Materials and Structures*, vol. 14, pp. 98–103, 2005.

59 V. Raghupathi and K.K. Sangeetha, 'Automatic DAQ for intrinsic optical fiber PH sensors using wireless sensor network,' *International Journal of Advanced and Innovative Research*, vol. 4, no. 4, pp. 151–158, 2015.

60 B. Zhou, S. Yang, T. Sun and K.T. Grattan, 'A novel wireless mobile platform to locate and gather data from optical fiber sensors integrated into a WSN,' *IEEE Sensors Journal*, vol. 15, no. 6, pp. 3615–3621, 2015.
61 A. Deivasigamani, A. Daliri, C.H. Wang and S. John, 'A review of passive wireless sensors for structural health monitoring,' *Modern Applied Science*, vol. 7, no. 2, pp. 57–76, 2013.
62 S. Caizzone and E. DiGiampaolo, 'Wireless passive RFID crack width sensor for structural health monitoring,' *IEEE Sensors Journal*, vol. 15, no. 12, pp. 6767–6774, 2015.
63 S. Caizzone, E. DiGiampaolo and G. Marrocco, 'Wireless crack monitoring by stationary phase measurements from coupled RFID tags,' *IEEE Transactions on Antennas and Propagation*, vol. 62, no. 12, pp. 6412–6419, 2014.
64 B.S. Tripathi, M.K. Shukla and M.K. Srivastava, 'Performance enhancement in wireless sensor network using hexagonal topology,' in *Communication, Control and Intelligent Systems*, Mathura, 2015.
65 A. Nayebi and H.S. Azad, 'Optimum hello interval for a connected homogeneous topology in mobile wireless sensor networks,' *Telecommunication Systems*, vol. 52, pp. 2475–2488, 2013.
66 M. Abdelhakim, J. Ren and T. Li, 'Mobile access coordinated wireless sensor networks—Topology design and throughput analysis,' in *IEEE Global Communications Conference*, Atlanta, 2013.
67 M. Li, Z. Li and A.V. Vasilakos, 'A Survey on topology control in wireless sensor networks: Taxonomy, comparative study, and open issues,' *Proceedings of the IEEE*, vol. 101, no. 12, pp. 2538–2557, 2013.
68 C.Y. Lee, L.C. Shiu, F.T. Lin and C.S. Yang, 'Distributed topology control algorithm on broadcasting in wireless sensor network,' *Journal of Network and Computer Applications*, vol. 36, pp. 1186–1195, 2013.
69 E.K. Lua, J. Crowcroft, M. Pias, R. Sharma and S. Lim, 'A survey and comparison of peer-to-peer overlay network schemes,' *IEEE Communications Surveys & Tutorials*, vol. 7, no. 2, pp. 72–93, 2005.
70 E. Gelal, G. Jakllari, S.V. Krishnamurthy and N.E. Young, 'Topology management in directional antenna-equipped Ad Hoc networks,' *IEEE Transactions on Mobile Computing*, vol. 8, no. 5, pp. 590–605, 2009.
71 P. Sethi, D.D. Juneja and D.N. Chauhan, 'A mobile agent-based event driven route discovery protocol in wireless sensor network: AERDP,' *International Journal of Engineering Science & Technology*, vol. 12, pp. 8422–8429, 2011.
72 H.F. Rashvand and J.M.A. Calero, *Distributed sensor systems: practice and applications*, Chichester: John Wiley & Sons, 2012.
73 N.J. Harvey, M.B. Jones, S. Saroiu, M. Theimer and A. Wolman, 'Skipnet: A scalable overlay network with practical locality properties,' *Networks*, vol. 34, pp. 1–36, 2003.
74 D. Wentzlaff, C. Gruenwald, N. Beckmann, K. Modzelewski, A. Belay and L. Youseff, '*A unified operating system for clouds and manycore: fos*,' MIT Computer Science and Artificial Intelligence Laboratory, Cambridge, 2009.
75 M. Villari, A. Celesti, M. Fazio and A. Puliafito, 'Alljoyn lambda: An architecture for the management of smart environments in IoT,' in *2014 International Conference on Smart Computing Workshops*, Hong Kong, 2014.

76 P. Costa, G. Coulson, R. Gold, M. Lad, C. Mascolo and L. Mottola, 'The RUNES middleware for networked embedded systems and its application in a disaster management scenario,' in *Fifth Annual IEEE International Conference on Pervasive Computing and Communications*, White Plains, 2007.
77 M.M. Wang, J.N. Cao, J. Li and S.K. Dasi, 'Middleware for wireless sensor networks: A survey,' *Journal of Computer Science and Technology*, vol. 23, no. 3, pp. 305–326, 2008.
78 T. Ahlborn, R. Shuchman, L. Sutter, C. Brooks, D. Harris and J. Burns, '*The state-of-the-practice of modern structural health monitoring for bridges: A comprehensive review*,' National Academy of Sciences, Washington, DC, 2010.
79 C.J. Lissenden and J.L. Rose, '*Structural health monitoring of composite laminates through ultrasonic guided wave beam forming*,' NATO Applied Vehilce Technology Symp. on Military Platform Ensured Availability Proc, Pennsylvania, 2008.
80 L. Borkowski, K. Liu and A. Chattopadhyay, 'Fully coupled electromechanical elastodynamic model for guided wave propagation analysis,' *Journal of Intelligent Material Systems and Structures*, vol. 24, no. 13, pp. 1647–1663, 2013.
81 R.V. Kranenburg and A. Bassi, 'IoT challenges,' *Communications in Mobile Computing*, vol. 1, pp. 1–5, 2012.
82 J. Liu, J. Li, W. Li and J. Wu, 'Rethinking big data: A review on the data quality and usage issues,' *ISPRS Journal of Photogrammetry and Remote Sensing*, vol. 115, pp. 134–142, 2016.

19

Error Manifestations in Industrial WSN Communications and Guidelines for Countermeasures

Filip Barac[1], *Mikael Gidlund*[2], *Tingting Zhang*[2] *and Emiliano Sisinni*[3]

[1] *Business Unit Network Products, Ericsson AB, Sweden*
[2] *Department of Information Systems and Technology, Mid Sweden University, Sweden*
[3] *Department of Information Engineering, University of Brescia, Italy*

19.1 Introduction

Industrial automation is one of the areas where wireless communication is yet to reach full maturity. The main obstacle in this sense is the inadequate reliability of standard industrial wireless sensor networks (IWSNs), a result of their inability to cope with the physical and electromagnetic properties of industrial environments. Safety-critical applications, such as closed-loop control, typically require communication reliability in excess of 99.999% and data delivery with latency on the millisecond scale. On the other hand, the dynamics of wireless propagation conditions at industrial sites are severe, often leading to extensive packet loss, as well as long communications blackouts, which these demanding applications cannot tolerate.

Industrial wireless is today standardized only for process control and, despite the existence of several proprietary IEEE 802.11-based solutions (such as iWLAN from Siemens [1]), IEEE 802.15.4-2006 [2] is the de facto standard for industrial wireless. The physical (PHY) layers of all three major IWSN standards – WirelessHART [3], ISA 100.11a [4] and WIA-PA [5] – are compliant with 2.4-GHz flavour of IEEE 802.15.4-2006 specification. While in the future one can expect a rethinking and eventual convergence of the three standards, the first step towards bridging the gap between expectations and reality is a thorough analysis of error properties. The convergence will most likely deal with higher layers, because higher layers are where the three standards are the most divergent. Nonetheless, even if the convergence introduces drastic changes in the PHY layer, the statistical properties of bit- and symbol-error footprints will remain essentially identical to the ones that today's technology is exposed to, because these properties are inherent to the error sources, rather than to any particular technology.

The target area of industrial wireless standards – automation – comprises three sub-areas with diverse communication requirements: process automation, factory automation and building automation [6]. This diversity is probably the reason why certain communication features are only briefly outlined in the IWSN standards, sometimes even at the level of a recommendation. A positive implication of this coarseness is the fact that it leaves plenty of room for improvements.

Wireless Sensor Systems for Extreme Environments: Space, Underwater, Underground, and Industrial.
First Edition. Edited by Habib F. Rashvand and Ali Abedi.

The two most desirable properties in IWSN communication are maximized communication reliability and swift reactions to communication failures. This is, however, a common feature for all communication domains, but the line of reasoning in critical WSN applications is fundamentally different from that for conventional WSN. The mission-critical mindset stipulates maximizing the probability of successful transmission at the first attempt. Conversely, an obvious problem with retransmissions in industrial WSN standards, which are time-division multiple access (TDMA) based, is that a retransmission is often not possible in the very next timeslot. That is, unless the node conquers the medium in the contention-based part of the superframe, the retransmission must wait for the next slot allocated to the transmitter in question, in which case it is plausible to send a new sensor measurement instead of insisting on retransmission.

This chapter has two main goals. The primary aim is to illustrate the low-level effects of the environment on industrial wireless communication at 2.4 GHz, and the bit- and symbol-level implications for IWSN signals are discussed. Second, a methodology for bringing performance closer to those demanded is illustrated in several examples. Because signal waveforms and other PHY-layer properties are strictly defined by WSN communication standards, the solutions for higher reliability are to be found in the design space of the data link layer (DLL) and its medium access control (MAC) sublayer. Before exemplifying the influence of industrial environments on communication quality, the compromising factors of industrial wireless must be discussed individually.

19.2 Compromising Factors in IWSN Communication

19.2.1 Physical Factors

Industrial environments come in a variety of configurations: from underground mines to factories. The focus in this chapter is on the most frequently encountered type of facilities: conventional factories. Radio channel conditions in factories significantly differ from those in office environments, as there are several factors that complicate wireless communication in industrial surroundings:

Sensor placement: Industrial facilities are designed exclusively to suit the industrial process. Apart from the possibility of placing the gateway at a prominent spot in the factory hall, there is next to no regard to communication issues. The placement of wireless sensor devices is subject to zero degrees of freedom, since the industrial process in question must be sampled at strictly defined points. Consequently, it is often impossible to establish line-of-sight (LOS) communication between the sensors. If the topology is such that a sensor node is unacceptably far from the rest of the network, introduction of relays must be considered, at a cost of additional timeslots in a superframe.

Reflective surfaces: Industrial machinery and factory walls are mostly encased in or made of metal, an excellent reflector of electromagnetic waves. This abundance of reflective surfaces has both positive and negative implications. While signal attenuation is a direct consequence of the law of conservation of energy, multipath fading stems from signal reflections off surrounding objects. The presence of multiple signal copies at the receiver usually leads to large variations of received signal strength.

Figure 19.1 A highly absorbing industrial environment: a paper roll warehouse at Hyltebruk paper mill, Sweden.

Multipath propagation is, in fact, both an enabler of and a showstopper for wireless communication. On the upside, it makes communication possible in the absence of LOS, while on the downside, the destructive superposition of multiple signal copies with different delays and phases at the receiver can push the received signal power below the reception threshold. Conversely, the properties of highly absorbent industrial environments (such as paper warehouses; see Figure 19.1) are such that even multi-antenna techniques, such as diversity and MIMO, bring little or no improvement in the absence of LOS [7].

Open-space layouts: Even though the absence of walls is generally favourable for indoor wireless communication, the purpose of an open-plan layout in industrial environments is to accommodate production tools and machinery of overwhelming dimensions, such as boilers, which can span multiple floors and occupy significant portions of the factory's volume. This is exemplified in Figure 19.2, where a wood crusher and central boiler occupy about 30% of the horizontal cross-section of a hall in a paper mill. Such large objects (in other words, obstacles) give rise to extensive shadowing and, due to the lack of sharp edges, diffraction is not always possible.

Moving obstacles: Staff and mobile machinery such as radio-controlled cranes, forklifts and trucks, often have a significant effect on wireless propagation. The influence of moving clutter has two contexts. First, due to the presence of a large number of signal components at the receiver, movements in the environment far away from the transmitter and/or receiver do not affect the power delay profile significantly [8]. However, if a large obstacle appears too close to the transmitter or the receiver on a link, outages caused by shadowing can occur for extended periods. Manual production lines are an interesting case in this respect, because the proximity of humans to wireless sensors causes both intensive small-scale fading and link outages [8].

19.2.2 Electromagnetic Interference

External sources of electromagnetic interference can be classified as intentional radiators (such as wireless communication systems) and unintentional radiators (such

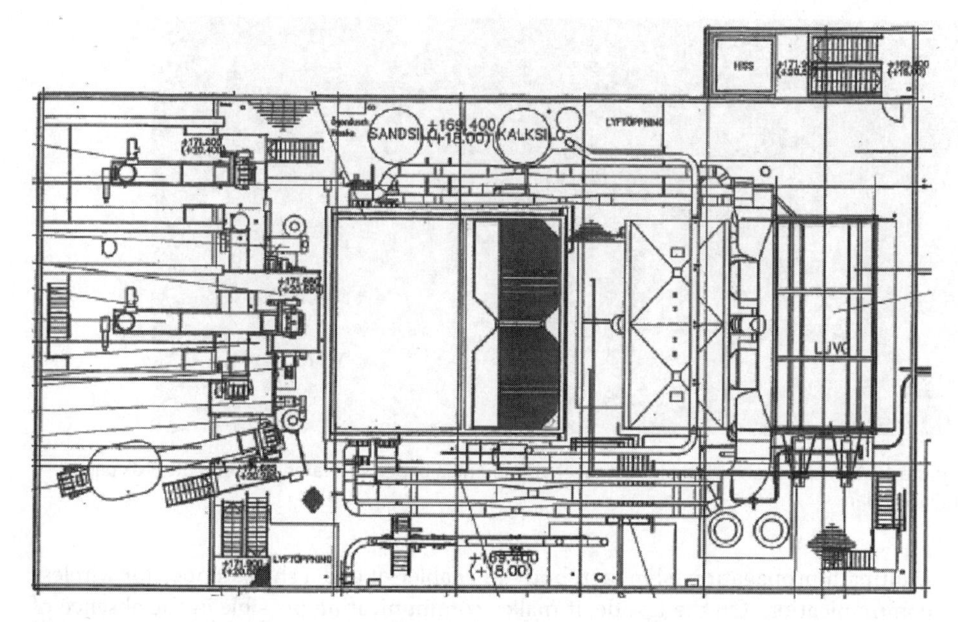

Figure 19.2 The floor plan of a paper mill in Borlänge, Sweden.

as industrial machinery producing spurious electromagnetic emissions). Contrary to common belief, previous research has shown that most unintentional radiators do not influence wireless communication in the ISM band, because most of the emissions are located below 1.5–2 GHz [9]. One notable exception is microwave ovens, which are used in certain branches of industry (such as the rubber industry), generating strong impulsive interference throughout the 2.4-GHz band. Table 19.1 shows the frequencies of interference generated by typical industrial tools.

Table 19.1 Frequencies of interference generated by typical industrial machinery.

Type of tool	Frequency band (MHz)
Frequency converter	<200
Punch press	<1600
CNC cutter	<400
Laser cutter	<1700
Weaving machine	<2000
Arc welder	<50
Spot welder	<150
Frequency converter	<0.2
Switchgear	<5
Relays	<100
Drives	<0.02
Induction heater	<5

Sources [10, 11].

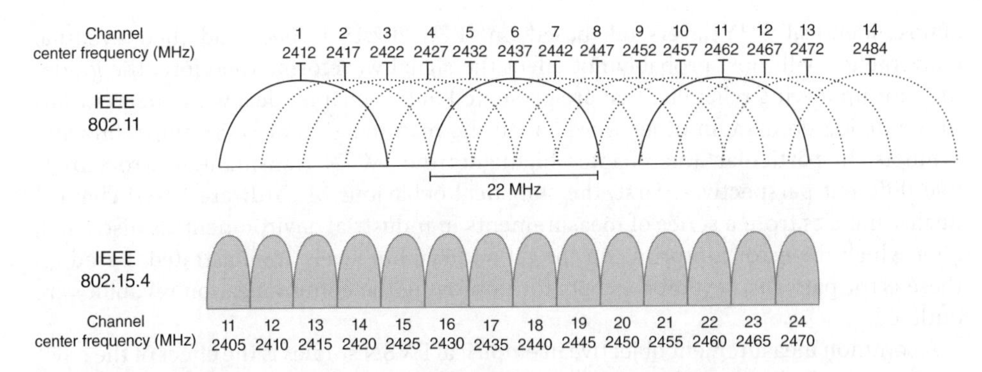

Figure 19.3 The overlap of IEEE 802.15.4-2006 and IEEE 802.11 spectra at 2.4 GHz [12].

The main sources of co-channel interference to IWSNs are the coexisting wireless communication systems in the 2.4-GHz band, most notably WLAN and Bluetooth networks. Bluetooth technology employs hopping over 79 available channels 1600 times per second, leaving WLAN networks as the biggest threat to IWSN communication. The maximum transmit power of WLAN devices (100 mW) exceeds its IWSN counterpart by a factor of 10, while one WLAN channel spans four IWSN channels (see Figure 19.3). An additional problem is the fact that WLAN interference can even leak from the neighboring facilities, so shutting down WLAN communication at the facility of interest might not suffice. Finally, the internal sources of interference are not considered in this chapter, since they are inherent to any type of contention-based communication, rather than the IWSN communication in particular.

19.2.3 Manifestations of Signal Distortion

The propagation environment primarily affects the signal waveform, after which the distortion spills over to the chip-, bit- and byte-level. In case of IEEE 802.15.4-2006 communication at 2.4 GHz, the distortion translates into incorrect direct sequence spread spectrum (DSSS) chip sequence decoding via the OQPSK demodulator, ultimately resulting in bit errors.

The research on IEEE 802.15.4 waveform distortion in industrial environments has been in the spotlight since WSN communication was first standardised, about a decade ago. The study objectives in the related literature are path loss modeling and channel impulse response measurements, often through the prism of root mean square delay spread, coherence bandwidth and coherence time. Signal spreads of the order of tens or hundreds of nanoseconds are typically observed in industrial environments [13], with log-normally distributed large-scale fading and small-scale fading closely following a Rician distribution [8]. Despite Rician- and Rayleigh-like behaviour of fading having been experimentally verified, studies reporting fading dips of 30–40 dB [6] suggest that fading distributions in certain environments are, in fact, heavy-tailed.

Nevertheless, waveform distortions are not the focus of this chapter, for two reasons. First, the modulation and pulse-shaping are strictly defined by the IWSN standards, and any alterations in that sense would imply breaching the compliance with the standards. Such information might be of use for signal processing design, which would likely increase the price of hardware. Secondly, the three major IWSN standards have almost identical PHY layers (all based on IEEE 802.15.4-2006) and their eventual convergence will most probably not affect the signal waveform. Therefore, the guidelines for improving reliability that are presented in this chapter deal with interventions in the digital domain; in other words, with the manipulation of DSSS chips, bits and symbols. In particular, this chapter highlights the IWSN transmission errors from two different perspectives. First, the statistical behaviour of hardware-based channel quality metrics from a series of measurements in industrial environments is discussed, after which the error footprints on the bit and symbol levels are illustrated. Based on these error patterns, several concepts for improving the communication reliability are outlined.

A common measurement objective in empirical IWSN studies is the effect of the environment on the quality of service parameters, such as the number of failed pollings, alarm latency, cycle time and round-trip time [14]. However, it is important to note that such packet-level tests are useful only for evaluation of the proposed solutions; they are of little value as an input to the design of packet recovery schemes, since it is impossible to extrapolate the results of packet loss measurements in a particular experiment to general conclusions. In other words, packet delivery ratio measurements should be the last stage in the iterative design process (that is, the evaluation stage), rather than the first step (that is, the input to the design process). Conversely, the result space in terms of bit- and symbol-level errors is reasonably small, and observations of error patterns are more easily generalisable, meaning that packet recovery algorithms should be designed based on the error patterns observed inside corrupted packets. This is why packet loss rates in this chapter are discussed only with respect to performance evaluation of the proposed solutions, rather than in the context generic propagation effects.

19.3 The Statistics of Link-quality Metrics for Poor Links

The IEEE 802.15.4-2006 standard stipulates that all compliant devices must provide two hardware-based channel quality metrics for every received packet: the received signal strength indicator (RSSI) and link quality indicator (LQI). The idea with this dichotomy is to obtain two complementary hardware quality metrics, whose synergy would give a comprehensive idea of channel quality. The two indicators are complementary in the sense that RSSI should quantify received signal power, whereas LQI should be a measure of signal purity; in other words, the amount of distortion inflicted by the channel.

The RSSI and LQI are the only quantities directly observable from IEEE 802.15.4-2006 receivers, with an additional advantage that they are available on a per-packet basis, unlike the packet delivery ratio, which is calculated by averaging over a number of transmissions and provides a rough idea about the cause of packet loss. Furthermore, RSSI and LQI require no additional computations nor are they dependent on time synchronization. These properties make them particularly attractive for applications requiring fast estimation of channel quality.

Assessing the trustworthiness of the two hardware-based metrics under bad channel conditions is of high interest for IWSNs, because their purpose is to accurately quantify the channel state and, as such, they are used in a deluge of WSN functionalities, ranging from channel access and localization to routing and transmit power control. The spotlight of this chapter is on the channel influence on the values of the two indicators and whether or not they are sufficiently trustworthy to provide a fast and precise insight into channel quality.

19.3.1 The Received Signal Strength Indicator

The RSSI is defined as the crude power of the received signal averaged over a number of symbols of the incoming packet. The main drawback of RSSI is the fact that the RSSI measurement module in commercial off-the-shelf (COTS) transceivers does not distinguish between useful signal and interference. Hence the RSSI value captures the total energy present on the channel at the sampling instant. This is potentially a problem, since many RSSI-based WSN protocols assume that channel quality is proportional to RSSI. Consequently, an RSSI measurement on a heavily interfered link can return a very high RSSI value, which is frequently overlooked in the literature on channel quality indicators. On the other hand, in the absence of interference, the RSSI can indeed accurately characterise good channels, provided that the link is not in the 'gray zone': the state of instability.

19.3.2 The Link Quality Indicator

The IEEE 802.15.4-2006 standard does not prescribe the calculation method for LQI, and leaves the concrete implementation in the hands of device manufacturers. The standard even allows the use of RSSI in LQI calculations, which seemingly breaches the idea of their mutual complementarity. One of the most widespread WSN transceiver chips, the Texas Instruments (previously Chipcon) CC2420 [15], as well as a number of newer transceiver chipsets, calculates the LQI in the form of CORR: the amount of correlation between a received chip sequence and its estimate. The CC2420 datasheet suggests expressing the LQI as $LQI = (CORR - a)*b$, where a and b are to be empirically derived, based on the reliability requirements of the user. The CORR calculation method in CC2420 is as follows:

1. Before the 250-kbps bit stream is OQPSK-modulated, each 4 bits are grouped into a symbol and mapped to one of 16 possible 32-chip (DSSS) sequences.
2. At the receiver side, each received symbol is correlated with all 16 possible chip sequences and the closest match is chosen for decoding.
3. The CORR is then expressed as the chip error rate calculated over the first 8 symbols after the preamble and start of frame delimiter (SFD), with respect to the closest match estimate.[endnl]

CORR is a unitless value in the interval [30,108], where the value 108 corresponds to the best measurable channel quality. In the remainder of this text, the CORR will be referred to as LQI.

19.3.3 The Ambiguity of RSSI and LQI Readings

A measurement study conducted at several locations in two paper mills [16] analysed the behaviour of RSSI and LQI on poor IEEE 802.15.4-2006 links. The behaviour of the two indicators on a link exposed to strong WLAN interference (Figure 19.4) reveals that RSSI values from unacceptably many corrupted packets (dark dots) have the RSSI values typical for correctly received packets (light dots), indicating no interference. The

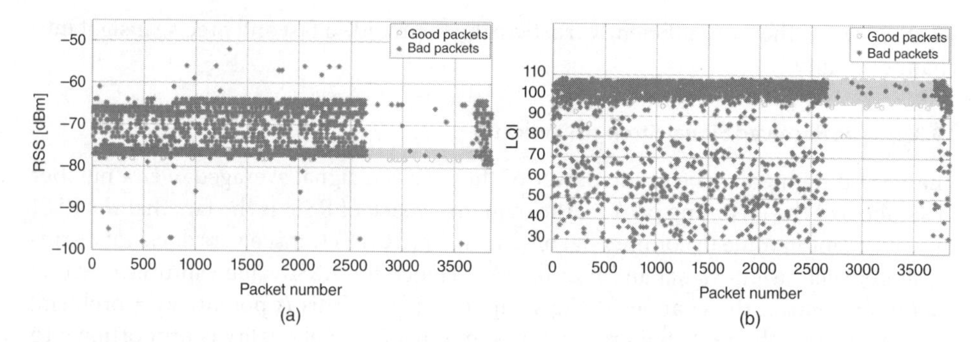

Figure 19.4 The misleading behaviour of quality metrics on a link exposed to WLAN interference: (a) RSSI; (b) LQI [16].

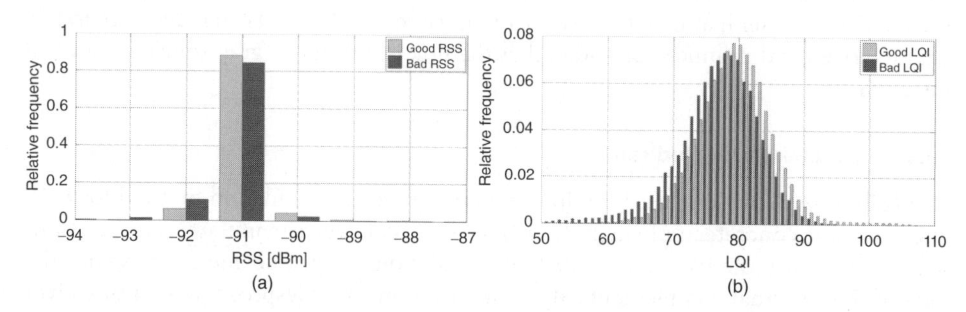

Figure 19.5 The overlap of RSSI/LQI probability density functions derived from correct and corrupted packets on a link exposed to MFA in a paper mill: (a) RSSI; (b) LQI [16].

situation is even more drastic in the case of LQI, where the majority of corrupted packets attain extremely high LQI values, which are typical of excellent channels. The conclusion is that the two indicators cannot be used in their original form for rapid channel quality assessment, which was in fact supposed to be their most plausible property. On the other hand, Figure 19.5 shows the RSSI/LQI distributions from a link exposed to multipath fading and attenuation (MFA). It is noticeable that the distributions of RSSI/LQI for corrupted and correctly received packets almost completely overlap, instead of being distinctively different, which is a clear sign of ambiguity.

What are the underlying reasons for this degree of underachievement? As mentioned earlier, the presence of an interfering signal can lead to RSSI boost, but this does not necessarily have to be the case. For instance, a weak interfering signal can corrupt an IEEE 802.15.4 packet, causing an RSSI increase of 1–2 dB, which will not necessarily be indicative of interference, because standard deviation of RSSI is about 2–3 dB, even on stable links. Meanwhile, LQI is, in most platforms, calculated based on the chip error rate. The main shortcoming of LQI in that respect is that the chip error rate is calculated with respect to the estimated chip sequence (that is, the closest match), and not necessarily with respect to the actually transmitted (true) sequence. The critical issue is that the channel can alter the transmitted chip sequence beyond recognition, resulting in the Hamming distance between the received and true sequences not being the minimum one. In that case, a heavily corrupted symbol may be very close (in the Hamming sense) to another valid sequence, resulting in an extremely high LQI. This opens the door to several problems, because a device is supposed to gain channel quality awareness through a channel quality metric the trustworthiness of which is, in turn, compromised by these same channel conditions.

Perhaps the most prominent shortcoming of RSSI/LQI calculation, at least when it comes to the CC24xx transceivers and a number of other popular chipsets, is the fact that the two are calculated over only eight symbols of the packet following the SFD. Such a short sampling time is far from sufficiently informative, considering that the duration of WLAN packets is significantly smaller than that of their WSN counterparts. For instance, the maximum transmission time of an IEEE 802.15.4-2006 packet at 2.4 GHz is 4.256 ms, compared to 1.906 ms and 0.542 ms in the IEEE 802.11-b and -g standards, respectively. Bearing in mind that CC2420 is by far the most widespread WSN transceiver, represented in the popular WSN platforms such as MicaZ, TelosB and TMote Sky, it is clear that the problem of short RSSI/LQI sampling periods has a significant impact on the performance of commercial WSN hardware. Furthermore, the problem of insufficiently long RSSI/LQI calculation periods persists even in the successors of CC2420, such as CC243x and CC253x series. One exception is the EM2xx and EM3xx platforms by Silicon Labs, which calculate the RSSI on the first eight symbols, but LQI is the chip error rate calculated over the entire packet.

There is a degree of awareness about RSSI/LQI deficiencies in the WSN community and comparative studies of RSSI accuracy of different platforms have been undertaken since as early as 2006 [17]. Conversely, the solutions proposed by authors that are unaware of this problem are mostly simulator-based and do not take into account the existence of interference. A commonly suggested workaround for the problem of RSSI/LQI trustworthiness is collecting several dozen [18] or even hundreds [19] of RSSI/LQI readings and filtering the values prior to application in higher-layer protocols. Unfortunately, this is not in line with the expectations of industrial wireless, where extremely short reaction times are imperative.

Extending the calculation period of RSSI/LQI from eight symbols to the entire packet would enable the receivers to grasp more channel-state information. This would, however, not entirely solve the previously mentioned problem of LQI derivation, because a corrupted packet is not a valid reference for calculating the chip error rate. On the other hand, a priori known bytes in a packet could be a valid reference for channel diagnostics. Intuitively speaking, channel scanning outside of packet reception times would also contribute to a clearer picture about the channel state, but a difficulty in that respect is determining when the scanning should take place, considering that all IWSN standards employ TDMA. The current channel used by a network can only be scanned during inter-packet intervals, which is only a small fraction of the time (a few milliseconds in a 10-ms timeslot). A potential solution is that devices could, from time to time, breach the channel-hopping pattern and listen to the channels that are not used by the network in the current timeslot; it should be noted, however, that in the case of an incident, the devices performing out-of-band listening would not be able to participate in the delivery of emergency traffic.

19.4 The Statistical Properties of Bit- and Symbol-Errors

The statistical properties of communication errors should, by all means, be considered in communication protocol design. In the ideal case, the analysis of error properties would reveal the concrete modeling decisions that can result in better communication reliability. In that respect, bit- and symbol-level errors offer more subtle channel-state information than packet loss and delay, which are the most commonly observed parameters in related studies [14].

Barac et al. [20] analysed IEEE 802.15.4-2006 bit- and symbol-errors in industrial settings. The statistical properties relevant for modeling decisions were derived from error patterns collected during 14 days of measurements at three industrial environments. A clear distinction between errors caused by MFA and errors inflicted by WLAN interference was observed, where the root cause was the difference in generic error patterns. The authors asserted that bit errors (and hence symbol errors) are sparse and randomly placed inside the packets corrupted by MFA, while in the case of WLAN corruption, they are more densely packed, interrupted by occasional appearances of correct bits. Typical bit-error patterns for the two cases are shown in Figure 19.6.

The error traces collected in the Barac study were analysed and used to evaluate the performance of Reed–Solomon (RS) channel code, in conjunction with several types of interleaving. The measurable of interest is the packet salvation ratio (PSR), which is the fraction of corrupted packets that the coding scheme was able to correct. It was found that the generic differences between the two patterns have a number of practically relevant implications:

Short bit-error bursts: Signal delay spreads in industrial environments (×100 ns) are of the order of one OQPSK symbol duration (1 ms in 2.4-GHz IEEE 802.15.4-2006) and consecutive errors are a common sight on poor IWSN links. As shown in Figure 19.7, the 95th quantile of bit-error burst length was 3 and 4 bits, for MFA and WLAN interference, respectively. Therefore, bearing in mind that symbol size in forward error correction (FEC) is proportional to decoding time, the coded symbol size of 4 bits is a compromise between performance and complexity.

Optimal interleaving method: Contrary to common belief, randomising symbol shuffling by sophisticated interleaving techniques is not always justified. In particular, simple matrix symbol interleaving outperformed helical, block and random interleaving in every experiment of the measurement campaign [20]. The underlying reason is

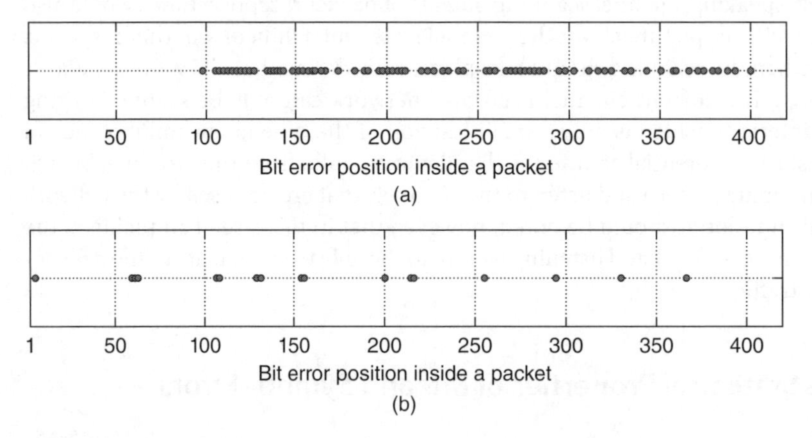

Figure 19.6 (a) The compact error footprint of WLAN, (b) The sparse error footprint of MFA [20]. The dots denote the positions of corrupt bits.

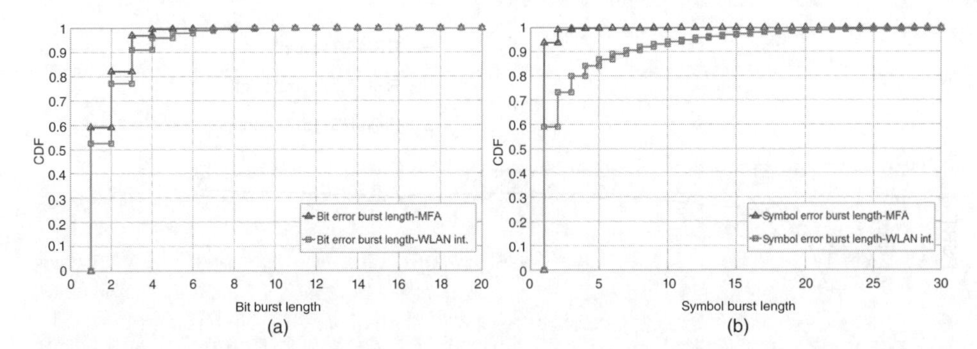

Figure 19.7 The CDF of bit- and symbol-error burst lengths from a 14-day measurement campaign at three industrial environments: (a) MFA-inflicted errors; (b) WLAN-inflicted errors [20].

that, in case of interference (that is, dense errors), matrix interleaving equalizes the error density throughout the packet and improves the ability of FEC code to cope with them. Meanwhile, the interleaving has no effect in case of sparse errors (such as MFA), because shuffling of sparse errors has little impact on error density (see Figure 19.8a).

Symbol vs bit-interleaving: Symbol-interleaving is more efficient than bit-interleaving because it distributes errors over fewer codewords. In other words, one erroneous m-bit coded symbol (with $\leq m$ corrupted bits) subject to symbol-interleaving will 'pollute' only one codeword, while bit-deinterleaving of such a symbol can 'pollute' up to m different codewords. The supremacy of symbol- over bit-interleaving is verified in Figure 19.8 for both types of scenario.

Optimal interleaving depth: The optimal interleaving depth under interference equals the codeword length because this depth achieves the biggest separation of symbols that were adjacent in the air. This can be observed in Figure 19.8b, where the peak of PSR for RS(15,7) code is at the 15-symbol (that is, 60-bit) interleaving depth.

These observations can be applied in packet-recovery schemes, as well as channel-diagnostic algorithms. The remainder of this chapter is dedicated to a number of solutions that are based on these observations.

19.5 Guidelines for Countermeasures

The high demands of safety-critical applications dictate the following dichotomy in IWSN communication design: a proactive component of the protocol suite should ensure as little packet loss as possible, while the reactive one should quickly recognise the cause of any packet loss and take appropriate action. The three industrial WSN standards include a number of countermeasures to packet loss, which are inherently reactive and unfortunately not particularly lightweight. For instance, the WirelessHART standard employs pseudorandom channel hopping and blacklisting in order to avoid interfered channels, while redundant path routing is expected to address the problem of link blockage. Unfortunately, the idea of channel diagnostics for selecting an adequate countermeasure has not been developed and there is no significant cooperation between different layers for this purpose. On the other hand, it is obvious from the error-pattern studies discussed in Section 19.4 that efforts towards

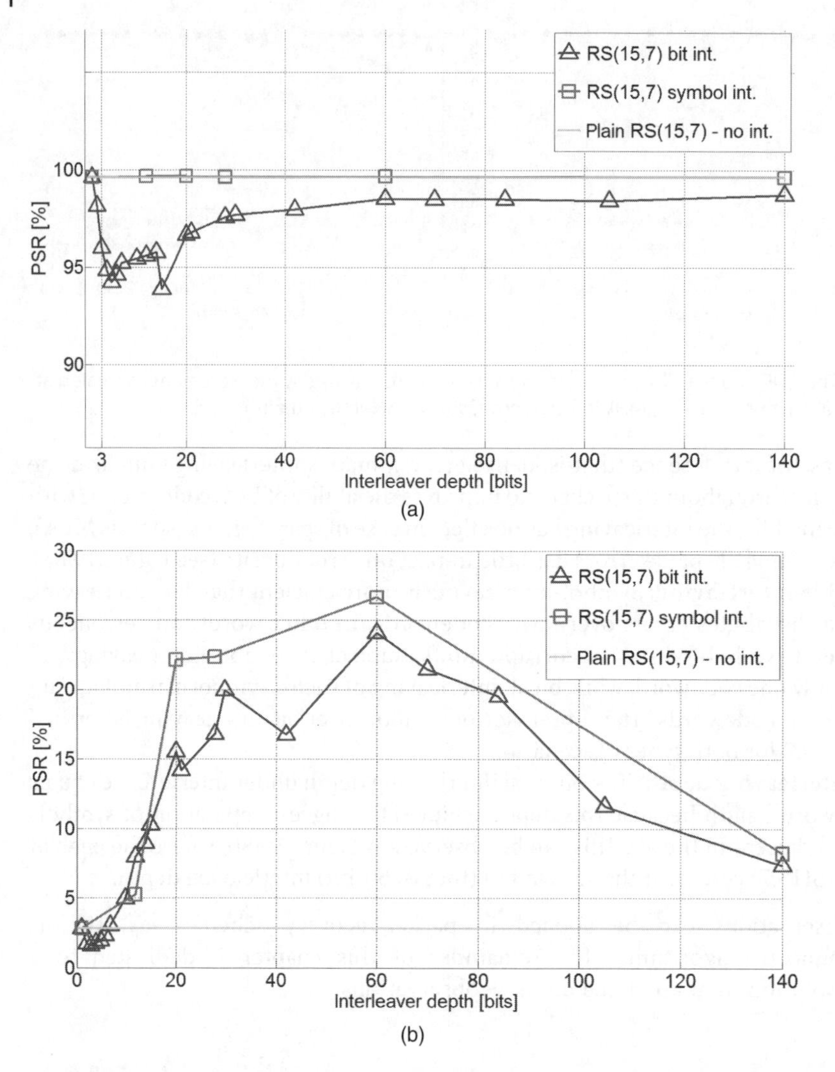

Figure 19.8 The performance of bit- and symbol-interleaved and non-interleaved RS(15,7) code on error traces: (a) MFA; (b) WLAN [20].

improving IWSN communication reliability should use contributions from different layers in the protocol stack. Bearing in mind the safety-critical nature of the target applications, the countermeasures must be brisk and closely fitted to the cause of packet corruption/loss. In particular, a desirable reliability framework should have the following properties:

- A PHY layer should act preemptively against packet corruption, preferably already on the DSSS chip level. This has two benefits: better reliability and facilitation of channel diagnostics.
- The data link layer should perform corrupted-packet forensics and diagnose the channel state (that is, the presence of MFA or interference). The channel diagnosis should be established even if no packet has arrived at the receiver.

- Based on the channel diagnosis, higher layers should reroute the traffic (if errors are caused by MFA) or adapt the channel hopping pattern (if interference is present).

As mentioned earlier, signal waveforms are strictly defined by the standard and they are not to be altered if compliance with the standards is to be maintained. Therefore, the design space for improving communication reliability boils down to DSSS chip- and bit-level manipulations. The following subsections adhere to this line of reasoning and explain several relevant techniques in more detail.

19.5.1 Forward Error Correction and Interleaving

Techniques such as retransmissions and partial packet recovery are inadequate for mission-critical WSN applications, where transmission of a new sensor reading is more sensible than retransmission of an old one. Instead, proactive approaches to error mitigation are desired, and the concept of FEC is in line with this philosophy. The adequacy of FEC has been widely recognised in the WSN community and its retrofitting into existing WSN standards has often been proposed. Moreover, even though the energy consumption in IWSNs is less important than reliability, FEC is, in fact, also beneficial from the energy-efficiency perspective, because transceiver modules are by far the largest energy consumers in sensor nodes [21]. Previous research has shown that it takes 2700 times more energy to send one bit than to execute a microcontroller instruction [22], meaning that energy investment in computation is reasonable if it means avoiding retransmissions.

Unfortunately, the restrictions imposed by the low computational power of COTS devices limit the number of utilisable coding schemes in WSNs. In particular, if the IEEE 802.15.4-2006 standard is considered, the decoding time constraint equals the *macAck-WaitDuration*, which is 0.864 ms. A general consensus in the WSN community is that block codes, such as RS, are a compromise between complexity and performance. Implementations on COTS devices show that these codes require decoding times below 1 ms for certain block lengths [23]. The capacity-approaching channel codes, such as turbo and LDPC codes, cannot be used in the industrial domain because of the long execution times involved, which breach the timing requirements of IWSNs. A reasonable solution in anticipation of more powerful low-cost WSN hardware is to increase the reliability by applying simple and computationally inexpensive tweaks to FEC. Three approaches that exploit the error statistics and the determinism in industrial wireless for boosting communication reliability are discussed below.

19.5.2 DSSS Chip-level Manipulations

Waveform deviations in IEEE 802.15.4-2006 translate into DSSS chip-level errors. Consequently, if a 32-chip sequence is altered beyond recognition, it will be mapped onto a wrong four-bit symbol. The idea of counteracting channel errors at the DSSS chip-level is a promising tool for better communication reliability in IWSNs. Unfortunately, this idea has received next to no attention in the WSN community, for two main reasons. First, none of the COTS IEEE 802.15.4-2006 platforms allow tampering with the PHY layer, which is why the possibility of using software-defined radio platforms, such as USRP [24], is the de facto prerequisite for any DSSS chip-level experimentation. Second, bearing in mind that every IEEE 802.15.4-2006 bit corresponds to eight chips,

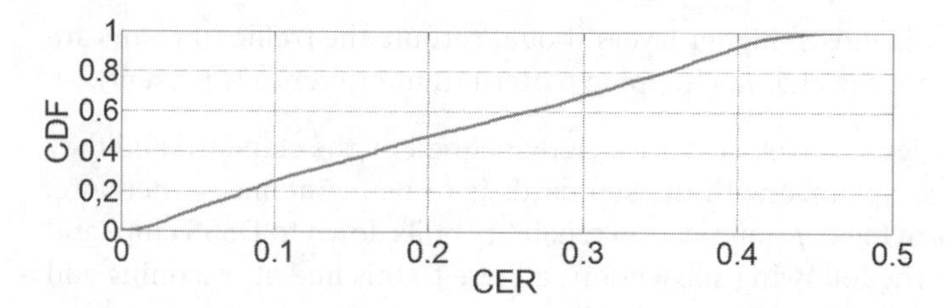

Figure 19.9 The CDF of chip error rate on WLAN-interfered links in an industrial workshop [25].

there is a concern that the induced computational load will be overwhelming for COTS devices. On the other hand, it is intuitively plausible that the error avalanche should be confronted as close as possible to the waveform–chip interface. Typical chip error rates (CER) in packets corrupted by WLAN interference can be as high as 50%, which is exemplified by the CDF of the CER acquired on an IEEE 802.15.4-2006 link exposed to WLAN interference in an industrial shop, shown in Figure 19.9 [25].

One of the loose ends of IEEE 802.15.4-2006 PHY with significant room for improvement is the design of chip sequences, which is currently far from optimal. Instead of being orthogonal, sequences have a high degree of similarity: each sequence can generate seven other sequences by mere circular shifts, which implies a high probability of wrong decoding. The motivation for such a design decision is likely to be the urge to facilitate the hardware implementation.

The problem of chip sequence design is addressed by the CLAP protocol [25], which calls for a redefinition of IEEE 802.15.4-2006 PHY by applying rudimentary manipulations on DSSS chips. CLAP introduces a redefined set of IEEE 802.15.4-2006 chip sequences, shown in Table 19.2. Each 32-chip sequence proposed by CLAP is a 7-repetition code representation of the corresponding 4-bit symbol, terminated by a 4-chip delimiter. The resulting sequences are more robust to corruption than the ones stipulated by the standard. In the transmitter chain, the sequences are subject to matrix chip interleaving, after which conventional DLL FEC can be applied. Figure 19.10 shows the performance of CLAP, which, on top of chip manipulations, uses conventional RS(15,7) code on the DLL. According to the evaluation on real error traces from an industrial workshop, the error-correction capability of packets corrupted by WLAN

Table 19.2 The redefined IEEE 802.15.4-2006 symbol-to-chip mapping [25].

4-bit symbol	IEEE 802.15.4 chip sequence	New chip sequence	4-bit symbol	IEEE 802.15.4 chip sequence	New chip sequence
0x0	0x744AC39B	0x0000000F	0x8	0xDEE06931	0x88888887
0x1	0x44AC39B7	0x1111111E	0x9	0xEE06931D	0x99999996
0x2	0x4AC39B74	0x2222222D	0xA	0xE06931DE	0xAAAAAAA5
0x3	0xAC39B744	0x3333333C	0xB	0x06931DEE	0xBBBBBBB4
0x4	0xC39B744A	0x4444444B	0xC	0x6931DEE0	0xCCCCCCC3
0x5	0x39B744AC	0x5555555A	0xD	0x931DEE06	0xDDDDDDD2
0x6	0x9B744AC3	0x66666669	0xE	0x31DEE069	0xEEEEEEE1
0x7	0xB744AC39	0x77777778	0xF	0x1DEE0693	0xFFFFFFF0

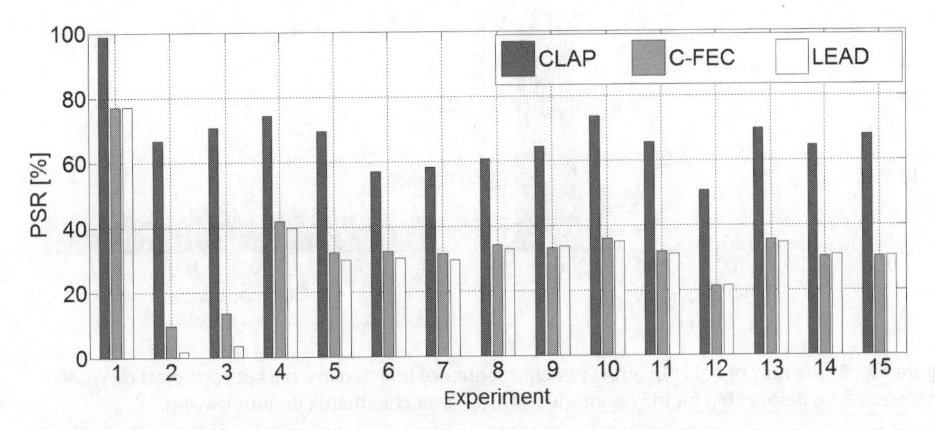

Figure 19.10 Comparison of CLAP, conventional DLL FEC (C-FEC) and a state-of the-art approach (LEAD) in terms of PSR.

interference is boosted by 78–588% with respect to conventional DLL RS(15,7) and another state-of-the-art packet recovery scheme, LEAD [26]. The large gains are a result of counteracting the errors at an early stage; that is, at the chip level.

To evaluate the individual contributions of three CLAP components – chip sequence redefinition, chip interleaving and conventional FEC – their individual contributions were normalised by the correction capability of the fully fledged solution. According to the results shown in Figure 19.11, mere matrix chip interleaving (IM), without any additional manipulations, corrects almost as many corrupted packets as conventional RS(15,7) DLL FEC (C-FEC). Meanwhile, the redefined sequence set combined with matrix chip interleaving (IM + SM) outperformed the DLL RS(15,7) code in each experiment. This is a surprising and encouraging result, having in mind the high complexity of RS(15,7), compared with simple matrix interleaving. The performance gain induced by CLAP outweighs the additional computational overhead, which is found to be a mere 10% of the execution time of OQPSK demodulator, a mandatory module in every IEEE 802.15.4-2006 receiver.

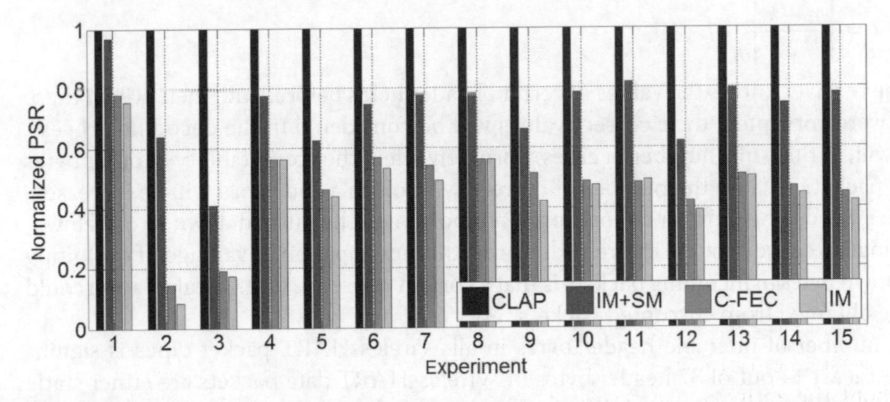

Figure 19.11 The contributions of different CLAP components, relative to the fully fledged solution [25].

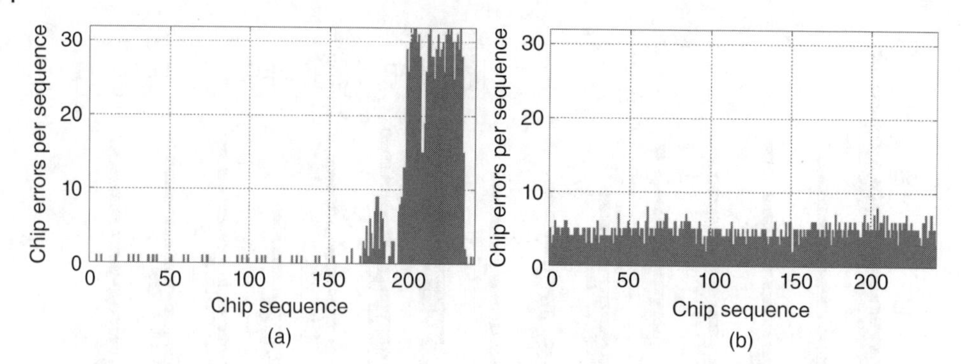

Figure 19.12 The number of chip errors per chip sequence inflicted in a packet corrupted by WLAN interference: (a) before chip matrix de-interleaving; (b) after chip matrix de-interleaving.

The high contribution of matrix chip interleaving can be understood by taking a microscopic look at the corrupted packets. Figure 19.12 shows the influence of interleaving on the number of corrupted chips per 32-chip sequence, where IEEE 802.15.4-2006 sequences can withstand 5–10 chip errors. This particular packet was hit by WLAN inteference at its rear end, which produced high error rates in a number of sequences. Figure 19.12b shows that matrix chip deinterleaving distributes the error load throughout the packet, eventually recovering the packet.

19.5.3 Exploiting Determinism in Industrial Wireless

Another way to boost the performance of FEC is to leverage the traffic properties of the application in question. Industrial wireless communication is inherently deterministic, with strictly predetermined transmission schedules, graph routing and predominantly TDMA channel access. Consequently, a significant fraction of header byte values in transmitted packets is a-priori inferable by the receiver. This feature can be a catalyst of FEC in the following way. Channel codes have a certain error tolerance, expressed as the number of corrupted symbols per codeword that the code is able to correct. The RS code with k information symbols in an n-symbol codeword can correct at most t corrupted m-bit symbols, where t depends on n and k as follows:

$$t = \left\lfloor \frac{n-k}{2} \right\rfloor$$

If the receiver knows the values of certain header fields beforehand, then even if these fields were corrupted, their correct values will be considered in the decoding process. This, will, in turn, in a number of cases, positively affect the correctability of other bytes in the packet, because the number of correct symbols in a codeword will be increased. In other words, the number of corrupted symbols might be pushed down to or below t, resulting in recovery of the entire codeword. Another apparent advantage of exploiting the known bytes in incoming packets is that a correct reference for LQI calculation could also be obtained from corrupted packets.

The number of inferable header bytes in all WirelessHART packet types is significant. At least 14 out of 37 header bytes in WirelessHART data packets are either static throughout the entire network lifetime or they are predictable by the receiving peer at almost any given point in time. For instance, since medium access in IWSNs is a

Table 19.3 The number of inferable bytes in WirelessHART packet types.

Packet type	Number of inferable bytes	RSM (%)
Data	14 out of 37 header bytes	11–35
Acknowledgment	9 out of 19	47
Advertisement	14 out of 26	54
Keepalive	9 out of 16	56
Disconnect	7 out of 10	70

hybrid of TDMA and contention-based access(where TDMA occupies most of every superframe), the DLL source and destination address fields in the TDMA part of a superframe are known in advance, because both the transmitter and receiver(s) are predetermined in every timeslot. Another example is the *absolute slot number* (ASN) field, which all the nodes participating in the network must know at any time. Table 19.3 shows the percentage of inferable bytes for all types of WirelessHART packets, with respect to the entire packet length, termed the relative size of the mask (RSM). While the percentage of inferable bytes in data packets is 11–35%, for WirelessHART control packets this percentage ranges from 47–70%.

PREED [27], a solution based on the concept of IWSN determinism, has been compared to conventional bit- and symbol-interleaved DLL RS(15,7) code, as well as the packet recovery scheme LEAD [26]. LEAD was originally proposed for IEEE 802.11 networks, and it has the same essential idea as PREED, but the interleaving applied in LEAD is bit-level, which is inferior to symbol-interleaving, as discussed earlier. The comparison of PREED to the other approaches was done using real error traces from industrial environments. According to the results shown in Figure 19.13, PREED improves packet correctability by 42–134% over to the second best approach, which is DLL symbol-interleaved RS(15,7) in most of the cases.

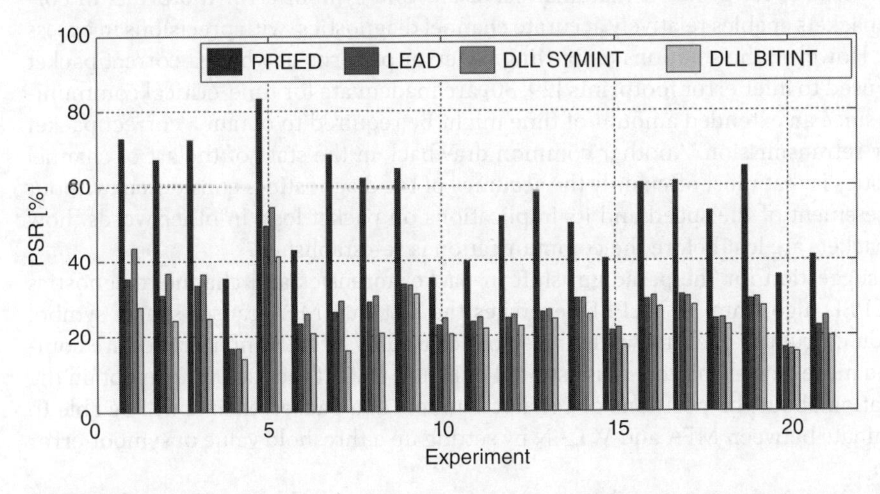

Figure 19.13 The boost in packet correctability induced by PREED on links exposed to strong WLAN interference [27].

The PREED concept described above can be used during most of the network lifetime. The exceptions are the join phase, the contention-based part of the superframe, and the interrupts due to an update of a header field value that the receiver would not be able to predict. These situations can be accounted for by introducing a dedicated flag (for example, the unused bit in the *frame length* byte) and a mechanism for updating the header byte knowledge.

19.5.4 Channel Diagnostics and Radio Resource Management

For a network to properly react to packet loss and communication outage, it is essential to determine the cause of packet loss because there is no single remedy for both MFA- and interference-related losses. For instance, packet loss caused by link blockage and shadowing should be tackled by rerouting the traffic by activating redundant paths, a mandatory feature in IWSN standards. However, rerouting will not help in case of interference, where the only appropriate measure is to blacklist and avoid the interfered channel(s). WirelessHART suffers from several problems in the domain of radio resource management [28]. First, the standard lacks a framework for channel diagnostics. Second, the channel blacklisting is a manual, rather than an automatic operation. Third, the procedure of reporting bad channels by individual nodes is undefined, so it is unclear how the network manager should maintain the blacklist. Finally, adding a new channel to the blacklist causes perturbations in pseudorandom hopping calculations and communicating this change throughout an operating network is much more challenging than distributing, for example, a routing table update. The ISA100.11a and WIA-PA standards allow for use of the same channel over several timeslots, meaning that these challenges could be addressed more easily. This chapter deals with the first of the issues, by discussing one possible approach to channel diagnostics.

As discussed earlier, RSSI and LQI should not be used as the primary tools for channel diagnostics, due to their apparent ambiguity [16]; other means of learning the channel conditions must be considered. In that sense, the generic differences between the error footprints of MFA and interference can be used for diagnosing the causes of packet loss. Previous studies have shown that analysis of bit- and symbol-error footprints in corrupted packets enables relatively accurate channel diagnostics, with precisions in excess of 90%. However, the solutions employing bit-error patterns in which a correct packet copy is used to infer error footprints [29, 30] are inadequate for time-critical communication, since an extended amount of time might be required to obtain a correct packet copy by retransmission. Another common drawback in the state of the art of channel diagnostics is that most often only the accuracy of the diagnostic is considered, without any assessment of the speed and its implications on packet loss; in other words, how many packets are lost before the communication is re-established.

One suggestion for the paradigm shift in packet forensics and channel diagnostics is the LPED algorithm [31]. LPED leverages the fact that MFA causes sparse symbol errors on corrupted packets, while the errors caused by WLAN interference are compact and more densely placed, as illustrated in Figure 19.6. If the receiver can obtain the information about the positions of corrupt symbols in a packet, then it will be able to discriminate between MFA and WLAN by setting up a threshold value of symbol-error density.

The key issue in this context is how a receiver can determine the error density but at the same time avoid the time-consuming retransmissions needed to obtain a correct packet copy. To calculate the error density, LPED employs the RS channel code in a novel way: in addition to error correction, the channel code is used for identifying erroneous symbol positions inside corrupted packets (which is one of the stages in conventional decoding), making it possible to infer error patterns and the causes of packet loss without obtaining a correct packet. After estimating the error density, a threshold is used to discriminate between MFA and WLAN interference. It should be noted that a channel-diagnostics algorithm must be able to set up a channel diagnosis even in the absence of packet; in other words, when the packet was not even acquired by the receiver.

Figure 19.14 shows the results of live tests at a mineral processing facility, where communication links were exposed to severe WLAN interference or MFA, and the number of packets lost before communication was re-established was measured. A simple RSSI sampling algorithm was added on top of LPED, to account for cases when the packet preamble was so corrupted that no packet was received at all. To avoid the pitfalls of network-wide packet-loss averaging, all presented results refer to individual links. According to the results, in the worst of the observed scenarios, 2.3 packets were lost on average until LPED managed to re-establish the communication by rerouting (when MFA was diagnosed) or channel hopping (WLAN interference diagnosed). Note that the lost packet count requires the loss of at least one packet. Considering that very small deviations in results are observed, the performance is in line with the common practice in control systems, where the loss of up to two out of three consecutive packets is tolerated before the fail-safe mode is invoked [6].

The issue of simultaneous MFA and interference is a legitimate one, although such occurrences are relatively infrequent (once in 10 000 cases in one study [31]). Since WLAN-inflicted symbol errors are more densely placed than those caused by MFA, it

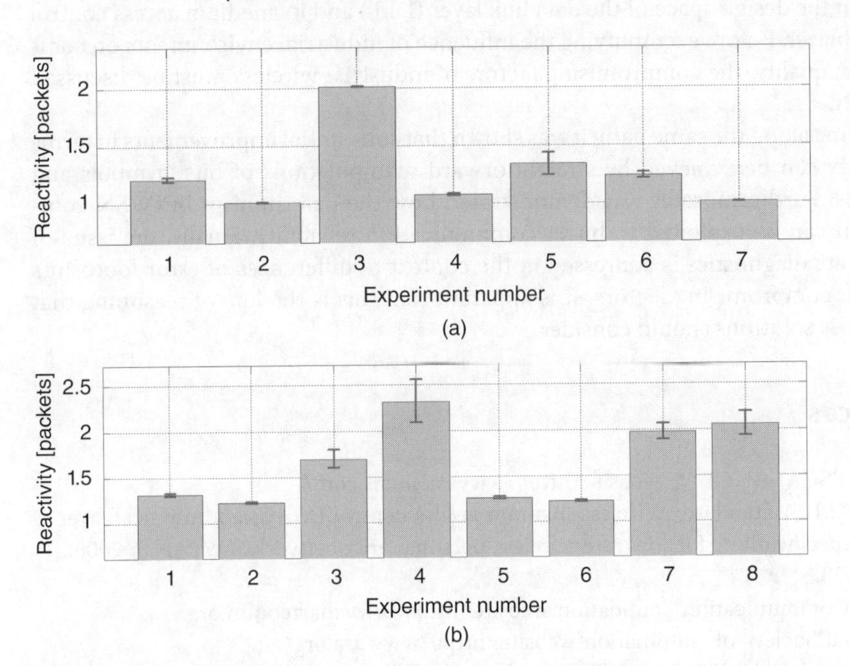

Figure 19.14 The mean number of packets lost before communication is re-established on a link under: (a) MFA; (b) WLAN interference [31].

follows that superposition of the two error patterns on a single packet would most likely result in a WLAN diagnosis. Furthermore, it should not be forgotten than in this situation both rerouting and channel hopping are necessary, hence the two factors (MFA or WLAN) will be diagnosed one-by-one and both addressed in a sequential manner. If the diagnostics algorithm includes a 'no decision' diagnosis, the maximum allowed number of such consecutive decisions should be limited to two or three, after which a countermeasure should be randomly selected.

19.6 Summary

The current IWSN standards cannot meet the high reliability expectations of underlying applications, due to their inability to cope with the harsh propagation conditions in industrial environments. The first step in the design of lightweight and resilient solutions is a thorough study of the nature of errors. In that context, this chapter takes a close look at the error footprints at bit-, symbol- and DSSS chip-level of IWSN signals. A distinction is made between error footprints inflicted by WLAN interference and those caused by MFA.

After a brief overview of the compromising factors in industrial environments, this chapter discusses the bit- and symbol-level implications on IWSN signals, after which a methodology for bringing the performance closer to those demanded is illustrated in several examples. Because signal waveforms and other PHY layer properties are strictly defined by the WSN communication standards, the solutions for higher reliability are to be found in the design space of the data link layer (DLL) and its medium access control (MAC) sublayer. Before exemplifying the influence of industrial environments on communication quality, the compromising factors of industrial wireless must be discussed individually.

Continuing along the same path, it was shown that substantial improvements in terms of reliability can be achieved by straightforward manipulations of bits, symbols and DSSS chips. Furthermore, it was demonstrated how the determinism in IWSN communication can be exploited to boost communication reliability. Finally, the issue of channel-state diagnostics is addressed in the context of differences of error footprints of different compromising factors. It is apparent that this is the line of reasoning that future IWSN solutions should consider.

References

1 Siemens SCALANCE W website: http://www.siemens.com/
2 IEEE 802.15.4 Standard: Wireless medium access control (MAC) and physical layer (PHY) specifications for low-rate wireless personal area networks (WPANs), 2006, pp. 1–323.
3 HART Communication Foundation website: http://www.hartcomm.org/
4 Industrial Society of Automation website: http://www.isa.org/
5 Chinese Industrial Wireless Alliance website. http://www.industrialwireless.cn/

6 J. Åkerberg, M. Gidlund, T. Lennvall, K. Landernäs, M. Björkman, Design challenges and objectives in industrial wireless sensor networks, in *Industrial Wireless Sensor Networks: Applications, Protocols and Standards*. CRC, 2013, pp. 79–97.

7 J. Ferrer Coll, P. Ängskog, J. Chilo, P. Stenumgaard, 'Characterisation of highly absorbent and highly reflective radio wave propagation environments in industrial applications,' *in Communications, IET*, vol. 6, no. 15, pp. 2404–2412, 2012

8 Tanghe, E., Joseph, W., Verloock, L., et al., 'The industrial indoor channel: large-scale and temporal fading at 900, 2400, and 5200 MHz,' in *Wireless Communications, IEEE Transactions on* , vol. 7, no. 7, pp. 2740–2751, 2008.

9 O. Staub, J.-F. Zurcher, P. Morel, A. Croisier, 'Indoor propagation and electromagnetic pollution in an industrial plant,' in *Industrial Electronics, Control and Instrumentation, IECON, 23rd International Conference on*, 1997, pp. 1198–1203.

10 'Coexistence of wireless systems in automation technology.' Zvei Automation whitepaper, April 2009.

11 F. Leferink, F. Silva, J. Catrysse, S. Batterman, V. Beauvois, A. Roc'h, 'Man-made noise in our living environments,' in *Radio Science Bulletin*, no. 334, pp. 49–57, 2010.

12 National Instruments website: http://www.ni.com/

13 J. Ferrer Coll, Channel characterization and wireless communication performance in industrial environments. PhD thesis, KTH Royal Institute of Technology, Stockholm, 2014.

14 M. Bertocco, G. Gamba, A. Sona, S. Vitturi, 'Experimental characterization of wireless sensor networks for industrial applications,' in *Instrumentation and Measurement, IEEE Transactions on*, vol. 57, no. 8, pp. 1537–1546, Aug. 2008

15 CC2420 Texas Instruments Datasheet, 2007.

16 F. Barac, M. Gidlund, T. Zhang, 'Ubiquitous, yet deceptive: hardware-based channel metrics on interfered WSN links,' in *Vehicular Technology, IEEE Transactions on*, vol. 64, no. 5, pp. 1766–1778, 2015.

17 A. Flammini, D. Marioli, G. Mazzoleni, E. Sisinni, A. Taroni, 'Received signal strength characterization for wireless sensor networking,' in *IEEE Instrumentation and Measurement Technology Conference, IMTC. Proceedings of the IEEE*, pp. 207–211, 24–27 April 2006.

18 C.A. Boano, T. Voigt, A. Dunkels *et al.*, 'Poster abstract: Exploiting the LQI variance for rapid channel quality assessment,' in *Proceedings of the International Conference on Information Processing in Sensor Networks IPSN*, 2009, pp. 369–370.

19 K. Srinivasan and P. Levis, 'RSSI is under appreciated,' in *Proceedings of 3rd Workshop EmNets*, 2006, pp. 1–5.

20 F. Barac, M. Gidlund, T. Zhang, 'Scrutinizing bit- and symbol-errors of IEEE 802.15.4 communication in industrial environments,' *in Instrumentation and Measurement, IEEE Transactions on*, vol. 63, no. 7, pp. 1783–1794, 2014.

21 E. Björnemo, 'Energy constrained wireless sensor networks: Communication principles and sensing aspects,' Ph.D. thesis, Uppsala University, Sweden, 2009.

22 J.H. Kleinschmidt and W. da Cunha Borelli, 'Adaptive error control using ARQ and BCH codes in sensor networks using coverage area information,' in *Proceedings of IEEE 20th International Symposium on Personal, Indoor Mobile Radio Communications*, 2009, pp. 1796–1800.

23 M.K. Khan, K. Mulvaney, P. Quinlan, et al., 'On the use of Reed-Solomon codes to extend link margin and communication range in low-power wireless networks,' in *Proceedings of 22nd Irish Signals System Conference (ISSC)*, 2011, pp. 124–130.

24 P. Ferrari, A. Flammini, E. Sisinni, 'New architecture for a wireless smart sensor based on a software-defined radio,' in *Instrumentation and Measurement, IEEE Transactions on*, vol. 60, no. 6, pp. 2133–2141, 2011.

25 F. Barac, M. Gidlund, T. Zhang, 'CLAP: Chip-level augmentation of IEEE 802.15.4 PHY for error-intolerant WSN communication,' in *Vehicular Technology Conference (VTC Spring), IEEE 81st*, 2015, pp. 1–7.

26 J. Huang, Y. Wang, G. Xing, 'LEAD: leveraging protocol signatures for improving wireless link performance,' in *MobiSys, Proceedings of ACM*, 2013, pp. 333–346.

27 F. Barac, M. Gidlund, T. Zhang, 'PREED: Packet recovery by exploiting the determinism in industrial WSN communication,' in *Distributed Computing in Sensor Systems (DCOSS), IEEE International Conference on*, 2015, pp. 81–90.

28 D. Chen, M. Nixon, A. Mok, *WirelessHART: Real-Time Mesh Network for Industrial Automation*. Springer, 2010.

29 F. Hermans, O. Rensfelt, T. Voigt, E. Ngai, L.-Å. Norden, and P. Gunningberg, 'SoNIC: Classifying interference in 802.15.4 sensor networks,' in *Proceedings of 12th International Conference on Information Processing in Sensor Networks (IPSN)*, 2013, pp. 55–66.

30 T. Huang, H. Chen, Z. Zhang, and L. Cui, 'EasiPLED: Discriminating the causes of packet losses and errors in indoor WSNs,' in *Proceedings of IEEE Global Communication Conference (GLOBECOM)*, 2012, pp. 487–493.

31 F. Barac, S. Caiola, M. Gidlund, E. Sisinni, T. Zhang, 'Channel diagnostics for wireless sensor networks in harsh industrial environments,' *Sensors Journal, IEEE* , vol. 14, no. 11, pp. 3983–3995, 2014.

20

A Medium-access Approach to Wireless Technologies for Reliable Communication in Aircraft

Murat Gürsu[1], Mikhail Vilgelm[1], Eriza Fazli[2] and Wolfgang Kellerer[1]

[1] *Chair of Communication Networks, Technical University of Munich, Germany*
[2] *Zodiac Inflight Innovations, Weßling, Germany*

20.1 Introduction

Communication in the aircraft has always been critical for a safe flight. The sense of flying safely is not intuitive for human beings even though accident reports prove that it is a lot safer than ground vehicle traffic [1]. This misconception about safety forces aircraft manufacturers to make any kind of failure almost impossible.

At the start of the era of manned flight, the control of aircraft was fully mechanical, and any input given by the pilot would be transferred to the related control with an increased force. Analogue electronics then replaced mechanical inputs in the shape of fly-by-wire, first installed a Concorde designed by Aerospatiale [2]. Digital electronics now mean that any kind of input can be converted to a digital message and can trigger the action required anywhere in the system.

Digital control systems brought the first communication systems in aircraft, in which the medium was shared and different messages are interpreted at different applications. This approach increased the functionality of a normal aircraft from the 1980s, one example being the A310. The removal of any need for mechanical inputs opened up the possibility of automated flights, but the communication infrastructure of copper wires became a huge burden for aircraft, so these had to be replaced with aluminum wires in order to make the communication infrastructure weigh less.

As the number of applications in aircraft continues to grow, we will see the problem of weight more and more often. On the other hand, the physical constraints of wires decrease system flexibility, since every new application requires new wire planning. On top of that, the planning time for the wiring infrastructure is another cost for the aircraft. The introduction of a wireless communication system can easily solve this problem. Ultra-reliable wireless for safety applications as part of the coming 5G and industrial automation technologies can become the ultimate communication solution for aircraft. In the future, many types of communication are expected to be wireless, so an investigation of the capabilities of the current candidate technologies is important. In this chapter, we aim to extract important aspects of current technologies and use these to provide a road map for future research on the reliability for wireless communications.

Wireless Sensor Systems for Extreme Environments: Space, Underwater, Underground, and Industrial.
First Edition. Edited by Habib F. Rashvand and Ali Abedi.

The replacement of a widely deployed communication system such as the wiring in an aircraft requires in-depth investigation. Reliability is the most important issue for us, but there are many other requirements too. Different applications have a wide range of requirements. With the introduction of wireless, every element in the aircraft will require extra power for signal transmission and so the energy consumption of every technology should be controlled. While some applications may require low-power communications, others have low latency as a more critical requirement. The maximum amount of data that can be carried in the fastest packet is a parameter that defines the applicability for some other applications. Node density will be an issue too, because of the large number of sensors that will be introduced for passenger and application monitoring. Initialization time is another important aspect of wireless technologies, because if a node detaches because of an error, it has to attach back in the shortest time possible.

Reliability is also an important requirement for factory automation. Wiring is one of the factors limiting the mobility of actuators and sensors. To overcome this problem, wireless technologies are already frequently used in factories and the technologies involved are therefore important candidates for use in aircraft:

- The Wireless Sensor and Actuator Network for Factory Automation (WSAN-FA) is a Bluetooth-based automation standard published by ABB. It started as a proprietary technology, but has since become open.
- WirelessHART is the wireless version of the HART standard, which is already used in factory automation and based on the IEEE 802.15.4 physical layer. For comparison purposes, we also introduce the non-modified structure of the IEEE 802.15.4.

Technologies from outside the world of factory automation can also be considered:

- LTE is one of the most potent candidate technologies, since it is the focus of most of the research and investment. 5G research is already beginning to replace the 4G standard.
- The ECMA-368 standard, with high data rates over short communication ranges allows for a really dense network with the ultra-wide band physical layer it uses.
- The widely established Wi-Fi IEEE802.11 standard is adaptable for many applications.

Even though there are many surveys of wireless sensors [3], personal area networks [4] and sensors in aircraft [5], none of them covers all of the critical aspects of a aircraft communication system. The closest surveys to our requirements are the ones on industrial automation. However, they do not give the full view required for aircraft applications. For example, Islam et al. cover security in detail but only briefly touch on reliability [6]. Gungor and Hancke [7] take commercial of-the-shelf (COTS) hardware as the focal point; Lee et al. [8] take technologies as they are, without specific application-related evaluations; the survey of Willig et al. [9] has the same perspective as Lee et al. but some of the possible candidate technologies are missing and the text is also outdated. Most importantly, none of these surveys has as a perspective the reliability framework that is a must for the aircraft communications and against which we want to provide a detailed assessment of each candidate technology for the applications that we summarised in an earlier study [10].

The structure is as follows. In Section 20.2 the reliability assessment framework is introduced with the fault-tree analysis to provide a starting ground for the safety criteria.

In Section 20.3 the Performance Metrics are introduced before going into deep investigation of wireless technologies. In Section 20.4 candidate wireless technologies are presented. In Section 20.5 the presented candidates are evaluated against the performance metrics with respect to the reliability framework. In Section 20.6 a brief summary concludes the discussion.

20.2 Reliability Assessment Framework

In probabilistic evaluations, the addition symbol + represents the logical OR function, while the multiplication symbol × represents the logical AND function.

Reliability in our evaluation means guaranteeing a certain availability of the system. Furthermore, the reliability of wireless communication in an aircraft can be better understood if separated into multiple layers.

The fault-tree analysis in Figure 20.1 helps us to break down the components of a failure of an example application. We use this model to introduce our assumptions, which lead from a system safety level to a maximum tolerated communication failure level. Here, we consider a passenger heat sensor application (PHSA). We aim to cover all possible aspects of this application, which can then be generalised to any other application. The first level of the failure rate comes from the different subsystems forming the application:

- power system, P
- sensor system, S
- control system, C
- communication system, Comm.

The system failure can be represented via the individual probabilities for failure of the subsystems as:

$$P_{\text{PHSA}} = PF + SF + CommF + CF, \tag{20.1}$$

where the probability of communication failure (*CommF*) can be broken down further. There can be many contributors to communication failures: communication hardware failure (*CHWF*), failing to deliver a message within a deadline required by a certain application *i*, T_W (P_{app_i}) and security failure (*SecF*). The multitude of messages delivered independently from each other adds another failure possibility. This results in a *CommF* of

$$CommF = CHWF + SecF + \sum_{i=1}^{j} P_{app_i}, \tag{20.2}$$

where j is the total number of packets required for a PHSA during a flight. This can be broken down to single packet transmission:

$$P_{app_i} = \prod_{k=1}^{Np} P_{comm_k^i}, \tag{20.3}$$

where the possible N_p transmissions within a T_W, k is the index of a packet transmission.

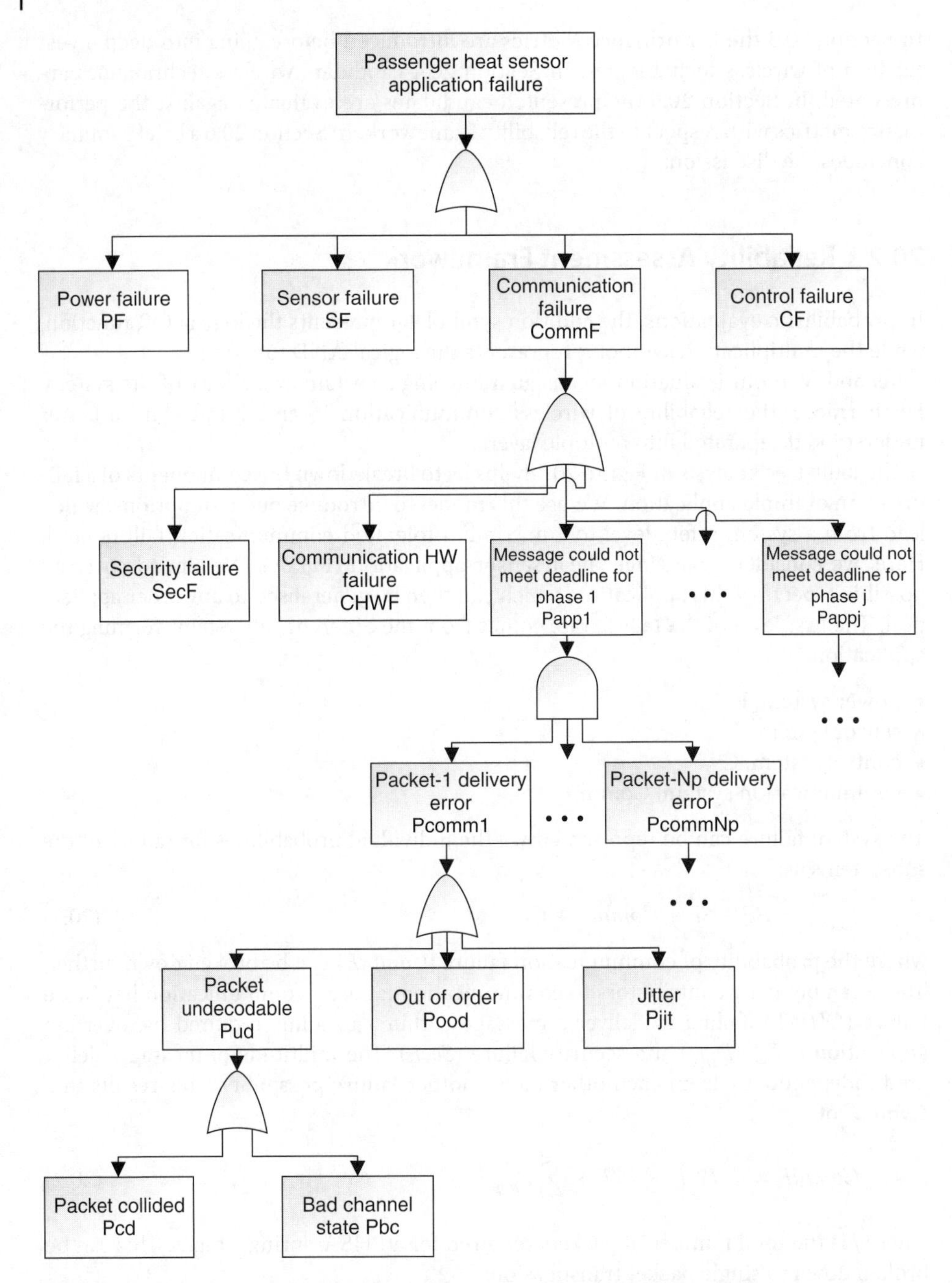

Figure 20.1 Application fault-tree analysis: heat sensor application failure probability.

The complete picture of the fault tree is

$$P_{PHSA} = PF + SF + CHWF + SecF + CF + \sum_{i=1}^{j} \prod_{k=1}^{Np} (P_{comm_k^i}), \tag{20.4}$$

In our analysis we will only focus on the communication failure probability *CommF*. This simplifies the error rate of our application to

$$P_{PHSA} = \sum_{i=1}^{j} \prod_{k=1}^{Np} (P_{comm_k^i}). \tag{20.5}$$

If the communication failure probability does not change within one time window we can write the AND operation without the product, and if P_{comm} doesnt change between different flight phases the OR operation can be simplified to,

$$P_{PHSA} = \sum_{i=1}^{j} P_{comm_k^i}{}^{Np} = \left(1 - (1 - P_{comm}{}^{N_p})^j\right) \tag{20.6}$$

On top of that, if the packet failure rate is constant between all of the flight phases *j* and with the linear approximation that $(P_{comm})^{N_p}$ is two orders of magnitude smaller than $\frac{1}{j}$ then we have,

$$P_{PHSA} \approx j(P_{comm})^{Np}. \tag{20.7}$$

Having the fault-tree analysis, we can now tackle the problem from a different angle by separating the reliability into three different layers for better visualisation of the assumptions we have made:

- *Transmission*: This layer covers the transmission of a single packet between two nodes.
- *Medium access*: This layer covers the medium access control and resource scheduling.
- *Safety*: This layer is the introduction, by the application, of a reliability requirement on the communication.

The transmission layer (TL) requires analytical modeling of the success of a packet transmission. In order to avoid such environment-dependent analysis we will use statistical data as an input for medium access layer (MAL) dimensioning. Finally, we will assume a certain reliability requirement from the application and see how the packet transmissions relate to this required reliability. In the following sections each of these layers will be investigated in order to provide a full picture of the reliability.

20.2.1 Transmission Layer

P_{comm} in Eq. (20.7) is the transmission layer in the fault-tree analysis, which can be interchanged with more detailed models to provide communication reliability designs with wider applicability. The parameters that can affect this layer are coding, modulation scheme, frequency band, co-existence, transmission power, hardware quality and many more.

20.2.2 Medium Access Layer

The medium access in a network can be optimized by setting reliability as a goal. A delay model can be built on top of this medium access to extract the number of possible transmissions in total before the delay budget is reached. Normally, delay is used as a statistical parameter that is within a percentage of a certain value. However, in our evaluation we will assume the maximum delay as a budget for transmissions, and derive our packet delivery reliability from packet transmission statistics. If a reservation-based medium access technique is assumed, the delay of a packet can be broken down as,

$$T_{delay} = T_{W1} + T_{W2} + T_{SE} + T_p, \tag{20.8}$$

where T_p is the propagation delay, T_S is the sending time of a packet from the radio, T_{W2} is the maximum medium access waiting time for a packet and T_{W1} is the buffer waiting time for a packet. Since we are considering short-range communications, we assume that T_p can be neglected. Let us define the buffer delay T_{W1} against deterministic R and Poisson arrivals as,

$$T_{W1} = \begin{cases} 0, & T_R > \dfrac{1}{\epsilon} \\ \dfrac{\frac{N\cdot\lambda}{\epsilon^2}}{1 - \frac{N\cdot\epsilon}{\lambda}}, & \dfrac{1}{\lambda} > \dfrac{1}{\epsilon} \\ \infty, & T_R, \dfrac{1}{\lambda} \leq \dfrac{1}{\epsilon} \end{cases} \tag{20.9}$$

where T_R is the deterministic inter-arrival time and λ is the Poisson inter-arrival rate for the application and ϵ is the mean serving rate for a packet. If T_R is larger than $\frac{1}{\epsilon}$, then we will always have the buffers empty. If we have a probabilistic arrival then we can calculate the expected buffer waiting time. In the third case, however, the system cannot serve all the incoming packets and the delay will grow to infinity. For the following applications we will assume that the system is well-dimensioned and has deterministic arrivals, so $T_R > \frac{1}{\epsilon}$. The maximum waiting time for medium access T_{W2} can be broken down as:

$$T_{W2} = T_{WC} + (N-1)T_{SE} \tag{20.10}$$

with N users, where the T_{WC} is the waiting time before the medium access is granted to the user, and T_{SE} is the use time of the medium by one user. The maximum T_{WC} is T_{SE} if a packet is generated just as another packet has started to be sent. In the worst case, we use this value instead of T_{WC}. So the formula changes to:

$$T_{W2} = N \cdot T_{SE}. \tag{20.11}$$

On top of this, for short-range communication we assume that T_p is 0. Then the delay from Eq. (20.8) simplifies to:

$$T_{delay} = (N+1) \cdot T_{SE} \tag{20.12}$$

At this point we define $N \cdot T_{SE}$ as T_{cyc} as it is simplified as the use of medium by N (all) users sequentially. And as for retransmission the user have to wait another T_{W2} we will use this for retransmission delay calculations.

Furthermore the T_{cyc} can be broken down to include elements of the medium access scheme as in,

$$T_{cyc} = T_{cyc\_control} + T_{cyc\_data} \tag{20.13}$$

the reserved time is broken down to control and data time. In $T_{control}$ signaling elements are contained such as beacon time, reservation techniques or acknowledgment time. In T_{data} number of slots M and time slot size T_{slot} is included where $T_{slot} = T_{ack} + T_{synchtol} + T_{packet} + T_{code}$ includes in slot acknowledgment time T_{ack}, synchronization tolerance $T_{synchtol}$ packet transmission time T_{packet} and extra time used for different coding T_{code}. This general structure is:

$$T_{SE} = T_{control} + T_{ack} + T_{synchtol} + T_{packet} + T_{code}, \tag{20.14}$$

which can be used for extracting the T_{cyc} from any wireless reservation based medium access technology. Then we can extract N_p through T_{cyc}, as in Eq. (20.15), to give the required reliability for packet transmission.

The time budget for medium access T_{TW} dimensioning is the critical parameter of the MAL. The model we build here is based on the assumption that the loss probabilities of sequential packet transmissions are independent from each other. We start by calculating the number of packets N_p that can be sent in a time window [11]:

$$N_p = \left\lfloor \frac{T_{TW}}{T_{cyc}} \right\rfloor, \tag{20.15}$$

where N_p is also the redundancy level. Following that, the error probability is:

$$P_{app} = (P_{comm})^{N_p}. \tag{20.16}$$

20.2.3 Safety Layer

The safety layer is a bridge between the safety requirements of an aircraft application [12] and the reliability requirements of communication using the MAL. For now we will follow the assumptions of the fault-tree analysis and use Eq. (20.7). However doing this will require us to derive an availability requirement from the safety requirements, based on a failure rate rather than a failure probability. In order to do so we introduce the average flight duration T_{avg} and all of the time durations are given in hours. The number of T_{TW} in an average flight is derived as:

$$\beta = \frac{T_{avg}}{T_{TW}}. \tag{20.17}$$

We assume that the application error rate does not change during the flight. Thus the total expected number of failures F during an average flight can be given by:

$$F = \beta P_{app} \tag{20.18}$$

We can use the expected number of errors F to deduce the failure rate per flight hour:

$$F_{PHSA} = F_{flight} = \frac{\beta P_{app}}{T_{avg}} = \frac{P_{app}}{T_{TW}} \tag{20.19}$$

and this leads to:

$$F_{PHSA} \cdot T_{TW} = P_{comm}{}^{N_p}. \tag{20.20}$$

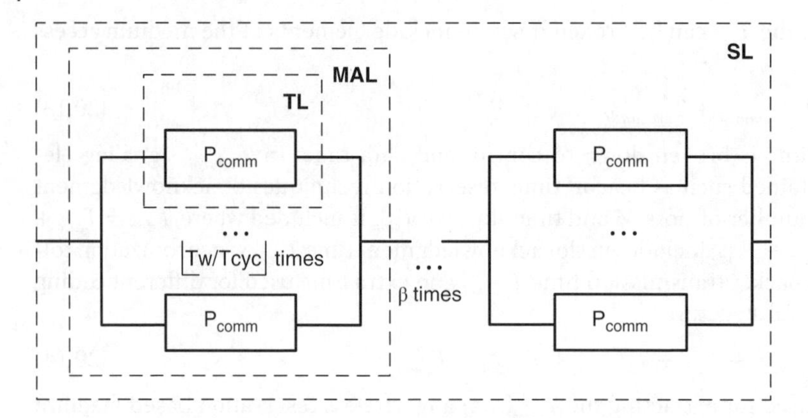

Figure 20.2 Reliability diagram of an average flight. Dashed lines mark the individual framework layers.

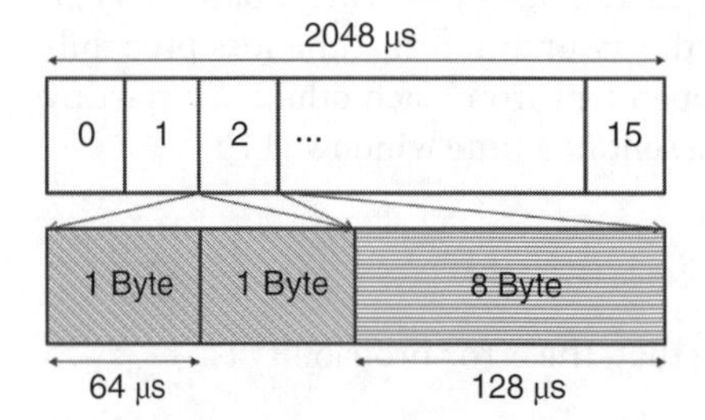

Figure 20.3 Superframe structure of WISA.

A simplified reliability diagram of an average flight with respect to the failure probability of its communication systems is presented in Figure 20.2. Parallel connections represent redundancy (AND operation on failure probabilities), and serial connections stand for dependency (OR operation on failure probabilities).

20.3 Metrics and Parameters

Following the reliability framework, we have to provide the performance metrics and design parameters for a clear comparison of the technologies. These performance metrics cover all of the applications but they will be used to assess technologies used in specific applications. Thus our aim is to provide a full comparison of technologies in a general way.

20.3.1 Design Parameters

20.3.1.1 Cycle Length and Packet Size

The time window and many other parameters will be equal for different technologies. However, the cycle time with a certain packet size will define the redundancy that is possible with the actual scheme. How the packet size and the cycle time are calculated will be detailed in Section 20.4. Still, the calculation is built around the overhead that each cycle has. The cycle type is detailed for each technology and possibilities of fitting more users into the cycles is discussed. Fair scheduling is always considered and due to the cycle limits the number of users served in a time slot is limited. However, with a fixed number of users we can have a constant cycle length since the users can have multiple slots in a cycle or at maximum one. If there is only one slot per user in a cycle, then the limit of the cycle is reached.

20.3.1.2 Node Density

The deployment of wireless sensors also brings limits on the communication network in question. The number of devices located in a small area is expected to only increase in future. This means that firstly the communication system should be able to handle the high density of nodes required by today's communications while also having enough capacity for future applications that might be deployed in the aircraft. It is important to note that the node density will not be the number of supportable users for a technology but rather how dense the communication can be so that the given number of users is supported. Hence, it is an advantage to have smaller cells to support denser communication. The frequency allocation for cell planning is another problem to consider in case of denser cells.

20.3.2 Performance Metrics

20.3.2.1 Power Consumption

As some of the devices in an aircraft communication system will be wireless sensors, it is possible that they lack a power cable and instead operate on batteries. For such applications, in order to have decreased maintenance times, it is important that the communication uses a small portion of the stored energy. On the other hand, not all applications in an aircraft operate on batteries.

The power consumption mostly depends on the frequency used. Higher frequencies require more energy to drive their more complex RF circuits. One other aspect of power consumption is the requirement for synchronization, which therefore prevents use of sleep mode: such devices will have the RF signal turned on all the time.

Another aspect of energy management is the complexity of a device's network roles. If a device has to switch between multiple roles to fulfil different tasks, this increases the power consumption as the devices must be configured in a mesh network, to allow them to behave as router, slave or gateway. Meanwhile in a fixed star network, users that require low energy consumption can be kept as slaves while the access point deals with all complex tasks.

20.3.2.2 Initialization Time

In case of a system restart or an emergency, a boundary on the time for all wireless devices to re-attachment to the communication system is required to prevent unacceptable delay. There are two different ways of joining a network. The network can be designed beforehand and all of the users given dedicated slots. Through use of a synchronization method, the user will align its message to its dedicated slot. The other way is for the network to use random access slots for users, who can get a dedicated time slot through contention access. In this regard, each of the technologies will be investigated to determine the worst-case initialization times.

20.3.2.3 Reliability

Using the delay as a budget rather than a metric, we can invert the calculations to obtain the reliability of a message delivery against a certain delay constraint. Doing the analysis in reverse opens up different possibilities for investigating high-reliability communications.

In the following section we will provide general information about technologies under consideration.

20.4 Candidate Wireless Technologies

The candidate technologies are mostly based on COTS systems. The reason for this is the expected decreases in design time if a technology is selected for future use. Before starting the discussion of each technology, the reason for selecting only contention-free technologies is set out.

The factory automation standard gives reliability, with high medium access granularity and low latency. Considering these aspects, the requirements for factory automation are similar to those for communication in an aircraft cabin.

20.4.1 WISA & WSAN-FA

The Wireless Interface for Sensors and Actuators (WISA) is a factory automation standard introduced by ABB. It comes with an energy harvesting system. It is built on the PHY layer of Bluetooth and takes advantage of frequency hopping on the shared spectrum. It was a proprietary protocol but ABB has now decided to have it standardised as WSAN/FA, also improving it on some aspects.

One of the strengths of the WSAN/FA is the small downlink timeslots, which are dimensioned to be available on a Time Division Multiple Access (TDMA) basis for every user registered on the system (see Figure 20.3). In this way, the latency of reaching the system is limited by the superframe length of 2.048 ms. It has a frequency division duplex scheme supporting different frequencies for uplink and downlink. For redundancy, retransmission on four different frequencies is possible, extending the cycle time to 8 ms. The payload considered for each time slot is 1 byte and the number of users is limited to 120 [13]. Alternatively, user numbers can be halved for a drastic increase in payload: 60 users at 10 bytes [14]. It uses COTS Bluetooth hardware for the PHY layer and 1 Mbit/s data rates are expected.

As it is a sensor and actuator network protocol, it has more uplink channels than downlink channels, which enables the sensors to report every cycle. The downlink channels are used for control of the actuators.

Figure 20.9 shows packet length versus latency. The superframe of 2048 μs of WSAN/FA supports two types of slot format: the small slot size of 64 μs, which allows 120 users with 15 frequencies and 4 redundant transmissions, and a bigger slotted MAC, which allows 60 users with a slot duration of 128 μs. One important thing to take from the figure is that the number of users is either 120 or 60 and the system does not allow multiple slot assignment to any of them. Decreasing latency through a smaller number of users is therefore not feasible. On top of that the low latency provided in the WSAN/FA vanishes fast with an increase in payload. Here, the latency is considered as the time to deliver a packet in an error-less environment, so basically it is the cycle length. With increasing payload, the efficiency of the protocol decreases due to large overloads.

20.4.2 ECMA-368

Published by the European Computer Manufacturers Association, the ECMA-368 due to its high carrier frequency has smaller cells with 30 m range, and thanks to the wider bandwidth, a higher data rate 480 Mbit/s [15]. A high carrier frequency of 3–10 GHz also come with its disadvantages, namely high fading and a high energy requirement due to the complex antenna structure needed. ECMA-368 has two MAC features:

- a distributed reservation protocol, where the users reach the channel with a guaranteed timeslot
- prioritised contention access, where there is a possibility to access some slots with the help of prioritisation on a contention basis.

There is no dedicated timeslot; instead there are medium access slots (MASs), which are allocated dynamically [16].

On the PHY layer it uses multiband orthogonal frequency division modulation, which enables 110 subcarriers in a superframe. For interference mitigation, frequency-domain spreading, time-domain spreading and forward error correction coding are available.

The superframe structure determines the actual data rate and worst-case latency, as can be seen from Figure 20.4. There are 256 MASs in a ECMA superframe. The first MASs are filled with beacons for each of the single devices attached to the system. These provide information about the structure of the upcoming superframe, so that each user knows how many slots it is able to use in the upcoming superframe. After the end of a superframe, a new superframe starts, with new settings that will be conveyed to the system via the beacons. The superframe structure creates a cycle time of 65 ms with 256 MASs of 256 μs. This cycle time limit can be overcome with the allocation of multiple slots in a superframe, but this will reduce the number of users in a timeslot. Due to the limit to the beacon frames to 96 at the start of the superframe and the requirement for each user to possess a beacon, the number of users in a cell is limited to 96 in a superframe [17]. In a synchronized case, the latency level depends on the number of users, as can be seen in Figure 20.10. One other advantage is the variable data length, which provides support for large payloads. With a decrease in coding rate, it provides highly reliable data transfer. The minimum packet length with the most reliable coding rate is 1.6 Kbyte, which is more than the packet length required for any sensor activity.

20.4.3 IEEE 802.11e

IEEE 802.11 provides contention-based access, with a request to send/consent to send scheme. However, in the 11e amendment to the standard, the option of contention-free

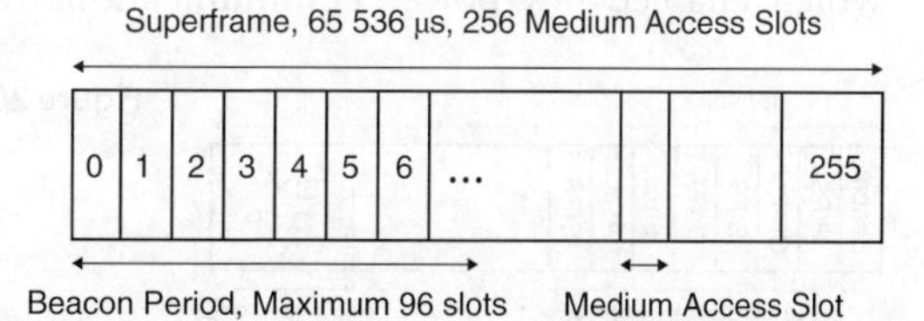

Figure 20.4 Superframe structure of ECMA.

access was added. A request can be placed, with access through the contention-based channel, to reserve upcoming contention-free slots. This enables dedicated time slots for users that can successfully reserve them. On top of that, there is another part in the time cycle which always has guaranteed slots for users. So all in all we can separate the cycle of 11e into three parts,

- full-contention access
- semi-contention access, with reservation being contested (enhanced distributed channel access, EDCA)
- full contention-free access (HCF control channel access, HCCA) [18].

As we have focused only on contention-free access on this survey, for a fair comparison we will consider HCCA.

The PHY layer of 802.11 provides high data rates within the ISM band. Interference of course is a problem due to the nature of the ISM band. Cell range is normally around 100 m, but can be increased and decreased via transmit power arrangements, which is more than enough for an in-cabin scenario. The reservation technique of the real time slots will limit the number of users in any single case. Normally, due to the contention-based access of the system, the number of users in a cell is not a problem. Figure 20.5 shows HCCA's performance. As it uses a polling mechanism through the access point, the access slot is reserved unless a packet error occurs during the polling. On the other hand one should also notice the amount of overhead required to guarantee such a real-time requirement, since the system itself is not designed for contention-free communication.

For transferring a data packet of 40 bytes, the total time required is 1.6 ms per user with the scheme provided in HCCA, with an 11-Mb/s data rate. If we choose a scenario with no packet loss and a service interval time of 100 ms, we can see how much the delay will be for each user before it is allowed to transmit another packet by the system. The service interval time is the time between two consecutive beacons, and with multiple users the contention-free period (CFP) can be extended up to that limit. Here, we will take advantage of this possibility in order to obtain lower latency levels with a higher number of users, as in Figure 20.11. In this figure we see that the number of users is capped at 60 and after that the 100-ms time-slot length is reached. Before that, the system benefits from allocating different controlled access periods (CAPs) to each user to decrease the latency. Latency here is defined as the time before next packet-sending opportunity.

20.4.4 IEEE 802.15.4

IEEE 802.15.4 is a standard designed for personal area network applications, but in a short time it has been applied to the Internet of Things, factory automation and many other areas. The modification that enabled it was the lightweight protocol stack, which enabled low-power communications with a decrease in latency. It uses the

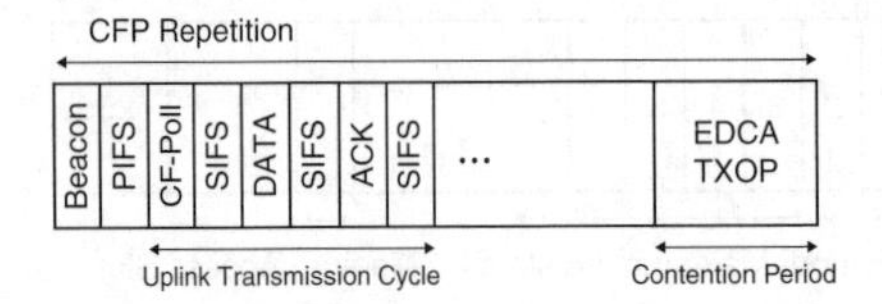

Figure 20.5 Reservation-based access cycle in 802.11e.

ISM band, like IEEE 802.11, and DSSS in order to avoid any interference. Due to the low-power nature of the chips, COTS equipment does not support a data rate higher than 250 kb/s but that is usually enough for the small chunks of data sent over sensor applications.

The standard, similar to IEEE 802.11, is not designed for high reliability because of that, it also lacks an optimized TDMA scheme. However, it has up to seven guaranteed time slots (GTSs) in order to serve users reliably in every time slot. The user limitation is a bound on the number of users in the cell, but the packet size can be enlarged with modification of the 'macsuperframeorder' (SO). As can be seen in Figure 20.6, the superframe is limited, with a contention access phase (CAP) and CFP. The CFP is only used if it is arranged in the previous slots for guaranteed communication. The minimum CAP is fixed in length at 440 bits. This limits the total number of bits available for CFP, out of the $960 * 2^{SO}$ bits in a superframe. All in all, it is not the packet size but rather the limit on the number of GTSs is a problem. The reason for not including the CAP is the assessment to give a fair comparison with the other standards in terms of packet load, number of users and latency.

Even though the number of users is limited, at the optimum conditions, latency levels as low as 15 ms are supported for a packet length of 20 bytes. With the modification of the SO parameter, a trade-off against latency can be seen in Figure 20.12. The SO parameter causes an exponential increase both in packet size and latency.

20.4.5 WirelessHART

WirelessHART is the specific protocol for industry automation. It uses the physical layer of IEEE 802.15.4, with enhancements to meet the needs of critical sensor networks. These have small and frequent chunks of data in the uplink, but less data in the downlink direction for the access point. The standard has really small timeslots arranged for parallel channel communication. On top of that, mesh networking is crucial in emergency cases in order to come up with the fastest routing path. In a major modification to the 802.15.4 standard, WirelessHART has a TDMA-based structure to give reliable service for all of the sensors.

It uses DSSS, like 802.15.4, to mitigate incoming interference on the 2.4-GHz band. Since WirelessHART supports up to 16 channels, 16 communication pairs is possible in a 10-ms timeslot [19]. The number of maximum communication pairs in the same area is an important feature, since this will be a defining parameter for cell communication if we are not considering a mesh network. On a mesh network, this can be used as a parameter of defining the frequency repetition range. In each timeslot, on top of the contention-free access guarantee, an ACK is also required to ensure no corruption against the bit error rate (BER), as can be seen from Figure 20.7. The MAC payload in a WHART packet contains 133 bytes. The ACK packet is 26 bytes, so the total data exchange in a WHART timeslot is 159 bytes, with a data rate of 250 kbps set by the physical layer [20]. So the timeslot structure is determined in WirelessHART but

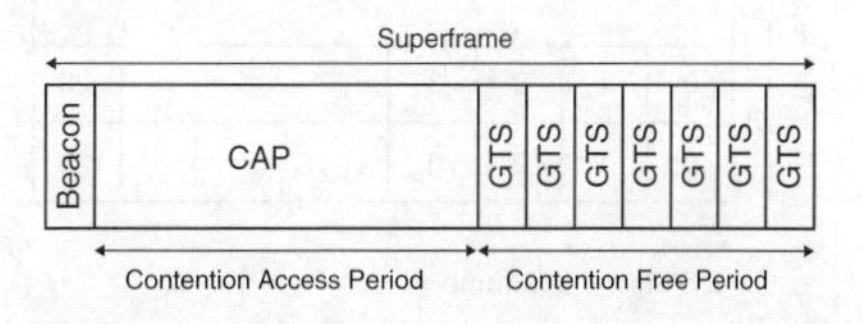

Figure 20.6 Superframe structure in IEEE 802.15.4.

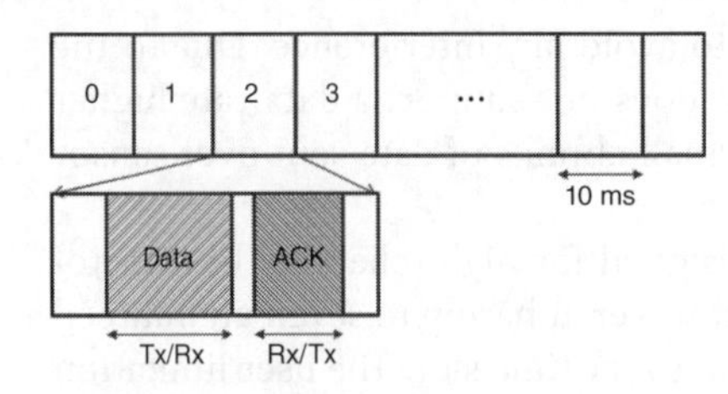

Figure 20.7 Time-slot structure in WirelessHART.

the reliability versus node density emphasis can be modified depending the number of simultaneous communications supported. WirelessHART supports both acknowledged and unacknowledged services.

In Figure 20.13 the tradeoff can be seen. Clearly, up to seven nodes can be supported, with double frequency allocation for interference mitigation where the delay for each node is limited to 10 ms. Of course, the predicted result of an increase in system population is a linear increase in latency. On a comparable level to previous protocols, 60 nodes can be served within 40 ms, with all the frequencies allocated for communication rather than having any redundancy.

20.4.6 LTE

The Third Generation Partnership Project Long Term Evolution (3GPP LTE) is the most recent standard for mobile networks. Due to the size of the market, it has developed to be one of the most efficient wireless communication technologies.

LTE uses multiple private bands, which is a huge advantage in terms of interference, compared to the previous technologies which all used the ISM band. LTE has resource blocks divided in terms of time and frequency, which allows for efficient scheduling. In Figure 20.8, the full frame structure of LTE is shown. A frame is divided into 20 slots, only 2 of which form a subframe because of the allocation of a minimum of two consequent resource blocks (RBs). A resource block is formed of 12 frequencies and 7 symbols and it is the minimum resource unit of the system. A resource block can transmit up to 40 bytes of payload data with a large overhead [21]. So a minimum transmission time interval will be 1 ms, with 50 RBs. This transmission interval can vary since the number of RBs allocated vary between 6 and 110.

On the other hand, we will use the uplink in order to assess the reliability of the LTE. The main strength of LTE lies in the OFDMA downlink channel, which is able to multiplex data from multiple stations on a single resource block, which is de-multiplexed with an inverse fast Fourier transform. This allows multiple subframes to be allocated

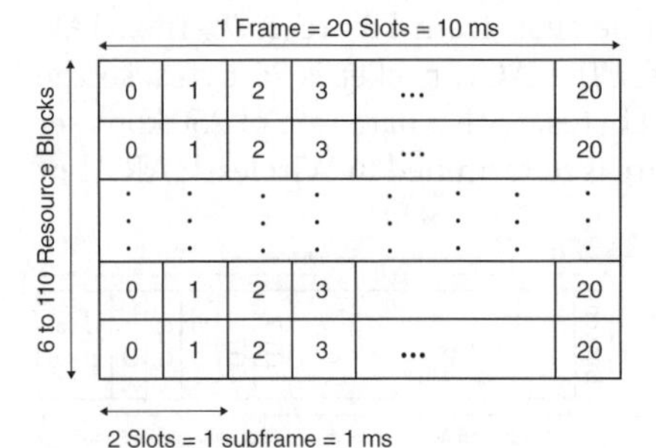

Figure 20.8 Frame structure of LTE [23].

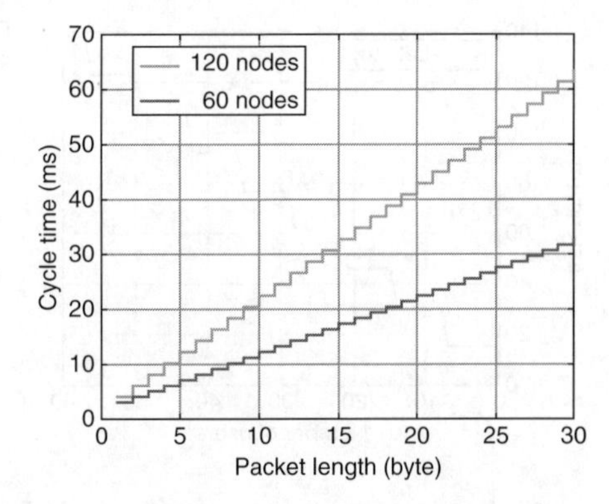

Figure 20.9 WSAN/FA cycle time vs payload.

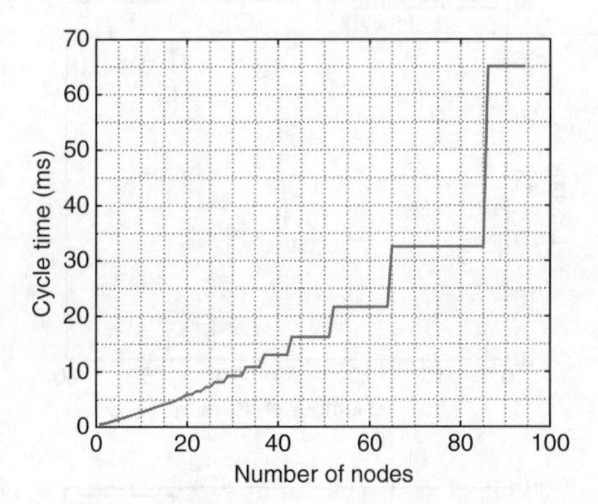

Figure 20.10 Cycle time vs number of users in the channel in ECMA.

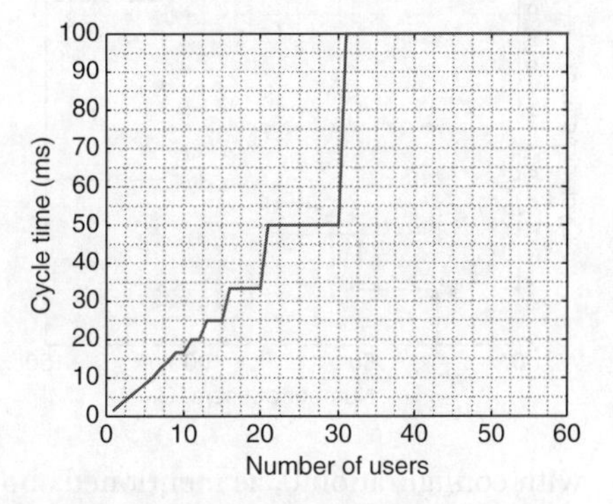

Figure 20.11 HCCA latency vs the number of users in the channel.

to the uplink where the downlink has a capacity increase. It is sufficient to have only a single subframe every 5 subframes. The uplink prioritization allows us to use it to our advantage via configuration 0 [21], where of the 10 subframes, 2 are downlink and 6 uplink, with 2 special subframes. The latter can be control information or a random access channel for synchronization. So random access is still a part of the technology due to the access requirement. One important remark to make here is that due to the random access nature of LTE, a re-initialization can be problematic when there are high numbers of users trying to reach to the system at the same time.

As presented in a previous study [22], we will try to calculate the worst cycle times in an LTE frame of 10 ms for varying numbers of users. We will use a 25 resource block size

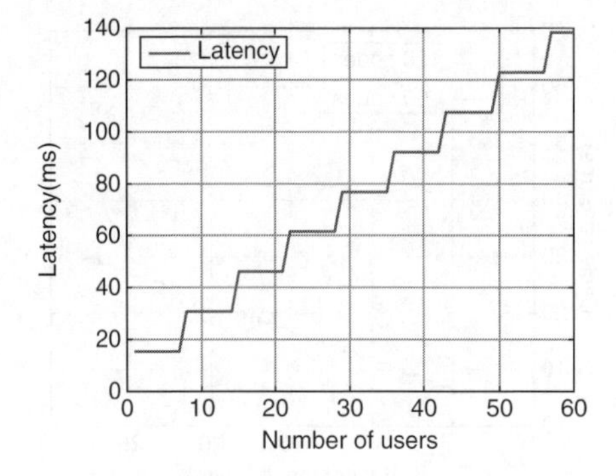

Figure 20.12 IEEE 802.15.4 GTS cycle time against SO parameter.

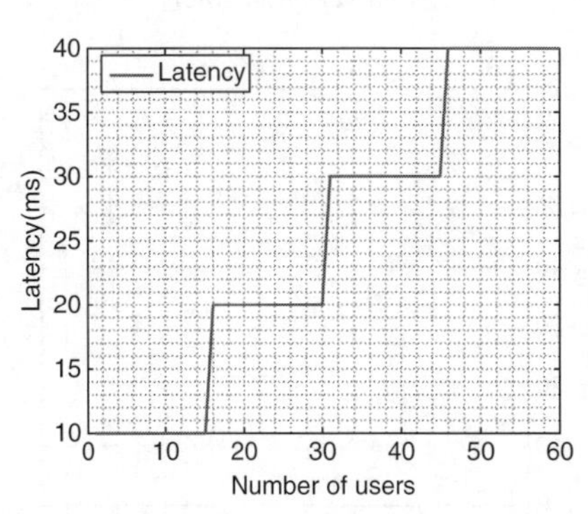

Figure 20.13 WirelessHART cycle time against network density with the number of frequencies available.

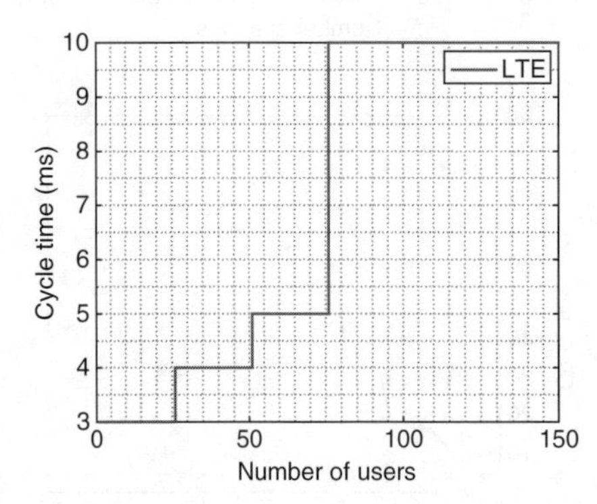

Figure 20.14 LTE cycle time with varying number of users.

with configuration 0, as mentioned above. We will not explore beyond 10 ms, because from that point the number of users and latency will tend to increase linearly and the data can be extrapolated to higher numbers of users. Because of the number of RBs available in a frame, a maximum of 150 users can be supported within 10 ms for a 40-byte data packet (see Figure 20.14), although 5 ms is possible with 75 users when the subframes are grouped into 3 and 3. The best cycle time achievable is 3 ms because of the downlink and the special frame time.

20.4.7 Comparison

In order to have a fair comparison, we used assumptions to investigate the most critical aspects of the technologies: the number of simultaneous communication links, the

Table 20.1 Comparison of protocols.

Technology	Cycle (ms)	Packet (bytes)	Power	Initialization	Node density
WISA WSAN/FA	2	8	Low	-	60
WLAN	100	40	High	Seconds	60
ECMA-368	22	1.6k	High	seconds	60
ZigBee	135	20	Very low	Seconds	60
WHART	40	133	Very low	Seconds	60
LTE	5	40	High	Milliseconds	60

packet length and cycle time T_{cyc}. The reason for the choice of these parameters is that, given the packet requirements of a certain application, the cycle length can be linearly extended depending on whether the requested packet fits the given packet length or not and the given or new calculated cycle time can be used in order to assess the reliability of the application with the given communication structure. Not all of the technologies have cycle lengths that scale linearly with the packet length but they are comparable.

To better visualise the technologies we have decided to fix one of those critical parameters and chose the number of simultaneous communications as 60. The reason of the choice is the structure of WSAN/FA which is limited up to 60 users.

With the complete reliability framework and full analysis of the protocol capabilities–a summary is given in Table 20.1 – we calculate the reliability of a worst case application in a cabin management system for each protocol.

In Figure 20.15, packet length ECMA-368 gives the best performance, with high data rates supported. The 2-ms latency of WSAN/FA is the worst. When both the packet length and the latency are compared at the same time, LTE represents the best trade-off. The problem with LTE, ECMA-368 and 802.11e is that while all of these standards are low power, they do not have special support for low-power applications. The disadvantage of WSAN/FA is its maximum of 120 users, which is a big problem for scalability. Scalability is also a problem for ECMA-368, where the number of beacons is

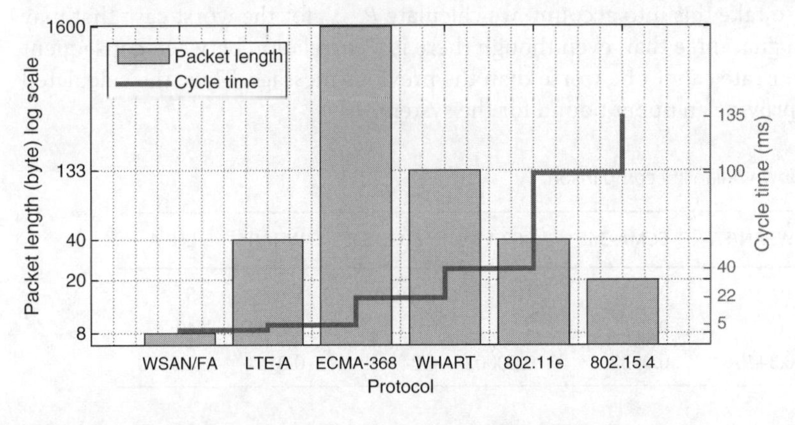

Figure 20.15 Cycle times and packet sizes of wireless technologies.

limited to 94. The limit is around 60 for 802.11e. Considering the future of the aircraft networks, with increasing density of devices, this lack of scalability will be a big problem.

On the other hand, WirelessHART has a limited [24] of 30 000 devices per network, while ISA100.11a uses IPv6 and does not have a limit. Nevertheless, due to the weight of traffic, the practical limit is a few thousands for both. The LTE network is limited of course to the number of subscriptions which should not be a problem in the near future.

20.5 Evaluation

With the complete reliability framework and full analysis of the protocol capabilities, we calculate the reliability of a worst case application in a cabin management system for each protocol. The application chosen is cabin lighting, which requires 36-byte messages, each of 200 ms. The total number of nodes in an aircraft is 250, so this value with multiple access points set in the aircraft will not be a problem. The time window is a lot larger than the propagation time, so it is neglected.

The first step for the reliability calculation is the adaptation of cycle times for the technologies in order to meet the required packet length. To accomplish this, we enlarge the cycle time of WSAN by a factor of 5 and of GTS by 2. We assume packet fragmentation does not come with an extra overhead. This results in updated cycle times of 10 ms and 270 ms respectively. We see that IEEE 802.15.4 is not able to meet the deadline in a worst case due to its cycle time being larger than the time window.

The next step is the calculation of N_p through Eq. 20.15, which results in P_{app} through Eq. 20.16. Since all share the same T_W, there are no comparable changes at the flight layer. The intuitive result from this, considering a reliability requirement of 10^{-5}, is that the technologies need to ensure a certain maximum communication error rate, as illustrated in Table 20.2. This is a good demonstration of how powerful retransmissions are for improving reliability.

This calculation assumes that transmission failures of consequent data packets are independent from each other. However, in reality they are not independent, since the nature of the air medium is not instantaneous but rather spans a certain amount of time. In order to take this into account, we calculate P_{comm} for the worst case that may occur. This will guarantee that, even though there is a correlation between consequent signals, the error rate cannot be worse than the previous message. Thus, the calculated error rate will provide an upper bound for the system.

Table 20.2 Top-down reliability comparison.

Technology	WSAN/FA	ECMA-368	802.11e	802.15.4	WHART	LTE-A
T_{cyc} (ms)	10	22	100	270	40	5
N_p	20	9	2	0	5	40
P_{comm}	0.3445	0.0937	0.00002	0	0.0141	0.5870

20.6 Summary

In this study, in order to have a generalisable approach, we have assumed fair scheduling for all of users in the medium. We have also evaluated all of the technologies against a simple reliability model without modifying the standard settings.

Initially we provided a reliability assessment framework model that takes its requirements from aircraft safety documentation and tries to merge these down to give the required packet transmission probability. Thereby framework we have broken down the assumptions that can be considered in an application availability evaluation.

Following this a delay model for a fair scheduling system was given. In order to do this we extracted the contention-free access parts of available wireless technologies and extracted the cycle times. At this point we assumed only re-transmissions are available for error control and dimensioned our reliability and delay accordingly.

Lastly we took a typical aircraft application for benchmarking purposes and compared the technologies in order to point out the unmodified capabilities. Moreover, we summarised additional critical characteristics of the technologies.

Our conclusions reflect the capabilities of COTS hardware for the evaluated technologies and a sample application. We believe that these results can be used to give insights to communication engineers for selecting the right technology for a given application. Moreover, the results can provide insights into new MAC scheme designs for reliability.

References

1 Locsin, A. (2008) Is air travel safer than car travel? *USA Today*. http://traveltips.usatoday.com/air-travel-safer-car-travel-1581.html/.

2 Traverse, P., Lacaze, I., and Souyris, J. (2004) Airbus fly-by-wire: A total approach to dependability, in *Building the Information Society*, Springer, pp. 191–212.

3 Alemdar, H. and Ersoy, C. (2010) Wireless sensor networks for healthcare: A survey. *Computer Networks*, **54** (15), 2688–2710, doi:10.1016/j.comnet.2010.05.003.

4 Latré, B., Braem, B., Moerman, I., Blondia, C., and Demeester, P. (2011) A survey on wireless body area networks. *Wireless Networks*, **17** (1), 1–18, doi:10.1007/s11276-010-0252-4.

5 Harman, R.M. (2002) Wireless solutions for aircraft condition based maintenance systems, in *IEEE Aerospace Conference Proceedings*, vol. 6, pp. 2877–2886, doi:10.1109/AERO.2002.1036127.

6 Islam, K., Shen, W., and Wang, X. (2012) Wireless sensor network reliability and security in factory automation: A survey. *IEEE Transactions on Systems, Man and Cybernetics Part C: Applications and Reviews*, **42** (6), 1243–1256, doi:10.1109/TSMCC.2012.2205680.

7 Gungor, V. and Hancke, G. (2009) Industrial wireless sensor networks: Challenges, design principles, and technical approaches. *IEEE Transactions on Industrial Electronics*, **56** (10), 4258–4265, doi:10.1109/TIE.2009.2015754.

8 Lee, J.S., Su, Y.W., and Shen, C.C. (2007) A comparative study of wireless protocols: Bluetooth, UWB, ZigBee, and Wi-Fi, in *Industrial Electronics Society, 2007. IECON 2007. 33rd Annual Conference of the IEEE*, pp. 46–51.

9 Willig, A., Matheus, K., and Wolisz, A. (2005) Wireless technology in industrial networks. *Proceedings of the IEEE*, **93** (6), 1130–1151, doi:10.1109/JPROC.2005.849717.

10 Gürsu, M., Vilgelm, M., Kellerer, W., and Fazlı, E. (2015) A wireless technology assessment for reliable communication in aircraft, in *IEEE International Conference on Wireless for Space and Extreme Environments (WiSEE).*

11 Bai, F.B.F. and Krishnan, H. (2006) Reliability analysis of DSRC wireless communication for vehicle safety applications, in *2006 IEEE Intelligent Transportation Systems Conference*, pp. 355–362, doi:10.1109/ITSC.2006.1706767.

12 US Department of Transportation, Federal Aviation Administration (2011) Advisory Circular on System Safety and Assessment for Part 23 Airplanes, *Tech. Rep.*.

13 Frotzscher, A., Wetzker, U., Bauer, M., Rentschler, M., Beyer, M., Elspass, S., and Klessig, H. (2014) Requirements and current solutions of wireless communication in industrial automation, in *Communications Workshops (ICC), 2014 IEEE International Conference on*, pp. 67–72, doi:10.1109/ICCW.2014.6881174.

14 Vallestad, A.E. (2012) WISA becomes WSAN–from proprietary technology to industry standard, in *Wireless Summit*.

15 Savazzi, S., Spagnolini, U., Goratti, L., Molteni, D., Latva-Aho, M., and Nicoli, M. (2013) Ultra-wide band sensor networks in oil and gas explorations. *IEEE Communications Magazine*, **51** (4), 150–160, doi:10.1109/MCOM.2013.6495774.

16 Fan, Z. (2009) Bandwidth allocation in UWB WPANs with ECMA-368 MAC. *Computer Communications*, **32** (5), 954–960, doi:10.1016/j.comcom.2008.12.024.

17 Leipold, D.F.M. (2011) *Wireless UWB Aircraft Cabin Communication System*, Ph.D. thesis, Technische Universität München. URL http://mediatum.ub.tum.de/doc/1079692/1079692.pdf.

18 Viegas, R., Guedes, L.A., Vasques, F., Portugal, P., and Moraes, R. (2013) A new MAC scheme specifically suited for real-time industrial communication based on IEEE 802.11e. *Computers and Electrical Engineering*, **39** (6), 1684–1704, doi:10.1016/j.compeleceng.2012.10.008.

19 Dang, K., Shen, J.Z., Dong, L.D., and Xia, Y.X. (2013) A graph route-based superframe scheduling scheme in WirelessHART mesh networks for high robustness. *Wireless Personal Communications*, **71** (4), 2431–2444, doi:10.1007/s11277-012-0946-2.

20 Petersen, S. and Carlsen, S. (2009) Performance evaluation of WirelessHART for factory automation, in *2009 IEEE Conference on Emerging Technologies and Factory Automation*, doi:10.1109/ETFA.2009.5346996.

21 Brown, J. and Khan, J.Y. (2012) Performance comparison of LTE FDD and TDD based Smart Grid communications networks for uplink biased traffic, in *2012 IEEE 3rd International Conference on Smart Grid Communications*, pp. 276–281, doi:10.1109/SmartGridComm.2012.6485996.

22 Delgado, O. and Jaumard, B. (2010) Scheduling and resource allocation in LTE uplink with a delay requirement, in *Proceedings of the 8th Annual Conference on Communication Networks and Services Research*, pp. 268–275, doi:10.1109/CNSR.2010.33.

23 Lioumpas, A.S. and Alexiou, A. (2011) Uplink scheduling for machine-to-machine communications in LTE-based cellular systems, in *2011 IEEE GLOBECOM Workshops*, pp. 353–357, doi:10.1109/GLOCOMW.2011.6162470.

24 Nixon, M. (2012) A comparison of WirelessHART and ISA100.11a, *Tech. Rep.*, Emerson Process Management. URL http://www.controlglobal.com/12WPpdf/120904-emerson-wirelesshart-isa.pdf.

21

Applications of Wireless Sensor Systems for Monitoring of Offshore Windfarms

Deepshikha Agarwal[1] and Nand Kishor[2]

[1] *Amity University, Lucknow, India*
[2] *Motilal Nehru National Institute of Technology, Lucknow, Uttar Pradesh, India*

21.1 Introduction

The use of wireless sensor networks (WSNs) spans a wide variety of applications, including automated monitoring of electrical systems. Electrical power is the backbone of modern industrial society and is expected to remain so for the foreseeable future. It is used almost ubiquitously, in applications ranging from entertainment, transport and communications to healthcare. The demand for this technology is constantly increasing, with new applications being developed for the convenience of society.

The existing power industry consists of three subsystems: generation plants, the transmission system and the distribution system [1]. Until now, the generation sector mainly used non-renewable resources like coal, oil and nuclear materials. The majority of generation is coal-fired. It is worth mentioning that coal is non-renewable and is fast depleting. The predicted year for total depletion is 2030, which is quite alarming.

The energy needs of the society are constantly increasing due to the introduction of new appliances that require power. According to the US Energy Information Administration [2], in 2014 non-renewable energy resources were 93.8% of the total generation, with coal having the highest share at 39.3%. Renewable resources contributed only 6.3%, with the largest contribution coming from wind energy, at 4.2% of the total. However, in the projections for electricity generation in 2030 it is suggested that the emphasis will be on renewable resources instead of non-renewable energy resources [3].

The burning of coal causes air pollution and emits significant amounts of carbon dioxide, a greenhouse gas [4, 5]. The extensive research carried out on these emissions has led to a belief that they result in global warming (a dangerous phenomenon, which is changing the weather cycle) [6].

WSNs can help in reducing these effects [7] by boosting the concept of alternative distributed energy resources [8]. Wind is present everywhere and in abundant quantities yet its intensity varies with the environment. In coastal areas, the intensity of wind energy is higher. In other areas, it varies according to the pressure and the vegetation present. Therefore, the most promising places for setting-up wind-based power plants is offshore and in coastal areas. Onshore wind power plants are still easier to maintain, monitor and control than offshore one. The offshore wind plants suffer from the corrosive effects of the surrounding water and humid air [9–12]. Also, detecting physical

Wireless Sensor Systems for Extreme Environments: Space, Underwater, Underground, and Industrial.
First Edition. Edited by Habib F. Rashvand and Ali Abedi.

damage to the structures and correcting them leads to long delays. Manual methods of error detection are outdated. WSNs are a boon to applications that require continuous and automatic monitoring. WSNs in offshore windfarms can aid remote monitoring of environmental conditions, for example the height of waves, sea-surface temperatures and the salt content of the water surrounding the turbines.

21.2 Literature Review

This section presents some important studies that focus on improvement of protocols and algorithms for WSNs, followed by recent literature on the issues related to WSN-based health monitoring of structures. WSNs use routing protocols for efficient transmission of sensed data to a destination sink node. In the Low-Energy Adaptive Clustering Hierarchy routing protocol (LEACH) [13], the sink-node is located close to the sensor nodes. However, the sensor nodes far from the sink node die sooner than those close by. To solve this issue, researchers have proposed energy-saving algorithms such as PEGASIS, H-HEARP and SEP [14–17]. However, with increase in field size, the performance of these algorithms is seen to degrade, suggesting that they are not equally suited to every application. Some applications – for example, weather monitoring – require the detection of data – temperature, movement, sound, light or the presence of specific objects and so on – from remote locations where human access is difficult. Here, application-specific WSNs need to be designed.

Solutions include adopting hierarchical structures [18] for better data aggregation, using heterogeneous nodes as cluster heads (CHs) for saving energy, fault-tolerance achieved through reconstruction algorithms, ZigBee WSNs for data collection [19], routing protocols based on mobility prediction and link-quality measurement [20], and centralised power-efficient routing algorithms [21]. Research on dynamic (or adaptive) selection of threshold values is extensive. Several techniques have been proposed, such as signal-histogram-based methods [22–24], and those based on spatial distributions of signal *k*-means [25].

Mutual information has been used in computer vision and machine learning for various applications, including data alignment [26], medical imaging [27], fusion of object detector outputs [28], feature selection for classifier training [29], contourlet-based adaptive threshold selection [30], variational background-based adaptive thresholding [31], geometrical thresholds for de-noising images [32], de-noising of satellite images [33] and fault prediction [34].

21.3 WSNs in Windfarms

WSNs can be used to autonomously monitor and control the operation of windfarms. The layout of the sensor network should be suitable for the design of the windfarm. Normally, the turbines are arranged in rows and columns with fixed distances between, but with the distance between rows twice the distance between columns or vice versa.

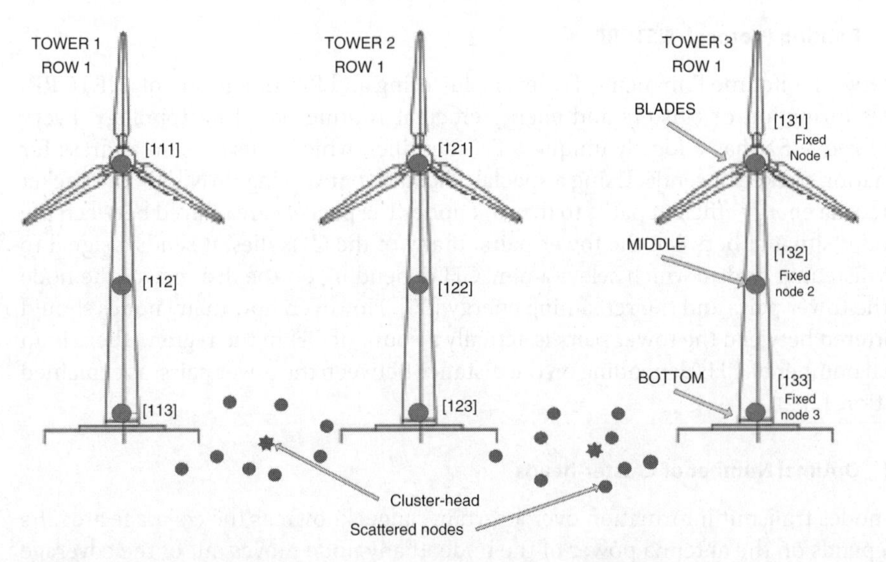

Figure 21.1 WSN deployed on a windfarm. Fixed row tower nodes (RTNs) (large circles) are attached to the towers. RTN node [123] represents the address of fixed node 3 for tower 2 in row 1. The CHs are represented by haxagons. The small circles represent scattered-nodes [35].

Such a topology requires deployment of sensor nodes in a fixed or hybrid topology depending on the communication distance or coverage area of the sensor nodes.

Figure 21.1 shows the proposed topology of a WSN in a windfarm. The WSN nodes are fixed onto the towers to collect the data required to monitor the working condition of the tower. The collected information is relayed to the nearby sink, which is connected to the last tower on the installation. The sink transmits the information to the base station (BS), which analyses the variation of the received parameter values from the threshold defined for normal working. Based on this analysis, suitable corrective or preventive measures are initiated from the maintenance and control center.

If the maximum range of data transmission is lower than the distance between the information-relaying nodes, there is a requirement for intermediate nodes. These must be deployed randomly between the towers to act as forwarding nodes. To save energy in the WSN nodes, a hierarchical routing approach is used.

The WSN network is divided into three layers:

- sensing nodes
- forwarding nodes
- CHs.

By coordinating the behaviour of these nodes, energy-efficient data transmission is possible in real-time. This system helps eliminate the need for staff for fault identification and reduces delays in troubleshooting.

The proper working of this system involves a suitable routing protocol and an efficient fault detection technique, as discussed in following sections.

21.3.1 Routing Protocol- NETCRP

The Network Lifetime Enhancing Tri-level Clustering and Routing Protocol (NETCRP) involves formation of clusters and energy-efficient routing based on topology. Every tower-fixed WSN has a locally unique RTN identifier, which it uses as an address for information packets it sends. Using a special calculation involving PAN [35], the packet is routed via energy efficient paths to the sink node. The packets are relayed between the CH nodes situated between the tower pairs. If any of the CHs dies, it sends a signal to nearby listening nodes, which select a new CH depending on the distance of the node from the tower pairs and the remaining energy [35]. However, how many nodes should be scattered between the tower pairs is actually a count of CH in the region. There is an optimal number of CHs depending on the distance between the tower pairs, as explained in Section 1.3.2.

21.3.2 Optimal Number of Cluster-heads

WSN nodes transmit information over a certain range, known as the coverage area. Its size depends on the antenna power of the node. If any node moves out of the coverage area, it is no longer able to communicate with other nodes. There should be some minimum (optimal) number of CHs (denoted OCH) that will allow communication across the entire windfarm. Depending on the distances between the rows and columns, OCH will vary. If the area increases, the optimal number of cluster-heads will increase and vice versa.

Let the number of rows be x and number of columns be y, and let z be the transmission range of a single WSN node in metres. The following cases present calculation of OCH:

Case 1: If the distance x between rows and the distance y between columns are both equal to z, then OCH = $(x-1) \times (y-1)$.

Case 2: If the distance x between rows and the distance y between columns are both equal to $2z$, then OCH = $(y+2) \times (x-1)$.

Case 3: If the distance x between rows and the distance y between columns are both equal to $0.5z$, then OCH = $(x-2) \times (y-1)$.

21.3.3 Adaptive Threshold

The fault-detection method involves several steps :

1. Find the adaptive threshold.
2. Find the distance matrix for the data sets.
3. Quantize.
4. Use combination-summation (CS) and flow direction (FD) to determine level of fault occurrence.

The most widely used value for setting a threshold in any parameter is the average method [36], which is non-adaptive to changes in environment. It is biased by the extreme values in the dataset and influenced to fall towards those values, giving an inaccurate picture of the actual scenario. Hence, by dynamically adapting the threshold value to cater for different environmental conditions, these limitations can be addressed.

Figure 21.2 Flowchart for threshold selection and quantization.

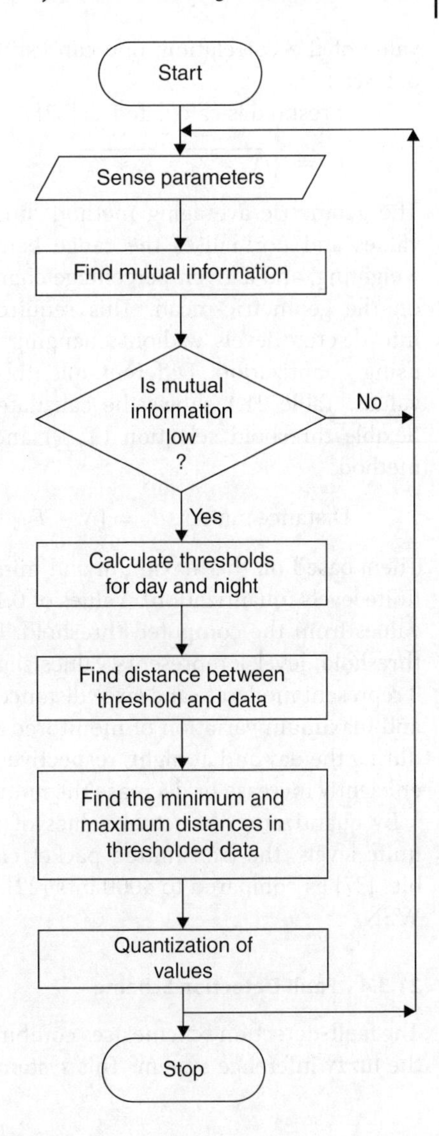

Figure 21.2 shows the flowchart for the threshold selection method. The formulae for mutual information, threshold and distance matrix [37] are all given below.

The mutual information between the two data sets $R(X_D, Y_N)$ is:

$$R(X_D, Y_N) = \frac{\sum_{i=1}^{N}(X_i - X_m) * (Y_i - Y_m)}{\sqrt{\sum_{i=1}^{N}(X_i - X_m)^2 * \sum_{i=1}^{N}(Y_i - Y_m)^2}} \tag{21.1}$$

If the correlation is 1, it means that the datasets correspond to the same period, whereas a value of 0 indicates that the datasets correspond to different periods. Based on the value of the correlation, one can estimate the degree of similarity between the two datasets.

The threshold is calculated as [37]:

$$T_X = \sqrt[N]{x_1 * x_2 * .... * x_N} \tag{21.2}$$

The geometric averaging method diminishes the influence of extreme low or high values and normalises the range being averaged, so that no range dominates the weighting, and a given percentage change in any of the properties has the same effect on the geometric-mean. This requires compression of the infinite sample values into discrete levels without changing the meaning of the information. This is done using quantization. Different quantization levels can contain unequal numbers of values. Table 21.1 shows the calculated threshold values for the proposed method, flexible threshold selection (T_{FTS}) and the static thresholding of the mean (T_{MM}) method.

$$\text{Distance matrix, } d_X = [X - T_X] \tag{21.3}$$

Then, based on the maximum and minimum values of d_X we bind similar values into finite levels (quantization). Values of 0,1,2,3 are used, depending on the distance of the values from the computed threshold. Level 0 represents values that are equal to the threshold, level 1 represents values slightly away from the threshold and levels 2 and 3 represent moderate and large distances. This is depicted in Figure 21.3. The minimum and maximum variation of monitored variables represents the data samples collected during the day and at night, respectively. Based on the variation of the values, one can efficiently increase or decrease the number of levels.

By quantization of actual values of the monitored data samples into discrete and finite levels, the transmitted packet carrying the information can be reduced to 23 bits [37] as compared to 4000 bits [22], and hence greatly enhances the lifetime of the WSN.

21.3.4 Fault Detection Scheme

The fault-detection scheme uses combination-summation and flow directions to design the fuzzy inference system. This system uses 'if-then' rules to evaluate a result against

Table 21.1 Threshold values.

Variation range	Sea surface temperature			
	Range values	FTS (T_{FTS})	FTS (T_{MM})	$T_{FTS}-T_{MM}$
Minimum	7.2	8.80863	16.6181	7.8095
Maximum	10.1			
Minimum	24.4	25.6094		−8.9913

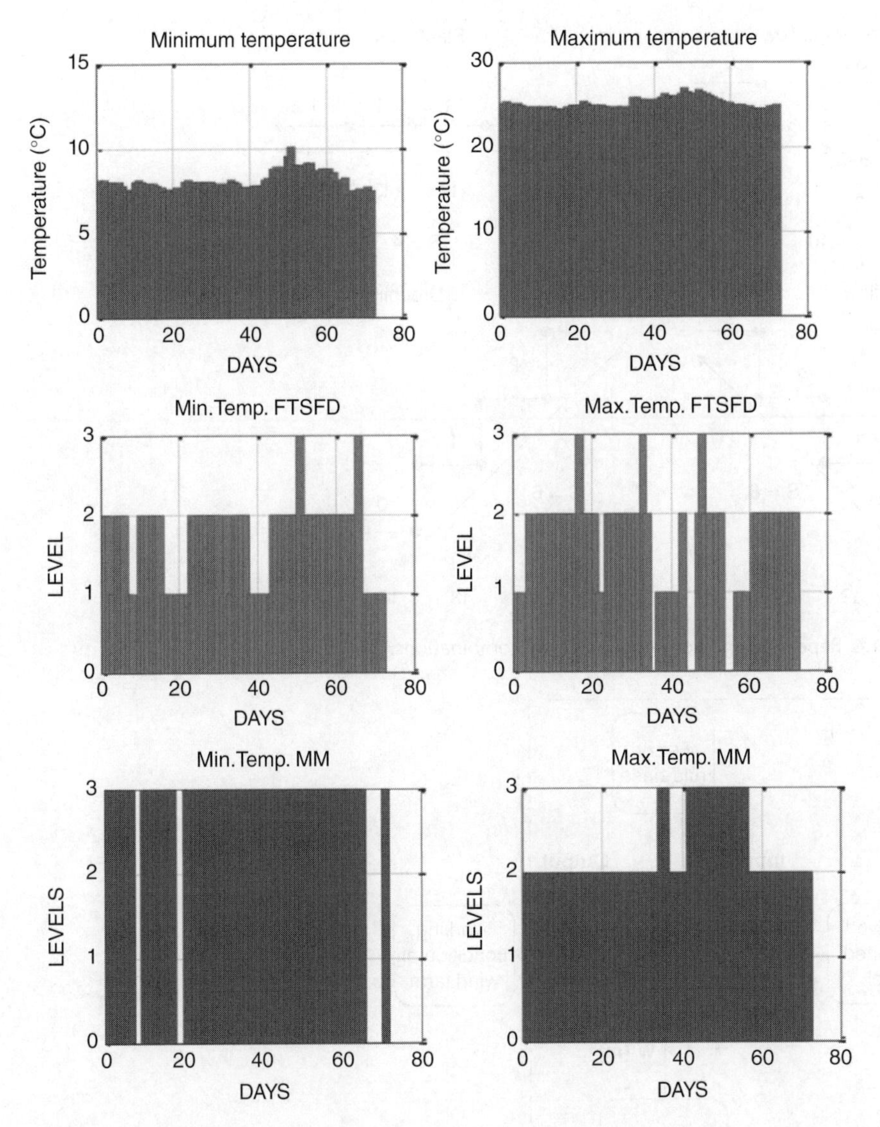

Figure 21.3 Quantization of sea-surface temperatures. Top, original data; middle, FTS quantization, bottom, MM quantization.

some inputs. Fault detection is the next step after quantization, and can be used to derive significant information from the quantized windfarm data about faults.

If the quantized levels received in time t are l_{t_1}, l_{t_2}, l_{t_3} and l_{t_4} then the sum of the levels is:

$$CS = \sum_{i=t_1}^{i=t_4} l_i = l_{t_1} + l_{t_2} + l_{t_3} + l_{t_4} \tag{21.4}$$

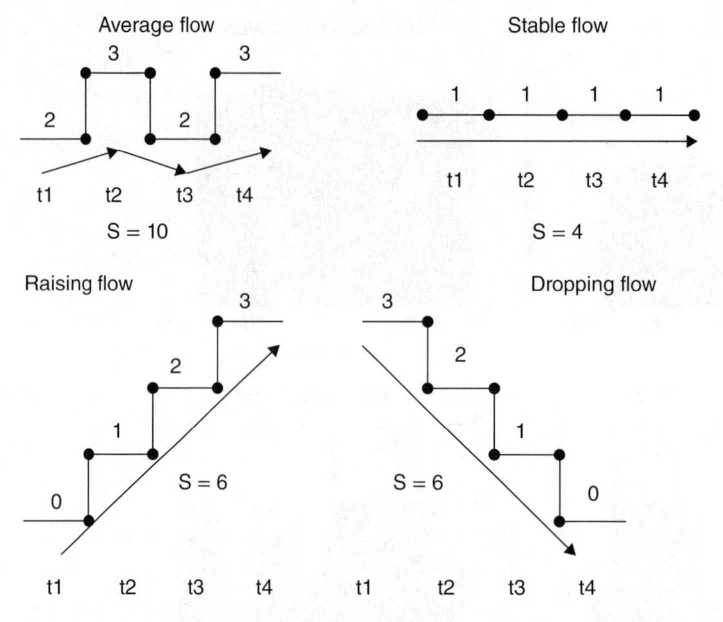

Figure 21.4 Repeating and non-repeating level combinations.

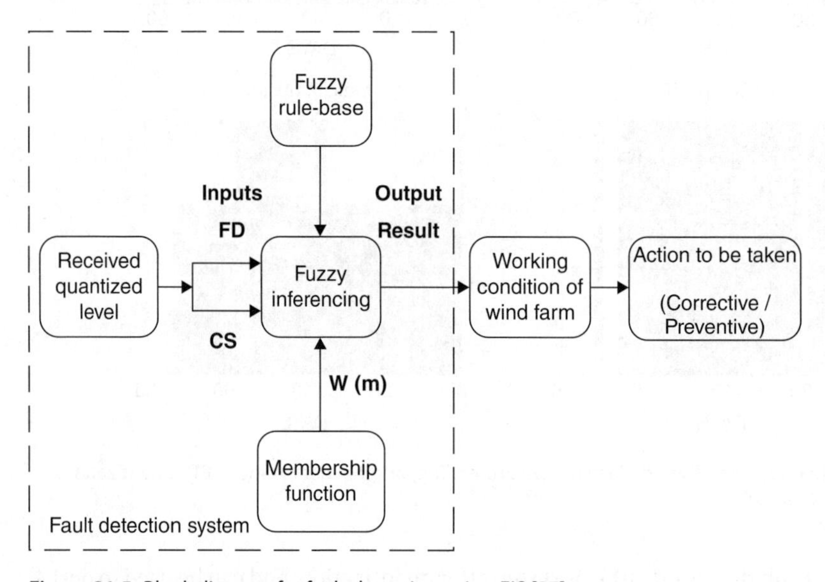

Figure 21.5 Block diagram for fault detection using FIS [37].

Flow direction means whether the received levels are increasing, decreasing, stable or varying constantly as shown in Figure 21.4. In this system (Figure 21.5), the fuzzy-set F can be described as:

$$F = \{\omega, m(\omega) | \omega \in U\} \text{ where } U = \{0 - 3{,}1\}$$
$$\text{and } m : \omega \rightarrow [\text{Normal operation, low risk, high risk}] \tag{21.5}$$

Table 21.2 Fuzzy rule-base for four levels.

Sl. No.	Flow direction	Combination-summation	Result (MF)
1	S	0	Normal
2	S	4	Normal
3	S	8	Low risk
4	S	12	High risk
5	R	6	High risk
6	D	6	Low risk
7	A	1	Normal
8	A	2	Normal
9	A	3	Normal
10	A	4	Normal
11	A	5	Low risk
12	A	6	Low risk
13	A	7	Low risk
14	A	8	High risk
15	A	9	High risk
16	A	10	High risk
17	A	11	High risk

S, stable; R, rising; D, dropping; A, average.

Table 21.3 Simulation parameters.

Sl. No.	Parameter	Value
1.	Area (m^2)	1000×1000, 4000×6000
2.	Number of rounds	100, 150
3.	Total number of nodes	416–1216
4.	Total number of fixed nodes	216
5.	Total number scattered nodes	200–1000
6.	Total number of turbines	72
7.	Packet size (bits)	23, 4000
8.	Assessment parameters	Energy efficiency Number of cluster-heads alive Total number of transmitted packets Round with first-deadnode Network-lifetime (all dead nodes)

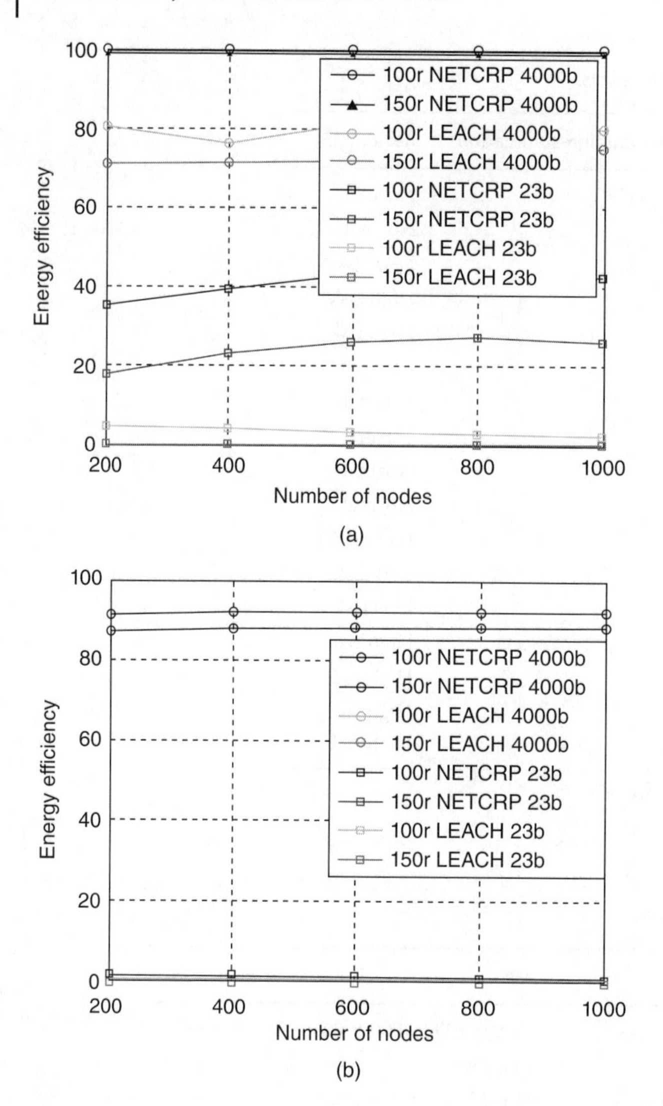

Figure 21.6 Energy efficiency: (a) 1000 × 1000 m; (b) 4000 × 6000 m.

where, ω is the combination summation and flow directions of the received levels by the remote-observer, $m(\omega)$ is the membership function for the received level and U is the universal set representing the set of all levels, as shown in Table 21.2. The membership function alerts the remote observer whenever the probability of fault occurrence becomes high.

The remote observer can derive meaningful information from these received quantized levels based on fuzzy-logic rules, as shown in Table 21.2.

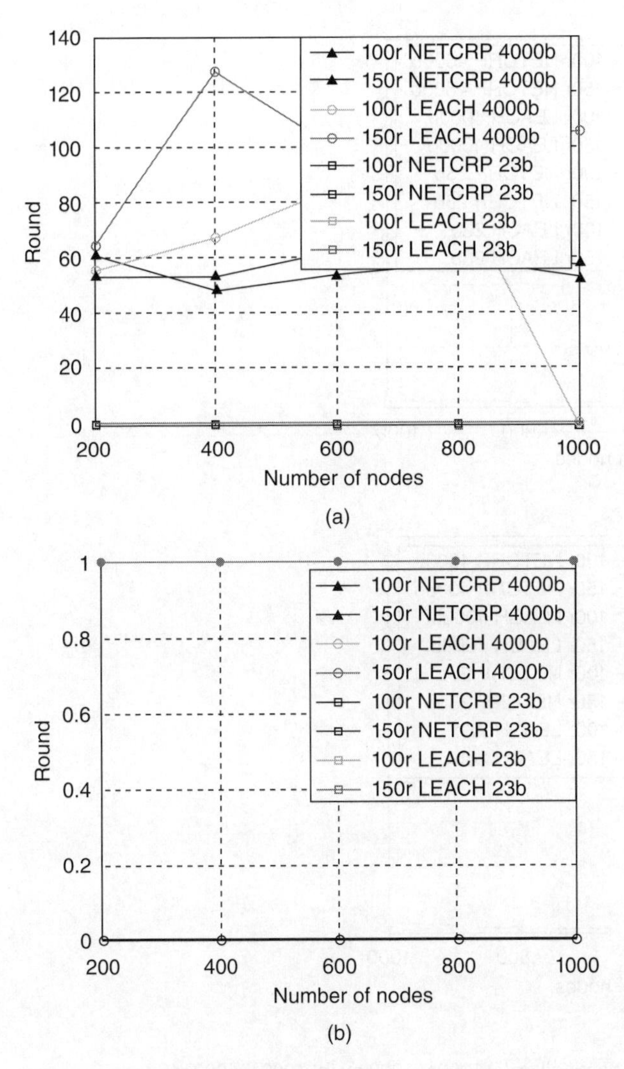

Figure 21.7 Round with first dead node: (a) 1000 × 1000 m; (b) 4000 × 6000 m.

21.4 Simulation and Discussion

This section presents simulation results of the fault detection method, prepared using NETCRP in MATLAB. The parameters used are set out in Table 21.3. The results of simulations of the flexible threshold method and the fault detection scheme are discussed in the following subsections.

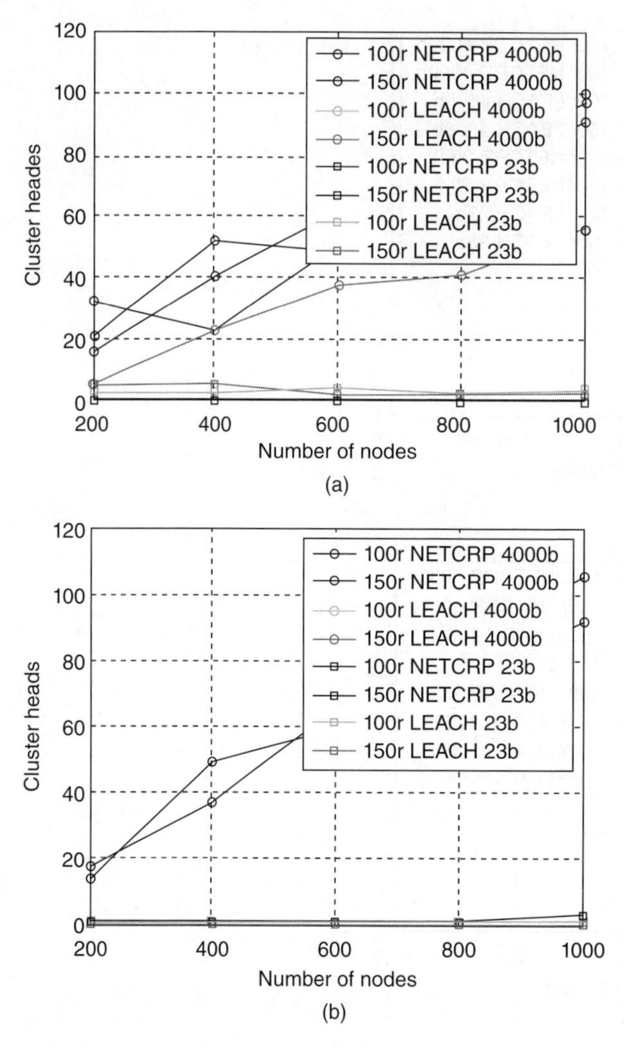

Figure 21.8 Total number of cluster-heads alive: (a) 1000 × 1000 m; (b) 4000 × 6000 m.

21.4.1 Flexible Threshold Method

The flexible threshold method is tested with the static threshold method of representing parameter values. Figure 21.3 shows the actual values for daytime and nighttime sea-surface temperatures, represented using bar graphs. The proposed method shows better sensitivity towards the changes in the values corresponding to different times of the day. On the other hand, with the static threshold, several changes of the values are neglected due to the method's non-adaptive nature.

21.4.2 Fault-detection Scheme

The protocol used a packet size of 23 bits to demonstrate the enhancement of the WSN lifetime. The initial comparison study, using packet sizes of 23 bits and 4000 bits in the NETCRP and LEACH [25] protocols is discussed elsewhere [24]. Figure 21.6 shows that

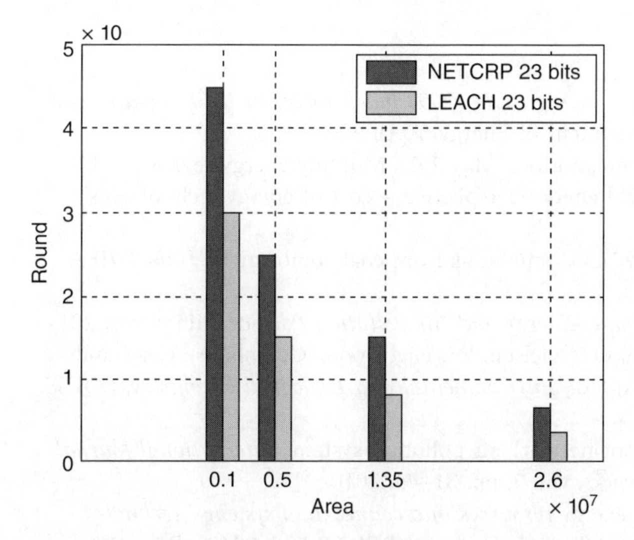

Figure 21.9 Round in which all the nodes become dead.

the energy-efficiency for reduced packet-size is higher for every area size and varying numbers of nodes. For large areas it becomes stable at nearly 90%. Moreover, the proposed protocol is more efficient when the number of rounds is 100 than it is with 150 rounds. Figure 21.7 shows that the first dead node does not occur in the proposed protocol, regardless of the area sizes of the nodes. This shows that the proposed protocol gives a longer lifetime than other protocols. Figure 21.8 shows a comparison of numbers of live CHs, and confirms that the proposed protocol has a higher live-node count for all area sizes and increasing numbers of nodes. Figure 21.9 shows the round in which all the nodes in the area become dead (an indicator of the network-lifetime). The count is found to be very high for the proposed protocol and more than 10 times the value in other protocols. This proves that the method yields an improved network-lifetime and better energy savings than earlier methods of threshold selection and fault detection. In summary, the method presented in this chapter is suitable for any area size, number of nodes and number of packets to be transmitted.

21.5 Summary

This chapter discussed the various application areas of WSN in windfarm monitoring. It emphasised on the greater usage of renewable energy sources for power generation instead of coal as it is not a pollution free resource. A unique application-specific routing protocol NETCRP which is energy efficient protocol along with a fault detection method is discussed. suitable improvements can be incorporated to further decrease the energy consumption and hence increase network-lifetime e.g., introducing sleep periods etc. Also, effects of increasing or decreasing the quantization levels can be studied and equations for optimal number of quantization levels can be established depending on the variations of the parameter values. Further, some method can be devised to automate the process of quantization. Moreover, this method can be applied to a wide variety of parameters e.g., wind velocity & humidity.

References

1 J.J. Messerly, *'Final report on the August 14, 2003 blackout in the United States and Canada'*, Report by US Department of Energy, April 2004.
2 US Energy Information Administration, 'May 2015 Monthly energy review', 2015.
3 World Energy Council, 'World energy perspective – cost of energy technologies', 2013.
4 International Energy Agency, 'CO_2 emissions from coal combustions', OECD/IEA, 2013.
5 G.C. Bryner *Integrating Climate, Energy and Air Pollution Policies*, MIT Press, 2012.
6 T.J. Wallington, J. Srinivasan, O.J. Nielsen, E.J. Highwood, 'Greenhouse gases and global warming', in Sabljic A, (ed.) *Environmental and Ecological Chemistry*, EOLSS, 2004.
7 K.K. Khedo, 'A wireless sensor network air pollution system', *International Journal of Wireless and Mobile Networks*, vol. 2, pp. 31–45, 2010.
8 N.R. Friedman, *'Distributed energy resources interconnecxtion systems: Technology review and research needs', Technical report NREL/SR-560–32459*, Resource Dynamics Corporation, Vienna, 2002.
9 US Department of Energy, 'Offshore resource assessment and design conditions – A data requirements and gap analysis for offshore renewable energy systems', 2012.
10 B.H. Bailey and D. Green, *'The need for expanded meteorological and oceanographic data to support resource characterization and design condition definition for offshore wind power projects in the United States '*, Technical report, American Meteorological Society, 2013.
11 British Wind Energy Association, *'Prospects for offshore wind energy'*, 1999.
12 A.W. Momber and P. Plagemann, 'Investigating corrosion protection of offshore wind tower'. *PACE*, Los Angeles, California, vol. 11, pp. 30–43, 2008.
13 W.B. Heinzelman, A.P. Chandrakasan, H. Balakrishnan, 'Energy- efficient communication protocols for wireless microsensor networks', *Proceedings of Hawaiian International Conference on Systems Science*, Wailea Maui, Hawaii, January 2000, **8**, pp. 1–10.
14 S. Lindsey, C.S. Raghavendra, 'PEGASIS: power-efficient gathering in sensor information systems', *Proceedings of IEEE Aerospace Conference*, Big Sky, Montana, March 2002, vol. 3, pp. 1125–1130.
15 M.G. Rashed, 'Heterogeneous hierarchical energy aware routing protocol for wireless sensor network', *International Journal of Engineering Technology*, vol. 6, pp. 521–526, 2009.
16 G. Smaragdakis, I. Matta, A. Bestavros, 'SEP: a stable election protocol for clustered heterogeneous wireless sensor networks', *Proceedings of International Workshop on SANPA*, Boston, MA, August 2004, pp. 251–261
17 M. Moazeni, A. Vahdatpour, 'HEAP: a hierarchical energy aware protocol for routing and aggregation in sensor networks', *Proceedings of Third International Conference on Wireless Internet*, Austin, USA, October 2007.
18 X. Chen, W. Qu, H. Ma, K. Li, 'A geography-based heterogeneous hierarchy routing protocol for wireless sensor networks', *Proceedings of 10th IEEE International Conference on High Performance Computing and Communications*, Dalian, China, September 2008, pp. 767–774.

19 Y. Song, B. Wang, B. Li, Y. Zeng, L. Wang, 'Remotely monitoring offshore wind turbines via ZigBee networks embedded with an advanced routing strategy', *Journal of Renewable and Sustainable Energy*, vol. 5, pp. 013110-1–14, 2013.

20 A. Malvankar, M. Yu, T. Zhu, 'An availability-based link QoS routing for mobile ad hoc networks'. *IEEE Sarnoff Symposium*, Princeton, NJ, March 2006, pp. 1–4.

21 Y. Wu and W. Liu, 'Routing protocol based on genetic algorithm for energy harvesting-wireless sensor networks', *IET Wireless Sensor Systems*, vol. 3, pp. 112–118, 2013.

22 N. Otsu, 'A threshold selection method from graylevel histogram', *IEEE Transactions on System Man Cybernetics*, vol. 9, pp. 62–66, 1979.

23 J. Kapur, P. Sahoo, A. Wong, 'A new method for graylevel picture thresholding using the entropy of the histogram', *Computer Graphics and Image Processing*, vol. 29, pp. 273–285, 1985.

24 P.l. Rosin, 'Unimodal thresholding', *Pattern Recognition*, vol. 34, pp. 2083–2096, 2001.

25 R.O. Duda, R.E. Hart, D.G. Stork, Pattern Classification, *2nd edn*. John Wiley & Sons, 2001.

26 P.A. Viola, 'Alignment by maximization of mutual information', Phd thesis, Massachusetts Institute of Technology, 1995.

27 P. Viola, M.J. Jones, D. Snow, 'Detecting pedestrians using patterns of motion and appearance', *Proceedings of IEEE International Conference on Computer Vision*, October 2003, **2**, pp. 734–741.

28 H. Kruppa and B. Schiele, 'Hierarchical combination of object models using mutual information', in *Proceedings of British Machine Vision Conference*, Manchester, September 2001,pp. 1–10 .

29 H. Peng, F. Long, C. Ding, 'Feature selection based on mutual information: Criteria of maxdependency, max-relevance, and min-redundancy', *IEEE Transactions on Pattern Analysis and Machine Intelligence*, vol. 27, pp. 1226–1238, 2005.

30 M. Kazmi, A. Aziz, P. Akhtar, A. Maftun, 'Medical image de-noising based on adaptive thresholding in contourlet domain'. *Proceedings of International Conference on Biomedical Engineering and Informatics*, Chongqing, China, October 2012, pp. 313–318.

31 F. Liu, X. Song, Y. Luo, D. Hu, 'Adaptive thresholding based on variational background', *Electronics Letters*, vol. 38, pp. 1017–1018, 2002.

32 Q.X. Tang and L.C. Jiao, 'Image de-noisingwith geometrical thresholds' *IET Electronics Letters*, vol. 45, pp. 405–406, April 2009.

33 V. Soni, A.K. Bhandari, A. Kumar, G.K Singh, 'Improved sub-band adaptive thresholding function for de-noisingof satellite image based on evolutionary algorithms', *IET Signal Processing*, vol. 7, pp. 720–730, 2013.

34 C. Xiong, 'Research on fuzzy fault prediction of the shell-feeding system', *Proceedings of IEEE International Conference on Quality, Reliability, Risk Maintenance and Safety Engineering*, Chengdu, June 2012, pp. 763–766.

35 D. Agarwal, N. Kishor, 'Network lifetime enhanced tri-level clustering and routing protocol for monitoring of off shore wind-farms', *IET Wireless Sensor Systems*, vol. 4, pp. 69–79, 2014.

36 Y. Ma and P. Guttorp, 'Estimating daily mean temperatures from synoptic climatic observations', *International Journal of Climatology*, vol. 33, pp. 1264–1269, 2013.

37 D. Agarwal, N. Kishor, 'A fuzzy inference based fault detection scheme using adaptive thresholds for health monitoring of offshore wind-farms', *IEEE Sensors Journal*, vol. 4, pp. 3851–3861, 2014.

Index

Wireless Sensor Systems for Extreme Environments: Space, Underwater, Underground, and Industrial.
First Edition. Edited by Habib F. Rashvand and Ali Abedi.